QUALITY ASSESSMENT OF PRINTED CIRCUIT BOARDS

QUALITY ASSESSMENT OF PRINTED CIRCUIT BOARDS

A Practical, Comprehensive Guide to the Inspection and Evaluation of Printed Circuit Board Quality

Preben Lund
Quality Manager,
Radiometer A/S,
Copenhagen, Denmark

The Innovators
Bishop Graphics, Inc.

Printed in the United States of America
First Printing, July 1985
ISBN 0-9601748-4-2
Library of Congress Catalog Number 84-073112
Reorder No. 10012

The Innovators
Bishop Graphics, Inc.
5388 Sterling Center Drive
P.O. Box 5007
Westlake Village, California 91359-5007

Contents

6.8.4 Nonplated-Through Boards, with Punched Holes 115

Preface

Considering the degree to which the electronics industry depends on printed circuit boards (PCBs), surprisingly little literature on quality assessment of PCBs can be found. Although a number of PCB magazines exist, they rarely contain articles on the occurrence of defects with their causes. When found, such articles address just one type of defect. A number of PCB books are in print, but they mainly address production techniques, and at the most, include a short chapter describing and assessing the most commonplace PCB defects.

It should, however, be emphasized that the IPC (The Institute for Interconnecting and Packaging Electronic Circuits) has contributed greatly by publishing a number of publications and standards. The most notable of these are "Acceptability of Printed Boards," IPC-A-600C and "IPC Test Methods Manual," IPC-TM-650.

In real life, the PCB inspector has a fairly small probability of finding help in the technical literature when establishing an incoming PCB inspection procedure. In the author's experience, an inspector should have a reasonably broad background knowledge of PCB defects and their causes. He should also have a qualitative knowledge of PCB production processes. Only then can he find the less obvious defects, assess their importance regarding reliability, and enter into a technical discussion with the PCB manufacturer.

It is the author's sincere hope that this book will fill a long-felt need within the electronics industry, and also contribute towards a more objective and functional quality assessment of PCBs than is the case today.

Another goal of the book is to draw the PCB designer's attention to good design practice in order to increase the producibility of the boards. This requires a broad background knowledge of fabrication methods as well as the faults attributable to an undesirable design. Since it is also possible to facilitate the inspection by modest design modifications, the author tries to give the PCB designer some insight into the problems a PCB inspector encounters in his daily work.

This book is based on the author's own experience, study of texts through many years, technical discussions with domestic and foreign PCB manufacturers, and exchange of experience within various PCB groups. It should be noted that all points set forth, including the illustrations and their interpretation, are the author's own responsibility and not necessarily shared by all the many experts, colleagues, and friends who have contributed to the book.

Acknowledgments

It is always difficult to write a book in a foreign language, and even when the author does his utmost to be linguistically correct, quite a few idiomatic blunders and incorrect sentence structures are bound to occur. The author wishes to thank Mr. Louis N. Jones, Santa Maria, California, for his painstaking efforts to correct the English, and for his valuable suggestions and constructive criticism when editing the manuscript.

The author also wishes to thank the publisher, Mr. Robert Meisterling, of Bishop Graphics, Inc., for publishing this book and for his personal interest in the project. Working together over a distance of 6000 miles requires much correspondance, so the author actually thinks of Bob more as a good pen-friend than as a business connection.

A serious problem was to illustrate rare defects, e.g. skip plating, but a number of PCB manufacturers and a metallographer kindly offered their assistance and provided boards and microsections exhibiting such hard-to-find defects. The author wishes to express his gratitude for all the help received. He also wishes to express his thanks for all technical discussions and valuable suggestions. It is, however, impossible to state all names, so the author wishes to single out the names of the following persons:

In the U.S.: Ronald A. Bulwith, Alpha Metals Inc.; David St. Andre, Colorado Springs Circuits; Ken Clasen, Delco Electronics Division, General Motors Corp.; Clyde F. Coombs, Jr., Hewlett-Packard Company; Louis J. Kane, Hughes Aircraft Co.; Len Vanderpool, Racal Dana Instruments, Inc.; Dieter Bergman, The IPC

In England: Anthony Cook, Tectonic (Electronics) Ltd.; Tom Hogan, Welwyn Electric Co.

In Scotland: Sine A. MacDonald, Norman Scott, Bepi Circuits Ltd.; John Pinkney, STC Exacta Ltd.; Roy Polson, Graham Fleck, Prestwick Circuits Ltd.

In Sweden: Fride Gunell, Cromtryck AB

In West Germany: Ulrich K. Voss, Schoeller & Co., Elektronik GmbH; Lothar Melchert, Melchert Elektronik GmbH & Co., KG; Alfred Märtens, Hans Kolbe & Co.

In Denmark: Ole Ravnsborg, Chemitalic A/S; Bjarne Olsen, Steen Flemming Kragelund, Cumatrix Electronics A/S; Poul Villien, Tom Jorgensen, Danish Research Centre for Applied Electronics; Finn Vesthede, Dansk Data Elektronik ApS;

Henrik Bjorn-Larsen, Dantek Elektronik A/S; H. C. Holmelund, PRI-DANA Elektronik A/S; Ove Ravn, Per Soetang Lauridsen, Jorgen Ib Johansen, Printca AS; Freddy Sorensen, Ruwel Danmark A/S

The author owes a debt of gratitude to Mr. Johan Schroeder, Director and General Manager of Radiometer A/S, for his kind permission to publish this book, which to a high degree is based on the author's experience of designing PCBs and assessing the quality of PCB deliveries over many years at Radiometer A/S.

Finally, my heartfelt thanks to Ellen, my dear wife, for her tolerant acceptance of an author's possibly excessive commitment of his spare time at the domestic writing-desk for five years, for affectionate patience with a messy study, and for wonderful company and good fellowship on study tours during summer holidays.

Preben Lund
Copenhagen
June, 1985

QUALITY ASSESSMENT OF PRINTED CIRCUIT BOARDS

Introduction

The author's experience indicates that the price of a bare printed circuit board (PCB) frequently equals 20 to 30 percent of the cost of the components which go on the board. Furthermore, with quantities comparable to those of large-scale integrated circuits such as microprocessors, printed circuit boards are often one of the largest classes of components purchased by an electronics manufacturer.

Because of the quantities and costs of printed circuit boards used in electronics equipment, manufacturers pay much attention to both the price and quality of these boards. In addition, rejecting an assembled board because of PCB defects can cause losses several times the board cost. To avoid these losses, many electronics companies perform incoming inspection on printed circuit boards.

One should not, however, expect the incoming inspection department to intercept all defective boards. Such thorough and comprehensive efforts would most likely cost more than they save in rejected assemblies.

Printed circuit boards have numerous characteristics to be inspected, usually visually. When every board of a batch is inspected (100 percent inspection) the inspector soon develops eyestrain and overlooks some defects. Screening can eliminate this problem while reducing the number of defects by 90 to 95 percent.

Obviously, a 100 percent inspection is too expensive for electronics companies in a competitive market. Such companies usually trade off the inspection costs against production losses caused by overlooked defects, as discussed in Part 1, and select the sampling inspection as the most economical method. When ultrahigh reliability is required, such as for space applications, the incoming inspection department inspects the lot a second time.

Sampling inspection, however, introduces a calculated risk that a defective batch will pass incoming inspection. The production and test departments must therefore expect defective boards on occasion.

Regardless of the type of inspection, the inspector must be familiar with all fault types in order to recognize the faults occurring in a batch. The inspector trying to study these faults finds little help in the technical literature, and probably no comprehensive guidance to performing actual inspections and evaluating the results.

The objective of this book is to provide the PCB inspector with a broad background knowledge of printed circuit board defects, inspection criteria, and causes of defects (discussed in Part 2). With such knowledge, the inspector is far better equipped to conduct realistic discussions with the PCB manufacturers.

This book emphasizes quality assessment on a functional basis. Such emphasis presupposes two conditions which the inspector should keep in mind:

1. Less weight is given to purely cosmetic faults which do not affect the function or reliability of the board. However, a reasonably attractive appearance projects a more positive image of the finished product.
2. Waivers for tolerances out of specification by insignificant amounts should be allowed.

These conditions deserve additional comments. Many boards are rejected for purely cosmetic faults because it is difficult or impossible to quantify many of the requirements regarding board appearance. Such requirements always contain subjective standards and frequently cause heated discussions between the PCB manufacturer and the inspector. Discussion frequently centers on the scratches easily inflicted on the soft tin/lead surface of the board.

The easiest and safest approach, which is to reject scratched boards, risks having the assembly line unable to start with the planned batch size. On the other hand, the inspector with a broad background knowledge soon learns to tell whether a scratch can affect the function or reliability. For example, a board with no short circuits between conductors caused by broken or overhanging tin/lead chips and with no copper visible at the botton of the scratch is not likely to fail.

The inspector can accept moderately scratched boards or issue a waiver. The scratches occurring in the customer's assembly line provide a standard for determining acceptable scratching.

Most tolerances are based on what is possible in normal production, not on what is actually required. A dimension which slightly exceeds its tolerance does not necessarily mean an unusable board or problems on the assembly line.

For example, consider a solder hole 0.031 in. (0.8 mm) in diameter. The specified tolerance of the diameter is frequently +0.006 in. (+0.15 mm) and −0.004 in. (−0.10 mm). If the diameter is slightly too small, say 0.026 in. (0.65 mm), it is not necessarily a cause for rejection.

The reduced diameter could be caused by excessive copper plating resulting from a missing or insufficient balancing of the pattern. The inspector can easily learn the diameter of the component lead that goes into the solder hole. If there is still sufficient clearance for the solder to fill the hole by capillary action during mass soldering, there is no reason to reject the board.

This consideration is valid only for manual component insertion. When components are inserted automatically, the inspector cannot routinely accept undersized solder holes because they could interfere with component insertion. In this case, the inspector should reject the board.

Evaluating minor tolerance violations especially requires the inspector to have a wide background knowledge in order to identify and consider possible side effects before issuing a waiver on an out-of-tolerance board.

A major theme of this book is that the customer, and especially the inspector, should treat the PCB manufacturer as a collaborator, with quality as the common goal. Printed circuit board users commonly give the PCB manufacturer feedback about the quality level as judged by the inspector so that the manufacturer can improve quality and thereby benefit the customer.

A formal inspection record preprinted with common defects aids this feedback. The inspector needs only to write the number of defects into adjacent blanks or boxes. This record also identifies whether a defect is a critical and major defect (A-defect) or a noncritical, minor defect (B-defect). Part 3, which deals with routine inspection, elaborates on the inspection record and its use. Preprinted inspection records have several advantages:

1. By presenting all information in a standard format, they reduce terminology problems between the customer and the PCB manufacturer.
2. By filling in the inspection record, the inspector formulates the feedback.
3. The inspection record lists all partial inspections, ensuring that all are completed.

Another important issue is cooperation between the inspector and the PCB designers, which also requires feedback from the inspector. Non-optimum patterns cause many recurring defects which will not be solved unless the designer knows of the problem.

Pattern improvements most likely will improve the PCB manufacturer's yield and, in the long run, reduce the costs of the boards. After all, the customer indirectly pays for manufacturing scrap through the PCB manufacturer's markup. Small modifications to the PCB pattern can also make the inspection much easier, but the inspector must inform the PCB designers of inspection problems before they make ease of inspection a design consideration.

This book frequently calls the PCB designer's attention to various modifications which can facilitate inspection. As a result, the designer can use the book to acquire a better understanding of PCB fabrication. Adapting a PCB design to the production process requires some understanding of the production procedures, especially of machining and the corresponding tolerances. For this reason the book describes those production tolerances achievable under ordinary conditions.

To avoid unnecessary complications, Part 3 discusses only routine inspections. Part 4 describes the more advanced inspection methods, such as the use of microsections for assessing the quality of plated-through holes. Many inspectors avoid preparing microsections because they believe that such methods require great expertise and expensive equipment. Part 4 describes a very simple means of preparing microsections, which the author hopes will encourage the inspector to gradually improve his methods by procuring more advanced equipment.

Quality assessment requires a reference, specifically the company's purchase specification for printed circuit boards. There are many varieties of purchase specifications, some more comprehensive than others. On the whole, the specifications contain similar requirements, such as the same nominal values and the same tolerance ranges.

To a large extent, this book is based on internationally accepted nonmilitary PCB specifications. On the other hand, military and nonmilitary printed circuit boards are sufficiently similar so that an inspector of military boards can use this book to advantage.

The inspector must of course tailor his quality assessment to his company's purchase specification. For example, a specification may mention the "75 percent rule" used in this book when assessing changes in the pattern. The rule requires that the actual value must not be less than 75 percent of the nominal value. This applies, for example, to conductor widths and insulation distances. Other specifications, including the military standards, specify an "80 percent rule." In practice, the rules differ very little—for example, only 0.0008 in. (0.02 mm) at a nominal insulation distance of 0.16 in. (0.4 mm).

Many companies have no formal purchase specifications. In such cases, the inspector should understand that all the inspection criteria given in this book relate to what normally can be expected from any professional PCB manufacturer. Therefore, the inspector may use these criteria and reasonably expect the PCB manufacturer to accept them. It is, however, strongly recommended that the company use a standard PCB specification—for example, published by the IPC—or else develop its own specification.

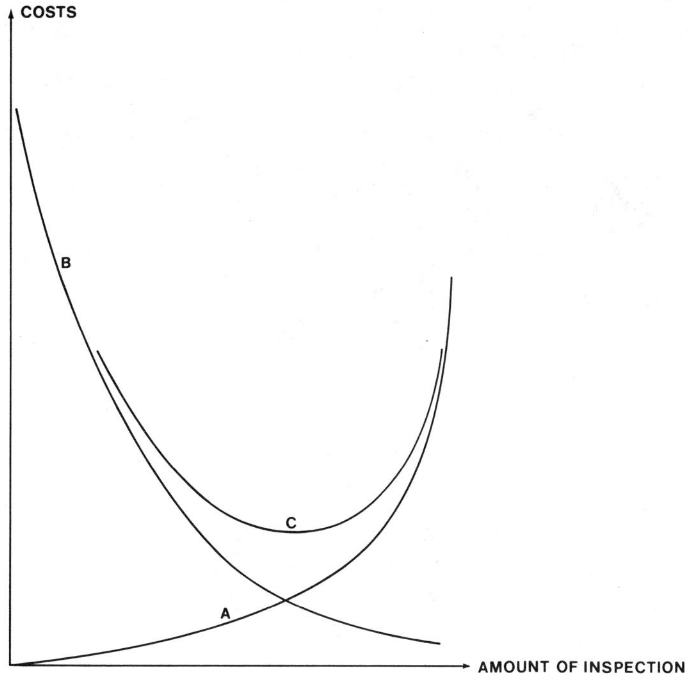

Part 1

Fundamental Concepts of Sampling Inspection

Chapter 1

Sampling Inspection of Printed Circuit Boards

This chapter deals with the theoretical background of an incoming inspection based on a sampling inspection by attributes. Such an inspection classifies a board as defective or good with respect to the customer's quality requirements. The latter part of the chapter describes a substantially simplified sampling system.

1.1 Introduction

1.1.1 The Literature

To some extent, the available literature describes various types of incoming inspections for printed circuit boards (PCBs), but on the whole this vexing problem receives limited treatment. Information on organized procedures, especially concrete descriptions of appropriate degrees of inspection, simply do not exist.

1.1.2 AQL Values

Many purchase specifications state AQL values and inspection levels (see Paragraph 1.2.5 for definitions of these terms), often giving one AQL value for major faults and another for minor faults. Unfortunately, an incoming inspection strictly based on such a system can become quite expensive. Nevertheless, it makes sense to specify AQL values in the purchase specifications in order to provide a legitimate basis for rejecting a delivery. Fortunately, this does not place any obligations on the customer himself with regard to the amount of routine inspection to perform, at least not in the case of PCBs for the normal civilian electronics market.

1.1.3 Purpose of Sampling Inspection

The goal of the PCB user is to get boards which do not cause production problems. In order to not disrupt production, he must consider exempting certain types of faults if the defective boards can be repaired without marring an otherwise good appearance. Such exemptions, however, should not become a regular event, especially not for the same type of fault; one otherwise might as well reduce the required quality level and demand a corresponding price reduction. Without doubt, rejection and return of several deliveries will stimulate the PCB manufacturer's quality consciousness, and, in the long run, should improve the delivery reliability as well as the general quality level. Should this not lead to more dependable deliveries, the only effective solution is to find another supplier with a better quality assurance system.

1.1.4 Effects of Rejections

Returning printed circuit boards to have them remade disrupts the PCB manufacturer's production no matter who is responsible. Rejections obviously affect both the customer, who produces more or less suitable documentation, and the PCB manufacturer, who produces the boards.

Most PCB manufacturers are well aware of occasional abuses in rejecting deliveries. An example is the customer who no longer needs the boards and subjects them to an extremely severe incoming inspection solely to reject them and receive credit. With the many parameters included in the quality specifications, not least the severe tolerances and the somewhat subjective assessment of certain quality characteristics, a skilled inspector can almost always find faults—usually nonfunctional—to justify rejecting the boards.

By accepting the customer's documentation package or, as unfortunately happens in some cases, by failing to express reservations against debatable documents, the PCB manufacturer has morally bound himself to deliver boards which fulfill the customer's specifications. Some PCB manufacturers prefer to take a certain risk than to criticize the customer's documentation.

1.1.5 Role of the PCB Designer

The PCB designer cannot be enjoined too often or too strongly that he has a heavy responsibility for the success of the production process. Whether the PCB design is generated manually or is a computer-aided design, it is important to consider all aspects of the production processes. An illustrative example is the great influence that the balancing of the PCB pattern has in meeting the diameter tolerances when plating plated-through holes; a small extra advantage is a faster incoming inspection because of the inherently greater reproducibility of the board.

1.1.6 Who Does the Inspection?

The incoming inspection of PCBs can be accomplished by any of four parties:

1. Usually by the customer's own inspection department
2. By an independent company or laboratory which performs an impartial inspection or a special investigation, most often because the customer's own incoming inspection department lacks the necessary measuring equipment or experienced personnel
3. By the PCB manufacturer by special arrangement
4. By the PCB supplier or agent by special arrangement.

In each case, a specification of the quality requirements, including the necessary inspection data and procedures, should ensure a uniform inspection independent of place and inspector.

1.1.7 Criteria for the Degree of Inspection

Because inspection costs increase rapidly as a function of quality assurance demands, certain criteria for the degree of inspection should be established. Theoretically, the company can find the best balance by finding the point where the inspection costs and the production losses caused by defective boards add up to the lowest cost.

In practice, however, it is extremely difficult to find this balance. The philosophy on which the following treatment is based is to keep the amount of inspection at a relatively low level, which implies efficient and prompt feedback from the production department. Should the amount of inspection be found to be too moderate, the inspection can be immediately increased in all areas which have caused criticism. This increase can be either general or selective, depending on whether the fault(s) is traced back to several suppliers or just one.

This philosophy is backed up by the common experience that, having chosen PCB manufacturers with a good technical background, modern production equipment, and a sound quality assurance system, the PCB customer usually receives satisfactory deliveries.

1.1.8 Causes of Faults

When faults occur, the reason is often one of the following:

1. A misunderstanding of the documentation, either because it is unclear or because it contains errors.
2. Over-specification demanding more than it is technically possible to meet with 100 percent confidence under normal production conditions.
3. Errors in production, inevitably occurring from time to time, not detected by the PCB manufacturer's own quality assurance system—for example, the sporadic occurrence of dewetting.
4. Raw materials defects not detected by the PCB manufacturer's own incoming inspection; for example, a base material which gasses when heated and therefore causes blowholes during mass soldering of the assembled boards.

In this connection it is emphasized that generally it is safer to inspect all deliveries to a moderate extent than to selectively and thoroughly inspect a few deliveries and not inspect the other deliveries at all.

These considerations are based on the assumption that the electronics products in question are intended for civilian applications, not for military or space applications where very thorough incoming inspection is prescribed.

The PCB designer may find receiving inspection fairly uninteresting, but all the same, he should devote some attention to this topic. In this way, he can learn some of the daily inspection problems with which the inspector has to contend. He can then modify his design practice to facilitate inspection. Small design modifications often cause the board prices to decrease rather than increase. For this reason a direct and efficient feedback from incoming inspection to the PCB designer should be promoted as much as possible.

1.2 Sampling Inspection

1.2.1 Introduction

1.2.1a Need for Inspection

In spite of the PCB manufacturer's efforts, an occasional delivery will contain defective boards. Thus, we have to establish an incoming inspection to verify that the batch quality is satisfactory. In order to quantify the concept of quality, we shall express the quality of a delivery as the percentage of boards meeting the quality requirements.

1.2.1b Determining the Degree of Inspection

A major problem of incoming inspection is the extent to which it is performed. The degree of actual inspection may range from merely verifying that the boards were received and are of the correct revision to a 100 percent inspection—that is, an extensive inspection of each board in the delivery. Obviously, the first extreme is inadequate to confirm the quality, and the second is usually uneconomical. Consequently, the latter would only be performed in special cases demanding extraordinarily high quality, such as boards intended for military or space applications. In such cases, the end user normally prescribes the inspection requirements.

1.2.1c Prerequisites of Quality

One prerequisite of satisfactory quality is high-precision artwork and adequate consideration of the production processes. Because a product is naturally more expensive when the end user imposes stricter requirements, a purchase specification should state only reasonable and well-considered requirements which any professional PCB manufacturer can meet at an acceptable price.

1.2.2 Price versus Quality

1.2.2a Quality Curves

To survive in today's highly competitive market, the PCB manufacturer must produce competitively priced boards of satisfactory quality and still make a profit, as shown in Figure 1.1. The X-axis indicates the general quality level, ranging from 0 to 100 percent; that is, from a delivery where all boards are rejected to a delivery where all boards are accepted. The Y-axis indicates the price or cost.

Curve A shows the price the market will pay as a function of quality. The curve flattens out at the higher quality levels because competition, among other factors, prevents further price increases.

Curve C shows the production costs. Straight line B shows the initial costs for manufacturing boards of quality q_1. The distance between curve C and straight line B represents the quality costs—that is, the extra costs necessary for increasing the quality. At first, the extra costs increase relatively slowly with increasing quality requirements, then more rapidly, and finish at a very high level. The distance between curves A and C represents

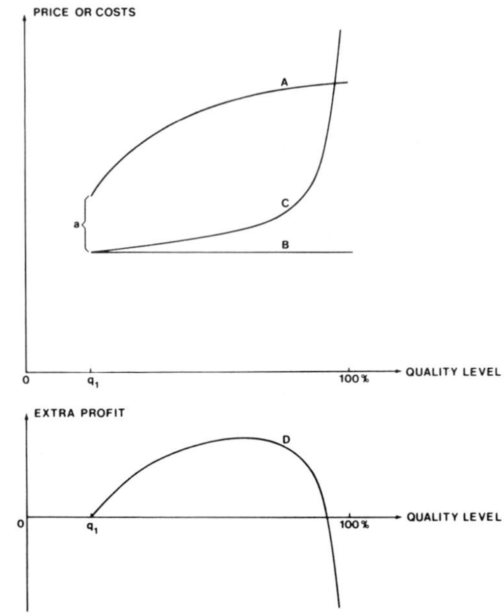

Fig. 1.1. Relationship between price, profit and quality level.
A: Market price. B: Basic costs of boards of quality q_1. C: Production costs. D: Extra profit.

the PCB manufacturer's profit. Curve D, the extra profit as a function of quality, is derived by subtracting curve C from curve A.

1.2.2b Maximum Profit Range

The shapes of curves A and B cause curve D to have a near-maximum over a wide range. For this reason, the PCB manufacturer earns a maximum profit by maintaining quality at an intermediate range. Consequently, the customer must anticipate PCB deliveries at less than 100 percent quality level; that is, a certain percentage of defects. It is important that the customer realize this, and not make impossible demands on the PCB manufacturer.

When very stringent quality requirements are specified, as in military applications, such boards fall outside the normal price assessment mechanism. Of

necessity, curve A will assume another shape in order to ensure the PCB manufacturer the same profit margin. The very steep end of curve C, which shows the production costs, explains the notoriously high price level typical of military boards.

For various reasons, an electronics company producing products for the civilian market wants PCB deliveries which are a reasonable compromise between price and quality. Figure 1.1 clearly shows that a high quality implies a high price which, within certain limits, is fully acceptable, because some cost reductions are gained in the assembly and testing of the boards due to the reduced waste and rework costs.

1.2.2c Costs of Undetected Faults

If defective boards reach the assembly line, they will cause large losses in the form of scrapped components and nonproductive labor or, alternatively, rework and consumption of new components. It should be noted that the PCB manufacturer is responsible for the boards only, not for lost work costs or scrapped components. To this is added the more intangible expenses arising from delayed production, breaking the natural production rhythm, late deliveries to the end user (lost goodwill), etc.

When the fault does not appear until the product is placed in use by the customer, very large recovery costs must be taken into account. Other extra costs are the result of product liability, either for a recall of already delivered equipment or, at worst, for the resulting lawsuits. The survey below shows the order of magnitude of the extra costs relative to the location where the faults are found:

In the assembly shop:	$1 per fault
In the test shop:	$10 per fault
In final inspection:	$20 per fault
At the end user:	$200 per fault

1.2.3 Optimum Degree of Inspection

1.2.3a Optimizing Inspection Costs

The preceding survey indicates that defective boards should be segregated as early as possible. This is normally achieved by setting up an incoming inspection department, responsible for ascertaining whether the boards comply with the quality requirements.

It was suggested earlier that in practice it is very difficult to optimize the inspection costs relative to the losses arising from inadequate incoming inspection. A qualitative consideration, however, clearly shows the economic consequences of just increasing the level of inspection.

1.2.3b Inspection Costs versus Assurance

Figure 1.2 compares the inspection costs (curve A) with the production losses (curve B). The inspection costs will inevitably rise with an increasing assurance against defective boards, which in turn is

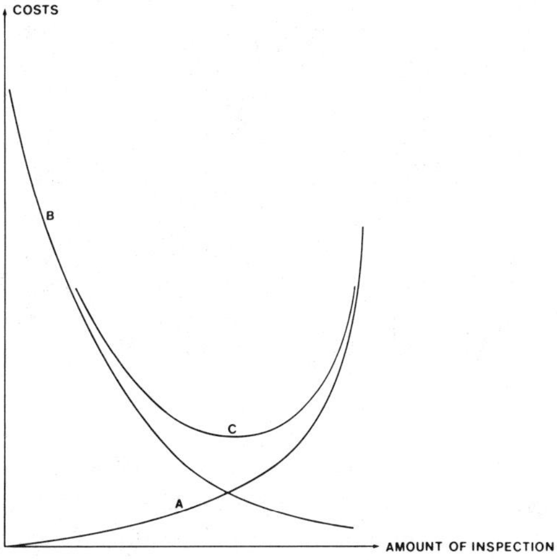

Fig. 1.2. Costs as a function of amount of inspection.
A: Inspection costs. B: Production losses.
C: Total costs.

reflected in decreasing production losses. The total costs are represented by curve C which is the sum of curves A and B. The minimum point of the total costs corresponding to the optimum level of inspection is found where curve C reaches a minimum, namely around the crossover point of curves A and B. This crossover is somewhat dependent upon the slope of the curves in this region.

As shown in Figure 1.2, the inspection costs increase disproportionately, with greater assurance against defective boards. Although production losses are somewhat reduced, the total costs will be very high, making it wiser to reduce the demand for full assurance and accept a sampling inspection.

An alternative is to spend a certain amount of the budgeted inspection costs on preventive measures: higher artwork quality, better vendor selection, or the PCB manufacturer's guarantee of a higher quality level. These choices, however, are somewhat difficult to make, because it is not easy to estimate the impact of the measures undertaken. Undoubtedly, a careful selection of PCB manufacturers, including regular factory visits, improves the quality level of the finished boards.

1.2.3c Role of Full Inspection

Full inspection of every board, usually called 100 percent inspection, should be carried out only in cases where, for security reasons, an undetected defective board is not allowed. Unless very special measures are taken, a 100 percent inspection is seldom fully effective. Owing to the many details to be inspected and the human fatigue factor, to which inspectors are also subject, the inspection must be repeated until is is quite certain that all defective boards have been found. Obviously, a 100 percent inspection is very expensive.

For the civilian market, it is not normally economical to perform a 100 percent inspection of the individual PCB deliveries. The user has to learn to accept the fact that a few defective boards will occasionally slip through incoming inspection, not to be detected until a later stage of production.

1.2.4 Principle of Sampling Inspection

1.2.4a Sampling Parameters

When a 100 percent inspection is out of the question, one has to perform a sampling inspection. Figure 1.3 shows the principle of single sampling. With a sampling plan, described in Paragraph 1.2.6, the following parameters are determined:

1. The sample size n as a function of the batch size N and the inspection level required.
2. The acceptance number c.
3. The rejection number r.

1.2.4b Accept–Reject Decision

If the number of defectives, p, found in the sample is equal to or less than the acceptance number c, the batch is accepted, whereas it is rejected if p is equal to or greater than the rejection number r. The case implies, of course, that an

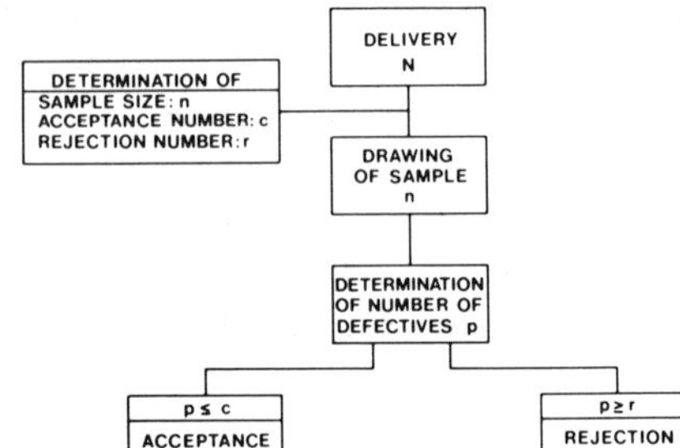

Fig. 1.3. Block diagram of single sampling.

agreement regarding the use of sampling inspection and the possible consequences was made beforehand with the PCB manufacturer. The rejection number r usually equals the acceptance number c + 1.

In order to perform the inspection we must be able to ascertain whether the individual boards conform to the specific requirements; that is, either to accept the boards as being satisfactory or to reject them as being defective. This is done by measuring physical characteristics, such as hole diameters, or by visually assessing such details as scratched surfaces. By counting all the faults detected during the inspection, the number of defectives, p, is ascertained, after which the fate of the delivery is decided.

1.2.4c Disposal of Rejected Batches

When the number of defective boards exceeds the acceptance number, the delivery or batch can be rejected and returned to the PCB manufacturer for screening, rework, or remake. As mentioned earlier, the customer does not always want to return the delivery because his production may stop. Instead, he may choose to screen the batch himself to remove defective boards. This implies a 100 percent inspection, but because a rejection is usually caused by a single characteristic, the inspection is primarily limited to this characteristic.

Defective boards may, in many cases, then be replaced by boards being stocked as surplus items from earlier deliveries. In other cases, the customer can choose to waive the defect if it is of a nonfunctional nature or easily remedied, and does not result in reduced board reliability.

A common misunderstanding is that the customer, by performing a sampling inspection, loses the right to return any defective boards sorted out, and have them replaced, or as an alternative, to require a credit from the PCB manufacturer. This is always possible, and, furthermore, the customer may claim payment for expenses in conjunction with screening and repairing defective boards. It is, however, a natural precondition that an agreement with the PCB manufacturer be made in individual cases.

A sampling inspection can also be performed as a double sampling or a multiple sampling. The principle appears in Figure 1.4, which shows a double sampling. By performing the sampling inspection successively it is possible to reach a conclusion earlier than by single sampling. This is because the first partial sample is smaller than the sample size prescribed for single sampling. In the case of double sampling the two-part sample sizes are equal; consequently one deals with only half the work if the first sample is either accepted or rejected.

We deal exclusively with single sampling because it is easier to deal with, requires less training, and yields the clearest information concerning the PCB manufacturer's quality level. Single sampling, furthermore, proves to be more economical when the actual percentage of defective boards falls around the acceptable

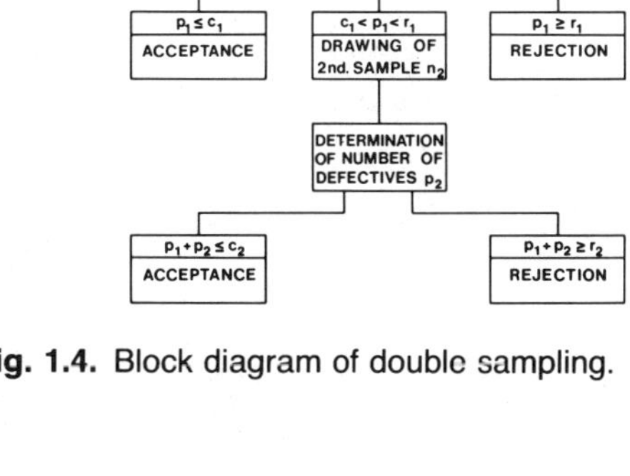

Fig. 1.4. Block diagram of double sampling.

quality level. This is exactly what we are aiming at when we determine the AQL value. See Paragraphs 1.2.5 and 1.2.6b.

1.2.5 Operating Characteristic Curve

We shall now assess the element of risk that a sampling inspection implies, since, by statistical means, it is possible to determine the probability of acceptance, P_a, as a function of the actual percentage of defectives, p, of the batch. We do not need to perform the actual calculations; instead, we can use the already available graphical presentations of the Operating Characteristic Curve, often abbreviated "OC curve."

Every sampling plan, which determines a certain set of sample size n and acceptance number c, has an OC curve as shown in Figure 1.5. Exceeding the acceptance number corresponds to the area above the curve. In case of a perfect batch ($p = 0$), we find a 100 percent probability of acceptance ($P_a = 1.0$). On the contrary, a completely defective batch ($p = 100$) has a probability of acceptance of 0. Usually, however, we will find ourselves in an intermediate range; that is, on the steep part of the curve.

It is, as a matter of fact, the steepness that determines the discriminating power of the OC curve; that is, its ability to distinguish between good and bad batches. An ideal OC curve would be a vertical line as shown in Figure 1.6. All batches having a defective percentage equal to or less than p_1 will be accepted, and rejected if the defective percentage is greater than p_1. This corresponds to a 100 percent inspection of the batch.

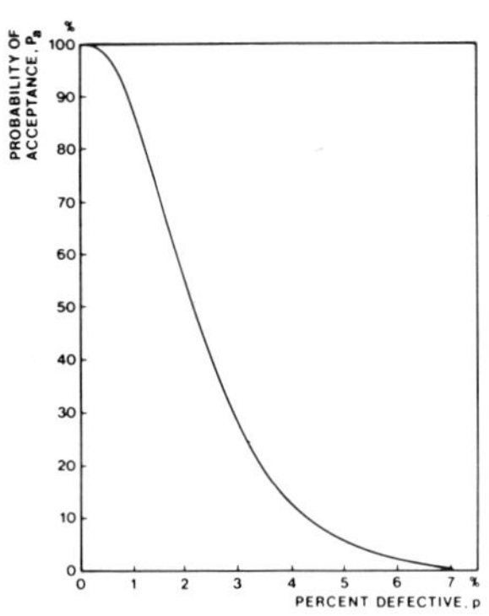

Fig. 1.5. Example of OC curve. $n = 125$. $c = 2$.

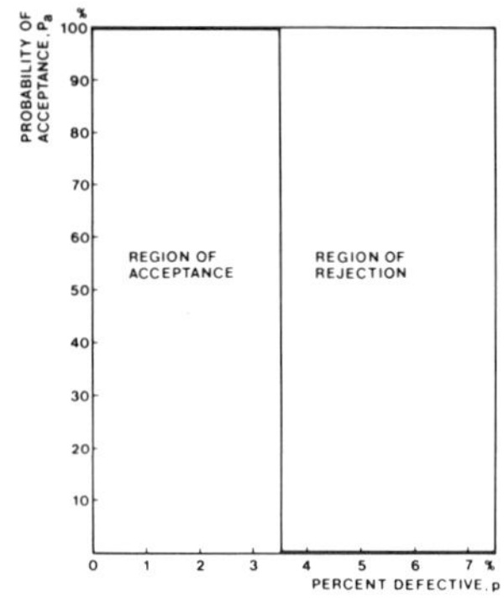

Fig. 1.6. Ideal OC curve.

In practice, we will never see an ideal OC curve, but curves that, in principle, correspond to the curve shown in Figure 1.5. Therefore, the problem is how certain we can be that the assessment is a correct one: We could perhaps accept a bad batch, which would be detrimental to ourselves, or reject a good batch, which would be detrimental to the PCB manufacturer, and, in the end, also injurious to ourselves when considering the reprocurement time. Therefore, various ideas have been introduced in order to define the risks of both the manufacturer and the customer, which are addressed below.

1.2.5a Acceptable Quality Level (AQL)

The Acceptable Quality Level (AQL) indicates the percent of defective boards for which there is a 95 percent probability of acceptance. See Figure 1.7. The AQL value can also be considered the greatest percentage defective which can be considered a satisfactory average of the

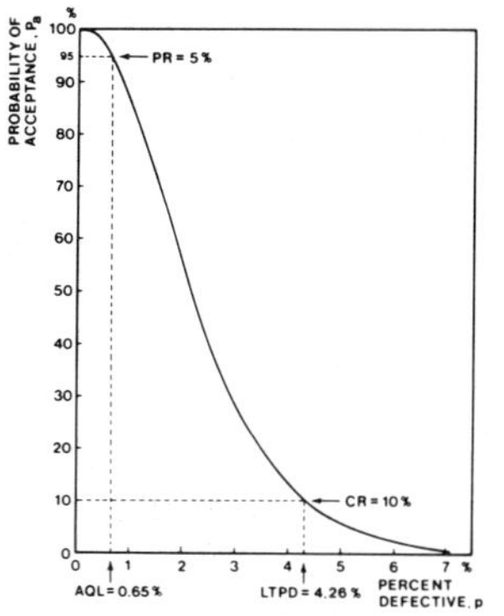

Fig. 1.7. Definition of risks connected with sampling inspection. PR: Producer's Risk. CR: Consumer's Risk.
AQL: Acceptable Quality Level.
LTPD: Lot Tolerance Percent Defective.

sampling inspection. To the PCB manufacturer, this means that the sampling plan will accept the vast majority of the delivered batches, on the assumption that the average percentage of defects in these batches is below the agreed AQL value.

Instead of characterizing the OC curve by the acceptance number, the AQL value can be used as in the sampling plans according to MIL-STD-105D. These plans are discussed in detail in Paragraph 1.2.6. Since the AQL values occurring in MIL-STD-105D form a standardized series, the AQL values do not always correspond perfectly to a 95 percent probability of acceptance, but will range from 88 to 99 percent. In practice, this is of no importance because the relationship between the PCB manufacturer and the customer regarding the acceptance conditions is defined in an unambiguous manner on the basis of the AQL value and the inspection level.

1.2.5b Producer's Risk

According to the above definition of the AQL value, the PCB manufacturer must accept the fact that a batch will be rejected 5 percent of the time, even if the requirements are met. Since he will suffer from this misjudgment, we speak of the Producer's Risk (PR), as shown in Figure 1.7. In many sampling plans this is stated as 5 percent.

1.2.5c Consumer's Risk

In exactly the same way the customer may accept a batch in a certain percentage of cases although the requirements are not met. The probability of this is called the Consumer's Risk (CR, Figure 1.7). Frequently the probability is stated as 10 percent, but it also happens that it is 5 percent, corresponding to the producer's risk.

1.2.5d Lot Tolerance Percent Defective

The percent defective corresponding to the consumer's risk is often designated as the Limiting Quality (LQ). This means an unsatisfactory quality level which should cause the sampling plan to reject the batch. However, owing to the inherent uncertainty of the sampling plan, such batches will occasionally be accepted.

When the consumer's risk is fixed at 10 percent, the corresponding percent defective is designated Lot Tolerance Percent Defective (LTPD), which is a special case of the Limiting Quality.

1.2.5e Discriminating Power of the OC Curve

The range between the acceptance probabilities 95 percent (PR) and 10 percent (CR) corresponds to an average quality level, and the steepness within this range determines the OC curve's discriminating power (ability to distinguish). The discriminating power varies with the percent defective of the batch, fortunately in such a way that the acceptance probability increases with a decreasing percent defective.

An important rule is that the efficiency of a sampling inspection is determined by the absolute size n of the sample and not by its relative size n/N, where N is the size of the batch from which the sample is drawn. This appears from the mathematical/statistical relations which form the basis of the sampling plans. It is, however, presupposed that the sample size n is no greater than 1/10 of the batch size N, and furthermore, that the percent defective (p) is less than 10 percent. This is usually the case with PCB deliveries.

1.2.6 Sampling Plans

Printed circuit boards are normally purchased in batches which, in the incoming inspection department, can be kept segregated from other PCB batches. Because of the homogeneous nature of the deliveries, the incoming inspection can be performed as a sampling inspection of the individual deliveries. The quality requirements should therefore be stated in the general purchase specification and indicate the AQL value, inspection level, and the type of sampling (single, double, or multiple).

The sampling system employed most internationally is MIL-STD-105D, "Sampling Procedures and Tables for Inspection by Attributes." It can be purchased from the Superintendent of Documents, U.S. Government Printing Office, Washington DC 20402.

Another publication, Handbook H53, "Guide for Sampling Inspection," which gives a very detailed description on how to use MIL-STD-105D, can also be purchased from the same source.

The sampling system defined in MIL-STD-105D is standardized within NATO, but is also used extensively in nonmember countries. Since it is very suitable for inspection of incoming goods, and furthermore is presumed to be familiar to all PCB manufacturers, it is made the basis of the following treatment.

1.2.6a Inspection Level

MIL-STD-105D operates at three inspection levels: General Inspection Level I, II, and III. These levels determine the degree of inspection (the sample size relative to the batch size). Inspection Level II is normally chosen as the starting point when performing an incoming inspection, whereas Inspection Levels I and III are used only in special situations, such as when the customer is trying to reduce the inspection costs.

The sample sizes of Inspection Levels I and III are about 60 percent less and 60 percent greater, respectively, than that of Inspection Level II. In addition to the inspection levels mentioned above, the MIL-standard has four Special Inspection Levels, S1 to S4, which can be employed if it is desirable to work with very small sample sizes and a greater risk of misjudging that the batch is acceptable.

In general, the Special Inspection Levels are satisfactory when the manufacturer has a reasonably good quality assurance system, and the goods are manufactured with processes that give a high reproducibility—for example, blanking and punching. On the condition that the PCB manufacturer has been chosen carefully and is regularly inspected, the above prerequisites are also valid for printed circuit boards. The vast majority of the individual processes utilized in PCB manufacturing give fairly uniform results from board to board. This is, however, not quite true in the case of very complex boards, for which reason a higher level of inspection is needed.

Although the purchase specification indicates a definite inspection level, usually General Inspection Level II, the customer is naturally not bound to use this level in routine inspection. He may inspect the boards according to his own standards to an extent which suits him, and he may freely choose one of the Special Inspection Levels, such as S3.

If an incoming inspection, however, shows such an unsatisfactory result that a rejection of the whole batch becomes necessary, the inspection should be repeated (or rather increased) by employing General Inspection Level II. The decision to reject a batch should always be based on the conditions stated in the purchase specification.

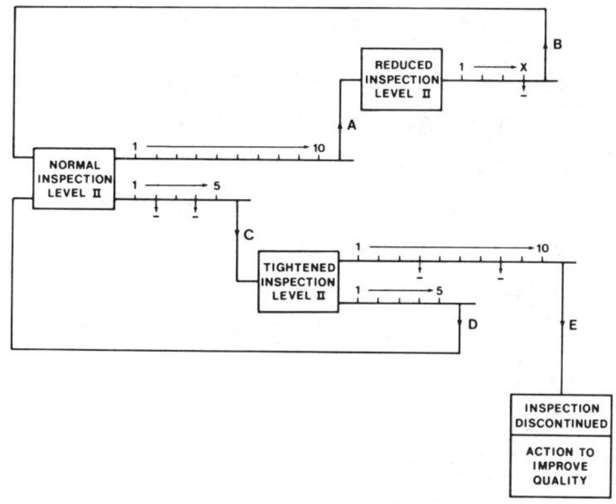

Fig. 1.8. Conditions for switching between different severity levels of inspection.

A: 1. No rejections within 10 preceding batches, and
2. Total number of defects is less than the maximum number of defects stated in Table VIII (MIL-STD-105D), and
3. Production is at a steady rate, and
4. Reduced inspection is specified.

B: 1. A batch is rejected, or
2. A batch is accepted according to Paragraph 10.1.4 (MIL-STD-105D), or
3. Production becomes irregular, or
4. Other conditions warrant that normal inspection should be instituted.

C: Two out of five consecutive batches are rejected.

D: Five consecutive batches are accepted.

E: Ten consecutive batches remain on tightened inspection; i.e., case D does not occur.

⊥ Batches to be inspected
↓ Rejection of batch

1.2.6b Severity of Sampling Plan

A prerequisite of selecting the AQL value is that the corresponding quality can be attained in practice. When the quality is high, the normal sampling plan safeguards against sudden changes in the quality level, while the screening operation in itself is less important. Consequently the discriminating power does not have to be particularly severe, since the customer can reduce the severity of the sampling plan in order to save inspection time and costs.

Conversely, when the quality is in general less than the quality level corresponding to the chosen or demanded AQL value, one needs a greater certainty in order to effectively identify bad batches. By tightening the severity of the sampling plan, one increases the discriminating power. For these reasons, the MIL-STD-105D provides two degrees of severity in addition to the one used for normal inspection: Reduced Inspection and Tightened Inspection.

The conditions for switching between the three severity levels are described in detail in MIL-STD-105D and illustrated in Figure 1.8 in a simplified form. Note B2, however, needs a more precise explanation: In the sampling plan valid for reduced inspection, the rejection figure does not always equal the acceptance figure $c+1$, but may equal $c+2$. If the number of defectives is $c+1$, the next sampling inspection shall, in turn, be performed at General Inspection Level II.

In practice, switching between the various severity levels obviously requires keeping a record of the individual suppliers. Administratively, such switching can be fairly complicated, and in normal civilian electronics production, MIL-STD-105D is usually not followed in this respect.

1.2.6c How to Use MIL-STD-105D

We shall now study the practical use of the sampling plans given in MIL-STD-105D. The most important sampling plans for single samplings are reproduced in Figures 1.9 to 1.12 with permission of the U.S. Government Printing Office, and their use is illustrated by means of some practical examples.

Example No. 1:

Let us inspect a batch of 125 boards, assuming General Inspection Level II and an AQL value of 1 percent. From Table I, Figure 1.9, the sample size code letter is found to be F, which is valid for batch sizes between 91 and 150. By entering Table II-A, Figure 1.10, at code letter F, we get a sample size of 20. After that, we move horizontally to the column marked 1.0 corresponding to the AQL value. The upward pointing arrow, however, indicates that no sampling plan is available and instead, the first sampling plan above the arrow is to be used. According to this the sample size is to be changed to 13, and an acceptance number (the first number) of 0 and a rejection number (the second number) of 1 are found. The sample is now inspected and if only one fault is found the batch is rejected and may be returned to the PCB manufacturer.

The sampling plan also has downward pointing arrows, which also indicate that no sampling plan is available. Instead, the first sampling plan below the arrow should be used, and the sample size changed accordingly. When the batch is very small the new sample plan may require a sample size exceeding the batch size. Consequently, a 100 percent inspection is performed.

Example No. 2:

A batch of 10 boards is to be inspected corresponding to an AQL value of 1 percent. The sample size code letter is B, and the arrow points down to a sampling plan corresponding to the code letter E, which has a sample size of 13. Obviously, the inspection level is 100 percent.

We shall now see what happens when the Tightened Inspection Plan is used instead of the Normal Inspection Plan.

Example No. 3:

Based on the same batch size and AQL value as in Example 1, we find the following: The sample size code letter remains unchanged and from Table II-B, Figure 1.11, we find a sample size of 20, and acceptance and rejection numbers of 0 and 1, respectively.

In other cases, we may find a reduction of the acceptance and rejection numbers. Both situations imply that the discriminating power is increased with respect to the Normal Inspection Plan.

Finally, we shall see what happens when using the Reduced Inspection Plan instead of the Normal Inspection Plan.

Example No. 4:

Again, we take the same batch size and AQL value as used in Example 1. As before, the sample size code letter is F, and by using Table II-C, Figure 1.12, we find an upward pointing arrow, which indicates that we have to use the first sampling plan above. Therefore, the sample size will now be reduced from 8 to 5, and the acceptance and rejection numbers are 0 and 1, respectively.

Lot or batch size	Special inspection levels				General inspection levels		
	S-1	S-2	S-3	S-4	I	II	III
2 to 8	A	A	A	A	A	A	B
9 to 15	A	A	A	A	A	B	C
16 to 25	A	A	B	B	B	C	D
26 to 50	A	B	B	C	C	D	E
51 to 90	B	B	C	C	C	E	F
91 to 150	B	B	C	D	D	F	G
151 to 280	B	C	D	E	E	G	H
281 to 500	B	C	D	E	F	H	J
501 to 1200	C	C	E	F	G	J	K
1201 to 3200	C	D	E	G	H	K	L
3201 to 10000	C	D	F	G	J	L	M
10001 to 35000	C	D	F	H	K	M	N
35001 to 150000	D	E	G	J	L	N	P
150001 to 500000	D	E	G	J	M	P	Q
500001 and over	D	E	H	K	N	Q	R

Fig. 1.9. Table I. Sample size code letters.

Sample size code letter	Sample size	0.010	0.015	0.025	0.040	0.065	0.10	0.15	0.25	0.40	0.65	1.0	1.5	2.5	4.0	6.5	10	15	25	40	65	100	150	250	400	650	1000
		Acceptable Quality Levels (normal inspection)																									
		Ac Re	Ac Re	Ac Re	Ac Re	Ac Re	Ac Re	Ac Re	Ac Re	Ac Re	Ac Re	Ac Re	Ac Re	Ac Re	Ac Re	Ac Re	Ac Re	Ac Re	Ac Re	Ac Re	Ac Re	Ac Re	Ac Re	Ac Re	Ac Re	Ac Re	Ac Re
A	2	↓	↓	↓	↓	↓	↓	↓	↓	↓	↓	↓	↓	↓	↓	0 1	↓	↓	1 2	2 3	3 4	5 6	7 8	10 11	14 15	21 22	30 31
B	3	↓	↓	↓	↓	↓	↓	↓	↓	↓	↓	↓	↓	↓	0 1	↑	↓	1 2	2 3	3 4	5 6	7 8	10 11	14 15	21 22	30 31	44 45
C	5	↓	↓	↓	↓	↓	↓	↓	↓	↓	↓	↓	↓	0 1	↑	↓	1 2	2 3	3 4	5 6	7 8	10 11	14 15	21 22	30 31	44 45	↑
D	8	↓	↓	↓	↓	↓	↓	↓	↓	↓	↓	↓	0 1	↑	↓	1 2	2 3	3 4	5 6	7 8	10 11	14 15	21 22	30 31	44 45	↑	↑
E	13	↓	↓	↓	↓	↓	↓	↓	↓	↓	↓	0 1	↑	↓	1 2	2 3	3 4	5 6	7 8	10 11	14 15	21 22	30 31	44 45	↑	↑	↑
F	20	↓	↓	↓	↓	↓	↓	↓	↓	↓	0 1	↑	↓	1 2	2 3	3 4	5 6	7 8	10 11	14 15	21 22	↑	↑	↑	↑	↑	↑
G	32	↓	↓	↓	↓	↓	↓	↓	↓	0 1	↑	↓	1 2	2 3	3 4	5 6	7 8	10 11	14 15	21 22	↑	↑	↑	↑	↑	↑	↑
H	50	↓	↓	↓	↓	↓	↓	↓	0 1	↑	↓	1 2	2 3	3 4	5 6	7 8	10 11	14 15	21 22	↑	↑	↑	↑	↑	↑	↑	↑
J	80	↓	↓	↓	↓	↓	↓	0 1	↑	↓	1 2	2 3	3 4	5 6	7 8	10 11	14 15	21 22	↑	↑	↑	↑	↑	↑	↑	↑	↑
K	125	↓	↓	↓	↓	↓	0 1	↑	↓	1 2	2 3	3 4	5 6	7 8	10 11	14 15	21 22	↑	↑	↑	↑	↑	↑	↑	↑	↑	↑
L	200	↓	↓	↓	↓	0 1	↑	↓	1 2	2 3	3 4	5 6	7 8	10 11	14 15	21 22	↑	↑	↑	↑	↑	↑	↑	↑	↑	↑	↑
M	315	↓	↓	↓	0 1	↑	↓	1 2	2 3	3 4	5 6	7 8	10 11	14 15	21 22	↑	↑	↑	↑	↑	↑	↑	↑	↑	↑	↑	↑
N	500	↓	↓	0 1	↑	↓	1 2	2 3	3 4	5 6	7 8	10 11	14 15	21 22	↑	↑	↑	↑	↑	↑	↑	↑	↑	↑	↑	↑	↑
P	800	↓	0 1	↑	↓	1 2	2 3	3 4	5 6	7 8	10 11	14 15	21 22	↑	↑	↑	↑	↑	↑	↑	↑	↑	↑	↑	↑	↑	↑
Q	1250	0 1	↑	↓	1 2	2 3	3 4	5 6	7 8	10 11	14 15	21 22	↑	↑	↑	↑	↑	↑	↑	↑	↑	↑	↑	↑	↑	↑	↑
R	2000	↑	↑	1 2	2 3	3 4	5 6	7 8	10 11	14 15	21 22	↑	↑	↑	↑	↑	↑	↑	↑	↑	↑	↑	↑	↑	↑	↑	↑

Fig. 1.10. Table II-A. Single sampling plans for normal inspection.

Sample size code letter	Sample size	0.010	0.015	0.025	0.040	0.065	0.10	0.15	0.25	0.40	0.65	1.0	1.5	2.5	4.0	6.5	10	15	25	40	65	100	150	250	400	650	1000
		Acceptable Quality Levels (tightened inspection)																									
		Ac Re	Ac Re	Ac Re	Ac Re	Ac Re	Ac Re	Ac Re	Ac Re	Ac Re	Ac Re	Ac Re	Ac Re	Ac Re	Ac Re	Ac Re	Ac Re	Ac Re	Ac Re	Ac Re	Ac Re	Ac Re	Ac Re	Ac Re	Ac Re	Ac Re	Ac Re
A	2	↓	↓	↓	↓	↓	↓	↓	↓	↓	↓	↓	↓	↓	↓	↓	↓	↓	↓	1 2	2 3	3 4	5 6	8 9	12 13	18 19	27 28
B	3	↓	↓	↓	↓	↓	↓	↓	↓	↓	↓	↓	↓	↓	↓	0 1	↓	↓	1 2	2 3	3 4	5 6	8 9	12 13	18 19	27 28	41 42
C	5	↓	↓	↓	↓	↓	↓	↓	↓	↓	↓	↓	↓	↓	0 1	↓	↓	1 2	2 3	3 4	5 6	8 9	12 13	18 19	27 28	41 42	↑
D	8	↓	↓	↓	↓	↓	↓	↓	↓	↓	↓	↓	↓	0 1	↓	↓	1 2	2 3	3 4	5 6	8 9	12 13	18 19	27 28	41 42	↑	↑
E	13	↓	↓	↓	↓	↓	↓	↓	↓	↓	↓	↓	0 1	↓	↓	1 2	2 3	3 4	5 6	8 9	12 13	18 19	27 28	41 42	↑	↑	↑
F	20	↓	↓	↓	↓	↓	↓	↓	↓	↓	↓	0 1	↓	↓	1 2	2 3	3 4	5 6	8 9	12 13	18 19	↑	↑	↑	↑	↑	↑
G	32	↓	↓	↓	↓	↓	↓	↓	↓	↓	0 1	↓	↓	1 2	2 3	3 4	5 6	8 9	12 13	18 19	↑	↑	↑	↑	↑	↑	↑
H	50	↓	↓	↓	↓	↓	↓	↓	↓	0 1	↓	↓	1 2	2 3	3 4	5 6	8 9	12 13	18 19	↑	↑	↑	↑	↑	↑	↑	↑
J	80	↓	↓	↓	↓	↓	↓	↓	0 1	↓	↓	1 2	2 3	3 4	5 6	8 9	12 13	18 19	↑	↑	↑	↑	↑	↑	↑	↑	↑
K	125	↓	↓	↓	↓	↓	↓	0 1	↓	↓	1 2	2 3	3 4	5 6	8 9	12 13	18 19	↑	↑	↑	↑	↑	↑	↑	↑	↑	↑
L	200	↓	↓	↓	↓	↓	0 1	↓	↓	1 2	2 3	3 4	5 6	8 9	12 13	18 19	↑	↑	↑	↑	↑	↑	↑	↑	↑	↑	↑
M	315	↓	↓	↓	↓	0 1	↓	↓	1 2	2 3	3 4	5 6	8 9	12 13	18 19	↑	↑	↑	↑	↑	↑	↑	↑	↑	↑	↑	↑
N	500	↓	↓	↓	0 1	↓	↓	1 2	2 3	3 4	5 6	8 9	12 13	18 19	↑	↑	↑	↑	↑	↑	↑	↑	↑	↑	↑	↑	↑
P	800	↓	↓	0 1	↓	↓	1 2	2 3	3 4	5 6	8 9	12 13	18 19	↑	↑	↑	↑	↑	↑	↑	↑	↑	↑	↑	↑	↑	↑
Q	1250	↓	0 1	↓	↓	1 2	2 3	3 4	5 6	8 9	12 13	18 19	↑	↑	↑	↑	↑	↑	↑	↑	↑	↑	↑	↑	↑	↑	↑
R	2000	0 1	↑	↓	1 2	2 3	3 4	5 6	8 9	12 13	18 19	↑	↑	↑	↑	↑	↑	↑	↑	↑	↑	↑	↑	↑	↑	↑	↑
S	3150			1 2																							

Fig. 1.11. Table II-B. Single sampling plans for tightened inspection.

Sample size code letter	Sample size	0.010	0.015	0.025	0.040	0.065	0.10	0.15	0.25	0.40	0.65	1.0	1.5	2.5	4.0	6.5	10	15	25	40	65	100	150	250	400	650	1000
		Ac Re	Ac Re	Ac Re	Ac Re	Ac Re	Ac Re	Ac Re	Ac Re	Ac Re	Ac Re	Ac Re	Ac Re	Ac Re	Ac Re	Ac Re	Ac Re	Ac Re	Ac Re	Ac Re	Ac Re	Ac Re	Ac Re	Ac Re	Ac Re	Ac Re	Ac Re
A	2	↓	↓	↓	↓	↓	↓	↓	↓	↓	↓	↓	↓	↓	↓	0 1	↓	↓	1 2	2 3	3 4	5 6	7 8	10 11	14 15	21 22	30 31
B	2	↓	↓	↓	↓	↓	↓	↓	↓	↓	↓	↓	↓	↓	0 1	↑	↓	0 2	1 3	2 4	3 5	5 6	7 8	10 11	14 15	21 22	30 31
C	2	↓	↓	↓	↓	↓	↓	↓	↓	↓	↓	↓	↓	0 1	↑	↓	0 2	1 3	1 4	2 5	3 6	5 8	7 10	10 13	14 17	21 24	↑
D	3	↓	↓	↓	↓	↓	↓	↓	↓	↓	↓	↓	0 1	↑	↓	0 2	1 3	1 4	2 5	3 6	5 8	7 10	10 13	14 17	21 24	↑	↑
E	5	↓	↓	↓	↓	↓	↓	↓	↓	↓	↓	0 1	↑	↓	0 2	1 3	1 4	2 5	3 6	5 8	7 10	10 13	14 17	21 24	↑	↑	↑
F	8	↓	↓	↓	↓	↓	↓	↓	↓	↓	0 1	↑	↓	0 2	1 3	1 4	2 5	3 6	5 8	7 10	10 13	↑	↑	↑	↑	↑	↑
G	13	↓	↓	↓	↓	↓	↓	↓	↓	0 1	↑	↓	0 2	1 3	1 4	2 5	3 6	5 8	7 10	10 13	↑	↑	↑	↑	↑	↑	↑
H	20	↓	↓	↓	↓	↓	↓	↓	0 1	↑	↓	0 2	1 3	1 4	2 5	3 6	5 8	7 10	10 13	↑	↑	↑	↑	↑	↑	↑	↑
J	32	↓	↓	↓	↓	↓	↓	0 1	↑	↓	0 2	1 3	1 4	2 5	3 6	5 8	7 10	10 13	↑	↑	↑	↑	↑	↑	↑	↑	↑
K	50	↓	↓	↓	↓	↓	0 1	↑	↓	0 2	1 3	1 4	2 5	3 6	5 8	7 10	10 13	↑	↑	↑	↑	↑	↑	↑	↑	↑	↑
L	80	↓	↓	↓	↓	0 1	↑	↓	0 2	1 3	1 4	2 5	3 6	5 8	7 10	10 13	↑	↑	↑	↑	↑	↑	↑	↑	↑	↑	↑
M	125	↓	↓	↓	0 1	↑	↓	0 2	1 3	1 4	2 5	3 6	5 8	7 10	10 13	↑	↑	↑	↑	↑	↑	↑	↑	↑	↑	↑	↑
N	200	↓	↓	0 1	↑	↓	0 2	1 3	1 4	2 5	3 6	5 8	7 10	10 13	↑	↑	↑	↑	↑	↑	↑	↑	↑	↑	↑	↑	↑
P	315	↓	0 1	↑	↓	0 2	1 3	1 4	2 5	3 6	5 8	7 10	10 13	↑	↑	↑	↑	↑	↑	↑	↑	↑	↑	↑	↑	↑	↑
Q	500	0 1	↑	↓	0 2	1 3	1 4	2 5	3 6	5 8	7 10	10 13	↑	↑	↑	↑	↑	↑	↑	↑	↑	↑	↑	↑	↑	↑	↑
R	800	↑	↑	0 2	1 3	1 4	2 5	3 6	5 8	7 10	10 13	↑	↑	↑	↑	↑	↑	↑	↑	↑	↑	↑	↑	↑	↑	↑	↑

Acceptable Quality Levels (reduced inspection)†

Fig. 1.12. Table II-C. Single sampling plans for reduced inspection.

If we, in Example 4, had used an AQL value of 4, we would have found a sample size of 8 and acceptance and rejection numbers of 1 and 3, respectively. In reality, this means an acceptance number of 2, but if two defective boards are found during the inspection, future deliveries from the manufacturer in question will be inspected at the normal inspection level.

In Example 4, the reduction of the inspection level is achieved by reducing the sample size by approximately 60 percent, whereas the acceptance and rejection numbers remain unchanged with respect to Example 1. In other cases we will find a simultaneous reduction of the sample size and the acceptance and rejection numbers.

1.2.6d Classification of Defects

The meaning of the term AQL has been explained in Paragraph 1.2.5a. It is, however, important to distinguish between defects of different relative importance, which makes it necessary to classify the faults. MIL-STD-105D defines three classes of defects: critical defects, major defects, and minor defects.

It may turn out, however, to be very difficult for an inspector to decide whether he has found a critical defect or a major defect, often because this implies a knowledge of the end product in which the board is used. In accordance with the philosophy expressed in this book, the number of defect classes should be kept as low as possible. We can simplify the defect classification considerably by using only two defect classes, namely A-defects and B-defects.

Incidentally, this philosophy is supported by MIL-P-55110, in which two defect classes are used: "major defects" equal A-defects, and "minor defects" equal B-defects. The standard states a number of typical A-defects and B-defects. Some examples are given below for the reader's information.

A-defects comprise lifting of copper foil, more than 20 percent reduction of conductor widths, voids covering more than 10 percent of the wall area of plated-through solder holes, formation of nodules in plated-through holes, etc.

B-defects are fewer in number and not specified with many details. They preferably concern surface faults—for example, cracks, scratches, and incorrect hole sizes (except solder holes).

1.2.6e Defect Class A

Defect Class A (A-defects) comprises all critical and essential major defects which, from experience or a technical evaluation, are likely to cause hazardous conditions for individuals, or incorrect performance, either during assembly and testing or when the board is installed in the end product and used by the customer.

A typical defect of this type occurs when the slot of an edge connector has been milled so wide that the board can be displaced in the fixed connector so that the connections fail. Another type of defect is pinholes in a gold-plated edge connector. If the board is subjected to a corrosive atmosphere, the underlying copper will be attacked and the corrosion products can spread over the contact surface, possibly creating an unacceptably large contact resistance.

1.2.6f Defect Class B

Defect Class B (B-defects) comprises all other types of defects, which therefore are characterized as relatively noncritical. In other words, they are immaterial for board functionality, even though they may degrade the board in one respect or another—for example, degrade its appearance because of surface scratches.

1.2.6g Selecting the AQL Values

Since the two defect classes are of different relative importance to the quality assessment of a printed circuit board, the two classes can be given different AQL values. The AQL values, however, cannot be determined on a technically well-founded basis. Unfortunately, one cannot predict the exact consequences of undetected defects, particularly the losses stemming from an unknown number of defective boards slipping through the incoming inspection. See Paragraph 1.2.3.

Several different AQL values are used in practical work. By way of example, BS 9762, which deals with double-sided, plated-through boards and which is published by the British Standards Institution, states AQL values ranging from 0.65 percent to 4 percent, depending on the characteristics to be inspected. Within the various electronics companies a wide range of AQL values is found, varying between 0.65 percent and 6.5 percent. There is a trend towards standardizing some of the AQL values. In the author's experience, a reasonable compromise in AQL values can be made as follows:

A-defects: AQL = 1 percent

B-defects: AQL = 2.5 percent

It is always possible to readjust the AQL values, depending upon the experience gained during the early phases of practical work in the incoming inspection department.

In order to simplify the total assessment of a board, it could be tempting to convert all B-defects to A-defects. This could be done by weighting the B-defects relative to the A-defects, so that, for example, three B-defects equal one A-defect. In this way, the inspector can use just one AQL value, and add the various defects directly. The most important advantage, however, would be the use of only one sample size. When using two AQL values, there may be two different sample sizes, one for the A-defects and another for the B-defects.

Assuming a batch size of 400 and a conversion of B-defects to A-defects according to the formula

$$A' = A + B/3$$

we find the results given in Table 1.1. In Example I, the boards are accepted when evaluated on the basis of two AQL values, because the number of A- and B-defects is less than the respective rejection numbers. By converting the B-defects to A-defects, the boards will be rejected, because A′ equals the rejection number.

Example II shows the reverse situation. Although the rejection number for B-defects is exceeded, the boards are accepted when evaluated on the basis of converted B-defects.

As demonstrated, conversion of B-defects to A-defects defeats the principle of classifying, which the inspector should realize if there are practical reasons to convert B-defects to A-defects.

Table 1.1. Assessment of boards based partly on one AQL value and conversion of all B-defects to A-defects, partly on two AQL values.

			Example 1		Example 2	
Type of defect	AQL	Rejection number	Number of defects	A′	Number of defects	A′
A	1%	2	1	2	0	1.67
B	2.5%	4	3		5	

It deserves notice that the customer has a free hand regarding how to arrange his incoming inspection, and many electronics companies have, in fact, chosen to weigh the A- and B-defects relative to each other. The most important thing for the PCB manufacturer is that a possible rejection of a batch be based on conditions agreed upon with the customer.

1.2.6h Different Sample Sizes for A- and B-Defects

When the sampling inspection is based on different AQL values for A- and B-defects, the sampling plans may indicate different sample sizes. The reason is that for certain combinations of AQL values and batch sizes (sample size code letters), no sampling plans are available, so that other sample plans have to be used as indicated by upward or downward pointing arrows. See Table II-A, Figure 1.10. The actual conditions valid for the AQL values 1 percent and 2.5 percent are shown in Tables 1.2 and 1.3. These are valid for normal inspection according to General Inspection Level II and Special Inspection Level S3, respectively.

In such cases, where different sample sizes are indicated, the largest sample is drawn first, and from this, the smaller sample is drawn. We shall now see two examples:

Example No. 1:

Here we inspect a batch of 100 boards. As indicated in Table 1.2, we draw a sample of 20 boards and from this a sample of 13 boards. The 13 boards are now inspected for A- and B-defects, and if we find one A-defect or two B-defects, the batch is rejected. If there are no A-defects, or no more than one B-defect, the remaining seven boards are inspected for B-defects. If the total number of B-defects equals two or more, the batch is rejected. A-defects which may be found when inspecting the remaining boards are disregarded, since the inspection for A-defects has been performed on the 13 boards.

Table 1.2. Sample size and acceptance/rejection numbers as a function of batch size and AQL value, valid for General Inspection Level II (Extract of Tables I and II-A of MIL-STD-105D).

	General inspection level 2					
	AQL = 1% (A-defect)			AQL = 2.5% (B-defect)		
Batch size	Sample size	Acceptance number	Rejection number	Sample size	Acceptance number	Rejection number
2–8	13	0	1	5	0	1
9–15	13	0	1	5	0	1
16–25	13	0	1	5	0	1
26–50	13	0	1	5	0	1
51–90	13	0	1	20	1	2
91–150	13	0	1	20	1	2
151–280	50	1	2	32	2	3
281–500	50	1	2	50	3	4
501–1200	80	2	3	80	5	6
1201–3200	125	3	4	125	7	8
3201–10000	200	5	6	200	10	11

Table 1.3. Sample size and acceptance/rejection numbers as function of batch size and AQL value, valid for Special Inspection Level S3 (Extract of Tables I and II-A of MIL-STD-105D).

	Special inspection level S3					
	AQL = 1% (A-defect)			AQL = 2.5% (B-defect)		
Batch size	Sample size	Acceptance number	Rejection number	Sample size	Acceptance number	Rejection number
2–8	13	0	1	5	0	1
9–15	13	0	1	5	0	1
16–25	13	0	1	5	0	1
26–50	13	0	1	5	0	1
51–90	13	0	1	5	0	1
91–150	13	0	1	5	0	1
151–280	13	0	1	5	0	1
281–500	13	0	1	5	0	1
501–1200	13	0	1	20	1	2
1201–3200	13	0	1	20	1	2
3201–10000	13	0	1	20	1	2

Example No. 2:

We inspect a batch of 250 boards. A sample of 50 boards is drawn first, and from this another sample of 32 boards. The smaller sample is inspected for A- and B-defects, and if we find two A-defects or three B-defects, the batch is rejected. If we find no more than one A-defect and two B-defects, the remaining 18 boards are inspected for A-defects before the batch can be accepted. If we now find a total of two or more A-defects, the batch is rejected. Possible B-defects found in the 18 boards are disregarded since the inspection for B-defects already has been performed.

The procedure is analogous to double sampling, discussed in Paragraph 1.2.4. In practice, however, the inspector presumably prefers to inspect the boards in the full sample size as prescribed for A- and B-defects, because a more rational work procedure is possible. This is addressed in Part 3.

1.2.6i Abuse of the Sampling Plan

Incoming inspection by sampling is a matter of confidence, and requires strict observance of the rules of the game. Unfortunately, most PCB manufacturers know of cases where a delivery has been rejected because the inspector has misunderstood the term "Customer's Risk," discussed in Paragraph 1.2.5c. He has therefore been hunting faults in the total delivery until finding the number actually valid for a limited part of the delivery, namely the sample itself, thereafter rejecting and returning the delivery.

Misunderstanding can be deliberate in special cases—for example, the customer for some reason may want to avoid a delivery. Obviously, this is difficult for a PCB manufacturer to prove. He will, of course, inspect the returned delivery very carefully, and if such rejections recur he soon can ascertain whether the rejection is justified.

There is every reason to emphasize that printed circuit boards should be assessed on their functional characteristics rather than their secondary characteristics, such as appearance. The inspector or the Material Review Board mentioned in Paragraph 3.1.2 should consider that slightly exceeding a tolerance seldom renders the board unusable. Tolerances are frequently based partly on experience about what is attainable by normal production practice, and partly on design requirements, which in the case of printed circuit boards may have a rather wide margin.

Chapter 2

Practical Level of Inspection

2.1 Prerequisites for a Sampling Plan

The information presented in Chapter 1 forms the basis for determining a practical level of inspection and developing an expedient sampling plan. A sampling inspection which closely adheres to MIL-STD-105D may be too expensive, considering the many parameters which must be evaluated in order to assess the exact quality of a batch of printed circuit boards (PCBs).

A company not manufacturing electronic equipment for military, space, or similar ultrahigh reliability applications tries to reduce the amount of inspection as much as possible. Obviously, such reduced inspection should not drop below that needed to be reasonably sure of detecting defective batches at incoming inspection, while holding cost to a minimum. The degree of reduced inspection depends on the following factors:

2.1.1 Knowledge of Manufacturers

Knowledge of the manufacturer's organization and production equipment is important. A normal objective is to maintain close contact between the PCB manufacturer and the customer, achieved by regular visits to the manufacturer. These visits give the customer insight into such organizational aspects as planning, process control, current production problems, changes in production methods, plant expansions, and new equipment procurement. Experience shows that introducing new processes, equipment, or production facilities usually causes quality to fluctuate. It is important, therefore, that the customer receive advance notice of such changes.

2.1.2 Manufacturer's Own Inspection

Knowledge of the manufacturer's finished-goods inspection is also important. If the manufacturer has inspected the batch, partly during the in-process quality control and partly during finished-goods inspection, the customer's incoming inspection department need only verify that a satisfactory quality level has been maintained, without duplicating the manufacturer's inspection. This requires that the customer first satisfy the requirement of Paragraph 2.1.1.

2.1.3 Manufacturer's Dependability

The manufacturer's long-term dependability concerning quality and promised delivery dates should be considered. The customer confirms such dependability by keeping records of each manufacturer's adherence to promised delivery dates, quantities, and qualities. These records must cover a relatively long span because PCB production encompasses many different disciplines and procedures, and occasional mishaps will affect a manufacturer's performance. If the records show a consistent problem, the customer must take some sort of remedial action.

2.1.4 Uniformity of Quality

As a rule, all boards in a batch containing a single part number will exhibit essentially the same quality, whether it is a satisfactorily high or an unsatisfactorily low quality level. This is especially true when automated processes are used or when all boards of a batch have been processed together or under identical conditions.

An example is the plating process. If one of the plating baths is slightly out of control, the metallization of the plated-through holes suffers. There is no acceptable reason for having a few bad holes while the process is under full control and the drilling itself is performed satisfactorily.

On the other hand, if the processes are carried out as separate operations, more or less dependent on the attentiveness of the operator, evaluation becomes less straightforward. For example, milling the profile of a board or of a polarization slot could be a manual operation, performed on individual boards and subject to human error. In such cases, generalizing about the uniformity of quality is impossible.

The reader can easily see that assessment in such cases requires a thorough knowledge of the processes used by the manufacturer in question. A quite different uniformity must be expected when the boards are blanked with a contour blanking tool or routed with a numerically controlled router, rather than with manual sawing or manual jig routing.

2.1.5 Feedback to Incoming Inspection Department

Prompt feedback from the assembly line to the incoming inspection department concerning defective boards which pass incoming inspection is a must for reducing the degree of inspection.

If the production department has automated test equipment and tests the boards immediately before or after assembly, the customer can receive rapid feedback concerning broken or short-circuited conductors, and avoid a detailed inspection of the pattern for such defects. It should be noted, however, that indentations or projections along conductor edges cannot be detected by automatic testing, so some visual inspection is still necessary.

The feedback information can result in design changes which improve producibility of the board. Often, only minor changes are needed to achieve trouble-free production.

2.2 Choosing a Sampling Plan

As emphasized earlier, the incoming inspection department has a free hand in choosing the amount of inspection and, as a result, the degree of assurance. This assumes, however, that any possible rejection and board return is based upon the agreed inspection level, usually General Inspection Level II (normal inspection). It is necessary to consider the supplier category and the nature of the order when choosing the detailed inspection level.

2.2.1 New or Unstable Suppliers

Deliveries from new suppliers and suppliers with variable quality levels must be inspected fairly carefully; for such deliveries we choose General Inspection Level II. While such suppliers are not approved for a lower inspection level, they can, of course, be approved after having delivered satisfactorily over a sufficient length of time.

2.2.2 Approved Suppliers

A supplier who has kept a stable and satisfactory quality level over an extended period of time can be approved for an inspection level substantially lower than General Inspection Level II. On the grounds listed in Paragraph 1.2.6a, we may choose Special Inspection Level S3, which implies a considerable reduction. The customer may, of course, choose the amount of inspection, but Special Inspection Level S3 has proven, in practice, to be a reasonable compromise between quality assurance and degree of inspection.

Part 2 shows that some PCB characteristics have such high reproducibility that it is unnecessary to inspect even to Special Inspection level S3. In many such cases, inspecting a single sample suffices.

2.2.3 Sampling Plan

Tables 1.2 and 1.3, valid for General Inspection Level II and Special Inspection Level S3, show sampling plans for AQL = 1 percent and 2.5 percent. For a given inspection level, the sample size and corresponding acceptance and rejection numbers are found in the tables as a function of the batch size. If a single type of defect causes the batch to exceed the acceptance level, it may be preferable—depending on the nature of the defect—to perform a 100 percent screening for that defect instead of returning the entire batch.

As emphasized earlier, rejecting an entire batch requires inspection according to Inspection Level II; an inspection performed according to Special Inspection Level S3 must be extended to General Inspection Level II before the batch is rejected. Again, the reader should note that screened boards may be returned to the PCB manufacturer whether or not the rest of the batch is accepted.

2.2.4 Recurring Defects

A recurring defect in a batch is usually caused by a process error. There may be, therefore, an increased risk that similar defects are present in other batches manufactured about the same time (within a few weeks of the date the defective batch was produced).

If such batches have passed through incoming inspection and perhaps been placed in stock, it may be wise to reinspect them for the same type of defect, either with an increased sample size or as a 100 percent inspection. Since this new inspection focuses on a specific type of defect, the actual additional inspection is fairly limited. The additional cost can be justified as extra assurance, offset by the savings of the lower initial inspection level.

2.3 Approval to a Lower Inspection Level

2.3.1 Conditions for Approval

MIL-STD-105D prescribes the conditions for switching between different degrees of severity in the sampling plans; see Paragraph 1.2.6b and Figure 1.8. The main prerequisite for switching to a lower level is that none of ten consecutive batches has been rejected.

2.3.2 Evaluation Period

Although switching to a higher level occurs as soon as a batch is rejected, no time limit has been prescribed for reclassifying a new and unknown PCB manufacturer to a lower level. Experience shows that, in order to reliably assess a manufacturer's stability regarding quality level, the evaluation period must include a time in which the process conditions actually have had an opportunity to deteriorate.

This consideration applies especially to plating lines, which usually have a long "time constant." Organic additives, which may be regarded as a controlled contamination of the plating bath, improve the quality of deposition. Unfortunately, these additives may decompose with time; the resulting products can reach a critical concentration after perhaps several months and unfavorably affect the deposition. The additional lower time limit mentioned below compensates somewhat for the notoriously larger risk connected with using very low inspection levels.

2.3.3 Approving to Special Inspection Level S3

We can now formulate the conditions for approving a PCB supplier to Special Inspection Level S3, provided that all the following conditions have been met:

1. The PCB manufacturer has been visited and approved as described in Paragraph 2.1.1.
2. A minimum of five consecutive batches have been inspected with no batches rejected.
3. Deliveries have been acceptably regular over a period of at least three months.

The inspector must decide whether deliveries have been sufficiently regular. The objective is to ensure that approval is not based, for example, upon four deliveries in the first two weeks and a fifth delivery in the last part of the third month. Because it is impossible to issue rules which are both simple and unambiguous, the inspector must assess and administer the switch.

2.3.4 Returning a Supplier to General Inspection Level II

The customer must understand and accept the fact that any PCB manufacturer can encounter production problems. With a safety net established in the form of prompt feedback from the production department, the customer should not need as rigorous an attitude as that expressed in MIL-STD-105D (Figure 1.8, case B) except in extraordinary cases. We shall, therefore, take a more lenient view, and set the following prerequisite for returning a supplier to General Inspection Level II:

> As soon as a batch is rejected, Special Inspection Level S3 is switched to General Inspection Level II, unless the PCB manufacturer can prove that a specific nonrecurring production fault caused the defect.

The rejection shall, as mentioned earlier, be based upon General Inspection Level II (normal inspection) and does not necessarily imply returning the whole batch. A screening could solve the problem of maintaining production.

2.3.5 Renewing Approval to Special Inspection Level S3

Since the incoming inspection department is naturally interested in returning to the lower inspection level as soon as possible, we present the following prerequisite:

> When five consecutive deliveries have been inspected according to General Inspection Level II and none has been rejected, Special Inspection Level S3 may be resumed.

Part 2

Defect Types and Inspection Procedures

Chapter 3

Initial Inspection

3.1 Introduction

This chapter contains a detailed discussion of quality criteria and the degree of inspection on which to base a practical PCB incoming inspection. Unlike most other items inspected by the incoming inspection department, printed circuit boards are subject to an immense number of different types of defects.

3.1.1 Examination Sequence

The inspector will not examine a board for one type of defect first, then for another type of defect, and so on, but assesses the board in general and observes possible abnormalities and defects. Therefore, he must have a broad background knowledge of the types of defects, so that he can identify any possible defects present in a given delivery of boards.

3.1.2 The Inspector's Authority

In some companies the inspector's authority is restricted to deciding whether the boards meet the required quality demands, which are generally tailored to the use of the end product. Especially in borderline cases, it may be difficult to decide whether a defective board must be rejected, repaired, or used as-is. Most specifications for printed circuit boards have been developed and based upon experience of what can be achieved under reasonable and normal production conditions.

In certain cases, departing from the specifications, even to a relatively high degree, is of no importance. The decision regarding the subsequent fate of the boards is therefore left to a committee, often called the "Material Review Board," acquainted with PCB technology as well as the user requirements. The committee decides whether the boards shall be discarded or whether they shall be accepted after the defects have been repaired, either by the PCB manufacturer or in the plant.

It is important for the PCB design department to be represented on the committee, because in this way the PCB designer gets direct feedback about the defects found. In many cases, a defect is caused by marginal artwork, so even small changes can preclude the defect's reappearing in later deliveries.

In other companies, the decision is left to the inspector, but wherever the decision is made, sufficient knowledge must be present to permit assessing the actual usability of the boards. In Part 2, the greatest importance is therefore attached to describing and assessing a wide spectrum of possible incidences of defects. This part may be regarded as a textbook which the inspector can study and refer to when in doubt.

3.1.3 Unjustified Rejections

Many PCB manufacturers know from bitter experience that, in certain cases, a delivery is rejected on an unprofessional basis. For example, attaching too much importance to cosmetic effects such as scratches on the surface, and insufficient regard to the board's utility.

Such decisions are usually due to the inspector's lack of knowledge. He dares not exempt the board and so rejects it. Hopefully, the discussion of the various types of defects and their importance relative to the use of the board will contribute towards a more professional quality assessment of printed circuit boards.

3.1.4 Inspection Guidelines

Part 3 describes the routine inspection and surveys an expedient working method. Special measuring equipment and seldom-used measuring methods are dealt with in Part 4.

The author is fully aware of the controversial aspects implied in the attempt to lay down clear guidelines for the incoming inspection of printed circuit boards, especially when it is a matter of establishing quality criteria and the degree of inspection. The PCB manufacturer and the customer will see these things a little differently.

The literature provides little help concerning PCB inspection. This seems somewhat peculiar, because the incoming inspection of printed circuit boards supposedly occurs in most electronics companies. The following description seeks to correct this deficiency by detailing solutions to the problem.

3.2 Flow of Goods Around the Incoming Inspection Department

For a somewhat clearer view of the incoming inspection department's place in the organization, we will follow the flow of goods from the moment the boards are received until they are delivered to components stock after having been inspected. The company can arrange the flow of goods in several different ways. The block diagram of Figure 3.1 shows a typical example.

3.2.1 Receiving Department

The Receiving Department fills out the paperwork and marks the package(s) before forwarding the boards to incoming inspection. The shipment may go directly to the incoming inspection department or to an intermediate storeroom controlled by the incoming inspection department. The containers should be marked to indicate that the batch has not been inspected, so that it will not be forwarded to components stock without inspection.

3.2.2 Incoming Inspection Department

The inspector draws the desired sample, or the whole batch in the case of a 100 percent inspection, and then performs the incoming inspection. In borderline cases which the inspector is not authorized to decide, he may turn to the Material Review Board for more precise instructions. The inspection will have one of the following results:

1. The boards are accepted, or exempted if unimportant defects occur or some of the specifications are insignificantly exceeded. When the batch is accepted, the package should be marked to indicate that the batch has been released (accepted), after which it is forwarded to components stock.

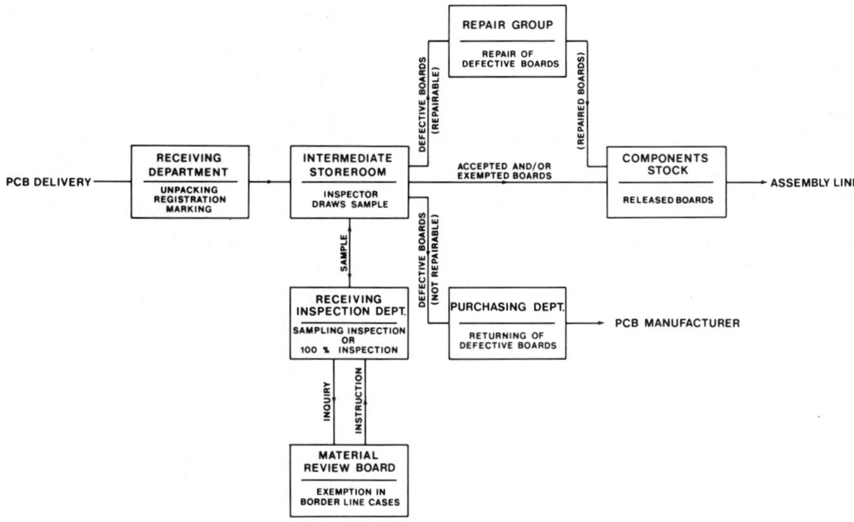

Fig. 3.1 Flow of goods from delivery to the assembly line.

2. The boards are rejected because of important defects, or some of the specifications are significantly exceeded. On rare occasions the whole batch may be rejected and returned to the PCB manufacturer for screening, remake, or rework. More frequently, a certain number of defective boards are rejected at the inspection, either from the sample or by a 100 percent inspection of the batch. The purchasing department usually returns the goods and requests that either a credit or replacement be given.
3. The customer repairs the boards. The customer with a pressing delivery deadline may make the repairs himself if this is possible. Modest defects may in special cases be remedied in the incoming inspection department, often at the same time as the inspection, but normally the incoming inspection department should not be responsible for repairs. In most electronics companies, defective boards are sent for repair to a special group within the production department, which is experienced in executing engineering change orders. See the remarks in Paragraph 1.2.4c concerning coverage of the repair expenses. After the repair has been completed, the boards are forwarded to the components stock.

3.2.3 Components Stock

In the components stock, the boards are counted and stored until needed. The purchasing department is informed of the number of boards received by components stock so that, in case of an incomplete delivery, the PCB manufacturer can be contacted.

3.3 Preparation for Inspection

Before actually inspecting a batch of printed circuit boards, the inspector must procure relevant information as illustrated in Figure 3.2 and discussed below. On the basis of this material only, he can decide whether the individual boards of the sample meet the quality requirements or not. The inspector enters the result of

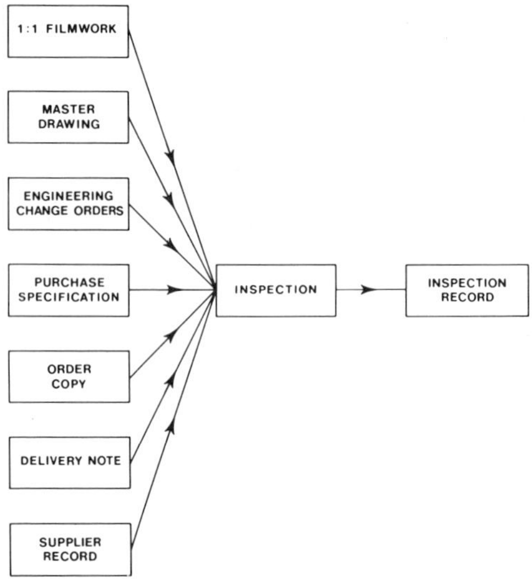

Fig. 3.2 Documentation necessary for the incoming inspection of printed circuit boards.

his inspection in an inspection record, which may be drawn up in many ways.

The simplest way is to let the inspector, in his own wording, note the defects found. One must be wary, however, of differences in terminology between inspectors. Another way is to use a preprinted inspection record form which lists the most common types of defects. If this is the case, the inspector has only to mark the appropriate defects found, which means that the PCB manufacturer receives feedback in a uniform way. Part 3 deals with the inspection record.

3.3.1 Filmwork and Master Drawing

3.3.1a Importance of File Maintenance

It is important that a routine be established to ensure that the inspection department's file of 1:1 filmwork and master drawings is automatically kept up-to-date. It is not particularly efficient for the inspector to rush about getting the necessary information after the boards have arrived. In this connection one should remember that smaller as well as larger changes may have been introduced after an order was placed, sometimes even after manufacturing began.

3.3.1b Response to Discrepancies

When an "organized" discrepancy between the delivered boards and the issued documentation has been demonstrated, the inspector should not immediately blame the PCB manufacturer. Instead, he should contact the PCB design department because, in all probability, he was not informed of recent design or requirement changes. For this reason, it is critical that the PCB design department identify all engineering changes, including small and apparently insignificant ones, in new revisions. The need to follow up all changes by automatically issuing revised documentation is obvious.

3.3.2 Purchase Specification

The purchase specification is a vital document because it itemizes the technical specification and the quality level expressed by the AQL value and the level of inspection. It is common practice for the PCB documentation to be decisive in the case of a discrepancy between the purchase specification and the PCB documentation. See Paragraph 4.2.1 of Reference A12 of the Reading List.

3.3.3 Order and Delivery Note

Other desirable documentation is a copy of the order and the delivery note of the batch. These papers show the ordered and delivered quantity of boards as well as the revision ordered.

3.3.4 Supplier Record

The supplier record is the basis for determining the inspection level. If kept according to the guidelines of Paragraphs

2.3.1 to 2.3.3, it will always show the supplier's correct category: new or unstable supplier (Paragraph 2.2.1) or approved supplier (Paragraph 2.2.2).

3.3.4a Contents of Supplier Record

The supplier record can contain information concerning previous deliveries, including the type of board, order number, number of boards and delivery dates (Paragraph 3.3.3), and the results of the inspection (acceptance, exemption, exemption with repair, or rejection). It should also record the number of A- and B-defects.

In this way the inspector can form an accurate picture of the supplier's general capability of production and delivery. More detailed information on the nature of the individual defect types can be obtained from the inspection record.

3.3.4b Maintaining the Supplier Record

The supplier record preferably should be kept by either the incoming inspection department or by the purchasing department, where most commercial information is kept. As the purchase orders are usually placed by the purchasing department, it may be advantageous for the purchasing department to keep the supplier record so that the buyers have immediate access to supplier information.

Maintaining the supplier record is often considered superfluous since the inspection records already contain the necessary details. The problem is that when there are many deliveries from multiple suppliers, the inspector has difficulty following the quality level of individual PCB suppliers.

An up-to-date supplier record sounds an early warning of a downturn in a PCB manufacturer's quality level, by recording when an abnormally high level of defects begins.

Incidentally, the supplier record can be kept electronically, possibly by a small computer in the incoming inspection department, making it easy to generate statistics, such as types of defects by PCB manufacturers. This information reveals the strong and weak points of the different PCB manufacturers.

3.3.5 Designating the Inspection Level

In Paragraphs 2.2.1 and 2.2.2 we defined the two inspection levels:

New or unstable suppliers:
General Inspection Level II

Approved suppliers:
Special Inspection Level S3

Paragraph 2.3.3 describes procedures for approving a supplier to Special Inspection Level S3; Paragraphs 2.3.4 and 2.3.5 deal with the return to General Inspection Level II and renewed approval to Special Inspection Level S3.

3.3.6 Determining the Sample Size

In Paragraph 1.2.6g we chose AQL values of 1 percent for A-defects and 2.5 percent for B-defects. We base the sample size on the size of the delivery and the inspection level selected according to Paragraph 3.3.5. It is more convenient to base the sample size on Tables 1.2 and 1.3 (Paragraph 1.2.6h) rather than MIL-STD-105D.

3.3.7 Inspection Record

To ease the inspector's record keeping, we can design the record as a register of nearly all probable defects, as described in Paragraph 3.3. Because the inspector fills in the appropriate boxes with the number of A- and B-defects found, the inspection record also becomes an efficient means of feedback to the PCB manufacturer and indirectly contributes to improved quality. Part 3 describes the practical design and use of the inspection record.

3.4 Delivery Reception

Here we deal with the degree of inspection and the quality criteria as they apply to routine inspection. Part 4 describes the more detailed inspections which can be used in case of doubt or with the qualification of test boards from new PCB suppliers.

3.4.1 Official Guides

There are no complete or easily accessible official directions for performing routine inspection of printed circuit boards, although IPC-A-600C (published by the Institute for Interconnecting and Packaging Electronic Circuits) and BS 9760 to 9763 (issued by the British Standards Institution) give some guidance. The inspector should consider the following description an attempt to provide some guidance in routine inspection.

3.4.2 Purpose of Guidance

The description mainly seeks to provide a reasonably wide information background so that the inspector can assess borderline cases not explicitly described. The author admits that, in some respects, these guidelines may arouse discussion or criticism, but he asserts that, with all its good and bad aspects, it has proven a usable basis for routine work. At any rate, it may serve as an inspiration to develop meaningful instructions for routine circuit board inspections, which is ignored all too often.

3.4.3 Recording A- and B-Defects

The number of A- and B-defects found during inspection are noted on the inspection record, which is filed to document the delivery quality. The record furthermore serves as a convenient guide for the inspector because all normal types of defects that occur are listed in the most practical sequence. It should be noted that defects of the same nature, such as voids in the hole barrel of a plated-through hole, are recorded as one defect per board if the number of plated-through holes with voids exceeds the acceptance level (Paragraph 7.3.3d).

3.4.4 Data Registration

The inspection record heading must be filled in with relevant data for the delivery in question, including the PCB part number and revision ordered, the order number, the PCB manufacturer, and the delivery number or job number as stated on the delivery note.

The inspection data and inspector's signature complete the inspection record.

3.4.5 Packing

3.4.5a External Damage

If the outer container is damaged, the fact should be noted on the inspection record. Printed circuit boards are commonly shipped in cardboard boxes. Because even a medium-sized box filled with boards is quite heavy, a weakly constructed box can fail during handling, exposing the contents to damage.

Even if the contents were not damaged, the inspector should report the box failure to the manufacturer and recommend that he either use a stronger box or improve his packing procedures.

3.4.5b Packing Specifications

The next step is to verify that the individual boards are packed according to valid purchase specifications. Various electronics companies may have different packaging requirements; some companies require that boards larger than a certain size, such as 4 by 6.3 in. (100 by 160 mm), the Eurocard size, be in separate plastic bags. Smaller boards may be required to be packed 25 per plastic bag, separated from each other by tissue paper.

Other companies do not demand individual packaging of large boards, usually for economic reasons. To preserve shelf life, the tissue paper must contain no sulfur, because sulfur will, in time, attack the board surfaces, with a drastic reduction in solderability. For the same reason, paper bags are not used. Only plastic bags, usually of polyethylene, are acceptable.

Opinions differ on the proper method of sealing the plastic bags. Folding the open end and securing it with a piece of tape allows the boards to breathe and dissipate moisture and other gases from the board. On the other hand, completely closed bags, whether by a zip-fastener or tape sealing, offer excellent environmental protection. Plastic welding provides the best protection, especially when used on laminated plastic bags. The welding must be done correctly to prevent forming gases which reduce solderability.

No matter which closing method is chosen, it is undesirable to store bare-copper boards for long periods, such as more than a few months, because the copper surface inevitably oxidizes and increases the soldering time to the point that mass soldering becomes impossible.

3.4.5c Handling Deviations

All deviations from the specified packing should be noted on the inspection record, but should not be included in the number of defects because the solderability will not have deteriorated measurably. The boards should be repacked, however, to prevent later deterioration.

3.4.6 Counting the Boards

3.4.6a Responsibility for Counting

Board counting may not necesssarily be the inspector's job. As mentioned in Paragraph 3.2.3, the stockroom personnel will count the boards because they are responsible for the number of boards received and delivered to production. Eventually, someone will check whether the delivered number of boards complies with the ordered quantity, within the tolerance stated in the purchase specification. As an example, the delivery tolerance may be −0, +2 pieces for batch sizes up to 20, and −0, +10 percent for batch sizes greater than 20.

3.4.6b Segregation of Production Runs

For all boards of the same part number to be inspected as a single lot, they must have been produced in the same production run. As long as any possible age limit is not exceeded, there is nothing wrong with receiving boards from different production runs when some of the boards are part of an overrun from an earlier order. If so, boards from different runs or batches must be segregated so their sources can be identified when the inspector draws the sample. Furthermore, the boards themselves should be marked with the actual manufacturing date.

3.4.6c Mixed Deliveries

Sometimes boards in mixed deliveries are not segregated. When excess boards are date-marked immediately before shipment, it becomes very difficult to distinguish between boards from different batches. If the user company suspects that this is happening, it must look for small dissimilarities which hint at a mixed delivery. These include inspection stamps, base material (color or watermarks), surface appearance, contour roughness, solder mask, etc. If the suspicion is justified, the customer should contact the PCB manufacturer.

A safe solution is to specify that the date marking be screen printed with the component notation, the date being the planned date of shipment.

3.4.7 Artwork Issues

The first thing to inspect is that the correct artwork issues for the solder side, component side, solder mask, insulation mask, and the component notation have been used. The identification system is usually organized so that the master drawing is the controlling drawing which defines the ordered boards by the issues or revision levels of the aforementioned artworks.

3.4.7a Master Drawing

The master drawing itself is characterized by a change letter/figure or revision letter/figure. Based on the revision level stated in the order, the inspector can identify the master drawing to be used, and from this the appropriate artwork revisions. See the Reading List, Reference A12, Paragraph 8.9.2.

3.4.7b Artwork Verification

The finished boards should always show the artwork revisions as etchings in the pattern or as markings in the masks and the notation. See the Reading List, Reference A12 Paragraph 8.5. Where this is done, the inspector can easily check whether the manufacturer has used correct revisions according to the master drawing.

If the artwork revisions are not printed on the boards, the inspector must use positive and negative 1:1 films to check the correctness of the pattern. The films are, in turn, laid over the pattern. With the positive film, the inspector verifies that no unwanted conductors or pads are present, and with the negative film that no parts of the pattern are missing. It should be noted that this investigation is not concerned with the general inspection of the pattern as regards the image transfer, etching, etc. Chapter 5 deals with this topic.

3.4.7c Extent of Pattern Check

When the boards are manufactured in a single batch, it is unnecessary to check more than one board for the correct pattern. Because of the step-and-repeat technique, the multiple film will show the same artwork for all the individual patterns, making it unnecessary to check more than one board. At this stage wc arc not interested in possible defects caused by the subsequent manufacturing processes.

3.4.7d Disposal of Incorrectly Patterned Boards

If the boards have been manufactured to the wrong revision, they are probably useless, which means that the defect is an A-defect. The registration system is based on the assumption that even the smallest change is reflected in new artwork levels. A new level may therefore cover many degrees of change from insignificant ones right up to large, functional circuit changes.

It is frequently possible to obtain an exemption when the changes are small. One reason is that they do not affect the function of the product or may only provide for easier assembly or testing. Another reason is that the situation, in many cases, can be remedied by cutting some conductors and soldering some components to the solder side of the board. This procedure may seem unprofessional, but it is followed by nearly all electronics companies.

Such changes, when accompanied by good workmanship, do not affect the board reliability. The inspector should, therefore, investigate, for example through the engineering department or the committee mentioned in Paragraph 3.1.2, whether it is acceptable to modify the delivered boards under the assumption that the extent is reasonably small. It is tempting to ask whether the boards in an obsolete revision would be discarded categorically if the company itself and not the PCB manufacturer were at fault.

3.4.8 Drawing the Sample

Having accepted the batch as delivered in the correct revision or exempted it for insignificant discrepancies, the inspector can draw the sample. The inspector has ascertained whether the batch is pure or mixed according to Paragraph 3.4.6c. In the case of a mixed batch, the inspector performs the sampling inspection as several separate inspections; in other words, he draws samples from the partial deliveries so that separate records can be kept for each.

In the following discussion, we assume that the delivery is pure, that is, the normal case. The sample size, determined according to Paragraph 3.3.6, is now noted on the inspection record.

In principle, the sample should be random; that is, drawn from different parts of the batch. As mentioned earlier, the boards are usually packed in a number of bags, each containing a certain number of boards. The safest method is for the inspector to draw samples from each bag and from different places in the individual bags.

In practice, such rigor is not necessary because the boards will be packed randomly due to the many processes to which they have been subjected during manufacture. It is sufficient that the inspector open a few of the bags and draw the necessary number of boards from these bags.

This procedure has two advantages: It simplifies sample selection and allows most of the delivery to be placed in stock unopened, providing the maximum shelf life. The bags do not need to be opened to allow the boards to be counted; this can be done by weighing the bags or measuring the stack height.

3.4.9 Board Markings

The next step is to verify all board markings. These can be a date marking; the PCB manufacturer's name, logo, or code; and certain certifications, such as a UL (Underwriters Laboratories, Inc.) flammability rating.

Because the markings vary among manufacturers, it is more practical to leave them out of the artwork and let the PCB manufacturer insert his markings on the artwork before photo reduction, or on the 1:1 master films, with transfer letters and figures. Since these markings will be applied automatically to the boards, inspection of more than one board is superfluous. The date marking can be stamped or preferably screen printed with the company notation. Possible faults or deficiencies are normally characterized as B-defects and entered as such in the inspection record.

A missing UL marking can be an A-defect, particularly if a nonrecognized base material is used. For this reason, a guarantee of having used UL-recognized base material of correct flammability rating should be procured from the PCB manufacturer. The inspector should furthermore realize that it is not sufficient to use a flame-retardant base material because the UL recognition implies the use of UL-recognized materials. (See Reading List, Reference A19.)

As the UL markings are frequently stated as codes, it is necessary that the inspector be provided with a list of approved codes. The author has experienced cases where insufficient information resulted in boards with too low a flammability rating passing unnoticed through incoming inspection, and later caused the rejection of a large number of assembled boards.

Chapter 4
Base Material Inspection

4.1 Incorrect Base Material

When the board is required to be flame retardant, for example of type FR4, the retardancy must be verified. It is, however, not necessary to perform burning tests because all flame-retardant base materials above 0.020 in. (0.5 mm), and characterized by the designation FR have red watermarks which normally correspond to the logo of the manufacturer of the base material. See Figure 4.1. International standards allow a maximum distance of 3 in. (75 mm) between the watermarks, so that even a relatively small board usually contains at least one watermark.

In principle, using the wrong base material is an A-defect, so the delivery is rejected. Nevertheless, in many applications it makes no difference at all whether FR4 is replaced by G10, which corresponds very closely to FR4 in all respects other than flame retardancy. The inspector should therefore ascertain whether it is possible to exempt isolated cases. He should, of course, also contact the PCB manufacturer and request the specified base material for subsequent deliveries.

The question of accepting FR4 instead of G10 is arguable as there can be situations, although extremely rare, in which the bromine in the FR4 materials poses a greater hazard than the flammability of G10. In such cases the PCB designer should specify the material to be "G10, not FR4" and not merely "G10" so that the PCB manufacturer clearly understands that he is not allowed to replace G10 with FR4.

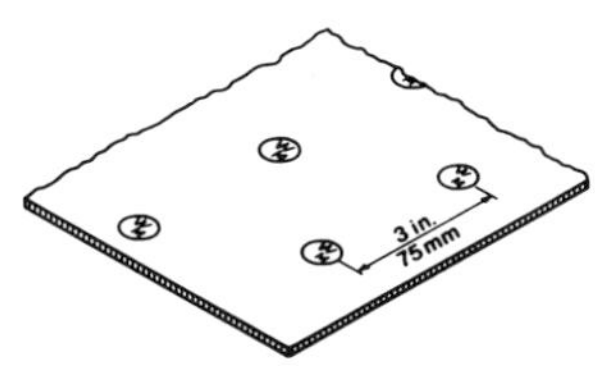

Fig. 4.1. Watermarks in base material.

Nonretardant base materials, such as G10, do not have red watermarks. They can have various colors, black or blue being the most common. Some base materials lack watermarks entirely. If the boards ordered were to be made of a G10 base material and they were made of FR4, the quality is just a little higher. Excluding the case of bromine hazard, there is no reason to comment on this in the inspection record, much less to complain. As time goes on, boards made of flame-retardant base materials are becoming more common so that economically it turns out to be more advantageous for the PCB manufacturer to stock only FR4 instead of both FR4 and G10.

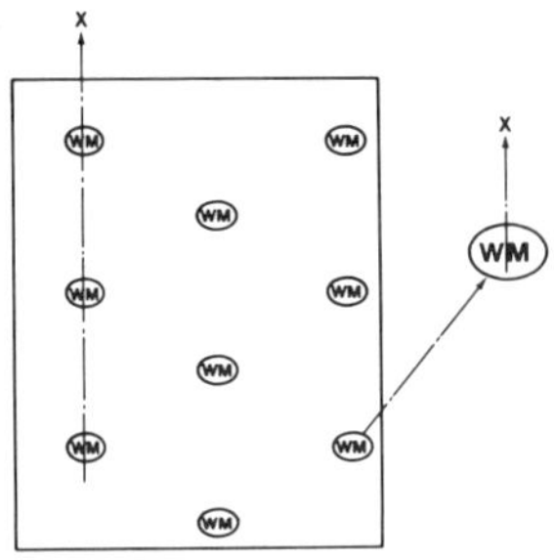

Fig. 4.2. Identifying lengthwise direction of a board.

4.2 Main Direction of Base Material

In exceptional cases it may be specified that the pattern of the board shall be placed in a certain way relative to the main direction of the glass cloth, so that the largest flexural strength and lowest coefficient of thermal expansion are attained. The main direction, also called the lengthwise direction or the X-axis direction, is usually indicated by means of the watermarks, either by a special part of the watermarks or by a line which can be drawn through the watermarks. Figure 4.2 illustrates these differences. The crosswise direction is also designated the Y-axis direction, and the direction perpendicular to the board plane is called the Z-axis direction. See Figure 4.3. The difference between the flexural strength and the thermal expansion in the X- and Y-axis directions is not particularly large. This appears from the numbers given below, which are valid for FR4 base materials:

Flexural Strength:
in the X-axis direction:
60,000 psi (4200 kg/cm^2)
in the Y-axis direction:
50,000 psi (3500 kg/cm^2)

Thermal coefficient of expansion, in./in. (mm/mm) per °C:
in the Y-axis direction:
1×10^{-5}
in the Y-axis direction:
1 to 1.5×10^{-5}

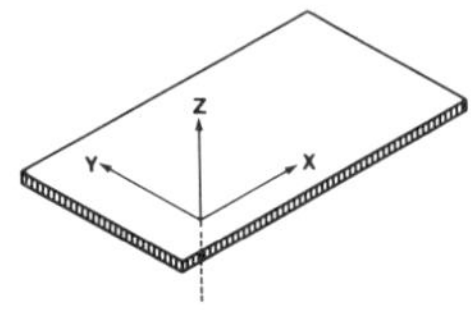

Fig. 4.3. X, Y and Z directions of a board.

Note: Sometimes the expressions "warp," "gill," and "weft" are found in the technical literature. Warp means the threads in the long direction, and gill or weft means the threads in the short direction, taken over and under the warp in weaving.

All boards of a batch are usually oriented in the same direction on the base material, but if the PCB manufacturer uses a computer program to minimize the waste of base material, some boards may be turned 90 degrees with respect to the others. Therefore, when a certain direction is specified, it is not enough to inspect one board, so a sample is drawn in the usual way. One reason for specifying a certain direction can be a wish to minimize the coefficient of thermal expansion in the long direction of boards intended for automatic component insertion.

Material movement during manufacturing, especially on large boards with small solder pads relative to hole sizes, and also on multilayer boards, can also be a factor in determining desirable warp and weft directions.

Failure to meet the requirement regarding a certain direction is an A-defect if this is a disadvantage in the use of the boards. The inspector should not overlook the fact that the board may be overspecified; the difference in the flexural strength and thermal expansion is insignificant in most applications. The inspector should therefore obtain a decision on whether the fault can be exempted, and possibly have the requirement canceled since it only adds to the cost of the boards.

4.3 Incorrect Thickness of Base Material

4.3.1 Effect of Thickness Variation

Except for boards having edge connectors intended for mating into a fixed connector, the thickness of the base material is usually of little functional importance. This is supported by the fact that many electronics companies in Europe specify a thickness of 1.5 to 1.6 mm (0.059 to 0.063 in.), not least because base materials in many cases are purchased by weight. It should be noted that a thickness of 1.5 mm (0.059 in.) is a special European thickness, not standard in the United States, which uses a standard thickness of 0.063 in. (1.6 mm).

With reference to the above considerations, we will consider out-of-tolerance thickness as a B-defect and, moreover, reduce the sample size to one or two pieces.

In general, it is sufficient to verify that the PCB manufacturer has used a base material of correct nominal thickness, and not a thinner or thicker base material. The thickness increments appear in Table 4.1, which shows thickness tolerances in accordance with various standards. As it is normally impossible to find base materials with other thicknesses than those given in the table, it is quite a simple matter to check the nominal thickness with a slide gauge. See Figure 4.4.

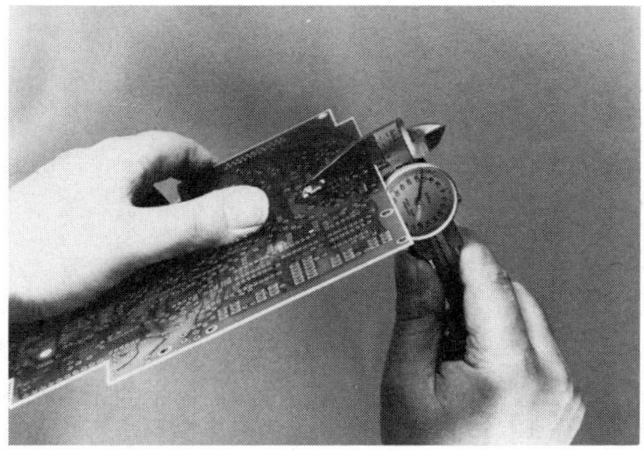

Fig. 4.4B. Measuring the nominal board thickness with a caliper equipped with a dial indicator.

4.3.2 Measuring the Thickness

When it is important to the use of the boards that the thickness tolerance be met, the measurement should be made with a micrometer gauge. See Figure 4.5. The measurement is made at four different places, such as the board corners, after which the average is calculated. In this case, the inspector should not measure on the board edge, but a small distance away from the edge, because profiling may affect the thickness.

4.3.2a Single-Layer Boards (Single- and Double-Sided Boards)

It deserves notice that the thickness tolerances include the copper foil, so when

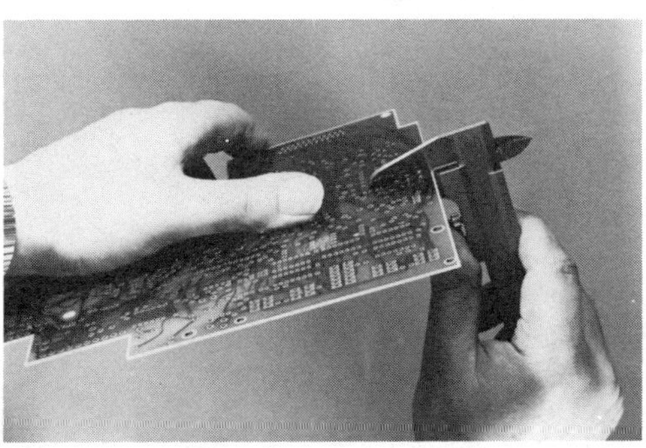

Fig. 4.4A. Measuring the nominal board thickness with an ordinary caliper.

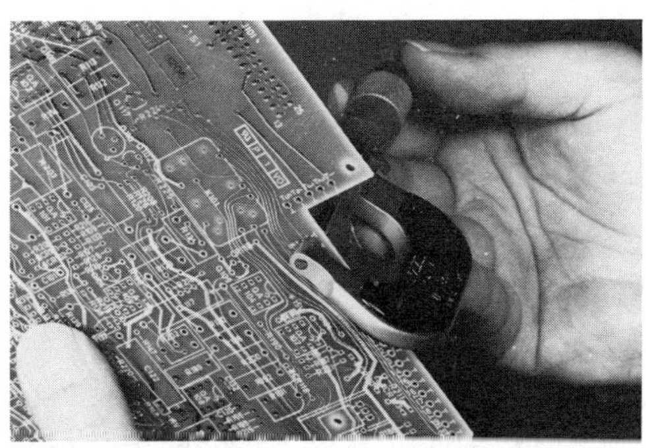

Fig. 4.5 Measuring the nominal board thickness with a micrometer.

Table 4.1. Thickness tolerances of base materials FR4/G10 according to various standards.

Thickness		NEMA-LI-1				DIN 40802				MIL-P-13949E					
		Class I		Class II		Normal		Tightened		Class I		Class II		Class III	
mm	in.	± mm	± in.	± mm	± in.	± mm	± in.	± mm	± in.	± mm	± in.	± mm	± in.	± mm	± in.
0.8	0.031	0.17	0.0065	0.10	0.004	0.15	0.006	0.09	0.0035	0.17	0.0065	0.10	0.004	0.08	0.003
1.0	0.040					0.17	0.0065	0.11	0.0045						
1.2	0.047					0.18	0.007	0.12	0.0047						
1.5	0.059					0.20	0.008	0.14	0.0055						
1.6	0.062	0.19	0.0075	0.13	0.005					0.19	0.0075	0.13	0.005	0.08	0.003
2.0	0.079					0.23	0.009	0.15	0.006						
2.4	0.093	0.23	0.009	0.18	0.007					0.23	0.009	0.18	0.007	0.10	0.004
3.2	0.125	0.30	0.012	0.23	0.009					0.30	0.012	0.23	0.009	0.13	0.005

the thickness is measured directly on the base material, it is necessary for double-sided boards to add twice the thickness of the copper foil. For the usual 1/2 or 1 oz. weight, the thickness is 0.0007 or 0.0014 in. (17.5 or 35 μm), respectively. When measuring on tin/lead-plated surfaces of double-sided boards, it is necessary to subtract twice the plating thickness. This is calculated as:

> 2 × [0.001 in. (25 μm) copper + 0.0004 in. (10 μm) tin/lead] + a plating tolerance of 50 percent or a total of approximately 0.004 in. (0.1 mm).

This measurement is somewhat unreliable for two reasons: plating tolerance and changes in the tin/lead deposit surface structure after reflowing. In addition, if solder and insulation masks have been applied, it is also necessary to correct for the thickness of the masks; that is,

> 2 × [0.0004 in. (10 μm) to 0.001 in. (25 μm)] or anywhere from 0.0008 in. (20 μm) to 0.002 in. (50 μm)

should be subtracted. In this case the measuring accuracy is so low that the method is useless.

The inspector can use a very moderate sample size of one or two pieces. Information on the standard and tolerance class to be used can be found on the master drawing or in the purchase specification. See Table 4.1. If the tolerance has been exceeded, for example because the manufacturer has incorrectly used a base material with a wide tolerance range (Class I), the inspection is extended to General Inspection Level II.

When it is necessary to inspect the board thickness with great accuracy, the PCB designer can facilitate the inspector's job by introducing, at appropriate places, apertures in the solder and insulation masks, possibly also in any ground planes to permit measuring the thickness of the bare base material with a micrometer gauge.

Where the thickness tolerance has been exceeded, we have an A-defect. The question now is whether the batch is to be rejected. The problem is that the thickness specification is often adapted from published standards and exceeding the tolerance does not necessarily mean that the boards cannot be used. The inspector should therefore, through appropriate channels such as the engineering department, get a decision as to how important it is to meet the thickness tolerance, so that the defect can possibly be exempted.

4.3.2b Multilayer Boards

In contrast to ordinary double-sided boards, the thickness specification of multilayer boards usually excludes the copper foil. The PCB designer should realize that most PCB manufacturers strongly desire that master drawings for multilayer boards, in addition to the thickness specification, also indicate how the thickness is measured; that is, including or excluding the pattern height (copper foil plus plating). The inspector should be aware of this detail.

4.3.2c Strip-Line Boards

In the case of strip-line boards, the thickness of the base material enters the calculations of the characteristic impedance. The inspector should note that the thickness of the copper foil is not included in the thickness of base materials intended for microwave applications. Usually there is a demand for very close tolerances for such materials (see Table 4.2) and violating the thickness tolerance should be considered an A-defect.

Table 4.2. Thickness tolerances of base materials for microwave applications according to MIL-P-13949E.

Thickness		Tolerance	
mm	in.	± mm	± in.
0.76	0.030	0.05	0.0020
1.52	0.060	0.05	0.0020
2.29	0.090	0.08	0.0030
3.05	0.120	0.09	0.0035
6.10	0.240	0.10	0.0040

4.3.2d Edge Connector Boards

A special problem arises with edge connector boards because manufacturers of receptacles set fairly strict demands on the board thickness within the edge connector area. Since the PCB manufacturer can produce only from base materials available on the market, he has to accept the thickness tolerances in force. He must therefore be given a certain level of relief.

The most common manufacturing method for gold-plated edge connectors is that the tin/lead deposit is stripped, after which nickel plating is applied followed by gold plating (see Reference A12, Paragraphs 3.7.4 and 4.9.3). The extra thickness of a single contact tab originates from 0.001 in. (25 μm) of copper plating plus a plating tolerance of +50 percent, 0.0002 to 0.00028 in. (5 to 7 μm) of nickel plating and 0.00004 to 0.00008 in. (1 to 2 μm) of gold plating, with a total of about 0.0018 in. (45 μm).

For a double-sided edge connector the extra thickness is therefore 0.0036 in. (90 μm) in addition to the thickness of the base material including the copper foil. In the case of a 0.063 in. (1.6 mm) printed circuit board, the thickness measured over the edge connector tabs must necessarily vary as follows:

Class I:

$0.063 + 0.0036 \pm 0.0075 =$
0.059 to 0.074 in.

$1.6 + 0.09 \pm 0.19 =$
1.50 to 1.88 mm

Class II:

$0.063 + 0.0036 \pm 0.0050 =$
0.062 to 0.072 in.

$1.6 + 0.09 \pm 0.13 =$
1.56 to 1.82 mm

Some connector manufacturers specify a working range of 0.056 to 0.070 in. (1.42 to 1.78 mm) for the thickness of the tab area. Obviously, in the case of base materials of Class I, and under worst-case conditions, the thickness of the edge connector area will have an excess of 0.004 in. (0.1 mm), for which reason base materials of Class II should be specified. In this case too, the thickness tolerance may be exceeded, but as the excess amounts to 0.0016 in. (0.04 mm) maximum, it may be neglected.

Since correct thickness of the edge connector tabs is usually quite important, we may choose to inspect the boards according to General Inspection Level II. With a micrometer gauge, the thickness across pairs of connector tabs is measured at several places as shown in Figure 4.6, after which the average thickness is calculated. If the thickness is outside the range of 0.062 to 0.072 in. (1.56 to 1.82 mm), the

Fig. 4.6. Measuring the thickness across a pair of connector tabs with a micrometer.

board should be rejected, unless the receptacle contacts have a larger working range than we have assumed here.

When the PCB manufacturer has a good record, the inspection level can be reduced to Special Inspection Level S3, or in certain cases, the inspection may possibly be omitted.

4.4 Warp and Twist

Warp and twist of the finished board must also be inspected because if these are excessive, they will make it difficult to assemble or solder the board, or to insert pluggable boards into receptacles.

4.4.1 Effects on Mass Soldering

When a board is to be mass soldered, it is suspended in a carriage which passes through the soldering machine, but the carriage may not correct the warp or twist enough for satisfactory soldering.

In extreme cases it may be necessary to introduce some support to the center of the board, either upward or downward, in order to reduce the warp. It helps with large boards, such as those larger than 8 by 8 in. (200 by 200 mm), which may exhibit a warp of up to 0.12 in. (3 mm), if the PCB designer introduces some fixing holes intended for a mechanical flattening of the board during mass soldering.

4.4.2 Mechanical Effects after Soldering

Another problem is the mechanical stress to which the components are subjected after soldering, when the board tries to resume the original warp or twist. The normally accepted values of warp and twist will not affect the components, but for safety it is recommended that one of the leads of fragile components, such as diodes, are provided with a stress relief loop.

4.4.3 Effects on Plug-In Boards

Inserting a plug-in board becomes more difficult the more it is warped, not so much because of the friction against the board guides, but rather on account of contact mating problems.

4.4.4 Causes of Warp and Twist

It may be a rather difficult matter to place the responsibility for excessively warped or twisted boards, because there are several different possible causes:

1. The base material may be stressed internally because of an unbalanced laminate buildup. This fault does not necessarily show up in the raw material itself, but is revealed after a number of manufacturing processes have been performed; for example, etching, where the mechanical tension in the copper foil is relieved.
2. The PCB manufacturer's processes may be too hard on the board. For instance, the temperature setting of the reflow machine may be too high, or the conveyor speed may be too slow.
3. The design may be unfavorable because the two sides of the board may differ substantially in circuit density. The PCB designer should always aim for a balanced pattern, including a relieving of large ground plane areas. See Paragraphs 6.6.3 and 6.6.4 of Reference A12, Reading List. In the case of multilayer boards, an asymmetrical buildup of the layers will constitute a serious cause of warping, so the PCB designer should always try for a symmetrical buildup.

4.4.5 Flattening Warped Boards

The PCB manufacturer may possibly succeed in flattening a warped board by applying pressure to bend it back, or by subjecting it to a heat treatment around the softening temperature of the base material. In the case of epoxy, the softening temperature is also called the glass transition temperature (248–257°F, 120–125°C). See Paragraph 8.2.1 for further details. This frequently proves to be a temporary flattening, because in time the board tends to revert to its original shape.

4.4.6 Establishing Flatness Requirements

If very strict demands are set on the board flatness, it is far better to discuss the problem with the PCB manufacturer, who may possibly choose quite different measures. He may, for instance, select another, possibly slightly more expensive, base material with less tendency to warp or twist.

Although it may be necessary to reduce the demands regarding other characteristics, such as machinability, the board's performance is not necessarily reduced. As already stated, the PCB designer can also contribute by designing the board so that it is less inclined to warp or twist. The aforementioned conditions imply that the company should avoid unnecessarily strict requirements on board flatness, since this will increase the cost.

4.4.7 Measuring the Warp and Twist

Before we more fully describe acceptable values of warp and twist, let us see how they are measured.

Warp is the ratio between the rise R of the curved side and the longest edge of the board and is expressed in in. per in. (mm per mm) or as a percentage. According to MIL-P-13949 and IPC-A-600C, the warp is measured by placing the board on a flat surface with the convex side up, as shown in Figure 4.7. The rise R is then defined as the diameter of the largest cylindrical feeler gauge which can be inserted under the board without lifting it.

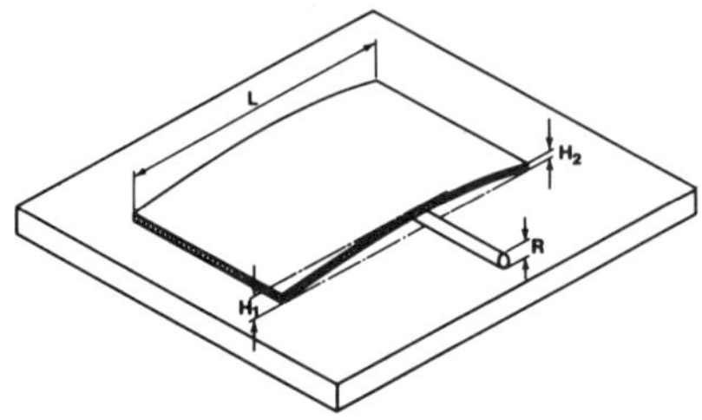

Fig. 4.7. Measuring warp with a cylindrical feeler gauge or a height indicator gauge.

The rise R can also be determined by measuring the heights H_1 and H_2 with a height gauge and then subtracting H_2 from H_1.

Great care should be taken to not depress the board when measuring H_1. According to British Standard BS 9760, the board is turned the opposite way, that is, with the concave side up, after which a light ruler is placed across the board as shown in Figure 4.8. The rise R is defined as the maximum clearance between the ruler and the board. In principle, all these methods give the same result.

Twist is defined in a similar manner to warp; that is, the ratio between the largest deviation from flatness and the diagonal of the board. Therefore the twist is also expressed in in. per in. (mm per mm) or as a percentage. In this case, it is most practical to place the board on a flat surface with the concave side up and, with a

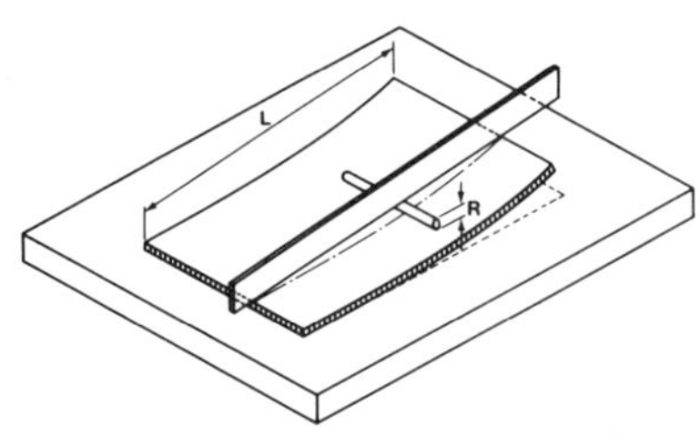

Fig. 4.8. Measuring warp with a ruler and a feeler gauge.

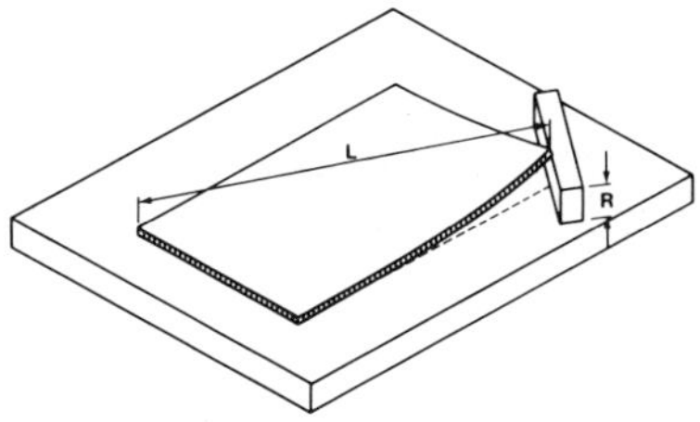

Fig. 4.9 Measuring twist with a feeler gauge.

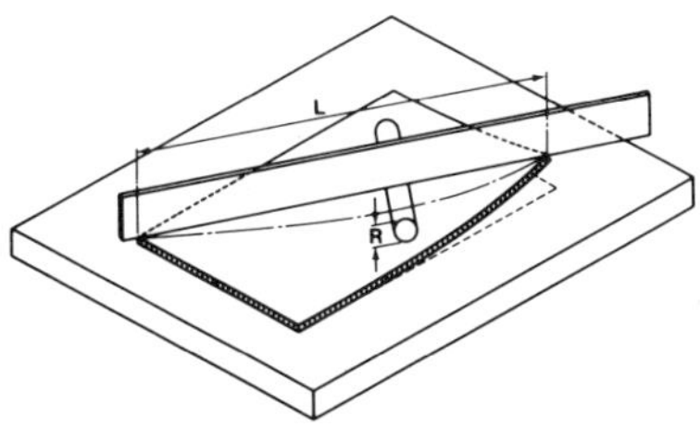

Fig. 4.10. Measuring twist with a ruler and a feeler gauge.

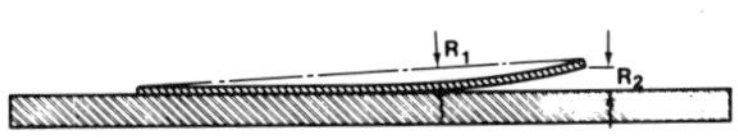

Fig. 4.11. Difference between the measuring methods illustrated in Figures 4.9 and 4.10.

feeler gauge, ascertain the largest deviation from the flat surface (see Figure 4.9), usually at one of the corners, or as the maximum clearance between the board and a light ruler placed across the corners (Figure 4.10). Figure 4.11 shows how a slightly larger value of twist results from the method shown in Figure 4.9 than that of Figure 4.10. The difference, caused by geometry, is of no practical importance because of the relatively small twist that actually occurs.

4.4.8 Specifications for Single- and Double-Sided Boards

There are many different specifications for warp and twist, official as well as unofficial. For general information, Table 4.3 shows the maximum acceptable values, taken from such specifications, valid for single-sided and double-sided boards of epoxy glass, whereas Table 4.4 shows corresponding values for multilayer boards.

By using Table 4.3 and adopting an adequately conservative attitude, we can lay down the following values for warp and twist for both single-sided and double-sided epoxy glass boards, regardless of size:

Thickness		**Warp or Twist**
in.	*mm*	*in./in.* *mm/mm*
0.031	0.8	0.015
0.063	1.6	0.015
0.094	2.4	0.010
0.126	3.2	0.010

The reason for not distinguishing between single-sided and double-sided boards is that a double-sided board exhibiting considerable imbalance between the two sides behaves rather like a single-sided board with respect to warp and twist.

The imbalance can, for example, be caused by an almost unrelieved ground plane on one side and a thin signal layer on the other side. Another example is an elongated board, with all conductors on one side running in the X-axis direction and on the other side in the Y-axis direction. By avoiding the differentiation between single- and double-sided boards we can relieve the inspector from subjectively assessing the degree of balancing.

When tighter limits of flatness are required, the company must reach agreement with the PCB manufacturer on an individual basis, and be prepared to pay an added cost. Tighter limits may, for instance, be required in the case of boards with edge connectors; for a 0.063 in. (1.6 mm) board, a maximum warp along the edge of the connector area of 0.010 in / in. (mm/mm) may be specified, possibly

Table 4.3. Warp and twist specifications for epoxy glass boards according to various standards. (S indicates panel size.)

Thickness			Warp or twist, mm/mm or in./in.				
			MIL-P-13949E			Companies A and B	IPC-D-300
mm	in.	Type of board	$S \leq$ 8 in. / 200 mm	8 in. / 200 mm $< S \leq$ 12 in. / 300 mm	$S >$ 12 in. / 300 mm	All sizes	All sizes
0.8	0.031	One side	0.015	0.015	0.020	0.015	0.015
		Two sides	0.005	0.010	0.010	0.015	0.01
1.6	0.062	One side	0.010	0.010	0.015	0.015	0.010
		Two sides	0.005	0.005	0.005	0.015	0.007

Table 4.4. Warp and twist specifications for multilayer boards, valid for all thicknesses, according to IPC-ML-950.

	Warp or twist, mm/mm or in./in.	
Class A	Stringent functional requirements	0.010
Class B	Normal functional requirements	0.030
Class C	Minimum functional requirements	No requirement

with a maximum absolute value of 0.059 in. (1.5 mm) regardless of the connector length.

If obtaining boards sufficiently flat along the connector area becomes a problem, a practical solution is to mount a stiffening rail just above the contact tabs as shown in Figure 4.12. The rail flattens the board in the edge connector direction and makes it possible to accept the general specification of 0.015 in./in. (mm/mm), which should solve all problems.

4.4.9 Specifications for Multilayer Boards

For multilayer boards, the following values of warp and twist are valid for all board thicknesses:

Class I, stringent functional demands: 0.010 in./in. (mm/mm)
Class II, normal functional demands: 0.03 in./in. (mm/mm)

The master drawing for the multilayer board must, of course, state the class desired.

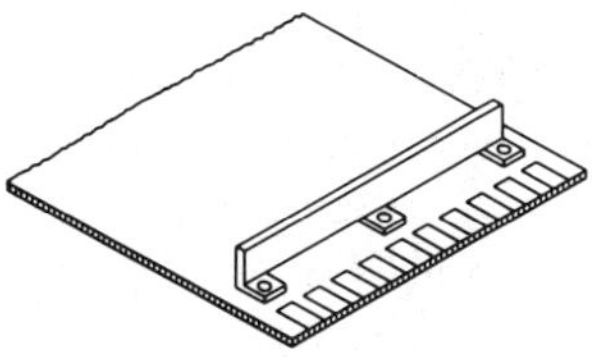

Fig. 4.12. Flattening of warped board along the edge connector with a stiffening rail.

4.4.10 Level of Inspection

Irrespective of the inspection level determined in Paragraph 3.3.5, we choose, for both supplier categories, to inspect according to Special Inspection Level S3 because warp and twist presumably depend more on the board design and base material than on the processes used by the PCB manufacturer. Exceeding the specified values of warp and twist is normally considered an A-defect.

4.5 Defects in the Base Material

In epoxy glass base materials a number of defects can occur:

- Measling
- Crazing
- Blistering
- Delamination
- Haloing
- Weave texture
- Weave exposure
- Incomplete curing
- Cracks
- Metallic inclusions
- Other inclusions

The inspector must always distinguish between the nature of the defects: Are they functionally important or merely cosmetic? Functional defects are A-defects, whereas local cosmetic defects can be disregarded with no remarks in the inspection record. There is, however, a limit to how many cosmetic defects can be accepted without a risk of seriously degrading the end product. Accordingly, if cosmetic defects occur all over the board, it should be rejected and the defect rated as an A-defect. By this time, it should be clear that laying down firm guidelines for assessing such defects is impossible; in the end, the assessment depends on the inspector's subjective judgment.

It is, however, essential that the inspector be familiar with the existence of various types of defects and their importance; for this reason, the following paragraphs describe them in detail. Here, we shall disregard the inspection level determined in Paragraph 3.3.5 and inspect according to Special Inspection Level S3 for both categories of suppliers.

4.5.1 Measling

Measling appears as white spots or crosses just below the board surface and is caused by small cavities where the fibers of the glass cloth intersect (Figure 4.13). Measle spots are tolerable if the following requirements are met:

1. Measle spots are widely scattered with an average density of no more than 50 spots per 16 sq. in. (about 100 cm^2). Figure 4.14 gives an impression of the accepted density.
2. No glass fibers are exposed on the board surface.

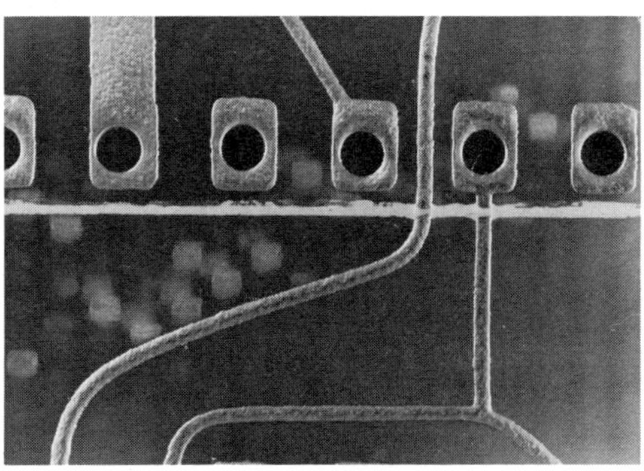

Fig. 4.13. Board with measling.

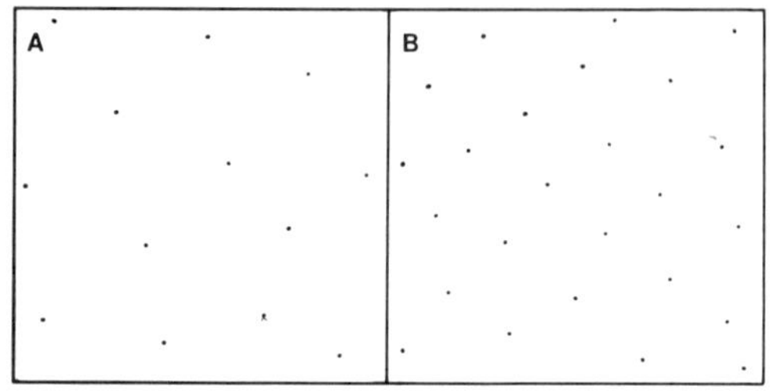

Fig. 4.14. Examples of measle spot density. Scale 1:1.
A: Acceptable density (50 spots per 16 sq.in. = 100 cm^2).
B: Unacceptable density (100 spots per 16 sq.in. = 100 cm^2).

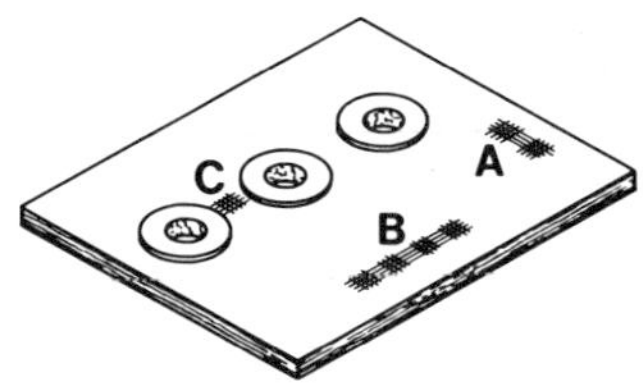

Fig. 4.15 Unacceptable occurrence of measle spots.
A and B: Two or more connected measle spots.
C: Solder pads connected by a measle spot.

3. No measle spots touch each other (Details A and B of Figure 4.15).
4. Solder pads and conductors of different polarity are not connected by one or more measle spots (Detail C, Figure 4.15).

If these requirements are not met, the board is rejected and the defect is an A-defect.

4.5.2 Crazing

Crazing, which can occur on or below the board surface, is best described as merging measle spots (Figure 4.16). Crazing usually occurs on machined surfaces and around larger drilled or punched holes. The porous crazed areas can absorb electrolyte from the plating baths, later lead-

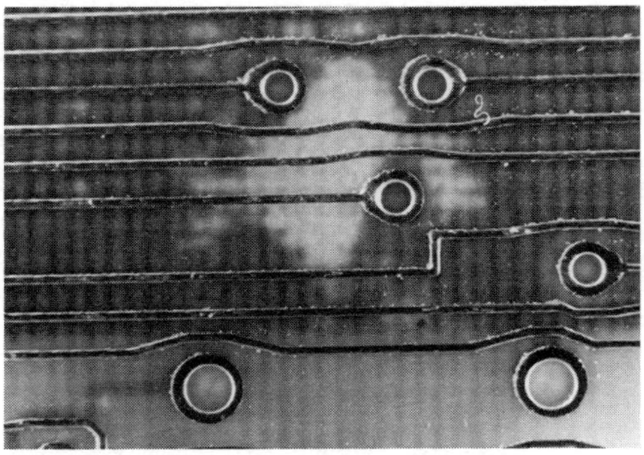

Fig. 4.16. Board with crazing.

ing to extensive damage by drastically reducing the insulation resistance.

For this reason, crazing is considered cause for rejection and rated as an A-defect. The only possible exemption exists when the area exhibiting crazing is located far away from electrically active parts of the circuit; that is, more than 0.4 in. (10 mm) from solder pads and conductors.

4.5.3 Blistering

Blistering is a local swelling and separation (delamination) of the base material, either between some of the glass cloth layers or between the base material and the copper foil (Figure 4.17). When evaluating blistered boards, the inspector should distinguish sharply between boards intended for hand soldering and those intended for mass soldering.

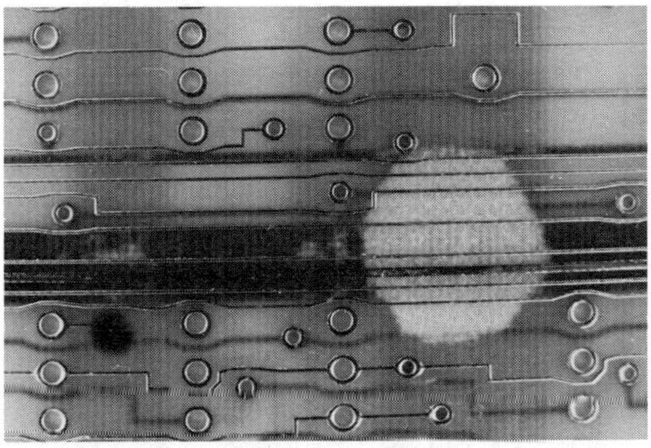

Fig. 4.17. Board with blistering.

4.5.3a Blistering in Hand-Soldered Boards

When a board is hand soldered, it may have two small blisters per 16 sq. in. (about 100 cm^2) of surface; small means having an area of 0.16 sq. in. (1 cm^2) per blister. Nevertheless, the blister must not lift the copper foil from the base material; if the foil is not lifted, there is no need to mention blistering on the inspection record.

When the copper foil is lifted from the base material (a real delamination) the board is unacceptable. It is cause for rejection and an A-defect (Paragraph 4.5.4).

4.5.3b Blistering in Mass-Soldered Boards

On boards intended for mass soldering, there is a risk that the blistering may spread during the soldering process because of the heat absorbed. In this case, blistering should not be accepted; any occurrence is cause for rejection. Consequently, blistering is always an A-defect when the board will be mass soldered.

Blistering is caused by gases developed within the base material itself, and is a result of heating. Because the board is exposed to thermal shock during mass soldering, the base material expands, primarily along the Z-axis. As a result, the internal bonding between the glass cloth layers or between the copper foil and the base material becomes so weak that the increasing gas pressure creates blisters or lifts the foil. The gas pressure is caused by the beginning of epoxy breakdown from the heat.

The tendency to blistering during mass soldering may be somewhat reduced by lowering the temperature of the solder bath slightly, say from 500 to 482°F (260 to 250°C). This lowering will also reduce warping caused by mass soldering.

4.5.3c Influence of Different Processes

Laminates of different makes may differ considerably in their thermal properties, mainly as a result of different degrees of cross-linking in the epoxy. The process conditions during the lamination also play a large part. It is, therefore, important that the PCB manufacturer use only such brands of base materials which experience has shown to be blister-resistant.

As an illustration, it can be mentioned that the time spent on developing the first blisters in FR4 material varies, at a solder bath temperature of 500°F (260°C), from 1 to 4 minutes, depending on the brand. The test is performed as a solder float test, in which the board is placed and floats on the surface of a hot solder bath.

In contrast, the time for developing blisters is from 10 to 20 seconds for base materials such as FR2 (phenolic paper) and FR3 (epoxy paper).

4.5.4 Delamination

Compared to blistering, delamination is a widespread separation between the individual glass cloth layers of the base material. When it occurs, it is usually easy to recognize as a gross defect because delamination shows as lighter areas in the board (Figure 4.18).

4.5.4a Effects of Delamination

Delaminated areas are unacceptable, partly because the separation between layers can spread over the board and partly because there is a strong risk that the cavities will retain moisture or electrolyte and drastically reduce the insulation properties of the board. Delamination is cause for rejection and an A-defect.

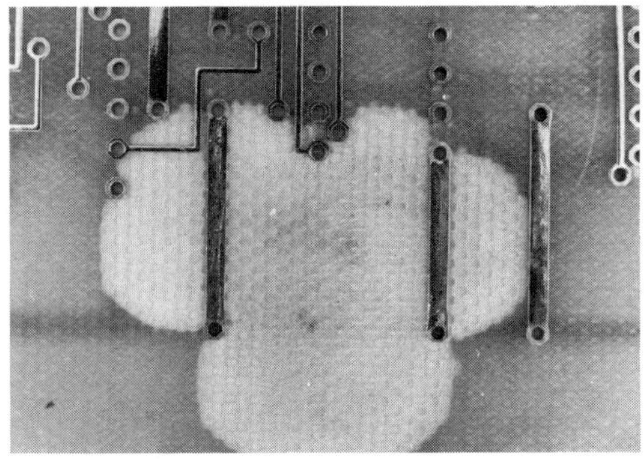

Fig. 4.18. Board with delamination.

Delamination in multilayer boards can be so small that it can be revealed only by microsectioning the board. When the delamination is so small that it cannot be seen with the naked eye and testing is destructive, it can be disregarded as a harmless fault.

4.5.4b Causes of Delamination

The reason for delamination is epoxy starvation in the glass cloth layers or an incomplete curing of the base material. It has been reported that some manufacturers of laminates have reduced the lamination press cycle a little in order to increase production.

In multilayer boards, visible delamination may be due to poor lamination of layers or insufficient bonding of prepegs to inner layer copper circuits. Here, the PCB manufacturer is in a way also the "manufacturer" of the base material. During an inspection visit the inspector should direct his attention to the capacity of the laminating press and also to the pressing and curing routines as well as to the quality assurance system.

4.5.4c Delamination after Assembly

Delamination may develop, however, in a board after assembly and soldering. There are several possible reasons for this:

Asymmetrical buildup of multilayer boards

Faulty soldering techniques, such as skipping the baking which removes retained moisture from the board. (A 1- or 2-hour baking at 230°F (110°C) immediately before assembly and soldering is highly recommended.)

Lack of preheating as the first step of mass soldering

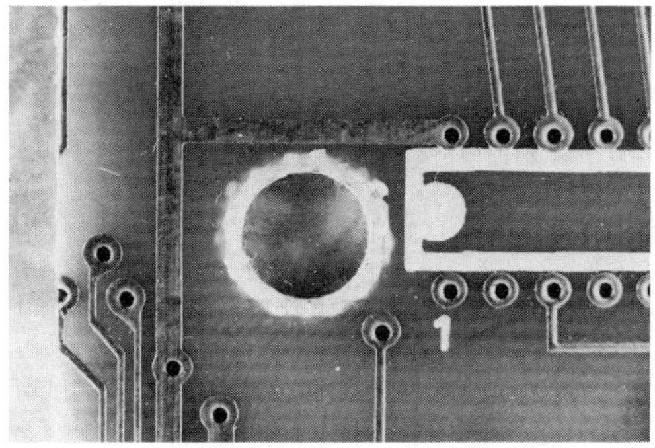

Fig. 4.19. Board with haloing around drilled hole.

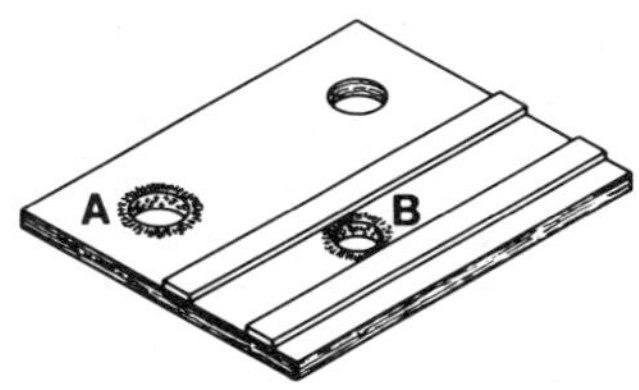

Fig. 4.20. Examples of acceptable and unacceptable occurrence of haloing.
A: Isolated occurrence is acceptable
B: Occurence between two conductors is rejectable

Delamination may also be caused by stressing factors arising from the insertion of a row of tightly fitting square terminal pins, or self-tapping screws in holes which are a little too small. Other reasons may be an unfortunate loading of the board during mass soldering, caused by especially heavy components such as large power transformers, or by mishandling of boards which are still hot after mass soldering while placing them in a rack. Heavy components will subject the boards in the rack to a bending moment and perhaps damage them.

If the production department lodges a complaint against a generally occurring delamination in a particular type of board, the problem may be self-inflicted; consequently, the production methods should be investigated before complaining to the PCB manufacturer. The inspector should ask the production people to check the soldering conditions or to solder-test a bare board.

4.5.5 Haloing

Haloing appears as a light area at machined areas of the board, either directly on the surface or in the base material—for example, as a light ring around a hole. See Figure 4.19. The cause is a local microsized crazing from the machining. In many types of punched board, haloing is unavoidable where the land between two holes is less than the material thickness, a fact that PCB designers should always consider. Another cause can be worn tooling, which can be remedied only by the PCB manufacturer.

Because of the possibility of absorption of electrolyte in the porous regions, the inspector must distinguish between the locations of such regions. When the defect is completely isolated as shown at A, Figure 4.20, it can be tolerated.

On the other hand, when the haloing occurs between two conductors or two solder pads as illustrated at B, Figure 4.20, there is a risk of reduced insulation resistance. The defect is cause for rejection and an A-defect.

4.5.6 Weave Texture

Visible glass weave is not considered a fault as long as it contains no broken glass

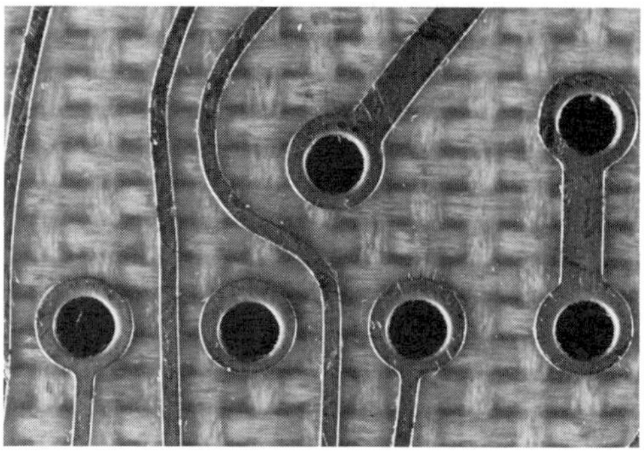

Fig. 4.21. Board with weave texture.

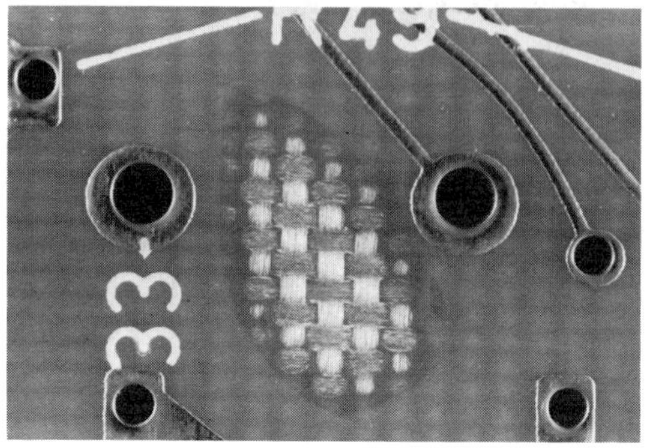

Fig. 4.22 Board with weave exposure.

fibers and is fully covered with epoxy. The fault is fairly easy to detect because the weave pattern is easily recognized. See Figure 4.21. This fault will not normally be recorded in the inspection record.

4.5.7 Weave Exposure

Uncovered glass weave in the board surface is worse than visible but covered glass weave, because such areas tend to be somewhat porous and can absorb electrolyte or etchant, which can cause extensive damage to the board later. See Figure 4.22. The fault is always cause for rejection and an A-defect.

4.5.8 Incomplete Curing

Fortunately, incomplete curing of the base material is a very rare defect. It is detected by the surface's being a little less hard than normal. In extremely rare cases, drilled holes can tilt slightly from the slight movement of the individual glass cloth layers, as shown in Figure 4.23.

Incomplete curing is cause for rejection for several reasons:

1. The board has not attained its full strength.
2. There is a strong tendency to form solder webbing during mass soldering. This means that certain areas of the board surface are covered with tiny threads of solder adhering to the surface, causing short circuits between the individual constituents of the pattern (Figure 4.24).
3. There is a risk of damaging the wall of plated-through holes due to the reduced stability of the glass cloth layers.

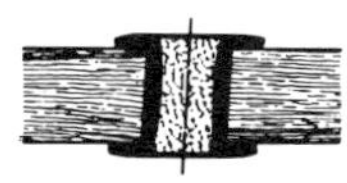

Fig. 4.23. Tilting of hole barrel from movement of glass cloth layers.

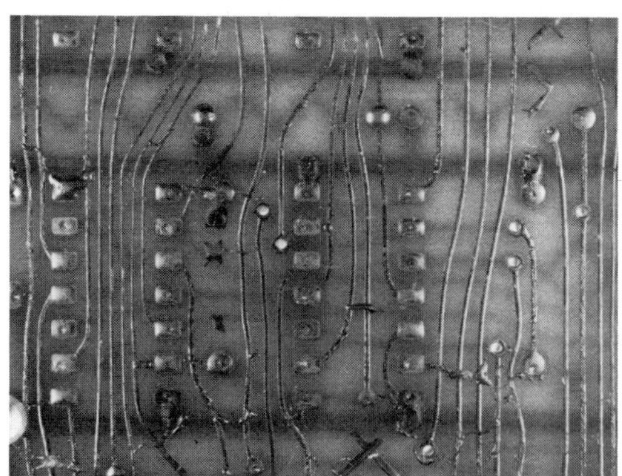

Fig. 4.24. Board with webbing after mass soldering.

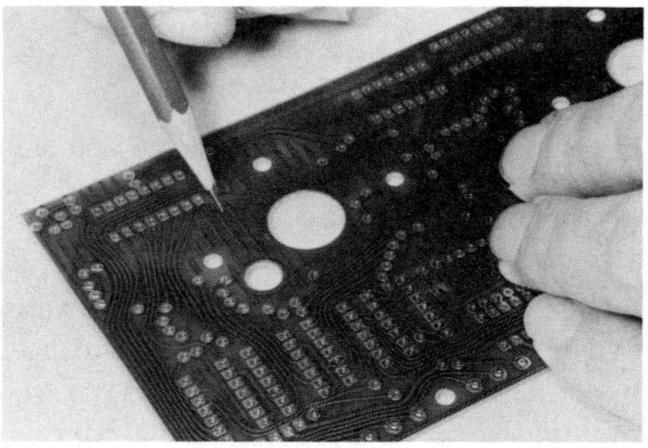

Fig. 4.25. Checking board curing with a pencil test.

Incomplete curing can be detected by a pencil test, in which a sharp-pointed pencil of H hardness is held firmly against the surface at a 45 degree angle and pushed with equal forward and downward pressure so as to make a 0.5 in. (13 mm) stroke. If the point penetrates the surface, the board is insufficiently cured. See Figure 4.25. The fault is rated as an A-defect and implies rejection of the board.

4.5.9 Cracks

The inspector must distinguish accurately between cracks extending through the base material from surface to surface and cracks present on only one side of the board. See Figure 4.26. A crack of the first type is acceptable only when it starts from a mounting hole near the board edge, and runs toward the edge of the board as shown at A, Figure 4.27.

It is, however, necessary that the area surrounding such holes be subjected to compression stress only. This limitation implies that the hole is used only for mounting; radial forces obviously are not permitted in such a hole. For example, the hole should not be used for mounting a board ejector because the hole wall probably will collapse in the direction of the board edge when the ejector is used. Unless it meets these criteria, a crack ex-

Fig. 4.26. Cross-section of board with cracks.
A: Crack extends through base material.
B: Crack is present on one side of board only.

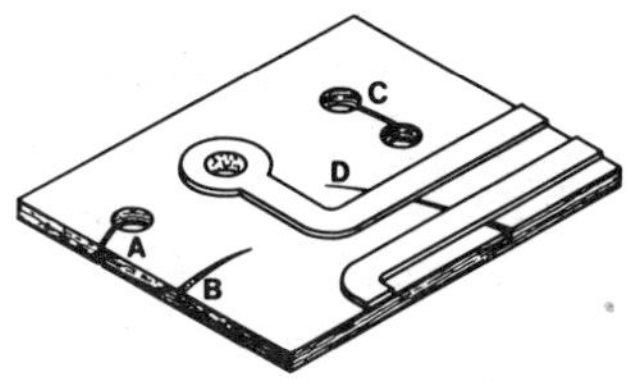

Fig. 4.27. Various types of cracks in a board.
A: Crack starting at a mounting hole.
B: Crack not extending through base material.
C: Crack extending between two holes.
D: Crack connecting two conductors.

tending through the base material shall be cause for rejection and rated as an A-defect.

A crack which does not extend through the base material is acceptable if:

1. The length of the crack is less than 0.5 in. (13 mm). See B, Figure 4.27.
2. The crack does not run between two holes. See C, Figure 4.27.
3. The crack does not run beneath any part of the pattern or connect any parts of the pattern, such as conductors and solder pads. See D, Figure 4.27.

If all the above conditions are not met, the defect is an A-defect and the board is rejected. The aforementioned limitations on cracks not extending fully through the base material also apply to cracks starting at mounting holes.

4.5.10 Metallic Inclusions

Metal particles encapsulated in the base material are extremely rare and are acceptable if:

1. They are smaller than 0.04 in. (1 mm) in any direction.
2. They are at least 0.02 in. (0.5 mm) from any active part of the pattern.
3. There are no more than two particles per 16 sq. in. (about 100 cm^2) of board area.

If the above conditions are met, no remarks are entered in the inspection record, but the entire batch should be 100-percent inspected if the sample contains several boards with metallic inclusions. If the conditions are not met, it is an A-defect and the board is rejected.

4.5.11 Other Inclusions

Some very special inclusions other than the metallic ones just described can be found in the base material on very rare occasions. For example, the author has personally seen an order slip encapsulated in the base material of a finished board. In another finished double-sided board, the buildup of the base material was wrong because an additional sheet of copper foil was placed in the middle of the laminate.

These faults, which should be regarded as curiosities only, are presumably due to a moment's distraction of the operator, and should not be treated as an expression of the general quality level of the base material. They are mentioned solely to show the inspector that a printed circuit board can exhibit unpredictable faults.

4.6 Surface Appearance

When evaluating the appearance of the surface of the finished board, the inspector should consider that it is very difficult, and always expensive, for the PCB manufacturer to procure base materials with a completely flawless copper foil surface. The risk of finding a few dents, pits, and scratches in the foil is ever present. See Figure 4.28.

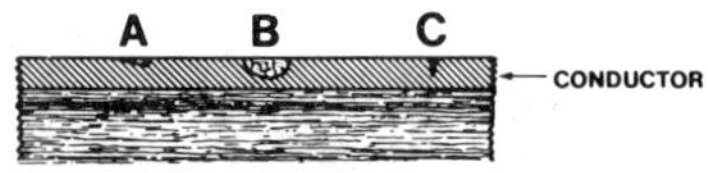

Fig. 4.28 Cross-section of board showing surface defects.
A: Dent. B: Pit. C: Scratch.

Such defects, to some extent, always can be found on the surface of the finished board. The production processes, by removing 80 to 95 percent of the copper, correspondingly reduce the probability of finding these defects on the remaining surfaces. Unless the user is prepared to pay a premium for a completely flawless surface, he must accept small defects in noncritical areas. The following paragraphs discuss the following defects which can occur on the board surface:

- Dents and pits
- Scratches
- Undesired metal particles
- Etching faults
- Contaminated surface
- Salt residues

4.6.1 Dents and Pits

Since defects, such as dents and pits in the copper foil, recur on the surface of the finished board, the reader should know which defects the PCB manufacturer must accept in his raw material. These are described in MIL-P-13949F.

A point system is used. The maximum total point count for dents and pits combined, in any 12 by 12 in. area (approximately 930 cm^2), shall be less than 30 for Class A. The point value for the individual defects depends on the size of the defects as shown in Table 4.5.

Translated into a practical example, a copper area of 4 by 6.3 in. (100 by 160 mm, Eurocard size) may contain a total of only five dents and pits, with the longest dimension no more than 0.01 in. (0.25 mm). If 80 percent of the copper is re-

Table 4.5. Point system for dents and pits. L indicates longest dimension.

Longest dimension of dent or pit		Point value
in.	mm	
$0.005 < L \leq 0.010$	$0.13 < L \leq 0.25$	1
0.020	0.50	2
0.030	0.75	4
0.040	1.0	7
0.040	1.0	30

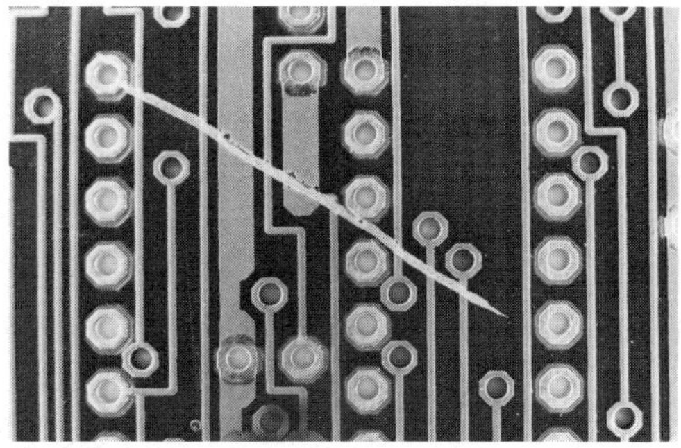

Fig. 4.29A. Board with unacceptable scratching.

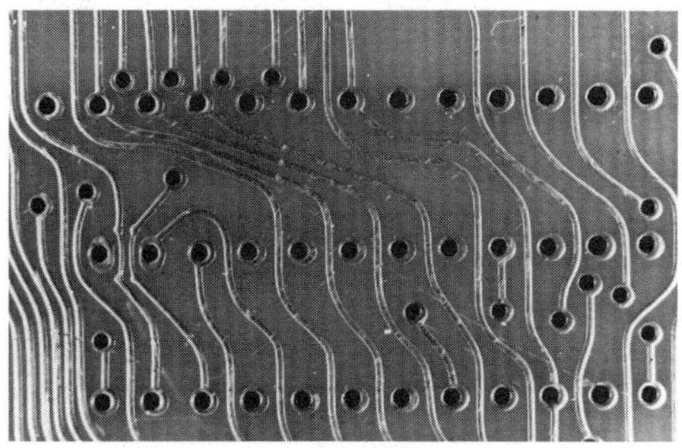

Fig. 4.29B. Board with acceptable scratching.

moved by etching, an average of only one defect (stemming from defects in the base material) per finished board should occur.

Table 4.5 is also valid for Class B, but here the longest dimension of the pits is 0.015 in. (0.38 mm). Within the same area as stated above, no more than 3 dents or pits are allowed, the largest dimension being from 0.005 in. (0.13 mm) up to 0.015 in. (0.38mm). No limit is stated for dents and pits smaller than 0.005 in. (0.13 mm). The restrictions stated imply, in this example, that only every tenth board should contain pits.

The PCB manufacturer normally uses a Class A base material; the first example clearly shows that the inspector must accept a number of irregularities in the delivered board.

Dents and pits in the board surface are usually of only cosmetic significance and have been mentioned only for the sake of completeness. These defects will normally occur to a very small extent and do not warrant remarks in the inspection record.

4.6.2 Scratches

It can be very problematic to evaluate scratches because it is nearly impossible to set adequate guidelines; over many years, the acceptable extent of scratching has been the subject of many discussions between customer and PCB manufacturer. The board surface inevitably receives some scratches during its manufacture, so the inspector must be prepared to accept boards with some degree of scratching.

Figure 4.29 can serve as a guide to acceptable and unacceptable scratching. It may be revealing to ask to what extent the customer is prepared to discard boards which he has scratched himself by careless handling during or after assembly.

Slight scratching of the tin/lead surface does not warrant remarks in the inspection record, because there are no functional problems and the cosmetic problems are negligible.

Widespread scratching in all directions indicates poor handling by the PCB manufacturer. This seldom causes functional problems; the fault is rated as a B-defect

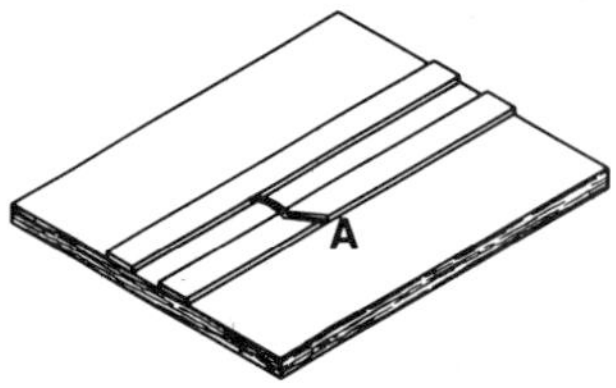

Fig. 4.30. A tin/lead chip caused by scratching forms a short-circuit to an adjacent conductor.

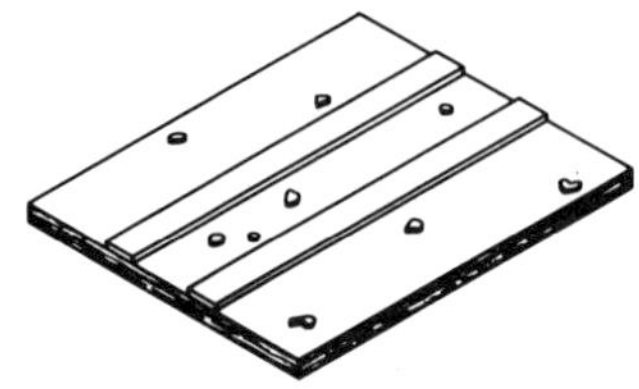

Fig. 4.31. Board with small metal residues on surface.

the board. The inspector should, however, note that in certain cases scratching of tin/lead plated boards can be dangerous because a tiny tin/lead chip may be drawn over the edge of a conductor, as shown in Figure 4.30. The chip can either cause a direct short circuit to an adjacent conductor, or it may break off, which also can cause a short circuit. When these defects are discovered during inspection, they shall be cause for rejection and rated as an A-defect.

Solder and insulation masks make it difficult to inspect the board surface; on the other hand, the encapsulation usually prevents the chips from breaking off.

Deep scratches exposing copper are normally cause for rejection and are rated as an A-defect. Exposed, unprotected copper can corrode, and the corrosion products may cause functional problems as they grow and spread over the surface, with the risk of short circuits between adjacent conductors. An exemption, if given, implies a repair where the scratches are covered by hand soldering or by sealing with epoxy. On the solder side, a repair is usually not necessary. The scratches will be covered after mass soldering if the copper has not oxidized beyond a level readily removed by the flux.

4.6.3 Undesired Metal Particles

This is to be understood as small metal residues (copper, sometimes covered with tin/lead) on the board surface, as shown in Figure 4.31. They are usually due to an incomplete image transfer (unwanted plating owing to pinholes in the plating resist, see Reading List, Reference A12, Paragraph 3.2.2) or an imperfect stripping of the plating resist (incomplete etching, see Reference A12, Paragraph 3.6).

4.6.3a Faults in Master Film

If the metal particles are present due to a definite fault in the master film, they will occur at the same location on each board. If the working film is a multiple film derived by a step-and-repeat process (see Reference A12, Paragraph 3.8.6) and having, for example, three patterns, the fault may occur in one pattern only, so that just one third of the finished boards will exhibit metal particles.

4.6.3b Randomly Distributed Particles

When the cause is not due to a fault on the master film, the metal particles can occur at random all over the board surface. The PCB manufacturer's process control or finished-goods inspection will normally detect such defects and ensure the removal of such particles, usually manually, with a scalpel. The scrape marks are usually visible after the particles have been removed, especially on boards without solder or insulation masks. If the epoxy surface of the base material is intact—the scraping did not expose the glass fibers—the removal will not cause functional problems and is not noted in the inspection record.

4.6.3c Multilayer Boards

In the case of multilayer boards, etching faults can, on rare occasions, occur in an inner layer, even when the PCB manufacturer takes great care and performs an extra inspection on the inner layers before they are laid up for lamination.

Etching faults can often be found with the naked eye, particularly when illuminated from below. Because it is difficult to distinguish between metallic and nonmetallic particles, full certainty of how critical the fault is can be achieved only by X-raying the defective area. Because PCB manufacturers producing multilayer boards always have an X-ray facility at their disposal, the inspector can have suspicious-looking boards X-rayed so that he can examine the X-ray photos.

4.6.3d Defect Assessments

When the sample contains boards with metal particles which have not been removed by the PCB manufacturer, the inspector must assess the extent of the occurrence. It is necessary to establish an upper limit to the size of metal particles; we will set the limit at 0.04 by 0.04 in. (1 by 1 mm). Metal particles larger than this in any direction are normally cause for rejection and rated as an A-defect.

If the metal particles occur in areas distant from active circuits, or within balanced areas (see Reference A12, Paragraph 6.6.3) considered to be passive areas, there is no need to be too rigorous, so the inspector can exempt the particles. In such cases metal particles larger than the aforementioned size could also be exempted.

It is conceivable that the PCB manufacturer has introduced limitations to the amount of board touch-up, restricting it to functionally important areas, that is, within active areas of the circuitry. The inspector should acquaint himself with the PCB manufacturer's policy in that respect

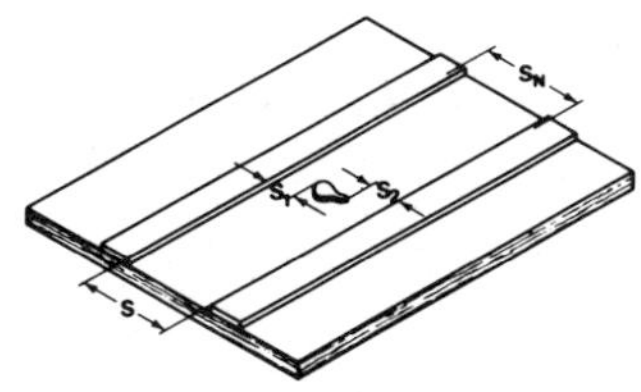

Fig. 4.32. Reduction of minimum spacing S caused by metal particle. $S_1 + S_2 \geq 0.75\ S_N$, where S_N = nominal spacing.

and, if necessary, define his position regarding the removal of metal particles.

Metal particles shall not inadmissibly reduce the minimum spacing between conductors and solder pads. This means that the reduction shall not exceed 25 percent of the nominal spacing S_N, which is measured on the original master film, as shown in Figure 4.32. A larger reduction shall be cause for rejection and rated as an A-defect.

In areas where power line voltage is present, the creepage directions shall strictly adhere to the requirements stated in the applicable PCB specification. On boards made to the directions stated in Reference A12, the effective minimum spacing is not less than 0.1 in. (2.5 mm). See Reference A12, Paragraph 6.5.2. This applies also when the board has solder and insulation masks.

4.6.3e Particle Removal

In principle, removal of undesired metal particles is permitted. It should not be done in the incoming inspection department, but left to the production department. Or, if time permits, the boards may be returned to the PCB manufacturer for rectification.

Metal particles need not be removed from nonfunctional areas. They must always be removed where they inadmissibly reduce the minimum spacing. If the board has solder and insulation masks, both inspection and rectification become rela-

because of the cosmetic depreciation of tively difficult. The surface will still show unacceptable traces even after the masks have been repaired.

4.6.3f Board Disposition

Having established the incidence of metal particles, the inspector takes the following steps. After counting the number of metal particles, he calculates the number per 16 sq. in. (100 cm^2) of board area. The counting shall always be performed on an area of at least 16 sq. in. (100 cm^2), which means that small boards must be laid together to obtain a sufficiently large area. Some general guidelines for disposition are:

1. When there are no more than 2 metal particles per 16 sq. in. (100 cm^2) of board area, no remarks are entered in the inspection record.
2. If there are between 3 and 10 metal particles present in this board area, one of the PCB manufacturer's processes has gone out of control. We will consider the fault as a B-defect and not reject the board. The inspector should carefully inspect all boards of the delivery for metal particles and order the removal of the most dangerous ones; that is, where the minimum spacing has been reduced by more than 25 percent.
3. If there are more than 10 metal particles per 16 sq. in (100 cm^2), there is a great risk of overlooking adversely located metal particles—after the board is assembled, it is almost impossible to find a defect on the component side under an IC package. Inspection and rectification become too difficult; we will, therefore, consider the fault as cause for rejection and rate it as an A-defect.

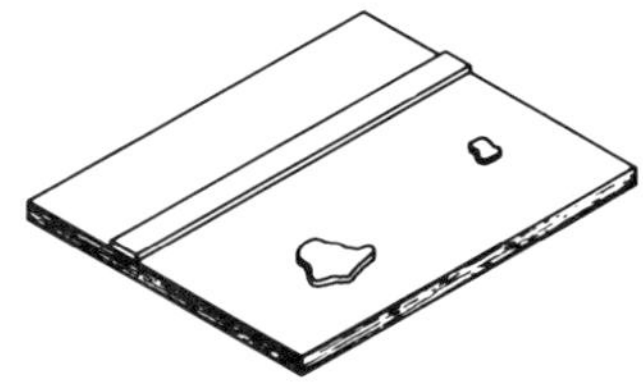

Fig. 4.33 Examples of etching faults.

4.6.4 Etching Faults

Etching faults appear in a manner similar to the undesired metal particles previously described, but with larger dimensions. In this paragraph we will deal only with nonfunctional etching faults, meaning that they occur in areas of the board where neither short circuits between conductors and pads nor reduction of minimum spacings are present. Chapter 5 describes functional etching faults.

On the whole, etching faults occur in the same way as described in Paragraph 4.6.3; Figure 4.33 shows typical examples. They are removed with a scalpel or a manually operated engraver as shown in Figure 4.34. The extent of etching faults is:

1. No more than two nonfunctional etching faults are permitted per 16 sq. in. (100 cm^2) of board area. The largest acceptable size of an etching

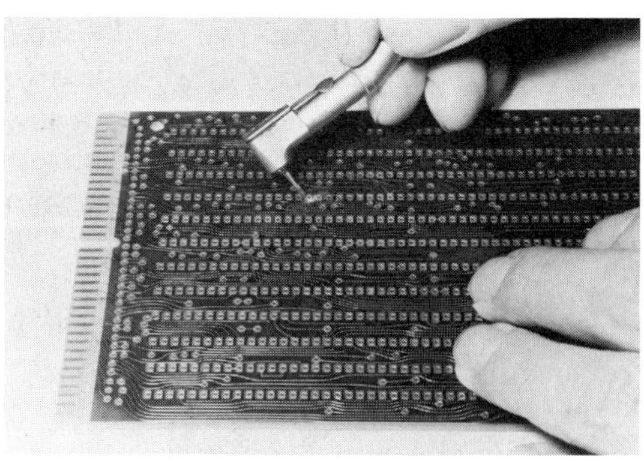

Fig. 4.34. Removing etching faults with a manually operated engraver.

fault is 1/8 by 1/8 in. (3 by 3 mm). Such etching faults must be found and removed by the PCB manufacturer before delivery, and are not recorded in the inspection record. If a solder or insulation mask is involved, it shall have been repaired; for example, by applying ink with a brush.

2. If the aforementioned etching faults have not been corrected, or there have been no more than five corrections per 16 sq. in. (100 cm^2) of board area, the fault is considered a B-defect.
3. Etching faults exceeding these limits are rated as an A-defect. The inspector must judge whether an exemption can be granted or whether the board shall be rejected.

4.6.5 Surface Contamination

The free-etched surface of the base material as well as the metal surface shall have no traces of oil, grease, organic film residue, dirt, or fingerprints. Furthermore, the surface shall not exhibit stains such as those caused by splashes of electrolyte or acid. Such defects are nearly always caused by careless handling during manufacturing. The main risks from such contamination are reduced solderability of the pattern and reduced surface resistance of the base material.

It should be noted that the tin/lead layer in very rare cases can have a bluish tint caused by tin oxides. They do not reduce solderability or warrant remarks in the inspection record.

4.6.5a Reduced Solderability

When the inspector doubts the board's solderability because of surface contamination, he can have a solderability test made. In the usual case, the company does not have the equipment to directly measure solderability, and the only practical solution is to pass an unassembled board through the mass soldering machine and evaluate the result. Chapter 10 describes solderability assessment. Unacceptable solderability is obviously cause for rejection and rated as an A-defect.

4.6.5b Reduced Surface Resistance

A reduced surface resistance can affect the board function because the insulation resistance between two adjacent conductors is reduced to an unacceptable degree. The manufacturer of the base material specifies a guaranteed minimum resistance of the base material. Typical values are:

phenolic paper (FR2): 10^{11} megohms
epoxy glass (FR4): 10^{12} megohms

Specific test conditions are usually stated in conjunction with the specified values; for example, exposure to a relative humidity of 90 percent and a temperature of 104°F (40°C) for 96 hours. The exposure is often followed by a drying period of 2 hours at a relative humidity of 75 percent and a temperature of 77°F (25°C).

It is difficult to get a reliable value of the insulation resistance between two conductors on the finished board. It must be remembered that the production processes can depreciate the electrical characteristics and that the insulation resistance is composed of the surface resistance and the volume resistance.

4.6.5c Surface Resistance Measurement

When the inspector suspects a reduced surface resistance, he can measure the insulation resistance between two closely spaced conductors with a megohmmeter, using a 100 volt test voltage. According to the IEC 326-3 and DIN 48303, the minimum insulation resistance value (R_I) can be calculated with the formula:

$$R_I = 160 \times R_M \times S/L, \text{ where}$$

R_M is a material constant, determined under specified test conditions,

S is the mutual distance between the conductors, and

L is the length of the conductors.

The insulation resistance measurement must be performed under the same conditions specified for the material constant. This requirement frequently complicates the measurement. Paragraph 28.5 discusses this in detail.

4.6.5d Factors Affecting the Insulation Resistance

Because many factors affect the insulation resistance, the value calculated by the above formula is mostly of theoretical interest. The most decisive factor is how well the board was rinsed and neutralized after etching, because salt residues can drastically reduce the insulation resistance (Paragraph 4.6.6). The presence of solder and insulation masks can also affect the resistance.

4.6.5e Rejection Value

Because of these conditions, as well as the uncertainties connected with the measurement of very high resistances, the author recommends the introduction of a safety factor, which in most cases can be set to 100. A rejection value R_R is therefore:

$$R_R = 0.01 \times R_I$$

If the conductor lengths are 4 in. (100 mm) and the spacing 0.012 in. (0.3 mm), the rejection value is:

$$R_R = 0.01 \times 160 \times R_M \times S/L$$
$$= 5000 \text{ megohms,}$$

R_M being 10^6 megohms. A value less than 5000 megohms indicates an unsatisfactory board, which should be rejected. The fault is, of course, an A-defect.

4.6.6 Salt Residues

The contamination discussed in the preceding paragraph, except splashes of electrolyte and acid, can be characterized as nonionic contamination. We now deal with an ionic contamination in the form of salt residues on the board surface (Figure 4.35).

4.6.6a Causes of Salt Residues

Salt residues are usually due to an insufficient rinsing after the plating process and/or an incomplete neutralization and rinsing after the etching process. After drying, such salt residues can be very difficult to remove. Similar conditions can arise after reflowing, not least if a flux containing glycol activators is used. In order to avoid ionic contamination of the board surface, the board must be cleaned immediately after reflowing; cleaning at a later stage may be extremely difficult.

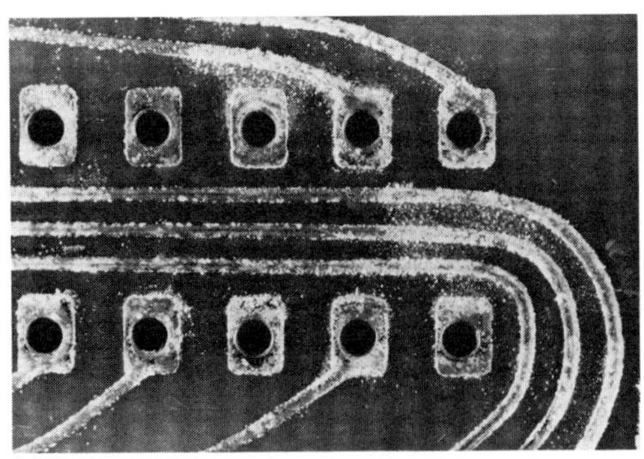

Fig. 4.35. Board with salt residues on surface.

4.6.6b Effect on Surface Resistance

Salt residues dramatically reduce the surface resistance of the board, especially when it is subjected to humidity. Under humid conditions, an electrolytic dissociation of the salt residues takes place. The salt residues dissociate into negative and positive ions, which carry an electric current, commonly reducing the surface resistance by a factor of 10^4 or 10^5. A disastrous corrosion of the pattern can also take place. The corrosion products—grey or white deposits—can interfere with circuit operation.

4.6.6c Detecting Ionic Contamination

The easiest way to demonstrate ionic contamination at the incoming inspection is to measure the insulation resistance as previously described. It may be expedient to moisten the board surface with distilled water, to expedite the dissociation of possible salt residues. If the board has solder and insulation masks, the water will penetrate the masks, so it is still possible to ascertain any ionic contamination.

Ionic contamination can be identified more accurately by various extraction methods. The principle is to extract the contaminating salts by means of an alcohol-water solution. By measuring the conductivity of the solution before and after the extraction, the inspector can ascertain the extent of the contamination. It is, however, outside the scope of this book to describe such methods in detail. See MIL-P-55110C, Paragraph 4.8.12.

4.6.6d Disposition

The occurrence of salt residues is an obvious cause for rejection and is rated as an A-defect. The boards are returned to the PCB manufacturer for remake. It is seldom necessary to routinely check for salt residues on boards from well-established PCB manufacturers.

Chapter 5

Pattern Inspection

5.1 Introduction

This chapter deals with the inspection of the pattern itself; that is, its conductors and solder pads. Hole drilling is covered in Chapter 6 and plating in Chapter 7.

In all discussions of circuit pattern inspection, the inspector must remember the requirements for minimum conductor width and minimum spacing between circuits, as expressed by the 75 percent rule. These requirements override the other criteria.

Solder pads and conductors represent the most important parts of a printed circuit board, thus they have to be inspected. For example, in a board with 1500 solder pads, the total length of the circumference of all solder pads is typically 10 yards (9 meters); the total conductor length can be surprisingly large, for example 50 yards (46 meters). Thus, the conductor edges will be 100 yards (91 meters), and the board will have a combined edge length of approximately 110 yards (100 meters).

There is, therefore, a great risk that different kinds of defects are present, partly along the edges of the solder pads and conductors, such as in the form of poor edge definition or small projections, and partly on the solder pads and conductors themselves, such as pinholes.

The quality of the finished pattern is usually determined by the processes of image transfer, plating, and etching. As these processes are fairly automated, and therefore somewhat operator-independent, the pattern quality will generally be uniform from board to board.

The inspector should not wrongly think it necessary to inspect only a single board. In spite of the prevailing degree of automation used by most larger PCB manufacturers, faults can easily occur. In order to give the inspector a wide background knowledge, the following subparagraphs describe how such faults occur.

5.1.1 Sources of Screen Printing Errors

Image transfer can be done by either screen printing or photo printing. In the first case, a fully automated screen printing machine is often used, which implies that the process, apart from the initial setting of the machine, is, on the whole, operator-independent. Nevertheless, errors can occur, as described below:

1. *Dust in the ink.* The reason for this is usually a lack of cleanliness and/or occurrence of dust due to inadequate flooring in the printing shop.
2. *Materials employed.* For example, the printing ink has incorrect viscosity, so that excessive flow-out occurs after printing.

3. *Process faults.* An incorrect machine setting, incorrect screen tightening, or improper ink cure.
4. *Screen pattern.* The pattern is blurred or damaged.
5. *Master film.* The film exhibits faults overlooked during inspection, and therefore not removed by touch-up.

5.1.2 Sources of Photo Printing Errors

Exposure and development of the resist used in photo printing take place automatically, but errors also can occur as described below:

1. *Dust on the film.* This is caused by electrostatic attraction. Dust-free working conditions are of prime importance.
2. *Materials used.* The dry film itself can contain pinholes.
3. *Process faults.* These include incorrect exposure setting and/or incorrect development time, used developer, problems in the vacuum system, clogged spray nozzles in the developing machine, incorrect temperature, etc.
4. *Pretreatment.* Incomplete cleaning of the production panels can cause lifting of the film resist during the plating process.
5. *Master film.* The film can contain faults overlooked during inspection, and therefore not removed by touch-up.

5.1.3 Sources of Plating Errors

Errors are also possible during plating, including:

1. *Plating baths.* Insufficient solution control can result in chemically unbalanced baths. Some examples are used-up bath solutions, organic contamination, or passive anodes.
2. *Plating current* not properly adjusted for the pattern density.
3. *Pretreatment.* Prolonged preplating storage can cause excessive oxidation, and improper cleaning can fail to remove oxidation.

5.1.4 Sources of Etching Errors

The finished pattern appears after etching, which usually is done on an automatic etching machine. In spite of the automation, there are some possible faults:

1. Process faults, such as prolonged etching time, incorrect etching temperature, and used-up etchant solution.
2. Posttreatment, including insufficient rinsing and neutralization after etching.

5.1.5 The Human Element

The aforementioned defects basically stem from human errors. Consequently, realistic quality costs include fairly regular inspection visits to examine the PCB manufacturer's quality assurance procedures.

5.1.6 Inspection Levels

Based on the preceding review, we can establish the following inspection levels: In the case of an approved supplier the inspection is carried out according to Special Inspection Level S3, whereas General Inspection Level II is used for the non-approved supplier.

5.1.7 Inspection Methods

An easy way to inspect the pattern is to lay a positive and then a negative film on top of the solder and component sides. By viewing the board, the inspector can now see possible short circuits between

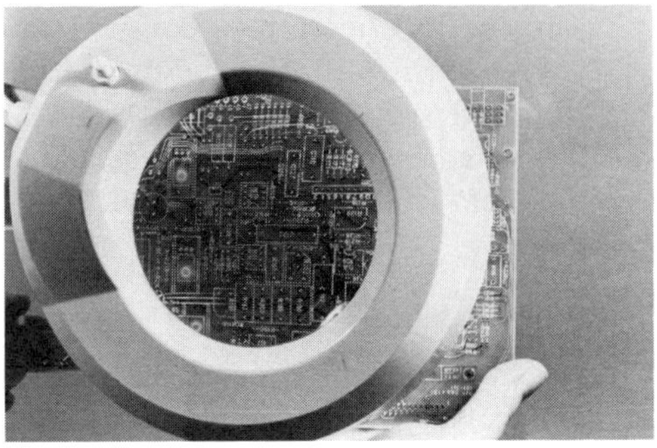

Fig. 5.1 Pattern inspection with an illuminating magnifier.

Fig. 5.2. Pattern inspection with a projector (courtesy Circuit Equipment Corporation).

conductors and solder pads when using the positive film, and missing pads, conductors, or parts of them when using the negative film.

The inspector should use a relatively large 2X or 3X magnifier with a built-in light source, and inspect the board systematically to ensure that all areas are scanned. See Figure 5.1.

Another way to inspect the pattern is to examine it through a magnifier without using the films. If the board is reasonably transparent, it can be placed on a light table. The inspector will soon be able to see irregularities such as short circuits or broken conductors; missing parts of the pattern, however, will be difficult to detect.

Finally, it is possible to use a special projector as shown in Figure 5.2. Various makes are available, but all are based on the same principle. The board is placed on a movable table and the pattern projected at a suitable magnification, frequently 10X, on a matte glass screen. The table is now moved in the direction of the X-axis, and after a full travel, shifted to the limit of the scanning field in the direction of the Y-axis. The table is then moved in the direction of the X-axis to achieve a systematic and complete scanning of the board.

One disadvantage of the projector is the slow inspection speed. Another problem is the relatively low light intensity on the screen, which requires a semidarkened room. This is especially the case when projecting light through the board, which has a low transparency, and even worse when the board has solder and insulation masks.

5.1.8 Defect Recording

As mentioned in Paragraph 4.6.3, nonfunctional pattern faults in balanced areas should not be recorded in the inspection record. The inspector shall primarily concentrate on the following major functional defects in the active circuit areas:

- Short circuits
- Broken conductors

Minor defects are also possible:

- Edge displacement of pattern
- Edge definition of pattern
- Projections and indentations
- Voids and pinholes
- Adhesion of pattern
- Width of annular ring
- Broken overhang

Each type of defect is entered in the inspection record as one A-defect or one B-defect per board, regardless of the actual number of each type on the board in question. It should be noted that a defect is entered only once in the inspection record even if it violates several different requirements. The defect is entered under the item most seriously violated. It is more informative to state, for example, that an annular ring is broken or that a minimum spacing has not been kept, than just to record poor edge definition.

5.2 Short Circuits

When dealing with short circuits between the different elements of the pattern (conductors and solder pads), the inspector should distinguish between two principal types: the common short circuits caused by an incomplete etching, and the rarer short circuits caused by metallic growth occurring after the board is manufactured.

5.2.1 Incomplete Etching

5.2.1a Systematic versus Random Short Circuits

Short circuits between conductors and solder pads as shown in Figure 5.3 can occur systematically or randomly. In the first case, the reason is usually a general fault in the image transfer process; for example, a short circuit between two conductors on the master film.

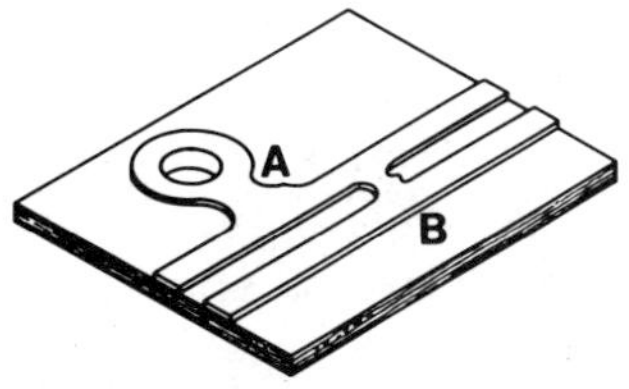

Fig. 5.3. Examples of incomplete etching.
A: Short-circuit between conductor and solder pad B: Short-circuit between two conductors

All boards in the batch will exhibit the same defect, which is not considered very dangerous because the inspector knows where to look for the defect on the individual boards. The inspector can choose between having the boards repaired by his company or returned to the PCB manufacturer. It is, however, wise to check one's own artwork or filmwork before complaining.

In the second case, the short circuits occur randomly and the defect can, for example, be attributed to a random pinholing of the applied plating resist. The unwanted spot is now plated with copper and later with tin/lead, which acts as an etch resist. In order to ease the inspection, the PCB designer should try to avoid interconnecting pairs of neighboring IC solder pads as shown in Figure 5.4. Executions A and B may resemble undesired short circuits.

The random short circuits can also be due to an incomplete removal of the plat-

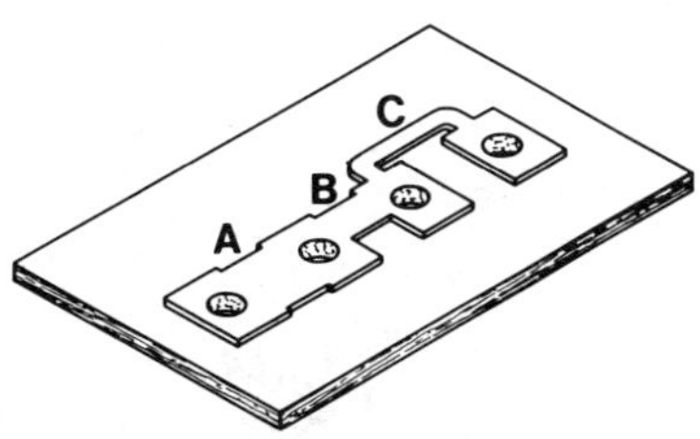

Fig. 5.4 Different methods of interconnecting IC pads
A and B: Some resemblance to etching faults
C: Preferred way of interconnecting IC pads

ing resist before etching; for example, between two closely spaced conductors. Even a very thin remaining film of resist can prevent etching at the spot in question which will therefore appear in copper. Owing to the color difference, the inspector can more easily detect this type of short circuit.

5.2.1b Effect of Spacing

It can be something of a problem for the PCB manufacturer to remove the plating resist between closely spaced, parallel conductors. Particularly when insufficiently thick resist is used, the resist is locked by the plated copper and tin/lead layers overhanging the resist as shown in Figure 5.5. This is sometimes called the "mushroom effect." See also Reference A12 of the Reading List, Paragraphs 3.5.5 and 3.6.

The PCB designer should always try to keep the spacing between parallel conductors as large as possible, preferably not less than 0.02 in. (0.5 mm). See Design Rule Number 2, Paragraph 6.5.1 of Reference A12. However, widths of 0.012 in. (0.3 mm) or even 0.010 in. (0.25 mm) are quite common today, so the PCB manufacturer is more or less compelled to use a dry film resist of sufficient thickness.

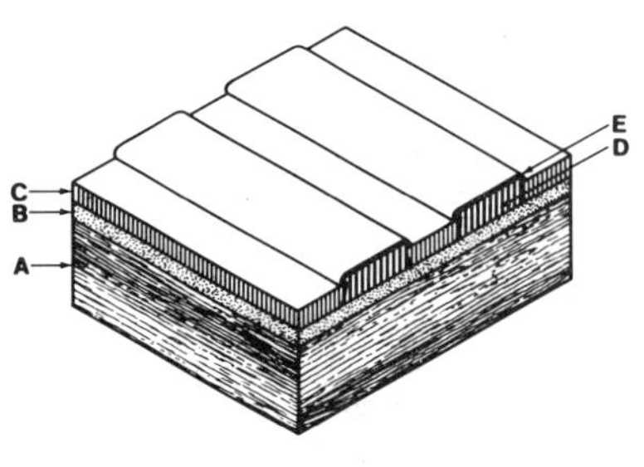

Fig. 5.5 Locking of plating resist by plating overhang.
A: Base material B: Copper foil C: Plating resist D: Plated copper E: Plated tin/lead

5.2.1c Board Disposition

In principle, short circuits are cause for rejection and consequently rated as an A-defect. Short circuits removed by the PCB manufacturer and leaving an acceptable board appearance are dealt with as follows:

1. Removed short circuits, up to 2 per 16 sq. in. (100 cm^2) or 4 × 4 in. (10 × 10 cm) of board area, do not warrant entries in the inspection record.
2. With up to 5 short circuits per 16 sq. in. (100 cm^2) of board area, the production processes are considered more or less out of control and, although the short circuits have been removed, the defect is rated as a B-defect.
3. If there are more than 5 short circuits per 16 sq. in. (100 cm^2), the defect is considered a cause for rejection and consequently rated as an A-defect.

With regard to an acceptable appearance, it should be noted that the largest acceptable dimension of a short circuit is 0.2 in. (5 mm). Conductors and solder pads shall not be lifted from the base material during short-circuit removal by the PCB manufacturer. When solder and insulation masks are damaged as a result of such removals, they must have been repaired, either by a renewed screen printing or by a brush touch-up.

When the boards have been rejected because of short circuits, but are so urgently needed that there is no time to return them, it is usually possible to repair them adequately with an engraver as shown in Figure 4.34. Obviously, the previously stated appearance requirements apply. The best way of repairing damaged solder and insulation masks is a brush touch-up. Out of consideration for pot life (useful life of the mixed ink), a one-pack ink should be used. A two-pack ink (epoxy plus resin hardener) starts curing as soon as it is mixed, but a one-pack ink can be used like paint, since it does not cure in the pot.

When the boards have been rejected for removal of more than 5 short circuits per 16 sq. in. (100 cm^2) of board area,the inspector may, depending on the nature and appearance, consider an exemption. Since the fault indicates poor manufacturing conditions, it should always be followed by serious discussions on quality with the PCB manufacturer.

5.2.1d Military Boards

It should be noted that MIL-P-55110C, which applies to military boards, does not permit repair by the PCB manufacturer (Paragraph 3.18).

5.2.2 Metallic Growth

Metallic growth, best known under the name "whiskers," is the very fine, monocrystalline threads growing out from metal surfaces. Such threads can grow long enough to cause short circuits to neighboring conductors.

The inspector probably will never see a board with whiskers because it can take months or even years before the growth starts, usually under the influence of elevated temperatures around 120°F (about 50°C) and a certain humidity. A polarization voltage, however, is not required, and the growth can occur during storage.

Figure 5.6 shows an example of whiskers, which can easily reach lengths of 0.02 to 0.04 in. (0.5 to 1 mm), enough to cause short circuits. The threads are so fine that they are difficult to see with the naked eye, so a magnifier is needed.

Tin/lead plated boards, fortunately, have little tendency to form whiskers, particularly when reflowed. Solder and insulation masks hinder the formation even more. Of course, the underlying laminate must be clean.

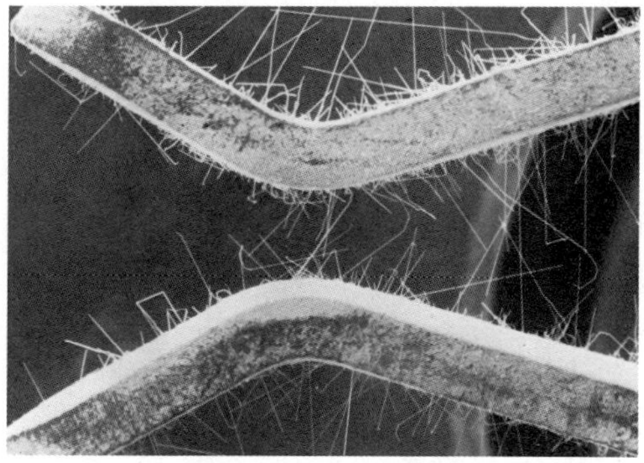

Fig. 5.6. Terminals with "whiskers" (courtesy Burndy).

Pure tin is more inclined to form whiskers, but this is not of great interest because tin plating is rarely used, mainly because of the high melting point of tin. Such boards cannot be reflowed without damaging the base material.

If the component leads are tinned—the case with certain IC packages—some care should be taken. However, if the tin is alloyed with a small percentage of lead, the tendency to form whiskers is eliminated.

Silver is very prone to grow, or to migrate as it is usually called, but this implies a dc potential and some moisture. Owing to the great tendency of silver to migrate, the general use of silver plating on printed circuit boards is not recommended.

Copper can form whiskers in a sulfurous atmosphere, but this tendency is so small that no problems occur in practice.

5.3 Broken Conductors

5.3.1 Systematic versus Random Breaks

Broken conductors (Figure 5.7) can occur either systematically or randomly. In the first case, the reason usually lies in the

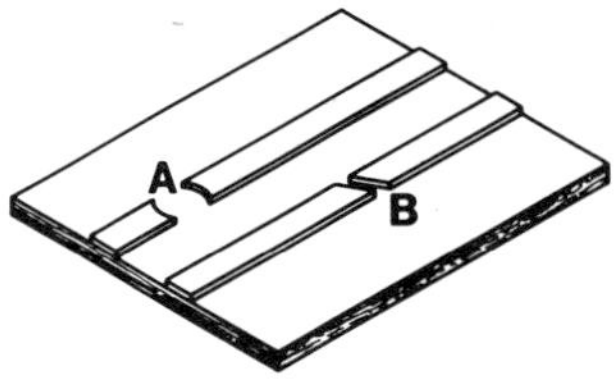

Fig. 5.7. Board with interrupted conductors.
A: Defect caused by hole in etch resist
B: Defect caused by a scratch in the etch resist

image transfer; that is, in the form of a broken conductor on the master film. All boards of the delivery will show the same defect, which therefore cannot be considered a very dangerous one from the inspector's point of view, because he knows where to find the defect on the individual boards.

In the second case, the interruptions occur at random. When a bare-copper board is involved, a hole in the etch resist or a scratch across a conductor causes a part of the conductor to be etched away, thus resulting in an interruption.

On plated-through boards, a conductor can be more or less blocked after the image transfer with no copper and tin/lead being deposited at the spot concerned. The missing etch resist causes the conductor to open when the board is etched. The basic cause of this defect can be an unwanted flow of the screen printing ink, incomplete development of the dry film, or contamination of the boards after image transfer.

5.3.2 Board Disposition

Broken conductors are an unquestioned cause for rejection, and are consequently rated as an A-defect. There are different opinions concerning the acceptance of boards with broken conductors which have been repaired by the PCB manufacturer.

5.3.3 Repair Methods

An often-suggested method is to weld a kovar strip to the conductor across the break. This method, however, requires a good welding routine to ensure correct adjustment of the welding equipment, and great care in preparing the spots where the welding is to take place. Obviously, as there can be some uncertainty regarding the quality of such repairs, the inspector himself or the Material Review Board must decide whether repair of broken conductors performed by the PCB manufacturer should be accepted.

Broken conductors can also be repaired by rendering the gap conductive; that is, with a metallic epoxy resin, then locally plating copper over the joint area.

When the boards are urgently needed, they can be exempted for up to 2 broken conductors per 16 sq. in. (100 cm^2) or 4 × 4 in. (10 × 10 cm) of board area. The breaks are repaired by soldering a piece of copper wire across the interruption, much like normal PCB modifications performed during assembly.

The repair work must be performed carefully by trained personnel. The component side is repaired before assembly, and the solder side after the board is mass soldered. The reader is referred to the repair instructions given in IPC-R-700B, "Suggested Guide Lines for Modification and Repair of Printed Boards and Assemblies," Paragraph 7.5.5.1 (Surface-Jumper Method).

5.4 Edge Displacement of Pattern

The pattern edges become slightly displaced because of growth or shrinkage, depending on the plating method: pattern plating or panel plating. The extent of displacement depends in turn upon the method of image transfer: screen printing or photo printing. It also depends upon the thickness of the copper foil and the etchant used. Details are given in Paragraphs 3.4.3, 3.4.4, 3.2.1, 3.2.2, and 3.3, respectively, of Reference A12 of the Reading List.

The result of this condition is either an outward- or inward-going edge displacement of the conductors and solder pads relative to the nominal location, which is determined on the basis of a correctly produced master film (correctly exposed and developed). In the case of an outward-going edge displacement, the width and diameter of the conductors and solder pads increase respectively, whereas all spacings decrease correspondingly. The reverse effect occurs when the displacement is inward-going.

Obtaining information about the processes used and the resulting edge displacement from the PCB manufacturer can be very inconvenient or impractical. Instead, the inspector should assess the edge displacement against the tolerances described in this paragraph. It is, however, difficult to find on the board itself a reliable reference against which to measure the edge displacement; the major cause of difficulty is that a possible misregistration of the pattern must not influence the assessment.

5.4.1 Measuring the Displacement

All edges of the pattern are equally displaced, either inward or outward. Therefore, it is possible to define the edge displacement ΔD as half the change in the width W of a conductor, the nominal width W_N of the corresponding conductor on the master film being used as a reference (Figure 5.8). The edge displacement ΔD is not affected by possible projections, indentations, and unevenness along the conductor edges.

The width of the selected conductor is measured with a 7X or 10X magnifier with a reticle; in the case of very fine-line boards—that is, with conductor widths of 0.004 in. (0.1 mm)—a somewhat higher magnification is recommended.

In the case of conductors with smooth edges, measurement should be no problem. When the conductor is uneven, the width is determined as shown in Figure 5.9; that is, as the distance between the two median lines of the crest-to-trough spans of the two conductor edge blemishes (Paragraph 5.5), disregarding isolated projections and indentations (Paragraph 5.6).

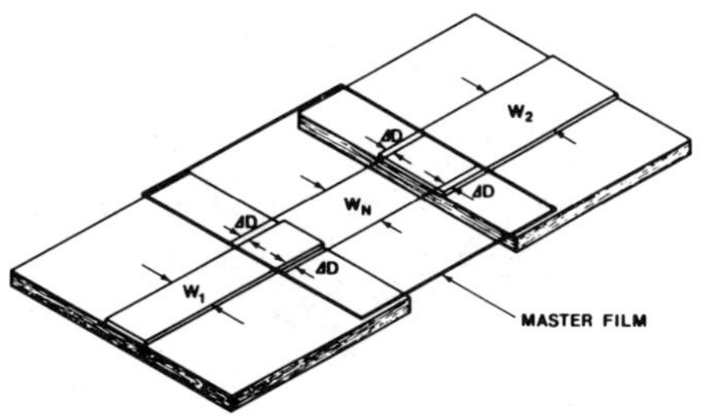

Fig. 5.8. Determination of edge displacement ΔD. $\Delta D = 1/2\,(W - W_N)$, where $W = W_1$ or W_2, and W_N = nominal width.

The measurement is somewhat subjective, but with some training the inspector can measure with satisfactory accuracy, although the conductor width is normally poorly defined because of the underetching.

Incidentally, the inspector should remember that the conductor width is, in reality, poorly defined owing to the underetching, normally called undercut (Figure 5.10). When measuring the conductor width, it is always the largest width which is determined, whether this is found at the base of the conductor (W_1) or at its top (W_2).

The edge displacement ΔD is defined as half the difference between corre-

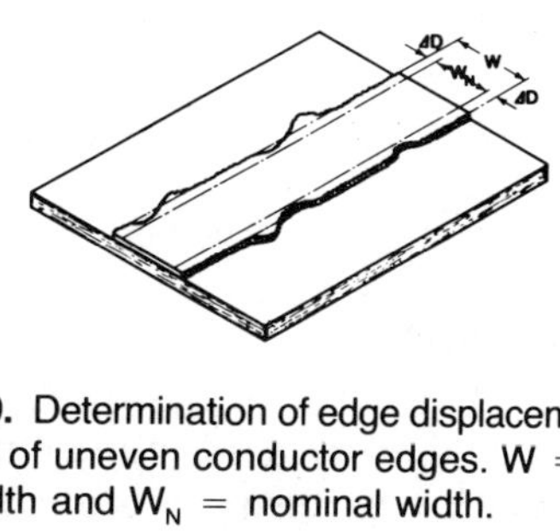

Fig. 5.9. Determination of edge displacement ΔD in case of uneven conductor edges. W = effective width and W_N = nominal width.

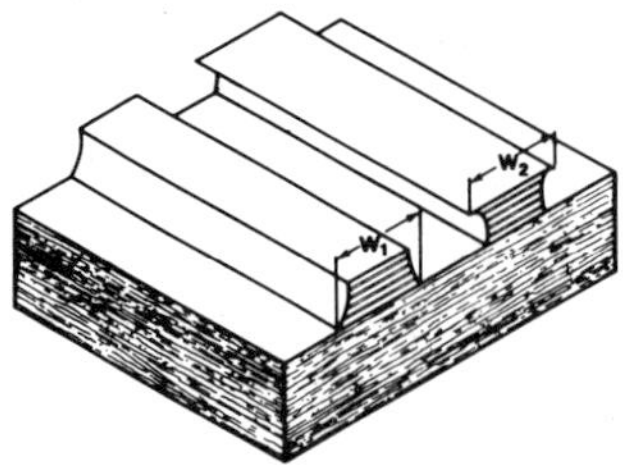

Fig. 5.10 Determination of conductor width. W_1 = conductor width at the base and W_2 = conductor width at the top.

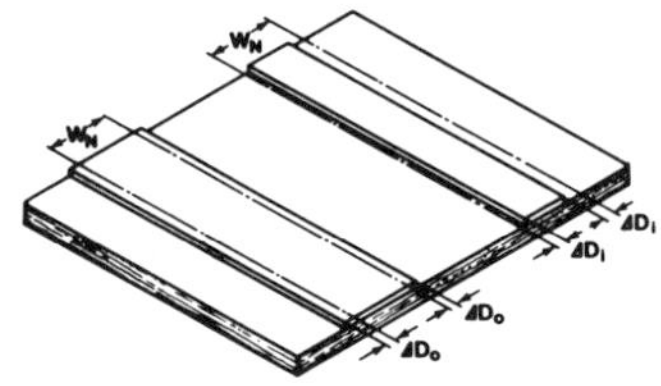

Fig. 5.11. Types of edge displacements. W_N = nominal width. ΔD_i = inward edge displacement, ΔD_o = outward edge displacement.

sponding measurements on the board and the master film; that is

$$\Delta D = \tfrac{1}{2}(W - W_N)$$

5.4.2 Tolerance Requirements

The edge displacement ΔD is now compared to the following tolerances, which imply a maximum thickness of the copper foil of 1 oz. = 0.0014 in. (35 μm):

1. In the case of plated-through, but not reflowed, boards, the edge displacement shall meet the following requirements:

$$\Delta D_i \leq -0.002 \text{ in. } (-0.05 \text{ mm}) \text{ inward,}$$
and
$$\Delta D_o \leq +0.004 \text{ in. } (+0.10 \text{ mm}) \text{ outward.}$$

See Figure 5.11.

2. In the case of reflowed boards as well as bare-copper boards, the edge displacement shall meet the following requirements:

$$\Delta D_i = -0.002 \text{ in. } (-0.05 \text{ mm}) \text{ inward,}$$
and
$$\Delta D_o = +0.002 \text{ in. } (+0.05 \text{ mm}) \text{ outward.}$$

See Figure 5.11.

For functional reasons it is not enough to verify that the edge displacements remain within the above limits. When deciding whether the actual insulation distance (spacing) or the actual width of a conductor is acceptable, it is necessary to include the combination of the following possible faults: unevenness, projections and indentations along the pattern edges, intervening metal particles within the insulation distance, and voids and pinholes in the pattern (for example, in the conductor). We can therefore formulate the following additional requirements which must also be met:

3. In the case of outward displacements, the spacing S to other parts of the pattern must be at least 75 percent of the nominal spacing S_N (insulation distance, Figure 5.12). This requirement shall be met at the most unfavorable combination of edge displacements, unevenness and projections along the pattern edges, and intervening metal particles.
4. In the case of inward displacements, the width W of a conductor

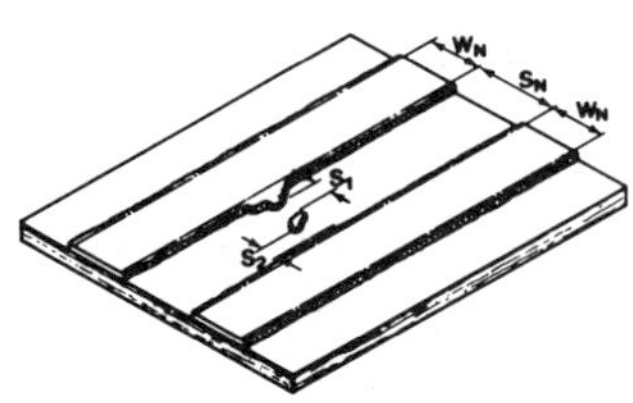

Fig. 5.12. Reduction of conductor spacing caused by pattern irregularities. $S_1 + S_2 \geq 0.75\ S_N$, where S_n = nominal specing, and W_N = nominal conductor width.

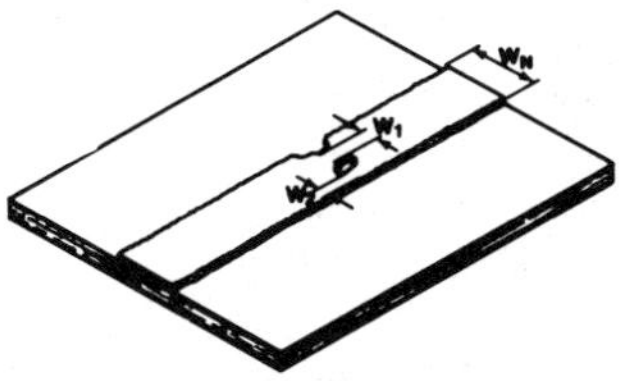

Fig. 5.13. Reduction of conductor width caused by pattern irregularities. $W_1 + W_2 \geqslant 0.75\ W_N$, where W_N = nominal width.

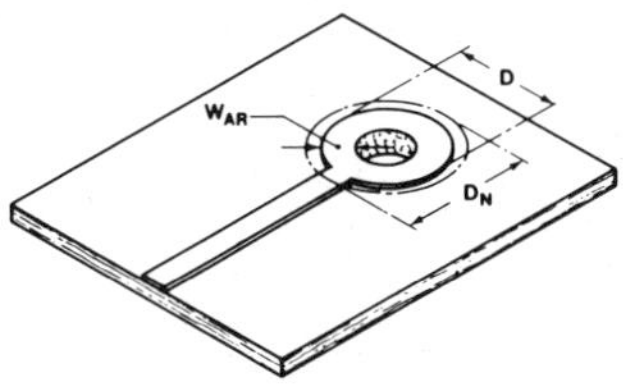

Fig. 5.14. Reduction of pad diameter caused by pattern irregularities. $D \geqslant 0.75\ D_N$, where D = pad diameter, D_N = nomnial diameter of pad. W_{AR} = width of annular ring.

> must be at least 75 percent of the nominal width W_N (Figure 5.13). Similarly, the diameter D of a solder pad must be at least 75 percent of the nominal diameter D_N, and the width of W_{AR} of the annular ring, as mentioned in Paragraph 5.9, must be met. See Figure 5.14. All requirements shall be met at the most unfavorable combination of edge displacements, unevenness and indentations along the pattern edges, and voids and pinholes in the conductors or solder pads.

The alert reader will note an apparent discrepancy between the accepted edge displacement and the requirement of not reducing certain dimensions of the pattern to less than 75 percent of the nominal value.

If the nominal value is 0.012 in. (0.3 mm), 75 percent represents 0.009 in. (0.225 mm). If we assume that the edges of two neighboring conductors are displaced by 0.002 in. (0.05 mm) outwards (that is, towards each other), the effective spacing becomes only 0.008 in. (0.2 mm). Similarly, we find that the effective width of a conductor becomes 0.008 in. (0.2 mm) when its edges are displaced 0.002 in. (0.05 mm) inwards. In either case the tolerance does not leave room for any of the irregularities previously mentioned.

5.4.3 Role of Processes

In conjunction with the above considerations, it should be emphasized that the PCB manufacturer must assess the nature of the pattern and adjust his processes to the actual conditions. Only in this way can he fulfill the requirements regarding the effective insulation distance or the conductor width, even when some of the irregularities mentioned above occur along the pattern edges. The PCB manufacturer must therefore be allowed to vary the parameters stated below:

- Copper foil thickness
- Image transfer method
- Plating method
- Etchant type

By combining the parameters in an appropriate manner (Table 6.2, Paragraph 6.5.1 of Reference A12), the PCB manufacturer can achieve an edge displacement within the range of −0.001 to +0.001 in. (−25 to +25 μm). This allows some room for such irregularities.

Frequently only certain areas of the pattern are so dense or of such a fine-line character that the above tolerance problem plays a decisive part.

On the whole, the tolerances regarding edge displacement can be met without much difficulty by any professional PCB manufacturer. If these requirements are violated, the fault should be rated according to Table 5.1.

Table 5.1. Rating of displacement faults according to requirements I to IV, Paragraph 5.4.

I or II		III and/or IV		
ΔD for nonreflowed boards	ΔD for reflowed boards	Spacing S	Width W Diameter D	
		(max. 25% reduction)		Rating
Observed		Observed		Accepted
Exceeded		Observed		B-defect
Observed		Exceeded		A-defect
Exceeded		Exceeded		A-defect

5.4.4 Board Disposition

Noncompliance with the remaining dimension of 75 percent of the nominal value shall be cause for rejection, whereas noncompliance with the specified value of the edge displacement will not, in itself, be cause for rejection.

It is emphasized that in areas carrying power line voltages, the creepage distance must always meet the safety requirements. In Paragraph 6.5.2 of Reference A12, this is defined as an effective insulation distance of at least 0.1 in. (2.5 mm), regardless of possible solder and insulation masks.

5.5 Edge Definition

Since the production processes do not distinguish between the edge of a pad and the edge of a conductor, the same degree of unevenness will be found along such edges. The edge definition of the pattern can therefore be defined as the maximum unevenness U, measured from crest to trough. This is illustrated in Figure 5.15, which is drawn oversized. The unevenness will normally be less than 0.004 in. (0.10 mm). In order to make it easier, we will only state below the size of the unevenness, implying the maximum crest-to-trough value.

5.5.1 Measuring the Unevenness

At solder pads, the unevenness is measured within 25 percent of the periphery, and along conductors within a length of 0.5 in. (13 mm). See Figure 5.16. All irregularities such as projections, indentations, and the like are disregarded.

A typical example of unevenness is shown in Figure 5.17. The fault is frequently due to a poor image transfer, especially in the case of screen printing, but just as common a cause can be an incomplete etching, usually due to an incomplete removal of the plating resist.

The measurement, made with a 7X or 10X magnifier with reticle, unavoidably becomes somewhat subjective, not least

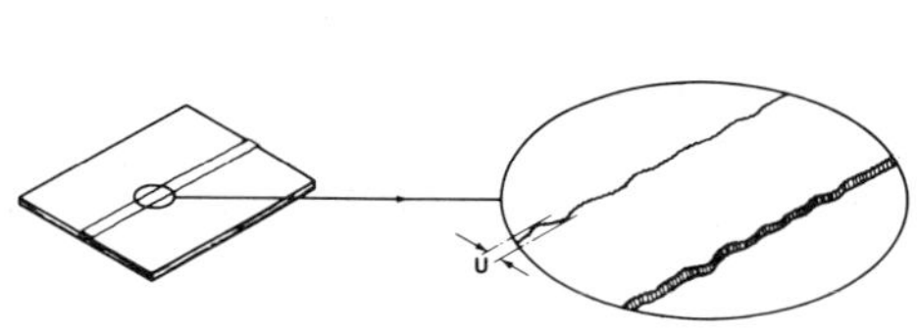

Fig. 5.15. Edge definition. Unevenness U is measured from peak to valley.

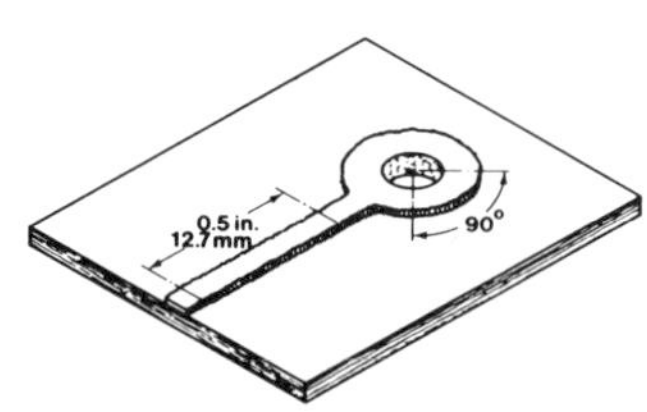

Fig. 5.16. Unevenness is determined within relatively short edge lengths.

Fig. 5.17. Board with uneven conductors.

at the solder pads owing to their rounding. A profile comparator with surface illumination facilitates the measurement, particularly at the solder pads, which can easily be centered by means of the bull's-eye on the screen.

It seems exaggerated to procure a profile comparator for the purpose of inspecting printed circuit boards only. It should be noted that an unevenness of 0.004 in. (0.10 mm) along the conductor edges is more a cosmetic question than a functional one, for which reason it does not require a very accurate measurement. An exception is high-power boards, where projections may cause arcing as described in Paragraph 5.6.1.

5.5.2 Board Disposition

When assessing unevennesses greater than 0.004 in. (0.10 mm), the inspector should consider how well the boards meet both the functional requirements, to which primary importance is attached, and the cosmetic requirements, to which only secondary importance should be attached.

The cosmetic requirements are fairly difficult to quantify, because the assessment will always be based on the subjective opinion of the inspector. In order to give the inspector certain guidelines, some judgment criteria are given below:

1. A generally occurring unevenness up to 0.004 in. (0.10 mm) is accepted without entering remarks in the inspection record. Unevenness up to 0.008 in. (0.20 mm) is rated as a B-defect, whereas unevenness greater than this shall be cause for rejection and consequently rated as an A-defect.
2. Unevenness between 0.008 and 0.020 in. (0.20 and 0.50 mm) is accepted as a B-defect on the condition that such uneven edges occur as an exception; that is, within an area corresponding to a maximum of 2 percent of the total board area, or with a total length of up to 4 in. (100 mm) per 16 sq. in. (100 cm^2) of board area, the length being measured along the pattern edges. If the unevenness occurs to a larger extent or is greater than 0.020 in. (0.50 mm), the defect is usually regarded as cause for rejection and rated as an A-defect in compliance with Requirement 1 above.

Both of the following functional requirements shall be met:

3. In the case of unevenness along the pattern edges, the spacing S to other pattern parts must be at least 75 percent of the nominal spacing S_N (insulation distance). See Figure 5.12. This requirement shall be met at the most unfavorable combination of unevenness, edge displacements, projections along the edges, and intervening metal particles.
4. In the case of unevenness along the pattern edges, it is a further prerequisite that the width W of a conductor be at least 75 percent of the nominal width W_N. See Figure 5.13. It is also required that the diameter D of a solder pad be at least 75 percent of the nominal diameter D_N,

Table 5.2. Rating of unevenness faults U along pattern edges according to requirements I to IV, Paragraph 5.5.

I	II	III and/or IV	
Generally occurring	Exceptionally occurring	Spacing S / Width W Diameter D (max. 25% reduction)	Rating
$U \leqslant$ 0.1 mm / 0.004 in.		Observed	Accepted
$U \leqslant$ 0.2 mm / 0.008 in.		Observed	B-defect
$U >$ 0.2 mm / 0.008 in.		Observed	A-defect
	0.2 mm / 0.008 in. $< U \leqslant$ 0.5 mm / 0.02 in.	Observed	B-defect
$U <$ 0.2 mm / 0.008 in.	$U >$ 0.5 mm / 0.02 in.	Observed Exceeded	A-defect A-defect

and also that the requirements regarding the width W_{AR} of the annual ring, as mentioned in Paragraph 5.9, are met. See Figure 5.14. All requirements shall be met at the most unfavorable combination of unevenness, edge displacements, indentations along the pattern edges, and also voids and pinholes in the conductors or solder pads.

The alert reader will again note an apparent discrepancy, similar to that in Paragraph 5.4, between the accepted size of unevenness and the requirement of maintaining certain dimensions of the pattern at 75 percent or more of the nominal value.

If the nominal value, as before, is 0.012 in. (0.3 mm), 75 percent represents 0.009 in. (0.225 mm). If we assume that an unevenness of 0.004 in. (0.10 mm) evenly distributed around the pattern edges, the width of a conductor becomes $0.012 - 2 \times 0.002 = 0.008$ in. ($0.30 - 2 \times 0.05 = 0.20$ mm), and correspondingly for the spacing. In both cases the tolerance leaves no room for any of the irregularities previously mentioned. The same considerations as given in Paragraph 5.4 also apply here, namely that the PCB manufacturer has to adapt his processes to the actual conditions.

The inspector can now inspect the boards according to these requirements. If the requirements are violated, the defect should be rated according to Table 5.2.

The inspector should be aware of the difference between nodules and unevenness. Nodules are outgrowths along the pattern edges, not only outwards but also upwards, and they are usually caused by overplating or by too large a current density during plating. See the extended discussion concerning nodules in Paragraphs 7.3.7 and 7.3.8.

5.6 Projections and Indentations

Projections and indentations along the pattern edges can be isolated so that the remainder of the edges are quite even,

Fig. 5.18A. Board with projections along pattern edges.

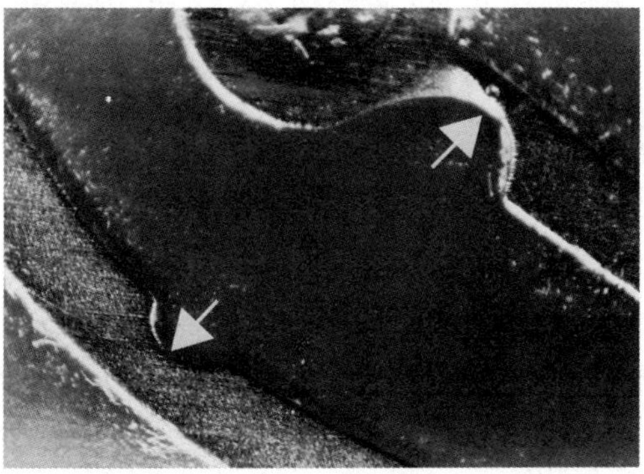

Fig. 5.18B. Board with indentations along pattern edges.

for the reasons mentioned in Paragraph 5.5. See Figure 5.18. These defects can occur along the conductor edges as well as the solder pads.

5.6.1 Hazards of Projections and Indentations

Projections are considered somewhat more dangerous than a general outward-going edge displacement—an increased conductor width—because spikes in the pattern increase the risk of arcing. Depending on their extent, projections and indentations can degrade the quality conception of the finished board. The following discussion primarily provides the inspector with a better background for assessing the importance of projections and indentations.

5.6.2 Projection and Indentation Tolerances

As described in Paragraph 5.5, the inspector should distinguish between cosmetic requirements and functional requirements. It is a little easier to assess cosmetic defects such as projections and indentations, than questions of edge definition. It is possible to state the following:

1. The size of the projection or indentation shall meet the following requirements: Lengthwise, the dimension L shall be less than 0.2 in. (5 mm), and crosswise, the dimension C shall be less than 0.04 in. (1 mm). See Figure 5.19.
2. A further limitation applies to solder pads, where the dimension L along the rim of the pad is limited to 25 percent of the periphery, independent of the shape of the solder pad. See Figure 5.20.

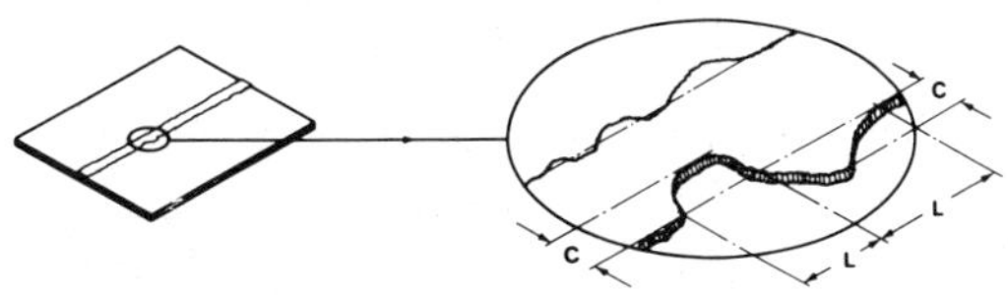

Fig. 5.19. Projections and indentations along conductor edges.
C = crosswise dimension of irregularity.
L = lengthwise dimension of irregularity.

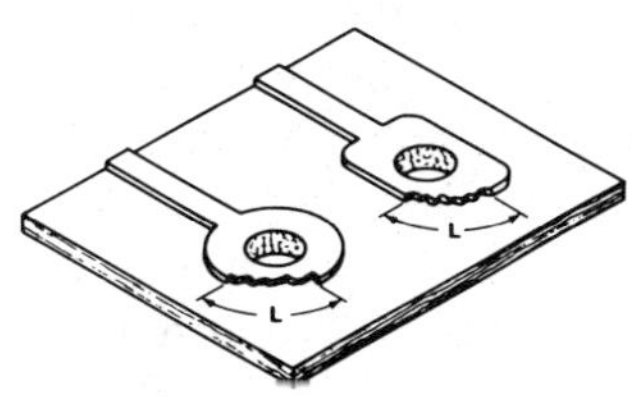

Fig. 5.20. Projections and indentations along rim of solder pad.

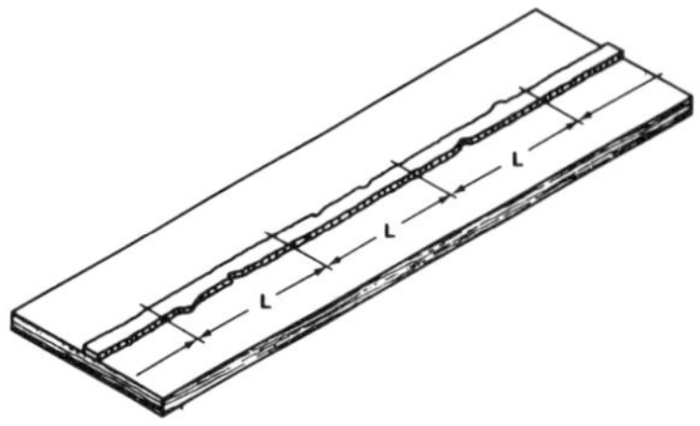

Fig. 5.21. Determination of average number of projections and indentations along pattern edges. L = 2 in. (50 mm).

3. The average number of defects (projections and indentations) along a pattern edge shall not exceed one defect per 2 in. (50 mm) length of edge. Thus, a quite random distribution is permitted. See Figure 5.21.

For functional reasons, the actual spacing (insulation distance), conductor width, or solder pad diameter must be at least 75 percent of the nominal value. The following two requirements shall both be met:

4. In the case of projections along the pattern edges, the spacing S to other pattern parts must be at least 75 percent of the nominal spacing S_N (insulation distance). See Figure 5.12.

   This requirement shall be met at the most unfavorable combination of projections, edge displacements, unevenness along the edges, and intervening metal particles.
5. In the case of indentations along the pattern edges, it is also a requirement that the width W of a conductor be at least 75 percent of the nominal width W_N. See Figure 5.13. Furthermore, the diameter D of a solder pad must be at least 75 percent of the nominal diameter D_N, and finally the requirements regarding the width W_{AR} of the annular ring, as mentioned in Paragraph 5.9, must be met. See Figure 5.14.

   All requirements shall be met at the most unfavorable combination of indentations, edge displacements, unevenness along the edges, and voids and pinholes in the conductors or solder pads.

Requirements 1 to 5 offer the PCB manufacturer very reasonable manufacturing tolerances; there should be no difficulty in fulfilling these requirements.

The expected size of projections and indentations as stated in Requirement 1 could seem fairly large, but Requirements 4 and 5 impose stricter limits on the size. For example, a projection of 0.04 in. (1 mm) is acceptable only if the nominal distance to a neighboring conductor (regardless of other possible defects such as edge displacements, unevenness, metal particles, etc.) is greater than 0.16 in. (4 mm).

On modern, densely packed circuit boards, there will be only a very few places where such large projections are present, and because of this, the importance of Requirement 3, which defines the number of acceptable defects, is reduced. Projections of the size mentioned may cause short circuits, which are dealt with in Paragraph 5.3. Analogous conditions are valid for indentations.

Therefore, accepting projections and indentations according to Requirements 1 to 3 is normally based on fairly coarse boards with large spacings and wide conductors; an alternative is open areas of otherwise dense boards. For this reason it seems sensible to adopt a functional assessment over a cosmetic assessment.

The inspector examines the boards according to Requirements 1 to 5. The measurement, which is carried out with a 7X or 10X magnifier with reticle and is a little subjective, still seems adequate. If the inspector finds a violation of the requirements, the defect is rated in agreement with Table 5.3 below.

Table 5.3. Rating of projection and/or indentation faults along pattern edges according to requirements I to V, Paragraph 5.6.

I		II	III	IV and/or V	
Size of defect			Average number N of defects	Spacing S / Width W Diameter D	
Lengthwise	Crosswise	Along pad rim			
L ≤ {5 mm, 0.2 in.}	C ≤ {1 mm, 0.004 in.}	L ≤ 25%	N ≤ 1 per {50 mm, 2 in.}	(max. 25% reduction)	Rating
Observed			Observed	Observed	Accepted
Observed			Observed	Exceeded	A-defect
I, II or III exceeded				Observed	A-defect

5.6.3 Board Disposition

When the inspector finds more or larger projections and indentations than allowed, the question of exemption or rejection arises. An intermediate possibility of exemption could be considered if the defects occur on component sides provided with insulation masks.

Unfortunately, it is not possible to guarantee that an insulation mask always fills the spacing between two parallel conductors right down to the base material when the spacing is small. Only from spacings of 0.020 in. (0.5 mm) and more is it to be expected that the ink completely encapsulates the conductors. See Figure 5.22.

This implies that the conductor thickness be less than approximately 0.004 in. (100 μm), namely 1 oz. = 0.0014 in. (35 μm) copper foil, 0.001 to 0.0014 in. (25 to 35 μm) plated copper, 0.0004 to 0.0008 in. (10 to 20 μm) tin/lead, and the rest covering changes in the surface structure of the tin/lead layer caused by reflowing. It is therefore rather risky to reduce the demand regarding the remaining insulation distance from 75 percent to, for example, 50 percent of the nominal value.

On the solder side of tin/lead plated boards, a reduction cannot be permitted because the solder mask is somewhat damaged after mass soldering. Because the underlying tin/lead layer melts and flows during mass soldering, the resist will at best wrinkle and at worst flake.

An exemption might be possible if the number of unacceptable projections is so small that all projections can be removed and the solder resist restored. Another possibility of exemption is to check whether the remaining insulation distance will be adequate for the particular application; that is, to carry the voltage applied. See Figure 6.11, Paragraph 6.5.2 of Reference A12. Such an exemption requires that the number of projections be limited. As these conditions are very similar to short circuits, the reader is referred to Paragraph 5.2.

In areas carrying power line voltage, the inspector shall very carefully verify

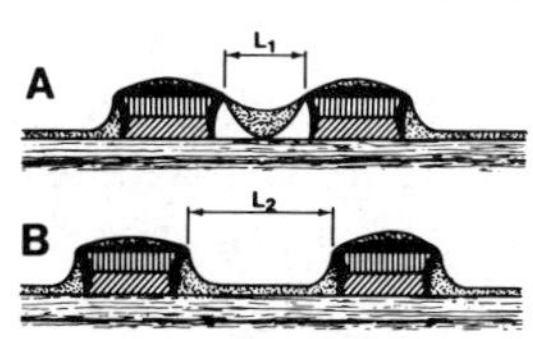

Fig. 5.22. Cross-sectional drawing of parallel conductors showing problems of achieving full filling of space between conductors.
A: Complete filling of space is not achieved (L_1 ≤ 0.02 in. = 0.5 mm). B: Complete filling of space is achieved (L_2 > 0.02 in. = 0.5 mm).

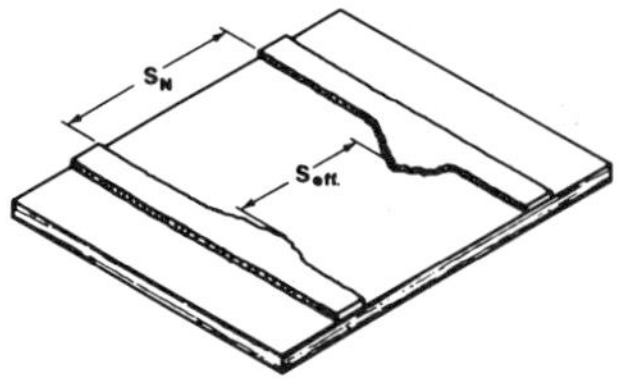

Fig. 5.23. Careful inspection of insulation distance is necessary in areas with applied power line voltage. The effective insulation distance should comply with national safety regulations. $S_{eff.}$ = effective insulation distance. S_N = nominal spacing.

that the insulation distance meets the safety requirements. On boards designed as described in Paragraph 6.5.2 of Reference A12, an effective minimum insulation distance of 0.1 in. (2.5 mm) is required. In Design Rule Number 3(B), stated in the referenced paragraph, the nominal safety margin is 0.12 in. (3.0 mm), so that there is a margin for certain irregularities along the edges.

If the sample shows a general occurrence of oversized projections all over the board, a 100 percent inspection of the batch should be performed in order to ensure that the safety requirements are met in areas carrying power line voltages. The inspector should be particularly alert to the possibility of projections occurring opposite each other between two neighboring conductors (Figure 5.23).

A safety distance of 0.1 in. (2.5 mm) is also valid even when the board is provided with solder and insulation masks. There is no guarantee that the insulation distance is completely covered (filled) with ink, and the solder mask will, as described above, be somewhat damaged during mass soldering.

The inspection is facilitated, and the security greater, if the PCB designer marks the master drawing to show those parts of the pattern which carry power line voltages.

Fig. 5.24. Board with pinholes and voids in pattern.

5.7 Pinholes and Voids

Due to processing faults, holes can occur in the pattern as shown in Figure 5.24. The holes are usually called pinholes and voids, but the technical literature makes no distinction between the two types of holes, which are both characterized by extending down to the base material.

In the following discussion, holes so small that they are difficult to see with the naked eye are called pinholes, and the upper limit of their size is set at a diameter of 0.006 in. (0.15 mm). Holes above this limit are called voids.

5.7.1 Hole Formation

If the etch factor is about one, the etchant attacks about the same in horizontal and vertical directions. The hole spreads sideways as hinted in Figure 5.25, which shows

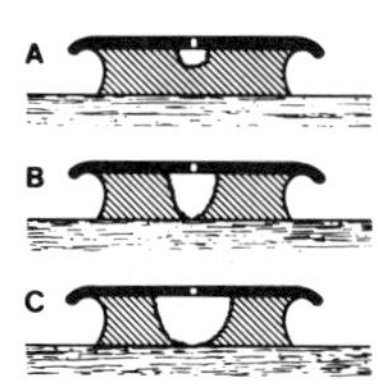

Fig. 5.25. Cross-sectional drawing showing development of a void or pinhole. A: Hole appears as a pit. B: Hole extends to base material. C. Hole grows sidewise.

Fig. 5.26. Cross-sectional drawing showing hole in bare-copper board.

a cross-section of a hole in a conductor. In (A) the hole has not reached the base material and is in fact a pit, whereas in (B) the hole extends to the base material. By prolonging the etching time the hole will grow as shown in (C).

In the case of a bare-copper board, the etch resist is removed and the hole assumes the appearance shown in Figure 5.26, which shows an easily visible hole. The conditions are a little different when the board is tin/lead plated. Here the etch resist is the tin/lead layer itself, which is not removed, and the cross-section of the conductor, before reflowing, has the appearance shown in A, Figure 5.27. It should be noted that the area around the hole in the tin/lead layer is undermined, for which reason it is almost irrelevalent to talk about the size of a pinhole or a void in a nonreflowed board, at least when the assessment is based on the surface appearance.

Let us assume that the etch factor (the ratio between the depthwise and sidewise etching) is 1.2. Let us furthermore assume that the thickness of the copper layer is 0.0024 in. (60 μm), namely 1 oz. = 0.0014 in. (35 μm) copper foil + 0.001 in. (25 μm) copper plating. The sidewise etching just beneath the tin/lead layer then becomes theoretically about 0.002 in. (50 μm) when the depthwise etching is 0.0024 in. (60 μm), which corresponds to free-etching the pattern.

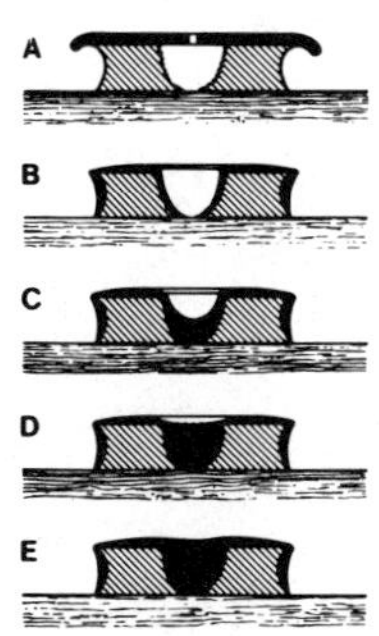

Fig. 5.27. Cross-sectional drawing showing different types of holes in a conductor on a reflowed tin/lead plated board. A: Appearance before reflowing. B: Hole appears as a pinhole or a void. C: Hole appears as a pit. D: Hole appears as a faint hollow. E: Hole is practically invisible.

In order to minimize the general undercut of the conductor, the etching is stopped as soon as the pattern develops. By assuming that the hole in the etch resist is 0.002 in. (50 μm) in diameter, the diameter of the undermined area becomes $2 \times 0.002 + 0.002 = 0.006$ in. ($2 \times 50 + 50 = 150$ μm), which means a large pinhole or a small void. A prolonged etching time enlarges the hole.

Tin/lead plated boards are usually reflowed, and the tin/lead layer which lies as a lid over the etched hole collapses because of this. Therefore, the cross-section can assume the appearance shown in A, B, C, D, or E of Figure 5.27, where the only difference is the degree of tin/lead filling of the hole. Consequently, the hole can be either a pinhole or a void (B) because the base material is exposed at the bottom, a pit (C), a faint hollow (D), or practically invisible (E).

5.7.2 Effects of Holes

Before discussing the effects of etched holes, it is expedient to see what happens when a component lead is soldered in a solder hole (Figure 5.28). For reasonably easy insertion of the component lead in the solder hole, and also for the filling of a plated-through hole with solder, the hole diameter is usually 0.008 to 0.020 in. (0.2 to 0.5 mm) larger than the diameter of the component lead (see Design Rule Number 5, Paragraph 6.5.4a of Reference A12). The electrical connection from the

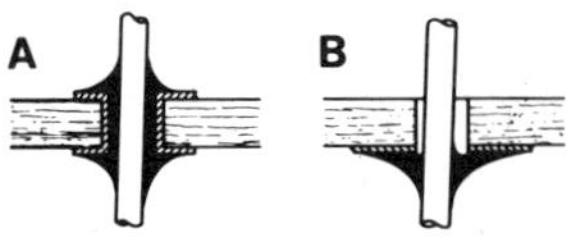

Fig. 5.28. Cross-sectional drawing showing a component lead soldered into a solder hole. A: Plated-through hole. B: Nonplated-through hole.

component to the solder pad passes through a solder bridge, which is universally accepted.

The above case can now be transferred to the actual problem of holes in a conductor. When the holes are closed as shown in C, D, and E of Figure 5.27, they can be regarded as solder spots in the conductors, and if they are not too large with respect to the conductor width, they may be regarded as harmless.

More or less closed holes are very difficult to detect during inspection, particularly when they are very small (less than 0.006 in. (0.15 mm)), especially if a solder or insulation mask has been applied. Usually quite a number of such holes will pass through the incoming inspection without being seen, or may be mistaken for pits.

When such holes occur in a bare-copper board (Figure 5.26) or in a nonreflowed tin/lead plated board (Figure 5.27), we cannot expect the holes to close during soldering; in fact, they will not close at all if the board is provided with a solder mask. Therefore, a hole acts as an effective width reduction of the conductor.

5.7.3 Inspection Requirements

The following guidelines are valid for the inspection of pinholes and voids in the pattern of bare-copper boards and reflowed tin/lead plated boards:

1. In solder pads the diameter of open holes (see Figure 5.26 and B of Figure 5.27) and of visible, but closed, holes (C and D, Figure 5.27) shall be less than 0.006 in. (0.15 mm).
2. Requirement 1 implies that the diameter D of a solder pad be at least 75 percent of the nominal diameter D_N (Figure 5.14), and also that the requirements regarding the width W_{AR} of the annular ring be met (Paragraph 5.9).
3. In conductors, the diameter of open holes or half-closed holes (Figure 5.26 and B and C of Figure 5.27) shall be less than 20 percent of the nominal width of the conductor. In the case of visible, but completely closed holes (D, Figure 5.27), the diameter shall be less than 40 percent of the nominal width of the conductor.
4. Requirement 3 implies that the width W of a conductor be at least 75 percent of the nominal width W_N in the case of open holes, and at least 50 percent of the nominal width W_N in the case of visible, but completely closed holes. See Figure 5.13.
5. Requirements 1 through 4 are to be met at the most unfavorable combination of holes, edge displacements, unevenness, and indentations along the pattern edges.
6. No more than one hole per solder pad shall be permitted, and none in the transition zone between the solder pad and the conductor. In plated-through boards, no hole shall be permitted at the rim of the solder hole. See Figure 5.29. On the average, no more than one hole per 2 in. (50 mm) of conductor length shall be permitted.

It should be noted that the invisible closed holes as shown in E, Figure 5.27, have not been considered, because unless the tin/lead layer is etched away, or a microsection is made, they usually cannot be detected. When the inspector suspects that

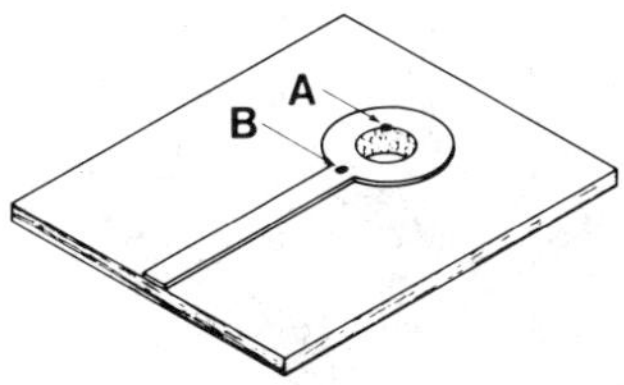

Fig. 5.29. Inadmissible locations of pinholes on solder pads. A: Pinhole occurs at rim of solder hole. B: Pinhole occurs in transition zone between solder pad and conductor.

such holes are present in a board, he can scrape the tin/lead layer away from the conductor surface with a scalpel. Possible holes will then appear as tin/lead-filled holes in the red copper.

5.7.4 Board Disposition

If these requirements are not met, the defect is cause for rejection and rated as an A-defect. It is possible to grant an exemption from requirement 6 regarding the number of holes, but this must depend on an assessment, among other things, of the size of the holes.

5.8 Pattern Adhesion

It goes without saying that it is very important that the pattern (solder pads and conductors) properly adheres to the base material. This applies not only just after the board has been soldered, but throughout the life of the board and the assembly. Lifted or loosened solder pads and conductors are an indisputable cause for rejection, for which reason the defect is rated as an A-defect.

5.8.1 Susceptible Areas

Very narrow conductors are liable to become loose due to the unavoidable undercut. For a total copper thickness of 0.0024 in. (60 μm) composed of 1 oz. = 0.0014 in. (35 μm) copper foil and 0.001 in. (25 μm) plated copper, and an etch factor of 1.2, the undercut is 0.002 in. (50 μm) at each side of the conductor. If the nominal width of the conductor is 0.010 in. (0.25 mm) there is only about 0.006 in. (0.15 mm) left to ensure the adhesion of the conductor to the base material.

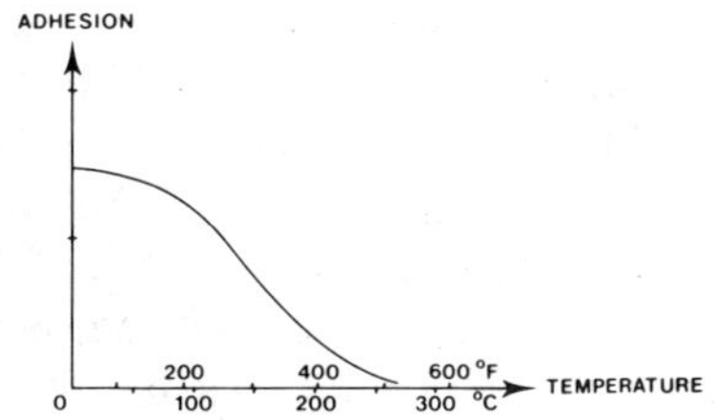

Fig. 5.30. Adhesion of pattern to base material as a function of temperature.

The inspector shall in particular be aware of points and sharp bends, as the greatest risk of lifted conductors occurs just here. The epoxy becomes very hot when the board is mass soldered and, as shown in Figure 5.30, the adhesion drops to nearly zero. Therefore, such conductors cannot withstand any external mechanical influence.

If the pattern is badly designed, such as with a sharp bend in a long and narrow conductor, it is possible, when mass soldering the board, that the difference between the coefficients of thermal expansion for the copper and the base material will cause large stresses in the conductor, so that the point of bend will lift as shown in Figure 5.31. This problem is created by the PCB designer, and no blame may be placed on the PCB manufacturer. The PCB designer must take the consequences and improve the board design; for example, by breaking all sharp corners with 45° bends. See Figure 6.32, Paragraph 6.6.2 of Reference A12.

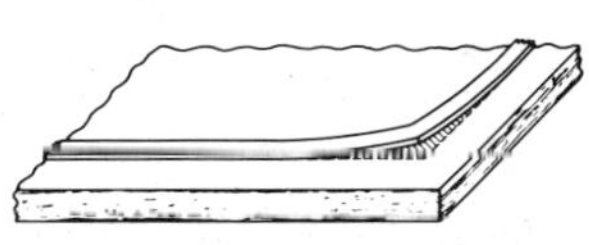

Fig. 5.31. Lifting of a conductor at a sharp corner.

Nonplated-through holes with narrow annular rings are also hazardous because the annular rings can loosen as a result of the mechanical influence of a hot solder iron, or even worse, from the recoil of a solder sucker used during the removal of a component. Here again it is unfair to blame the PCB manufacturer, and the PCB designer must take care to design his boards with adequately dimensioned annular rings. See Design Rule Number 7, Paragraph 6.5.4b of Reference A12.

5.8.2 Measuring the Adhesion

The adhesion of the pattern can be ascertained by the "Peel Test," with which the inspector measures the force necessary for peeling off a conductor of a specified width, the direction of force being kept perpendicular to the board within 2°. The shape of the conductors is frequently standardized as shown in Figure 5.32. Shape A is prescribed in IEC-249 and BS 9762, the only difference being the length. According to BS 9762, the test coupon comprises three conductors. Shape B is prescribed in MIL-P-13949. In this specification a test coupon comprises four wide and six narrow conductors and the test prescribes test coupons turned 90° with respect to each other so that the peel strength is determined in the lengthwise and the crosswise directions of the base material. (A test coupon is a separate printed circuit region on the production panel provided for various tests.)

It is also possible to use one of the sample boards, provided it comprises a conductor having a minimum width of 0.04

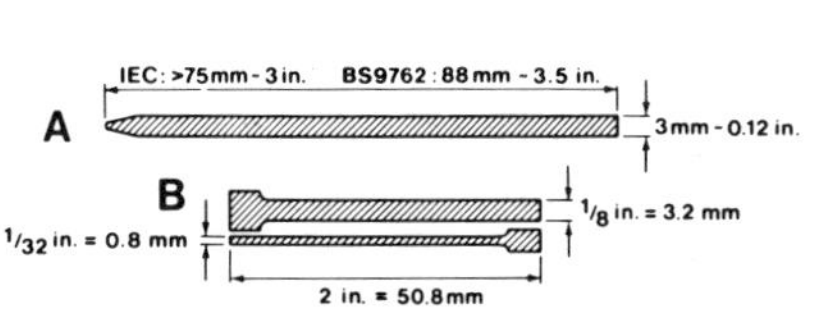

Fig. 5.32. Shape of conductors intended for peel testing.

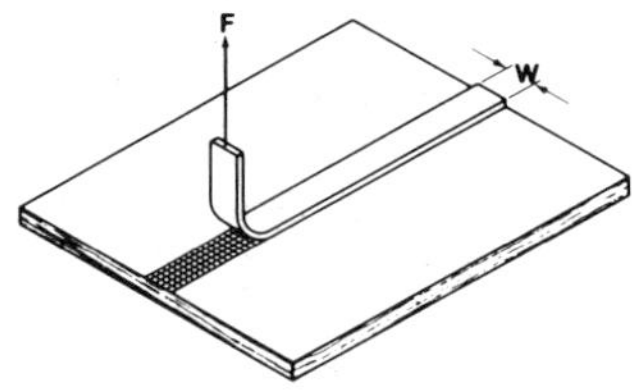

Fig. 5.33. Principle of a peel test. F: Force for peeling off the conductor. W: Width of conductor.

in. (1 mm), preferably 0.12 in. (3 mm), and a minimum length of 2 in. (50 mm). The conductor is detached from the base material for a length of 0.5 in. (13 mm), bent upwards, and gripped over its entire width with a clamp. Then the conductor is peeled off at a steady rate of 2 in. (50 mm) per minute. The required force is measured, in the simplest case, with a spring balance.

The peel strength is calculated as the ratio between the force F and the conductor width W. See Figure 5.33. In the U.S. and in England, the peel strength is normally expressed in pounds per in. of width (lb/in.). In the SI system, the force is measured in newtons (N) and the width in meters, so that the peel strength is expressed as kN/m or as N/mm. (1 N equals approximately 0.1 kg.) Multiplying the peel strength expressed in lb/in. by 0.18 converts the value to N/mm.

5.8.3 Peel Strength Requirements

The minimum peel strength required by several specifications is as follows:

Phenolic paper (FR2):
6 lb/in. $\simeq$ 1.0 N/mm

Epoxy paper (FR3):
7 lb/in. $\simeq$ 1.2 N/mm

Epoxy glass (FR4):
8 lb/in. $\simeq$ 1.4 N/mm

These values are valid for 1 oz. = 0.0014 in. (35 μm) copper and increase with increasing copper thickness because the adhesion area at the spot of peeling is a little larger due to the increased radius of curvature. By way of example, for 2 oz. = 0.0028 in. (70 μm) copper the values increase by 10 to 20 percent.

To more easily comprehend the tabulated values, consider a conductor 0.012 in. (0.3 mm) wide. On an epoxy glass board, the peel strength is about 1 oz. (28 g), the effective width of the conductor being taken into consideration. This is not a particularly large adhesion, and from the curve shown in Figure 5.30, it is easy to understand that a narrow conductor cannot withstand any mechanical influence when heated.

Of similar interest is the adhesion of the solder pads, and we will first consider the pull-off strength of solder pads with nonplated-through holes. The pull-off strength is measured as the force F, perpendicular to the board, required to pull a solder pad of a specified size off the base material. See Figure 5.34.

According to IEC-326, the diameters of the pad and the hole are 0.157 in. (4 mm) and 0.051 in. (1.3 mm) respectively, whereas according to BS 9760, they should be 0.15 in. (3.8 mm) and 0.039 in. (1 mm) respectively. As the difference in area is less than 7 percent, it is immaterial whether the first or second specification is adopted.

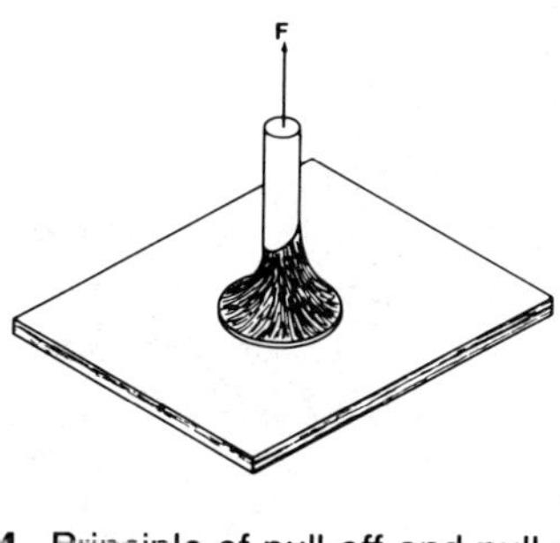

Fig. 5.34. Principle of pull-off and pull out tests of nonplated-through holes and plated-through holes, respectively. F: Force necessary for pulling-off/out.

A special test board with solder pads as specified above can be used, or the inspector can find corresponding solder pads on a board taken from the sample. In the latter case, the solder pads must be released from any conductors connected to the pads.

A 0.035 in. (0.9 mm) pretinned, hard-drawn brass wire is now inserted into the hole and soldered to the pad with a soldering iron at a temperature of 518 ± 18°F (270 ± 10 °C) and a soldering time of 4 ± 1 seconds. In this operation, the solder shall wet the solder pad completely. After cooling for 30 minutes, the board is placed in a tensile test machine. By increasing the force at a rate not exceeding 11 lb/s (50 N/s ≃ 5 kg/s), the inspector measures the force necessary to detach the solder pad from the base material. The following minimum values for pull-off strength, valid for one soldering operation, shall be verified:

Phenolic paper (FR2):
6.8 lb (31 N ≃ 3.1 kg)

Epoxy glass (FR4):
13.2 lb (60 N ≃ 6 kg)

Here also, it is true that the pull-off strength drops almost to zero when the solder pad is heated during soldering, for which reason a small mechanical influence will cause the solder pad to lift.

Finally, we will discuss the pull-out strength of plated-through holes. In principle, the measurement is carried out in the same manner as for solder pads; the diameters of the solder pads and the solder holes are 0.08 in. (2 mm) and 0.04 in. (1 mm), respectively. Having soldered the brass wire in place, the inspector performs five unsoldering and resoldering operations, then allows the hole and pad to cool for 30 minutes. It should be noted that in MIL-P-55110C no cooling period is prescribed (Paragraph 4.8.3). The sample is now ready for the pull-out strength test, where the following minimum value shall be verified:

Epoxy glass (FR4):
13.2 lb (60 N $\simeq$ 6 kg)

Owing to the plated-through barrel of the solder hole, the solder pad is better anchored to the base material, so that it withstands repeated unsoldering and resoldering operations far better than a nonplated-through solder pad.

A peel test is a rather complex, time-consuming and somewhat uncertain test, especially when the board is tin/lead plated and reflowed. If the board also has solder and insulation masks, it is almost hopeless to obtain reproducible results, even when the conductor is cut free of the mask along its edges. Therefore, there is no valid basis for peel-testing a board. Similar conclusions apply to a pull-off or a pull-out strength test for solder pads and plated-through holes, respectively.

5.8.4 A Simple Adhesion Test

It is, however, possible to assess the adhesion with a very simple, but also very subjective test, which requires only a scalpel. This test requires that some parts of the pattern, preferably nonfunctional, or a balancing network, not be covered by a solder or insulation mask.

The scalpel edge is placed against the accessible part of the pattern, which the inspector very carefully tries to lift by twisting the scalpel a bit. See Figure 5.35.

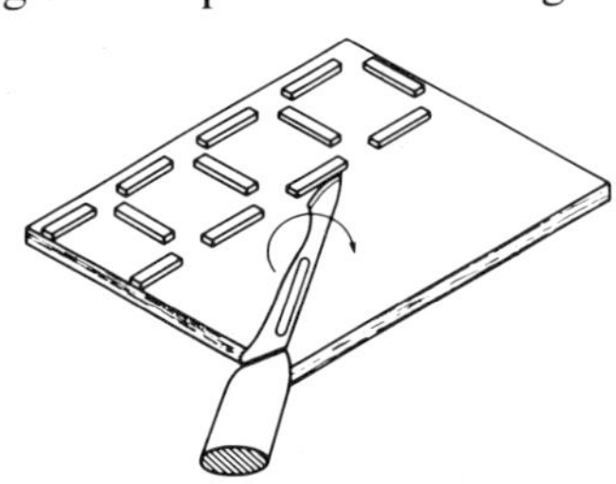

Fig. 5.35. Subjective assessment of the adhesion by means of a scalpel.

The inspector will soon learn how much to twist the scalpel in order to lift the copper, and on the basis of such a "standardized feeling," he can assess whether the pattern has sufficient adhesion.

5.8.5 Board Disposition

Where the adhesion is too low, information concerning lifted copper will quickly be fed back to the incoming inspection department from the assembly line. Before contacting the PCB manufacturer about the problem, however, it is advantageous to investigate whether the pattern design can be the reason for the lifting. An indication of this is that the lifting always occurs at the same location on a large number of boards.

Repair of a lifted conductor or solder pad on an assembled board can be performed according to the instructions given in IPC-R-700B, "Suggested Guide Lines for Modification and Repair of Printed Boards and Assemblies," Paragraphs 7.5 and 7.6. The PCB manufacturer should not be expected to repair lifted conductors or solder pads. This is cause for rejection and consequently rated as an A-defect.

5.9 Width of Annular Ring

Broken annular rings were once viewed as a fairly serious defect, but practice has shown it to be exaggerated. See Figure 5.36. Plated-through boards have not exhibited a reduced reliability due to soldering of pads with broken annular rings. This is also supported by the fact that, in recent years, many boards have been designed with intentionally broken annular rings. See Figure 5.37. In this way, it is possible to increase the diameter of the solder hole and still place a conductor between two IC solder pads.

The need for a large-diameter solder hole is nearly always caused by the use of

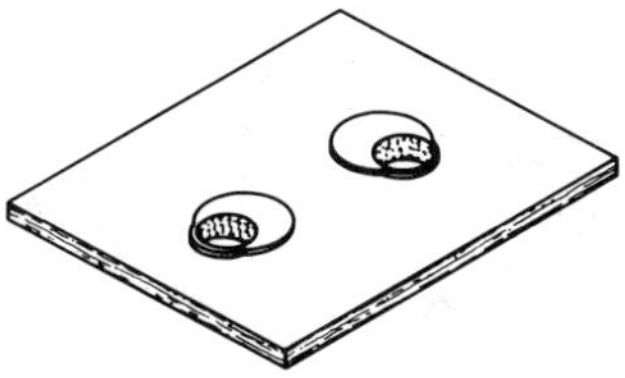

Fig. 5.36. Examples of broken annular rings.

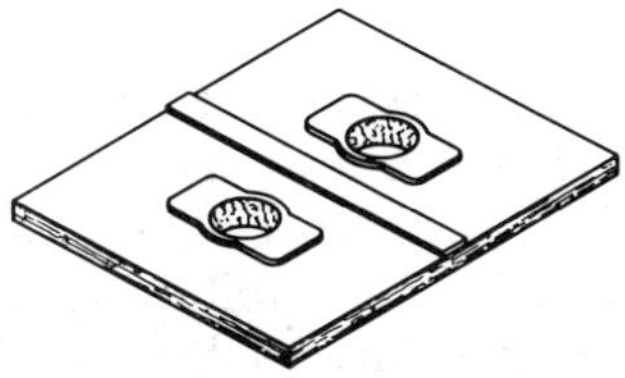

Fig. 5.37. Example of design with intentionally broken annular rings.

automatic component insertion. This technique sets great demands on the positional tolerance of the solder hole, and because the dimensional stability of the board is limited, one way to aid the automatic insertion is to increase the diameter of the solder hole. For example, the diameter of holes which could have been 0.031 in. (0.8 mm) might be increased to 0.040 in. (1.0 mm) or even to 0.043 in. (1.1 mm) when automatic component insertion is introduced. On the basis of the experience gained, it is considered reasonable to grant an exemption, to a certain extent, for broken annular rings.

5.9.1 Effect of Broken Annular Rings

In the case of nonplated-through boards, broken annular rings can prevent soldering of the solder pads when mass soldered. By giving such solder pads a shape as shown in Figure 5.38, the holes will not be closed. This is, as a matter of fact, utilized in some nonplated-through boards at places where it is required that certain empty holes remain open after mass soldering, which otherwise would render a

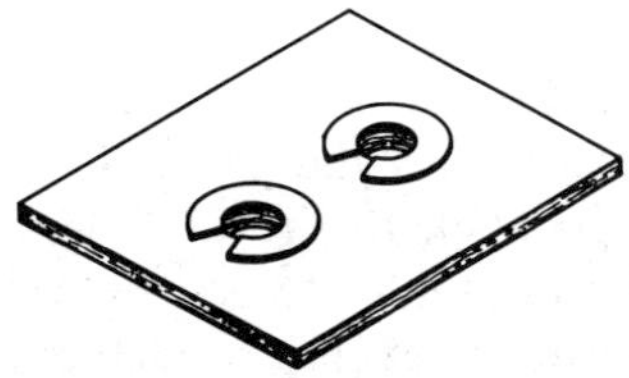

Fig. 5.38. Special shape of pads of nonplated-through holes prevents closing of holes during mass soldering.

later insertion and soldering of various components difficult. A practical example is certain types of precision capacitors which cannot pass through the mass soldering machine without changing value due to the heat.

If the customer is prepared to accept plated-through boards with a certain number of broken annular rings, it seems expedient to introduce this in the purchase specification. It can, for example, be stated that broken annular rings be accepted at 1 percent of the solder pads. This will, as a side effect, bring about a great reduction of the PCB manufacturer's need for communication with the PCB designer regarding any problems.

5.9.2 Causes of Broken Annular Rings

The artwork can cause broken annular rings; frequently the reason is a combination of misregistration of the solder pads of the two artwork sides, and the use of undersized solder pads relative to the solder holes. Other reasons can be found within the production processes, for example, inaccurate image transfer, or during hole drilling where the drill bit, for one reason or another, has drifted. The PCB manufacturer should have found all possible faults in the artwork during his audit of the documentation package received and have expressed reservations regarding possible broken annular rings.

If there are many broken annular rings, the inspector should check the accuracy of the film set; that is, the side-to-side registration beween the solder pads on the solder and component sides.

Where the registration is poor, he should go back to the PCB designer to have the original artwork checked. This may, however, exhibit a perfect registration, and the reason for the problem may be that one of the films has moved slightly, but sufficiently to cause broken annular rings.

The gelatin film of a photographic film can, under some conditions, flow a little and cause this fault. A commonplace fault is drying the developed film by hanging it in one corner with a single clip, which can cause film distortion.

Finally, it should be mentioned that the original set of master films can be correct, but that the PCB manufacturer's working films have changed a little. Many PCB manufacturers have stopped using photographic films and are now using diazo films. On very rare occasions the diazo film material can be insufficiently stabilized by the film manufacturer and incur small dimensional changes.

As mentioned in Paragraph 6.4.4f of Reference A12, there is a long tolerance chain from the films to the finished boards, and it is only possible to let the film inaccuracies use up a small part of the total tolerance.

In the following paragraphs we will assume a perfect side-to-side registration of the artwork. When, in spite of all precision and care, a misregistration is found, the reason is usually artwork rectifications having rendered it necessary to apply solder pads on the solder and component sides. Experience shows that it is nearly impossible to achieve an exact registration between such solder pads.

5.9.3 Annular Rings in Plated-Through Boards

1. In the case of eccentrically drilled solder holes, the width of the annular rings on the finished board shall not be less than 0.002 in. (0.05 mm). See Figure 5.39. This implies the use of the following minimum design values of the width of the annular rings:

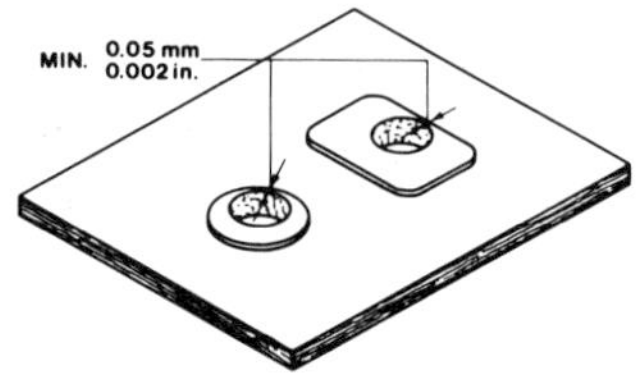

Fig. 5.39. Minimum width of annular ring (plated-through board).

   0.012 in. (0.30 mm) on manual artwork, and

   0.010 in. (0.25 mm) on photo-plotted filmwork
2. Broken annular rings are accepted at 1 percent of the solder pads. When the thickness of the base material is less than 0.031 in. (0.8 mm), broken annular rings are accepted at 2 percent of the solder pads.
3. Requirement 2 applies to board dimensions (length and width) up to 10 in. (250 mm). For each length of 4 in. (100 mm) of board dimension beyond 10 in. (250 mm), the acceptable number of breakouts generally should be doubled.
4. Maximum 25 percent or 90° of the periphery of the solder hole may be located outside the solder pad. See Figure 5.40.

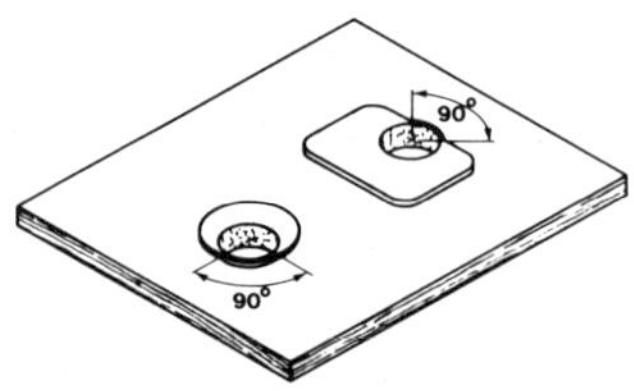

Fig. 5.40. Acceptable eccentricity of solder hole with respect to solder pad.

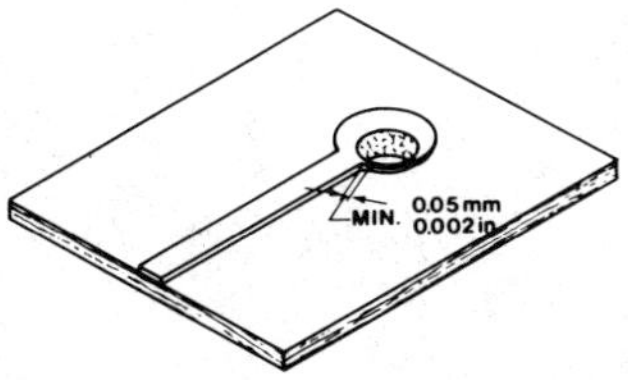

Fig. 5.41. Nearest admissible location of hole break-out with respect to connecting point of conductor.

5. The portion of the hole barrel which is outside the rim of the solder pad must not reduce the minimum distance to an adjacent pad or conductor.
6. Breakout shall not occur closer than 0.002 in. (0.05 mm) to the connecting point of a conductor. See Figure 5.41.

It should be noted that the design value of the width of the annular ring is a little lower in the case of photoplotted filmwork than in the case of manual artwork. The reason is, obviously, the high degree of accuracy which characterizes filmwork photoplotted directly to a 1:1 scale.

With a maximum of 10 percent, the occurrence of breakouts to a number twice the numbers given in Requirements 2 and 3 above is rated as a B-defect. A larger number of breakouts is rated as an A-defect, for which reason the board is rejected. This usually indicates a systematic fault caused by an inferior image transfer.

Regarding Requirement 5, the edge of the solder pad concerned may be already at such distance from an adjacent conductor that, in breaking out of the pad, the edge of the hole will be too close to the adjacent conductor.

5.9.4 Annular Rings in Nonplated-Through Boards

In the case of eccentrically drilled or pierced holes, the width of the annular

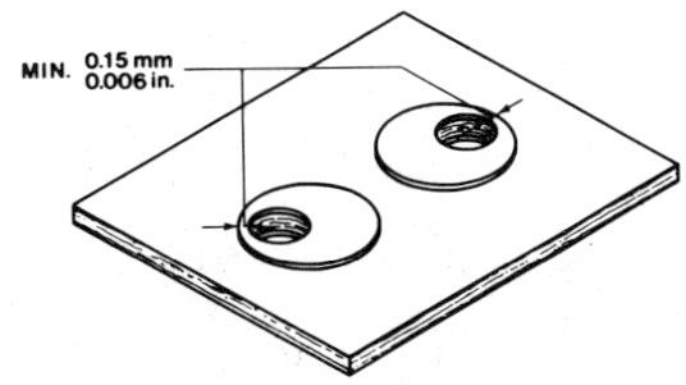

Fig. 5.42. Minimum width of annular ring (non-plated-through board).

rings on the finished boards shall not be less than 0.006 in. (0.15 mm). See Figure 5.42. This implies that the minimum design value of the width of the annular rings is 0.020 in. (0.50 mm), valid for manual artwork as well as for photoplotted filmwork.

Breakout of the annular rings is not accepted, partly because this can result in soldering problems and partly due to the tendency that broken annular rings have to lifting. Breakouts shall therefore be cause for rejection and are rated as an A-defect. The drilling tolerance is a little greater than in the case of plated-through boards, so that the risk of breakouts is considered fairly insignificant.

5.10 Broken Overhang

Removal, by reflowing, of the hazardous tin/lead overhang (Figure 5.43) along the edges of conductors and solder pads is dealt with in detail in Paragraph 8.4.

Not all tin/lead plated boards can be reflowed. Certain types of boards, such as strip-line boards for microwave applications, have to be made of a special base material in order to obtain well-defined high-frequency characteristics. Frequently, the material consists of a cross-linked polystyrene. Unfortunately, the

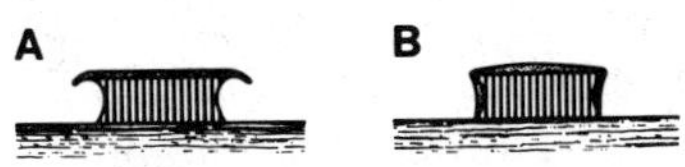

Fig. 5.43. Removal of tin/lead overhang by reflowing.
A: Before reflowing. B: After reflowing.

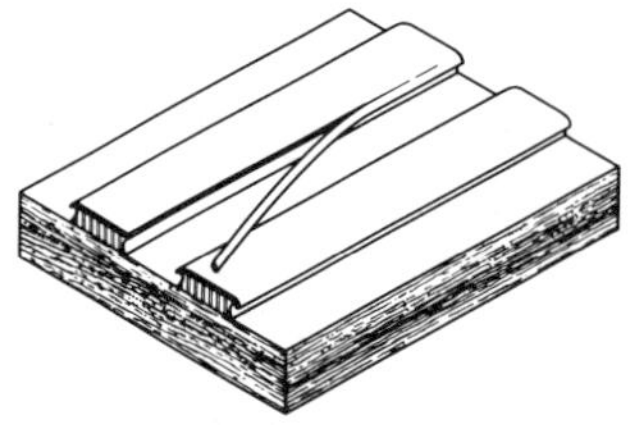

Fig. 5.44. Short-circuit between two conductors caused by partly detached overhang (board not reflowed).

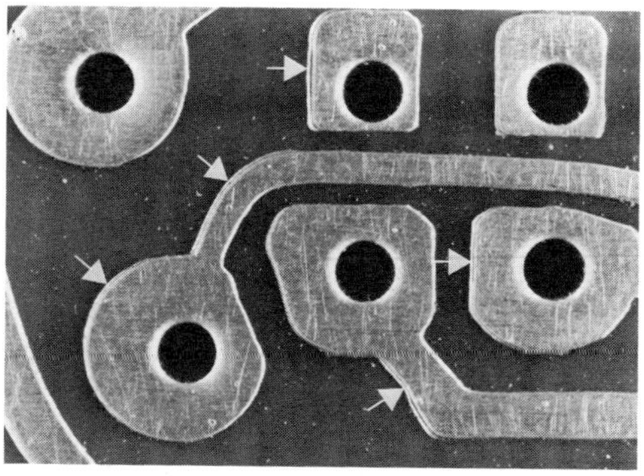

Fig. 5.45. Non-reflowed board beginning formation of cracks along conductor edges.

softening temperature is so low that such boards cannot withstand reflowing, just as they cannot withstand mass soldering.

It can also happen that a pure tin plating has been selected instead of a tin/lead plating. The melting point of pure tin is so high that the base material cannot withstand an increased reflow temperature.

The PCB manufacturer tries to achieve as small an overhang as possible by controlling the etching process or by suggesting a base material with a thin copper foil cladding; for example, ½ oz. = 0.0007 in. (17.5 μm) or thinner. In spite of such measures, the inspector should always inspect the boards very carefully for overhang. Nonreflowed boards imply a latent risk of a later functional failure in the form of very thin threads or slivers. See Figure 5.44.

Fig. 5.46. Execution of a tape test. The tape is being pulled off.

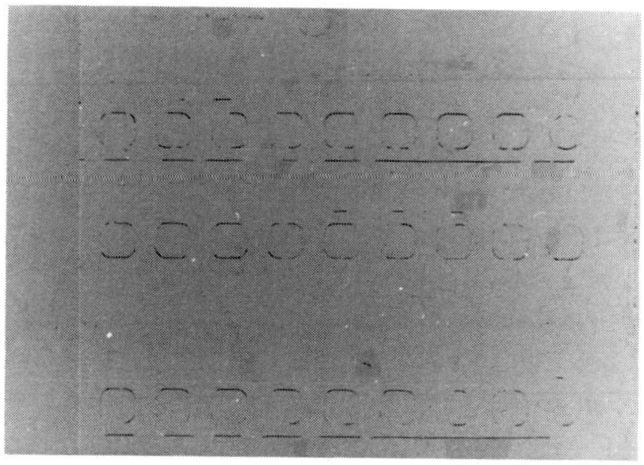

Fig. 5.47. Tape with adhering overhang (slivers).

It can be very difficult, or even impossible, to see an overhang or a beginning formation of cracks along conductor edges with the naked eye. Figure 5.45 shows an example of a 0.02 in. (0.5 mm) wide conductor viewed under a microscope. The crack is practically invisible to the naked eye, but quite clear in the microscope. Since it is a strenuous and slow procedure to check a board under a microscope, it is more common to perform a tape test.

A description of a tape test is given in MIL-P-55110C, Paragraph 4.8.2. Initially the board is degreased, after which a 0.5 in. (13 mm) wide tape 2 in. (50 mm) long plus a little to hold to, is pressed firmly onto the board so that any air bubbles disappear. With a rapid pull perpendicular to the board the tape is pulled off as shown in Figure 5.46. Any overhang adheres to the tape, which is examined with a magnifier or a microscope. An example of a tape with an adhering overhang is shown in Figure 5.47. The prescribed pressure sensitive cellophane tape may be difficult to obtain, but it can be replaced by a PVC tape, type 691 with a similar adhesion, obtainable from the 3M Company. Obviously, a fresh piece of tape should be used for each tape test.

If cracks form along the pattern edges or half-broken overhangs are found, the board is, in principle, to be rejected, and the defect is therefore rated as an A-defect. It is recommended that a 100 percent inspection of the batch be performed, and that the PCB manufacturer be contacted about the problem.

A remedy is difficult, although the worst overhang possibly can be removed by a light brushing with a rotating nylon brush or similar brush. This usually is discouraged because there is no guarantee of removing all the overhang. Furthermore, brushing adds the risk of tiny metal particles working down into the surface of the base material and reducing the surface resistance.

Chapter 6

Drilling Inspection

6.1 Introduction

A most important part of a board is its solder holes and mounting holes (fixing holes). This chapter deals with the quality of the drilling, particularly the location and the hole size, whereas the next chapter deals with the quality of the internal hole plating.

6.1.1 Effects of Incorrect Drilling

Incorrect drilling of a board can interfere with the assembly and soldering processes, and the inspector must therefore ensure that all holes are located correctly and all hole sizes are correct.

6.1.2 Sources of Error

Although the PCB manufacturer does a good piece of work to ensure a correct result, it may happen that during the digitizing of the hole pattern, the operator misplaces a hole or confuses two hole sizes. This is, unfortunately, not always discovered during the inspection following drilling.

As customers, we may expect that all the delivered boards have correct hole locations and sizes, particularly when the drilling is performed with a numerically controlled (NC) drilling machine. If the wrong drill bit has been inserted in one of the drill heads, this will appear during the first piece inspection, or at any rate, during the inspection of the diameter tolerance (Paragraph 6.8).

Another possibility of erroneous drilling is that the controller of the NC drilling machine is a little out of order; for example, sensitive to large noise pulses on the power line. This may cause a few holes to be drilled at random, and such boards may pass undetected through the sampling inspection. The risk of such spurious holes is usually so small that it can be discounted.

When the holes are punched, the boards will, of course, be identical, particularly if the base material is punchable at room temperature. Since a punch needle can break, the operator will inspect the first and last boards for missing holes. If the last board has a full complement of holes, all intermediate boards most likely have the required holes. It is very unusual that plated-through boards have punched holes because the hole wall would be too rough for good plating quality.

6.2 Inspection of Hole Locations

The significance of the preceding statement is that we can accept a relatively low inspection level when inspecting the hole locations, including cases where the batch is made by a new supplier. The inspector has to distinguish between solder holes for components and mounting holes for fixing the board. Owing to the difference in the positional tolerances required, the inspector must distinguish between solder holes for manual and automatic component insertion. This is dealt with in Paragraphs 6.3 and 6.4. The location of solder holes and mounting holes can be tested in different ways, and since it is important that the inspector have a deep insight in such matters, a detailed treatment is given below.

6.2.1 Hole Locations Based on the Master Film

6.2.1a Methods of Locating Solder Holes

Two different methods of digitizing the master film can be followed. In the first, the operator aims at the center of the solder pads. The PCB draftsman usually tries to locate the solder pad symbols so that their center holes lie symmetrically with respect to the intersections of the grid lines.

If the periphery of the solder pad symbols is not quite concentric with the center holes, a minor displacement of the drilled holes will occur. The eccentricity between the periphery of the solder pad symbol and its center hole amounts to maximum 0.002 in. (0.05 mm) in the case of recognized brands. By preparing the artwork in a 4:1 scale, the PCB draftsman reduces the eccentricity to about 0.0005 in. (0.013 mm), which is negligible.

With the second method, the operator aims at the periphery of the solder pads. In theory, the drilled holes are concentric with the periphery of the solder pad, which of course leads to a uniform width of the annular ring. In practice, the solder hole can be displaced a little, but not more than stated above.

In order to use the first method, the solder pads must have center holes, which obviously are not necessary in the second method. On the other hand, the latter method requires that the solder pad symbol be symmetrical with respect to the center. Solder pad symbols trimmed by the draftsman (because the minimum insulation distance to an adjacent conductor cannot be met otherwise) can present a problem. In some cases, the inspector can expect broken annular rings.

When working in a 2:1 or 4:1 scale, it can be difficult for the PCB draftsman to decide how far he can trim a pad without going too close to the drilled hole. Similar conditions are valid in the case of groups of pads; that is, for ICs of type T070 where the location of the solder hole is not symmetrical with the periphery of the solder pad.

In photoplotted filmwork, the solder pads are always located accurately, which means that they lie on a grid. This implies, of course, that the carrier disc holding the solder pad symbols be made very precisely and be correctly adjusted. A drill tape is frequently provided, but as elaborated in Paragraph 6.2.3, the PCB manufacturer often prefers to redigitize the hole pattern.

It should be noted that a number of components have their leads off the grid, such as subminiature D-connectors. Owing to the great resolution of most digitizing systems, such deviations are usually of no importance.

6.2.1b Methods of Locating Mounting Holes

The mounting holes can be located with solder pad symbols, and in such cases the inspector should observe the following conditions:

1. The solder pads can, on the master film, be smaller than the mounting holes themselves, for which reason they disappear when the holes are drilled.

2. The solder pads can, on the master film, be larger than the mounting holes, but they do not necessarily occur on the finished board. If, on a plated-through board, nonplated-through mounting holes are desired, the PCB manufacturer may use the "tenting technique."

   By retouching the master film after digitizing, so that the solder pads disappear, the dry film plating resist used for imaging the pattern covers (tents) the mounting holes and the surrounding areas. No copper and tin/lead will therefore be deposited at these places. Since this means that no etch resist occurs, the result after etching is nonplated-through mounting holes with no pads.

The PCB manufacturer can also choose to drill the mounting holes in a second drilling operation—that is, after the plating and etching processes have been completed. Today, the most common procedure is to drill such mounting holes using an NC drilling machine. If there are few mounting holes, and the demand for positional accuracy is rather low, the PCB manufacturer may choose to drill the mounting holes as small as solder holes; for example, 0.031 in. (0.8 mm). These holes are plated-through together with all the other solder holes, and redrilled at a later stage, frequently by a manual drilling operation.

The inspector should realize that a narrow annular ring has a tendency to show a reduced adhesion as a result of the mechanical influence and the temperature occurring during the drilling. A test of the adhesion can be made easily with a scalpel as shown in Figure 5.35 of Paragraph 5.8.

Mounting holes can also be located by center markings, frequently in the form of very small solder pad symbols which disappear on the finished board.

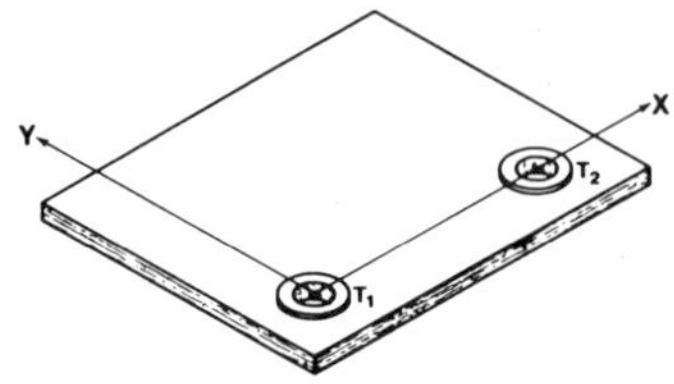

Fig. 6.1 Reference system for determining the holes of the board. T_1 = major tooling hole. T_2 = secondary tooling hole.

It is not uncommon that the master drawing also specifies the center-to-center dimensions of the mounting holes. This is actually a double specification, and in the case of manual artwork a certain discrepancy between the dimensions stated and the real center-to-center dimensions, as measured on the master film, exists.

When there is no reference system to lock the contour, the mounting holes, and the pattern together, the inspector should ignore the "floating" dimensions. He shall therefore inspect the location of the mounting holes with respect to the pattern; that is, relative to the master film.

6.2.2 Hole Locations Based on a Reference System

Here the location of the holes is specified by coordinates in a reference system as shown in Figure 6.1. The coordinates are frequently derived as multiples of the basic grid module (raster) which usually corresponds to 0.1 in. = 2.54 mm, or in some European countries, to 2.50 mm = 0.0984 in. Generally, the 2.50 mm grid is a cause of error in manufacturing as the worldwide industry is based on a 0.1 in. grid. Major problems occur in the case of long connectors if the deviation accumulates.

6.2.2a Zero Point

For the sake of future inspections, it is very important that the reference system is clearly defined regarding the location of its zero point (datum), and the direction of its axes. The datum is frequently located in the center of the solder pad of the major tooling hole (T_1), and the direction of the X-axis is determined by the center of the solder pad of the second tooling hole (T_2). In this connection, "center" refers to the virtual center of the pad and not to the center of the center hole of the pad symbol. This complies very well with digitizing as the operator usually aligns the bulls-eye with the rim of the pads.

6.2.2b Digitizing to the Periphery

Note the remarks in Paragraph 6.2.1a regarding digitizing to the center or the periphery of the solder pad symbols. The advantage achieved by digitizing according to the periphery of the tooling hole pads is that the datum of the reference system as well as the directions of the axes can be "reconstructed" on the finished board. This implies a reasonable edge definition on the board. Inward or outward edge displacements will be of no importance to the concentricity.

A practical advice to the PCB designer is that the pads of the tooling holes should be circular in order to increase the accuracy when digitizing to the periphery.

6.2.2c Digitizing to the Center

When digitizing by aiming at the centers of the solder pad symbols, the aiming point is not found on the finished board. Unless it is assumed that the drilled tooling holes lie concentrically with the aiming points or with the periphery, it can be difficult to define the reference system very accurately. However, these considerations are perhaps of academic interest only, as the difference between the two procedures is negligible.

6.2.2d Tooling Holes

When the inspector has no coordinate table, for example as shown in Figure 6.4 (see Paragraph 6.4), it is common practice to use the tooling holes as the basis for measuring the dimensions. This implies that the reference system is now supposed to be determined by the fictitious centers of the drilled tooling holes. Owing to inevitable small drilling faults, a measuring uncertainty amounting to about 0.002 in. (0.05 mm) is introduced, for which reason this procedure in principle is not advised. See Paragraph 6.4 for further details.

When tooling holes are to be utilized as the basis for measuring the dimensions, it is a prerequisite that the holes be drilled in the first drilling operation at the same time as the solder holes are drilled. It is also required that the hole diameter be reasonably large; for example, from 0.08 to 0.12 in. (2 to 3 mm). Owing to the rigidity of the drill bits, only a small deviation occurs during the drilling.

In plated-through boards, the tooling holes will normally be plated-through unless the tenting technique is utilized. Many PCB manufacturers, however, express great concern about plated-through tooling holes and suggest that they be drilled in a second drilling operation and, in some cases, do so without obtaining the customer's consent. Drilling in the second drilling operation (after plating and etching) always involves a risk of adding to the uncertainty of the location of the hole center. Since this addition can easily be larger than the uncertainty caused by plating and reflowing (uneven metal deposition and restructuring), plated-through tooling holes are preferred to drilling in a second drilling operation. The purchase specification should therefore state that internal scratching of tooling holes is acceptable. This makes it possible for the PCB manufacturer to use the tooling holes where desirable, without the fear of rejected boards due to internal scratching of the tooling holes.

6.2.2e Raster Digitizing

Instead of identifying the location of the solder holes by means of coordinates, the location can be determined by raster or grid-round-off digitizing. This procedure is based on an automatic locking of all holes to the grid, even when the pads are slightly displaced. It is therefore possible to achieve the same positional tolerance as when the coordinates of the holes are specified. It is, in both cases, a natural precondition that the solder pads on the artwork or the master film be accurately located. Otherwise, the finished board will exhibit a certain eccentricity between the solder pads and the solder holes, possibly with broken annular rings as a result.

6.2.3 Hole Locations Based on the Customer's Drill Tape

As a result of the increasing use of computer-aided design (CAD), PCB customers frequently wish to deliver drill tapes to the PCB manufacturer in order to save some of the tooling charges. Due to the fact that the drill tape is derived from the same data base as the plotter tape used for plotting the master film, a very high degree of coincidence between solder pad location and solder hole location should be achieved.

It is beyond the scope of this book to deal with all the problems a PCB manufacturer faces by receiving finished drill tapes. We will, however, mention the following main problems:

1. Compatibility between the drill tape and the NC drilling machine requires that the drill tape be prepared in close agreement with the programming rules of the drilling machine in question.
2. The PCB manufacturer usually has some programming rules of his own which also must be met.

For these reasons, it is almost impossible for the PCB user to prepare a drill tape usable by all PCB manufacturers without editing. Furthermore, it may prove very complicated for a CAD operator to prepare the tape so that it is optimal for PCB manufacturers; for example, with respect to achieving minimum travel of the NC drilling machine.

Likewise, the drilling sequence should take the hysteresis of the drilling machine into account. The drilling machine should, as much as possible, always move in the same direction and not have to move backwards and forwards any more than necessary.

As a consequence of the above circumstances, some PCB manufacturers prefer to prepare the drill tape themselves on the basis of the master film. This may sound absurd, but the digitizing takes place quite easily as the operator, for example, has only to digitize the first of the solder pads of an IC group, after which the remaining solder pads can be introduced automatically.

On the other hand, there are good reasons for dealing in more detail with the responsibilities regarding correct hole locations and hole sizes. Since it is the form of the documentation which is decisive for the distribution of the responsibility, the inspector should, in the case of complaints about drilling faults, check the documentation sent to the PCB manufacturer.

3. The drill tape is the only documentation. In this case the PCB manufacturer regards the drill tape as the decisive documentation and will not accept complaints about drilling faults which can be traced back to faults in the drill tape. The PCB manufacturer will normally always

check the drill tape; for example, by drilling a copy of the master film.

Based on the hole registration relative to the solder pads, it is easy to check the hole locations. Except for quite obvious drilling faults such as removing a pad by using too large a drill bit, it is, on the other hand, impossible to check the hole sizes.

The responsibility for wrong hole sizes, except for obvious drilling faults, must therefore rest with the tape supplier (the PCB customer or his CAD service bureau). On the other hand, the responsibility for correct hole locations rests with the PCB manufacturer.

4. A specification of hole sizes and locations exists, frequently in the form of a special film where the solder pads are replaced by symbols indicating the various hole sizes. Incidentally, this film can be derived from the data base of the CAD system, and is preferred to a manually prepared survey of the hole sizes, for which reason the responsibility for correct drilling lies with the PCB manufacturer. The material will also facilitate a possible redigitizing of the board.

6.3 Solder Holes for Manual Component Insertion

In the case of boards for manual component insertion (hand assembly), it is sufficient to check the hole locations relative to the solder pads. The easiest way of doing this is to visually inspect one of the board sides; for example, the solder side.

A perfect drilling operation has all solder holes concentric with the solder pads. In practice, however, there will always be a small amount of eccentricity.

Owing to the very narrow coherence between solder hole, annular ring, and solder pad, the following discussion corresponds closely to the treatment in Paragraph 5.9 of the width of the annular ring. In this and the following paragraph we address the drilling quality (the hole location) in relation to the use of the board, and not to the appearance of the annular rings as such.

6.3.1 Level of Inspection

On account of the manual insertion of components, it is not necessary to insist on a very high accuracy of the solder hole location, for which reason we can restrict the inspection to Special Inspection Level S3. This applies also to boards delivered by a new supplier.

If the inspector finds an unacceptable hole drilling, the inspection should be extended to General Inspection Level II, because an erroneous hole drilling can be of decisive importance for the use of the board.

6.3.2 Random Faults

If, by way of example, the erroneous drilling is caused by noise pulses on the power line which disturbs the NC drilling machine, the fault occurs at random. It is therefore necessary to perform a 100-percent inspection if it is not desirable to return the batch to the PCB manufacturer. Where the drilling fault does not interfere with the pattern, the hole is harmless, but obviously it can be missing in one of the solder pads. Repair possibilities are mentioned in Paragraph 6.6.

6.3.3 Digitizing/Programming Faults

When the drilling fault is caused by a digitizing fault or a programming fault, it appears systematically; that is, at the same place and on all boards of the batch. A quite common fault is a generally occurring displacement of the solder holes relative to the solder pads. Although the effect is the same, it is incorrect to interpret the fault as a drilling fault, because it is in all probability the PCB pattern which, during the image transfer, has been displaced a little with respect to the hole pattern. Since the image transfer process, to a certain degree, depends on the operator, the fault will not necessarily appear systematically on all boards.

6.3.4 Dimensional Changes

A visual inspection of the hole location relative to the solder pads is a sufficient basis for assessing the drilling quality, and it is not necessary to check whether a possible permanent dimensional change of the board has affected the center-to-center distances of the holes. Such dimensional changes, which are caused by wet and hot manufacturing processes, are so small that they are of no importance within an IC group of solder holes. Even for the insertion of a 40 pin IC package the changes are not noticeable. They may, however, affect the insertion of a long plug or socket; for example, a 120 pin connector.

6.3.5 Annular Ring Width

As a criterion for assessing the drilling quality we can, as stated in Paragraph 5.9, use the annular ring of the solder pads and require that it be unbroken. We will, however, accept broken annular rings in plated-through boards to a level of 1 percent of the number of solder pads. This implies a nominal design width of the annular ring of:

0.012 in. (0.30 mm) for manual artwork, and

0.010 in. (0.25 mm) for photoplotted filmwork.

The inspector, therefore, must obtain information concerning the current design practice. For manual artwork it can, for example, be 0.031 in. (0.8 mm) solder holes in 0.055 in. (1.4 mm) circular solder pads, corresponding to a width of 0.012 in. (0.30 mm) of the annular ring. In contrast, for a photoplotted filmwork it can be 0.031 in. (0.8 mm) solder holes in 0.051 in. (1.3 mm) circular solder pads, corresponding to a width of 0.010 in. (0.25 mm) of the annular ring. Many CAD systems use a 0.050 in. (1.27 mm) solder pad size, but the difference in the width of the annular ring is so small that it can be neglected.

In the event of wider annular rings, this criterion cannot be used as it would allow unreasonably large eccentricities. Therefore, the inspector must assess the eccentricity with respect to a maximum acceptable eccentricity of:

0.012 in. (0.30 mm) for manual artwork, and

0.010 in. (0.25 mm) for photoplotted filmwork.

An experienced inspector can visually assess whether the eccentricity is satisfactory. More precisely, the eccentricity E can be defined as half the difference between the largest and the smallest widths of the annular ring; that is,

$$E = \tfrac{1}{2}(a_1 - a_2)$$

See A, Figure 6.2. The measurement requires a 7X or 10X magnifier with a reticle having scale divisions of 0.005 in. (or, for a metric reticle, 0.1 mm).

In the case of rectangular solder pads

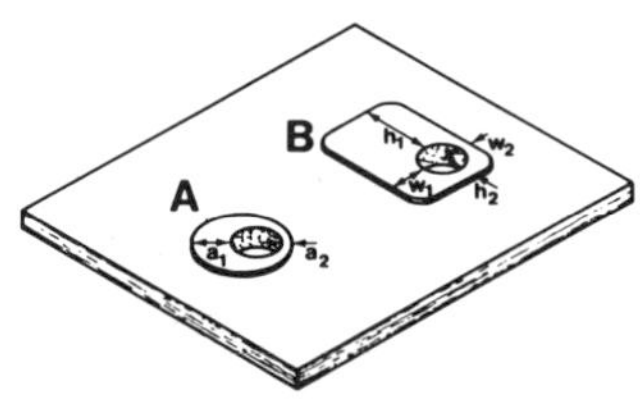

Fig. 6.2. Determining the eccentricity between hole and solder pad.

the above considerations do not apply because of the difference in height and width of the pads. We can, however, talk of the eccentricities in the two directions mentioned, and ascertain them as

$$E_H = \tfrac{1}{2}(h_1 - h_2) \text{ and}$$

$$E_W = \tfrac{1}{2}(w_1 - w_2)$$

See B, Figure 6.2. The largest value is then assessed relative to the relevant value of the maximum acceptable eccentricity stated above.

6.3.6 Inspection Considerations

The procedures outlined above may sound onerous; by using his background knowledge concerning the current design practice, the inspector can relax the inspection. In the vast majority of cases, the drilling of boards for manual assembly will be based on either:

1. A digitizing based on the actual location of the solder pads on the master film (manual artwork), or
2. A drill tape derived from the same data base as the plotter tape for the master film (CAD filmwork).

The result is a reasonable concentricity between all solder holes and solder pads. Consequently, the inspector can limit the inspection of drilling quality to running his eye over the board to get a general impression of the hole locations. If the annular rings meet the requirements regarding the occurrence of breakouts, the drilling quality can be considered satisfactory.

The specification, however, may demand drilling to the grid (raster drilling). A CAD board will present no problems, but if the board is hand taped, the artwork will have some inevitable sources of error, which appear after drilling:

3. The PCB draftsman cannot ensure that all solder pads are located with sufficient accuracy relative to the grid system.
4. The grid used on the light table may have altered due to changes in the relative humidity and temperature in the drafting room, or simply as a result of aging. The draftsman should realize that the additional heat from the light table can reduce the accuracy of the grid, on a short-term basis as well as on a long-term basis (aging).

From the outset, the board has some weaknesses caused by a more or less random displacement of the solder pads on the artwork, whereas the solder holes, as a result of the raster drilling, will be located very close to the exact grid positions. Consequently, an increasing degree of eccentricity along the board can be expected. The inspector should therefore assess the hole drilling somewhat more closely than in the first case, possibly by measuring the eccentricity at five to ten places along one of the board diagonals.

The fundamental problem connected with raster digitizing is that the customer demands a process which may not be justified by the PCB documentation. When the PCB manufacturer has performed his duty to assess the PCB documentation and is asked to do his best, it is not fair to blame the PCB manufacturer for the fault. If raster digitizing is not absolutely necessary, the inspector should try to have the specification changed so that the solder pads are digitized according to their actual position.

The inspector shall also inspect the board for undrilled solder pads. Although a solder hole is indicated on the master drawing (drilling drawing), the pattern can be designed in such a way that the hole marking disappears in a conductor. A, Figure 6.3 shows an example.

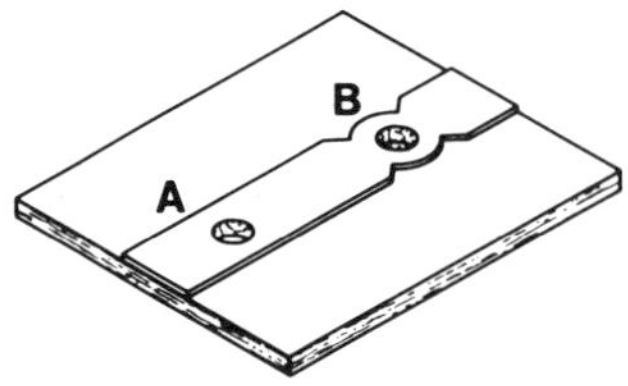

Fig. 6.3. Conductor with solder holes. A = marking of solder hole "disappears" very easily. B = preferred execution.

It may, furthermore, happen that in the rush to inspect and touch up the master film, the hole marking is interpreted as a pinhole which has to be closed. In order to achieve the greatest possible security against missing holes, the PCB designer should be persuaded to use execution B, Figure 6.3, instead of A, as B also implies certain soldering advantages. Incidentally, this is an example of how the PCB designer, in quite small ways, can influence the producibility in a positive direction.

6.4 Solder Holes for Automatic Component Insertion

Automatic component insertion demands precise positioning of the solder holes for reliable component insertion. The PCB designer must, within reasonable limits, take care to adapt the requirements regarding positional tolerance to the insertion equipment in question as well as the general PCB manufacturing processes. The design requirements are dealt with in Chapter 10 of Reference A12 of the Reading List. In the interests of the PCB manufacturer as well as the inspector, the tolerances should always be indicated on the master drawing.

6.4.1 Hole Diameter versus Positional Accuracy

Obviously, by increasing the diameter of the solder holes in excess of the diameter necessary for ordinary hand assembly of components with preformed leads, it is possible to compensate for an insufficient positional accuracy. However, there are certain limits for the maximum size of the solder holes with respect to the diameter of the component leads, if mass soldering is to yield satisfactory results. See Paragraphs 6.5.4a and 10.3.5e of Reference A12.

6.4.2 Basis of Positional Tolerance

It is not sufficient to inspect the position of the solder holes on the basis of the pattern as stated in the previous paragraph. In the case of boards for automatic component insertion, the inspector has to consider the permanent dimensional change.

Meeting a positional tolerance of ±0.003 in. (0.08 mm), as demanded by several manufacturers of automatic component insertion equipment, may seem rather questionable, at least when large boards are considered. Some PCB manufacturers, however, specify positional accuracies of ±0.003 to ±0.005 in. (±0.08 to ±0.13 mm), depending on the size of the board.

In the author's opinion, there is some insincerity in this tolerance as it can only be taken as an expression of the accuracy which can be achieved by the drilling process itself, and not for the positional accuracy of the holes in the finished board. The inspector should realize that, when the drilling accuracy is expressed as a positional tolerance, the tolerance of the center-to-center distance between two holes is twice this value.

6.4.3 Dimensional Changes

The permanent dimensional change has the largest influence on plated-through boards because of the wet and hot processes during the board manufacture, and also because of the stress relief in the board due to the etching of the copper foil.

For plated-through boards it is not unrealistic to anticipate a permanent dimensional change around ±0.05 percent, which is valid for most types of base material. Therefore, it stands to reason that on a large board, the uncertainty of the hole positions can easily become quite unacceptable for automatic component insertion. The PCB manufacturer is therefore compelled to use specially selected base materials with a greater dimensional stability, or to try to stabilize the base material by different heat treatments before the PCB manufacture.

Nonplated-through boards exhibit a far less permanent dimensional change, and when considering boards having the solder holes punched in one of the last operations, the positional accuracy can reach values of ±0.004 in. (±0.1 mm) in total. If the punching is carried out by means of several punching tools, an additional tolerance of ±0.003 in. (±0.075 mm) should be added per punching operation.

6.4.4 Effects of Digitizing the Hole Positions

In order to keep, with a reasonable certainty, the positional tolerance demanded, the PCB manufacturer must require that the insertion holes (solder holes) be determined by coordinates, a drill tape, or by raster digitizing. Digitizing the actual position of hand-taped solder pads seldom yields a satisfactory result. In turn, digitizing to the grid (raster digitizing) implies an increased risk of broken annular rings, which on the other hand should not be regarded as a manufacturing defect, but rather as the result of a design fault.

6.4.5 Effects of Changing Drill Bits

In the case of boards for automatic component insertion, the inspector should note a decisive factor in the drilling accuracy: If the drilling is performed by an NC drilling machine with no automatic drill bit change, and the operator does not change the drill bits manually for each board, small unwanted shifts between the various sets of hole sizes can occur.

The reason is that some PCB manufacturers prefer to drill all holes of one size in the total batch before changing to the next size. This procedure implies as many handlings of the production panels as there are hole sizes.

The PCB specification should state, in the case of boards demanding a very high positional accuracy, that drilling must be performed either in an NC drilling machine with automatic drill bit change, or, if this type of NC drilling machine is not available, the individual panels must be drilled in one turn. The point is that the production panels must remain in the drilling machine until all holes have been drilled.

6.4.6 Other Causes of Displacement

Even when a production panel is drilled under the same drill head every time, small displacements can occur. There are two reasons: small play in the pinning system, and perhaps more important, temperature variations in the drilling shop during the prolonged drilling sequence. The displacement will be further enhanced when the production panel is drilled under different drill heads, which easily can happen because of human error.

6.4.7 Level of Inspection

In pursuit of hole accuracy, the inspector has every reason to perform a more thorough inspection of boards for automatic component insertion than of boards for hand assembly. This is not least due to the fact that the extent of the permanent dimensional change is unpredictable, and in order to increase the inspection certainty, we will use General Inspection

Level II here. It cannot be taken for granted that the permanent dimensional change is the same all over the board due to the variation in the pattern density, among other things. It is therefore necessary to check the dimensions over long and medium distances, both crosswise and lengthwise.

6.4.8 Measuring the Positional Accuracy

The best way of carrying out the measurement is to use a coordinate table as shown in Figure 6.4. The board is placed on the table and aligned so that one axis of the reference system is placed in parallel with the X-axis movement. This is verified by finding the same Y-coordinates for the solder pads of the tooling holes which determine the direction of X-axis. See Paragraph 6.2.2. It is then easy to determine the coordinates of the selected solder holes, so that the center-to-center distances can be found quickly. In this case, the inspector shall use the fictitious centers by aiming at the rim of the holes and not of the solder pads. Instead of the visual adjusting device, the measuring equipment can be provided with a conical point as shown in Figure 6.5.

The master drawing can, in support of the inspector, state selected pairs of solder holes and the corresponding dimensions. For the sake of the measuring method mentioned below, the selected pairs of solder holes should lie on lines parallel to the X- and Y-axes of the coordinate system.

When the electronics company neither has a coordinate table at its disposal, nor wishes to invest in the relatively expensive equipment, it is possible to do nearly as

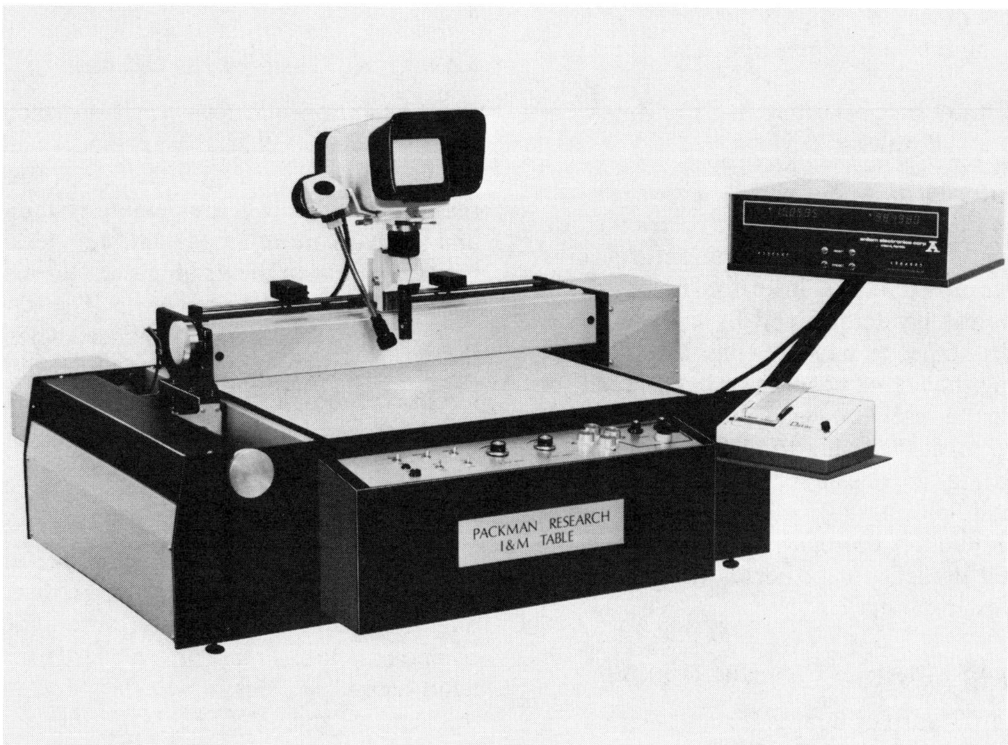

Fig. 6.4. Measurement of board dimensions by means of a digital coordinate table with visual alignment (Courtesy Packman Research Limited).

Fig. 6.5. Measuring board dimensions by means of a coordinate table as shown in Fig. 6.4, but with alignment carried out by means of a probe with a conical point (Courtesy Packman Research Limited).

well by means of a special slide caliper* as depicted in Figure 6.6. It can be seen from the illustration that the jaws of the slide caliper are provided with two cones so that the center-to-center distance between two solder holes can be measured directly.

Unlike the measuring method based on the coordinate table, it is expedient to measure in parallel with the X- and Y-axes of the reference system, since the measuring result corresponds directly to the difference between the X- or Y-coordinates of the solder holes. It might be feared that measuring on plated-through

*The slide caliper which has the type designation MULTIPRE DBP-PL, can be purchased in 10 in. (250 mm) and 20 in. (500 mm) versions from Friedrich J. Schall, D-7442, Neuffen/Württ., Postfach 80, West Germany.

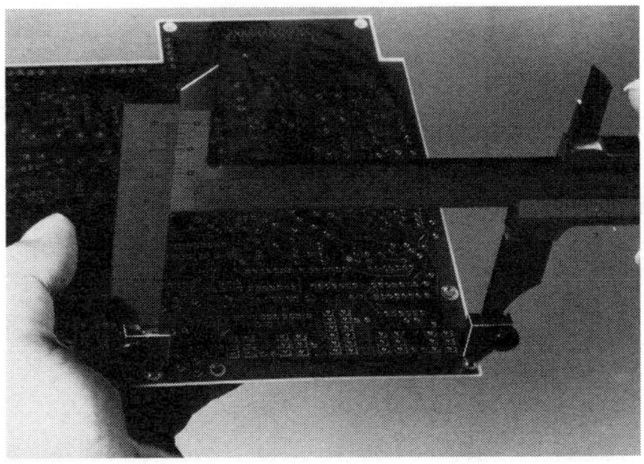

Fig. 6.6. Measuring board dimensions by means of a special slide caliper. Providing the jaws with cones makes the caliper suitable for measuring center-to-center distances.

holes, with an irregular tin/lead layer, would give an inaccurate result. The cones, however, rest only on the hole rims where the tin/lead layer, after reflowing, will always be very thin. See Figure 6.7. The axis through the cone will therefore coincide very closely with the axis through the solder hole.

A third method is to base the measurement on an accurate film having every

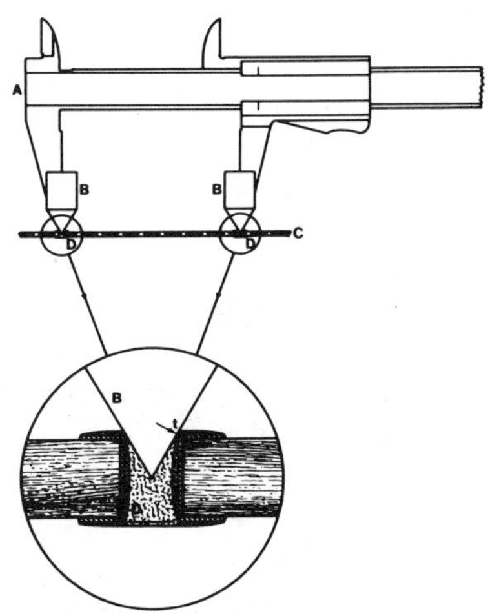

Fig. 6.7. Principle of measuring center-to-center distances with caliper provided with conical jaws. A: Slide caliper. B: Cone. C: Board. D: Plated-through hole. t: Thickness of tin/lead layer at rim of hole.

solder hole represented by a black circular pad. The diameter of such a pad must equal the minimum diameter of the solder hole minus the positional accuracy of the solder hole.

The board is now placed on a light table, and the film is overlaid and aligned with respect to the pattern. All the black pads should now be surrounded by a luminous ring. Where this is not the case, the corresponding solder holes are either too small or displaced. This method requires an accurate film with special sizes of the black pads, for which reason the method is best suited for CAD boards.

6.4.9 Board Disposition

Because exceeding the tolerances required can seriously affect the automatic component insertion, it is considered cause for rejection, and the fault is consequently rated as an A-defect. If this is the case, a screening of the total batch is recommended, and very quickly it will probably become apparent that it is only necessary to perform some very few measurements on the board to separate those that are defective.

It is not advisable to exempt the boards, which means that the rejected boards should be returned to the PCB manufacturer. In conjunction with the return of the boards, it is reasonable to discuss the tolerance problem with the PCB manufacturer so that suitable measures regarding the design and/or manufacture can be taken into account in order to secure acceptance of future deliveries.

6.5 Mounting Holes

Although the locations of the mounting holes are inspected according to the aforementioned guidelines, there are very good reasons to deal with certain details a little more thoroughly.

6.5.1 Tolerances

Unless otherwise specified in the master drawing (drilling drawing), larger displacements of the mounting holes up to ±0.012 in. (±0.3 mm) for boards with side lengths up to 16 in. (400 mm) should not be accepted. The PCB designer should have assessed the positional tolerances and the oversize of the mounting holes relative to the tolerances of the mechanical assembly in which the board is to be mounted or which the board is to carry.

6.5.2 Board Disposition

Exceeding the tolerances should normally be rated as an A-defect. If the mounting of a board in a chassis can be made difficult, or even be impossible, it is not always necessary to reject the board. A slight relocation of the mounting holes can easily be achieved by use of a round file or by reboring.

The PCB designer as well as the inspector should realize that in case of boards for automatic component insertion, the mounting holes are frequently used for mounting the board on the holding fixture of the automatic component insertion machine. The relationship between the plated-through solder holes and the mounting holes is of major importance, and the relocation mentioned above is dangerous.

To ensure correct component insertion, a tighter positional tolerance of ±0.002 to ±0.003 in. (±0.05 to ±0.08 mm) is frequently required. Failure to meet this tolerance is cause for rejection, and the fault is rated as an A-defect.

6.5.3 Inspection Methods

Various manufacturing methods for plated-through as well as nonplated-through mounting holes have been elaborated above and will not be repeated here. Actual conditions, however, may dictate the inspection method:

1. The mounting holes are determined by the pattern: The displacement is assessed on the basis of the pads as described in Paragraph 6.3. When no pads are found, the displacement is determined with respect to a positive film which is laid over the board and aligned relative to the pattern. The degree of eccentricity between the mounting holes and the hole markings on the film indicates the displacement.

   As mentioned earlier, the mounting holes can be dimensioned on the master drawing. When a reference system is not used, the dimensioning floats and is of no particular value to the inspector who has to inspect the absolute, and not the relative, location. In such cases, he should ignore the dimensioning and check the location relative to the master film.
2. The mounting holes are determined by means of a drill tape: In case there is no other documentation regarding the position of the mounting holes such as a table of their coordinates relative to a reference system, the master film is used as a reference. When the coordinates exist, measurement is performed as described in Paragraph 6.5, either with a coordinate table or the special slide caliper shown in Figures 6.4 and 6.6.

   There is a slight measuring problem when using the slide caliper in such cases, where the center line between the mounting hole and the tooling hole used as datum of the reference system is not parallel to one of the axes.

Fig. 6.8. Slide caliper modified with displaced jaw and one cone. The slide caliper is now suitable for measuring from edge to center of hole.

It is, fortunately, possible to modify the slide caliper by removing the cone from the fixed jaw which is then displaced transversely a little so that it can be set with the board edge. By using the board edge as a reference line it is possible to measure the distances to the desired holes, and find the individual center-to-center distances as the difference between the measuring results. See Figure 6.8.

The above method implies that the edge is smooth and parallel to one of the axes of the reference system. Usually, the error of parallelism is less than 0.1 degree, corresponding to a deviation of 0.0066 in. (0.17 mm) over 4 in. (100 mm). If the slide caliper is displaced 0.4 to 1.2 in. (10 to 30 mm) between the measurements to the zero point (the tooling hole) and the mounting hole, the measuring error amounts to 0.0008 to 0.002 in. (0.02 to 0.05 mm), which is negligible.

The inspector should ensure beforehand that the edges are parallel to the axes of the reference system, and for this purpose it is most practical to have three tooling holes as

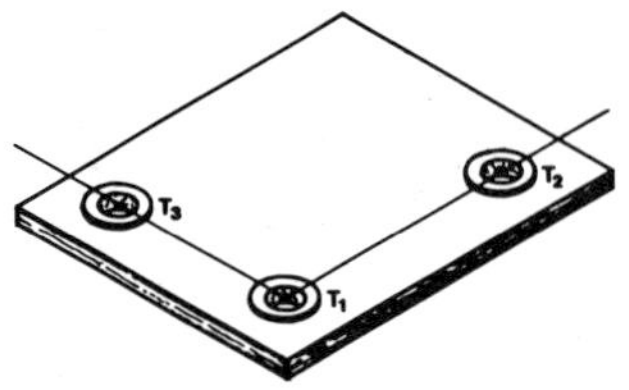

Fig. 6.9. Checking the parallelism between board edges and reference axes is facilitated by means of three tooling holes. T_1 = major tooling hole. T_2 and T_3 = secondary tooling holes.

shown in Figure 6.9. Here also the PCB designer can easily facilitate the inspection.

Theoretically, the directions of the axes of the reference system can only be determined on the basis of the rim of the solder pads and not the hole drilling. With the cone on the slide caliper, the measurement is based on the center of the drilled hole, but—as discussed in Paragraph 6.2.2d regarding drilling of tooling holes—the measuring error is considered negligible. However, the inspector should always verify that the holes are concentric with their pads.

3. The mounting holes are determined by coordinates in a reference system: The inspection is performed as described above.

6.6 Possibilities of Repair

In principle, drilling errors should be regarded as A-defects, but in many cases it is possible to repair a board as described below without impairing the reliability. In the event of isolated cases from an otherwise dependable PCB manufacturer, and where the assembly department needs the boards urgently (which is frequently the case) it can be necessary to exempt the fault and perform a repair.

A repair assumes that the fault appears only to a limited extent—for example, one or two erroneous drillings per board. It is furthermore assumed that the fault appears in a systematic manner, which means at the same location. Out of consideration for supplier statistics, such erroneous drillings should always be entered as A-defects in the inspection record, including exempted cases. To guide the inspector, immediate practical repair possibilities are stated below:

1. A plated-through solder hole is missing. If no conductors are connected to the pad on the component side, the hole can be drilled and left unplated. The missing through-plating is of no importance to the electrical function of the board.
2. In other cases the solder hole mentioned above also acts as a via hole, which means that both sides of the board have conductors connected to the solder pads of the hole. The missing hole can be drilled slightly oversized, then an eyelet is inserted and soldered on both sides. This technique is not always possible in dense boards because the annular rings are so narrow that they disappear during the drilling.
3. If Item 2 cannot be implemented, the solder hole can, in an emergency, be drilled to the correct size. After assembly, the component lead can be soldered on both sides of the board.

   If soldering of the component lead on both sides is impossible or inexpedient, a second hole is drilled to take a wire link to connect between the two sides. The advantage of the additional hole is that it can be placed beside the conductor and/or the solder pad at a convenient place. The hole needs to be only large enough to pass the wire link.

4. In nonplated-through boards, a missing solder hole is drilled to the correct size.
5. Erroneous drilling outside the solder pad is exempted if conductors or the like have not been interrupted. The fault is, however, rated as an A-defect in the inspection record.
6. Incorrectly drilled mounting holes are exempted as in Item 5, and redrilled at the correct location or relocated a little with a round file. The inspector should note the remarks in Paragraph 6.5.2 regarding the relocating of mounting holes in boards for automatic component insertion.

Drilling should preferably be performed by a high-speed drilling machine, using special PCB drill bits. In many cases it is advantageous to use a simple perspex drilling jig which is guided relative to the tooling holes by means of guide pins. In multilayer boards it is more problematic to drill a missing hole, particularly when connection to a pad in one of the inner layers must be established. This requires special equipment for either a selective plating of the hole or a special "surgery-in-depth" technique. However, it is outside the scope of this book to describe such methods.

It is only fair that the PCB manufacturer pays for all repair work he has caused, but an agreement should be reached beforehand in all individual cases. It is assumed, of course, that a careful account of the repair expenditure is kept.

6.7 Inspection of Nominal Hole Size

The initial inspection described below addresses only the nominal hole size (whether the holes have the intended size) and not the tolerances on diameter, which are dealt with in Paragraph 6.8. When choosing the inspection level, the inspector should act on his background knowledge regarding the drilling process.

6.7.1 Level of Inspection

Because the vast majority of boards today are drilled on NC drilling machines, of which the newer types have automatic drill exchange, the probability of drilling the holes in the right size is high. It is therefore reasonable to limit the inspection to a single board.

Since NC drilling machines have several drilling heads, frequently four, it is quite conceivable that a wrong drill bit could be used in one of the drilling heads so that, for example, 25 percent of the boards exhibit erroneous drillings. The probability of this happening should normally be quite low, among other things, because the manufacturer performs a special check regarding drill bit size.

However, should the fault occur in spite of all efforts, there is every probability that it will be found when the diameter tolerance is inspected as described in Paragraph 6.8.

For the same reason, when checking the nominal hole size of boards drilled in other ways, such as jig drilling, we shall base the inspection on only one board.

It is not intended to inspect every hole in the selected board for nominal hole size. Usually the board will show a large predominance in the number of holes of a certain size. On digital boards, perhaps 90 percent of the solder holes are intended for IC packages, frequently 0.031 in. (0.8 mm), whereas the next hole size can be 0.043 in. (1.1 mm) or 0.051 in. (1.3 mm).

Due to the fact that IC solder pads are located in a regular pattern, there is little risk that the digitizing operator will mistake them for other solder holes. The inspector is therefore fully justified in inspecting the nominal size of some few of the many IC solder holes—for example, 10 holes—so that he can concentrate on the remaining hole sizes. Here the diameter indications on the master drawing are compared with the actual diameter of the holes.

It is seldom necessary to check all holes within the various hole sizes, so the inspector can make do with one or two holes in each cluster. The holes most likely to be incorrect are the ones which occur singly all over the board, and all such holes should be checked.

The above considerations apply mostly to first issues or to raised issues. In case of repeat orders with no changes, the risk of erroneous drilling is extremely low, and the inspection is therefore limited to checking just a few holes within each hole size.

6.7.2 Measuring the Diameter

The checks can be performed with cylindrical plug gauges with a diameter corresponding to the lower tolerance limit of the hole size in question. See A, Figure 6.10. It is, however, much quicker to use a taper pin which can be inserted into the hole to a depth determined by the hole size. A calibrated dial shows the hole diameter directly. B, Figure 6.10, shows only the principle.

Measuring the diameter with a taper pin gauge is not very accurate, as it implies an absolutely cylindrical hole with no internal projections or nodules. This method is therefore best suited for drilled, nonplated-through holes where a cylindrical shape is taken for granted. Since plated-through holes in reflowed boards frequently will be somewhat irregular—having, for example, an hourglass shape—the measuring accuracy will be less. It is, however, only a question of determining the nominal hole size, for which this accuracy is considered adequate.

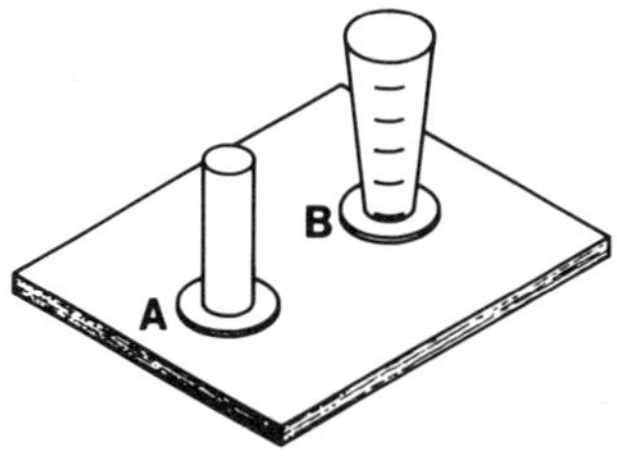

Fig. 6.10. Checking the hole diameters.
A = cylindrical plug gauge. B = tapered pin.

The inspector must apply the gauge with care to avoid damage to the internal hole plating—for example, by removing small projections or nodules, or by cracking the hole wall. On some boards, the copper plating is protected by a solder lacquer instead of tin/lead. Since this solder lacquer is relatively soft, the inspector may need to use an optical method. In the simplest case, he can use a magnifier with a reticle containing a circular scale, and in an advanced case, a profile comparator with surface illumination.

6.7.3 Board Disposition

An incorrect nominal hole diameter should, in principle be regarded as an A-defect, but in most cases it is easy to repair such boards. If it is an isolated case from an otherwise dependable supplier and the production department needs the boards urgently, it may be necessary to exempt the fault and perform a repair.

It is a condition that the fault only appears to a limited extent—for example, on the order of one or two defects per board. It is furthermore assumed that the faults appear systematically—that is, at the same locations on each board. In consideration for the supplier statistics, such erroneous drillings should always be entered as A-defects in the inspection record, including cases which are exempted. As guidelines for the inspector, immediate accessible repair possibilities are given below:

1. Solder holes are too small: They should be treated in accordance with the guidelines in Paragraph 6.6, Items 1, 2, 3, 4, and 6. In this context, drilling means redrilling of already existing holes.
2. If solder holes are too large, they can frequently be discounted. This implies that the annular ring has not disappeared and still has an acceptable width. Soldering can usually be carried out without complications although the clearance between component lead and hole is not optimum. In some cases manual post-assembly soldering may be necessary.
3. Generally, oversized mounting holes do not present any problems, except that it is necessary to perform a more careful location of the component or the board itself before tightening the screws. In some cases oversized mounting holes cannot be accepted—for example, if they are used to secure a mechanical unit on a dense board. During later service operations, there is a certain risk that the unit will be incorrectly positioned, possibly causing a short circuit. Similar considerations apply to boards for automatic insertion (Paragraph 6.5).

   Boards presenting problems should be rejected and returned. The PCB manufacturer can carry out a repair by epoxy casting the holes and redrilling to correct size. This fault is consequently rated as an A-defect.

6.8 Tolerance of Hole Diameter

For the sake of assembling and soldering the board, it is important that the diameter tolerances of the solder holes and mounting holes are met, particularly regarding the lower tolerance limit. The tolerances are stated below where d indicates the finished hole diameter.

6.8.1 Plated-Through Boards, Nonreflowed

$d \leq 0.04$ in.: ± 0.004 in.
$0.04 < d \leq 0.25$ in.: ± 0.005 in.
$d > 0.25$ in.: ± 0.008 in.

$d \leq 1.0$ mm: ± 0.10 mm
$1.0 < d \leq 6.3$ mm: ± 0.13 mm
$d > 6.3$ mm: ± 0.20 mm

6.8.2 Plated-Through Boards, Reflowed

$d \leq 0.04 \text{ in.: } \left\{\begin{array}{l} +0.006 \text{ in.} \\ -0.004 \text{ in.} \end{array}\right.$
$0.04 < d \leq 0.25$ in.: ± 0.006 in.
$d > 0.25$ in.: ± 0.008 in.

$d \leq 1.0 \text{ mm: } \left\{\begin{array}{l} +0.15 \text{ mm} \\ -0.10 \text{ mm} \end{array}\right.$
$1.0 < d \leq 6.3$ mm: ± 0.15 mm
$d > 6.3$ mm: ± 0.20 mm

6.8.3 Nonplated-Through Boards, with Drilled Holes

$d \leq 0.04$ in.: ± 0.002 in.
$0.04 < d \leq 0.25$ in.: ± 0.004 in.
$0.25 < d \leq 0.40$ in.: ± 0.008 in.
$d > 0.40$ in.: ± 0.010 in.

$d \leq 1.0$ mm: ± 0.05 mm
$1.0 < d \leq 6.3$ mm: ± 0.10 mm
$6.3 < d \leq 10.0$ mm: ± 0.20 mm
$d > 10.0$ mm: ± 0.25 mm

6.8.4 Nonplated-Through Boards, with Punched Holes

$d \leq 0.25$ in.: ± 0.004 in.
$d \leq 6.3$ mm: ± 0.10 mm

6.8.5 Determining the Tolerance of Hole Diameter

The inspection is based on cylindrical plug gauges which can be purchased in steps of 0.0005 in. (in the metric system, in steps of 0.01 mm) at reasonable prices. A set can consist of the following plug gauges:

0.02 to 0.06 in. in steps of 0.002 in.
0.06 to 0.25 in. in steps of 0.004 in.
0.5 to 1.5 mm in steps of 0.05 mm
1.5 to 6.3 mm in steps of 0.10 mm

Figure 6.11 shows an example of a set of plug gauges. The inspector is usually content with checking the lower tolerance limit of the hole since this is most important in assembly. In the event of plated-through boards there is, due to the reflowing, a somewhat greater risk that the holes become too small than too large. Incidentally, the inspector will soon learn how to assess the upper tolerance limit on the basis of the play he feels between the gauge and the hole. Figure 6.12 illustrates the practical measurement.

6.8.5a Level of Inspection

Let us assume a batch of 200 boards, each having 700 holes of 0.031 in. (0.8 mm). The inspection is carried out according to General Inspection Level II, and the AQL value is 1 percent. By this the sample size is determined to be 50 boards, and the problem is now how many holes to check.

It is quite impossible, and also unnecessary to check all 35,000 (50 × 700) holes of the sample. Instead we can perceive the batch as a delivery of a large number of holes, namely 140,000 (200 × 700) holes. Based on the sampling plan we find that we have to inspect 500 holes with acceptance and rejection numbers of 10 and 11, respectively.

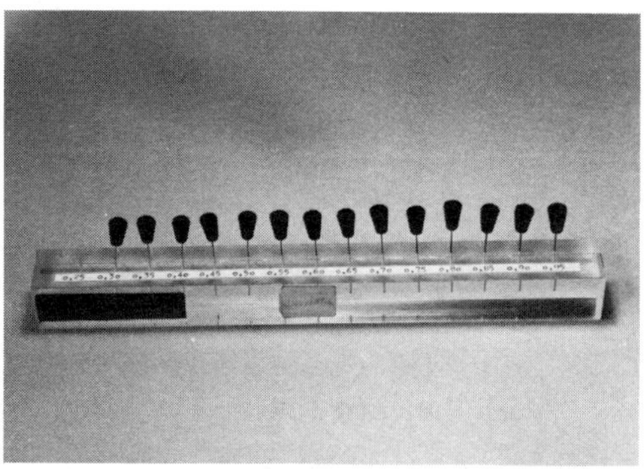

Fig. 6.11. Example of a set of plug gauges.

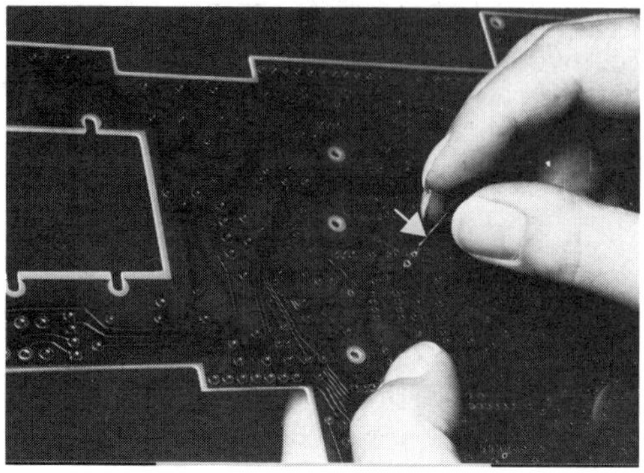

Fig. 6.12. Practical measurement of hole diameters.

The number of holes corresponds more or less to one board; however, it is not reasonable to check all these holes in one board only, so it is recommended to distribute the checking over the total sample.

Five hundred holes are quite a lot, and the checking will necessarily take some time. In practice, it is unnecessary to proceed in this way, especially when the holes are nonplated-through. When the base material is phenolic paper—that is, with a very homogeneous structure—there is no reasonable cause to assume that drilling or punching of the holes should result in large diameter variations.

Due to the glass weave, epoxy glass base materials are less homogeneous, and the inspector should expect a somewhat larger distribution of the hole diameters. In plated-through boards the most essential contributions to variations of hole diameters originate in the plating and reflowing processes.

The plating thickness in a hole depends on its immediate surroundings and its location on the board. In the case of a good balancing of the pattern, the current field lines in the plating bath have approximately the same density all over the board, and consequently, all holes will, in general, achieve the same plating thickness.

If large variations in pattern density are found, large variations from hole to hole may occur, in extreme cases with mutual differences of 0.005 in. (~0.1 mm) or more. The plating thickness of the solder pads varies in the same way.

It can be easily understood that, due to the varying amounts of tin/lead in the holes and on the solder pads, reflowing can cause large diameter variations and, at worst, blocked holes. Balancing the pattern is dealt with in detail in Paragraphs 6.6.3 and 8.3.11 or Reference A12, and it is strongly recommended to balance the pattern.

The case above illustrates clearly how the PCB designer, with some very few and cheap means, can improve the producibility of the board, and therefore reduce the price a little, and help the inspector a lot. A typical example of a well-balanced board is shown in Figure 6.13.

We will now assume that the PCB designer works in a professional manner and achieves a satisfactory balancing of the pattern of plated-through boards. Such boards therefore rank with nonplated-through boards with regard to the uniformity of the holes. Consequently, we can reduce the inspection level substantially by carrying out the inspection of the hole diameters on a few of the holes which the sample, according to the above considerations, otherwise should comprise. When the plated-through board has not been balanced, the number of holes to be inspected must necessarily be slightly larger.

The inspector should distribute the holes fairly evenly over a few boards, so that there are about 10 holes per board. Within the individual boards, the holes should be evenly distributed, and they can therefore be selected equidistantly along various lines—for example, along diagonals as shown in Figure 6.14, but preferably at different locations on the individual boards.

The inspector needs, of course, more specific guidelines than the above vague wording. It is reasonable to use Special Inspection Levels S3 and S4 in the cases mentioned. Sample size and acceptance and rejection members appear in Table 6.1. Instead of 500 holes in the basic example, the inspection comprises 50 and 80 holes, respectively, which is more practicable.

It is very important that the inspector, when setting the inspection level, avails himself of his background knowledge concerning the reproducibility which can be

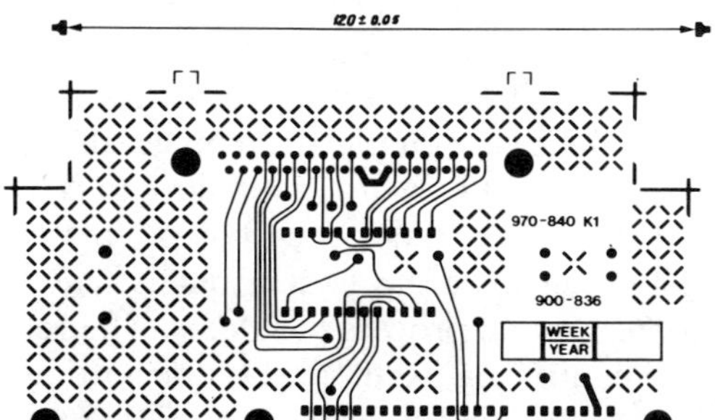

Fig. 6.13. Typical example of a well-balanced board.

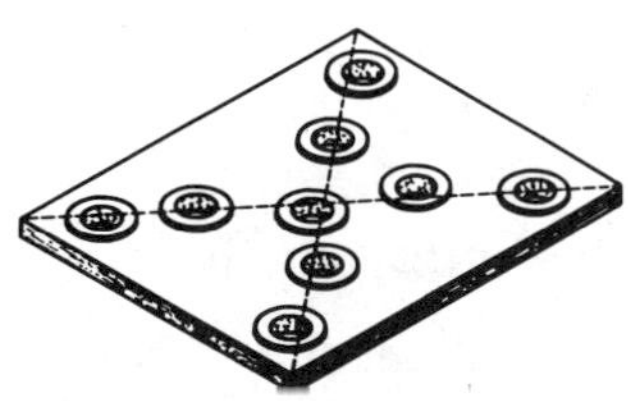

Fig. 6.14. Holes to be inspected can be selected along diagonals or other lines.

Table 6.1. Sample size and acceptance/rejection numbers to be used when performing a reduced inspection of the hole sizes of a PCB batch at an AQL value of 1%.

	Type of board			
	Plated-through and balanced Nonplated-through		Plated-through and nonbalanced	
	Special Inspection Level S3		Special Inspection Level S4	
Total number of holes n in PCB batch	Number of holes in sample	Acceptance number Rejection number	Number of holes in sample	Acceptance number Rejection number
$n \leq 10{,}000$	13	0-1	50	1-2
$10{,}001 < n \leq 35{,}000$	13	0-1	50	1-2
$35{,}001 < n \leq 150{,}000$	50	1-2	80	2-3
$150{,}001 < n \leq 500{,}000$	50	1-2	80	2-3
$500{,}001 < n$	50	1-2	125	3-4

obtained within the various processes. The art is, in this way, not to set up a long and expensive inspection program, but rather to keep it within a limited, but sound, framework.

Since this inspection level is very limited, the inspector can take precautions by utilizing his experience and focus on the more dubious areas—that is, where the solder pads are isolated from the remaining pattern.

The extent of the aforementioned inspection is valid for the most frequently occurring hole size, usually 0.028 in. (0.7 mm) or 0.031 in. (0.8 mm). The remaining holes, perhaps 20 percent of the total and distributed over various hole sizes, are inspected to a correspondingly limited extent—for example, on one or two boards. This can be justified on the assumption that these holes are often less critical with regard to the tolerance on diameter, and since they are larger than 0.31 in. (0.8 mm), they are less likely to block after reflowing.

6.8.5b Board Disposition

In principle, exceeding a hole tolerance is rated as an A-defect, but frequently it is of no practical importance, particularly if the hole is a little too large.

Holes which generally are too large can indicate insufficient copper plating. The author has experienced some cases where the copper plating was accidentally omitted in a limited number of boards corresponding to a few production panels. The only copper in the holes was the electroless copper, which carried the tin/lead layer. Consequently, the hole diameters were about 0.004 in. (0.10 mm) larger than usual. A safe method of revealing such faults is a resistance measurement as described in Paragraph 7.3.1d.

When the holes are too small, the component insertion may be difficult. Many times a small spare margin can be utilized for an exemption. The PCB designer normally specifies that the diameter of the solder hole be 0.008 to 0.020 in. (0.2 to 0.5 mm) larger than the diameter of the component lead. Paragraph 6.5.4a of Reference A12 discusses this in detail.

In the case of undersized holes, the inspector should ensure that failure to meet the tolerance range does not occur randomly all over the board, but is due to some general process problems. The reason could, for example, be holes drilled too small or excessive plating. Other reasons could be a design error, particularly an inadequate or missing balancing of the pattern.

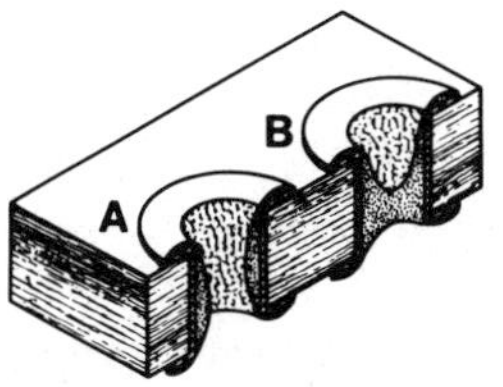

Fig. 6.15. Reflowed board. A: Blocked hole. B: Closed hole.

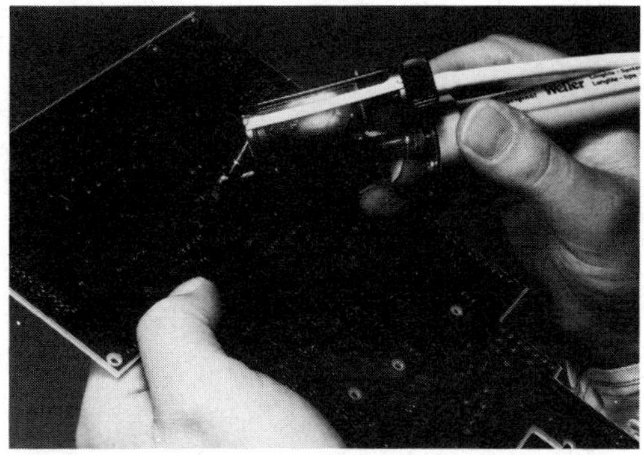

Fig. 6.16. Opening of blocked or closed holes by means of a vacuum extraction equipment.

In the above cases, the defect will be repeated from board to board, a fact which, of course, makes it easy for the inspector to make a decision with regard to an exemption. A reduction of the hole diameter by 0.002 to 0.004 in. (0.05 to 0.10 mm) will, in many cases, be of no practical importance, but the decision is the inspector's.

A special problem is presented by solder holes intended for insertion of square pins. The pin manufacturer frequently prescribes a ±0.0002 in. (±0.05 mm) tolerance on the diameter. In the case of plated-through and reflowed boards it is, however, quite unrealistic to demand such a tight tolerance.

Instead, the PCB designer can supplement the normal tolerance on the diameter of the finished solder holes with the required tolerance, but valid for the copper plating only. The soft tin/lead will not prevent the insertion of the pins, and the PCB manufacturer is able to control the drilling and copper plating processes to achieve a tolerance of ±0.002 in. (±0.05 mm), possibly ±0.003 in. (±0.075 mm). In practice, the latter tolerance proves sufficient to ensure a safe insertion of the pins.

When necessary, the inspector can perform a destructive inspection by etching away the tin/lead layer or by preparing microsections. If he normally receives prompt feedback regarding assembly problems, he does not need to perform these examinations.

When holes are blocked, or even completely closed by tin/lead after reflowing, as shown in Figure 6.15, a note is entered in the inspection record. This type of fault is cause for rejection and rated as an A-defect. Only in the case of a very limited number of blocked holes is it advisable to try to save some boards by solder extraction methods. This is conveniently carried out by means of a solder extraction equipment as shown in Figure 6.16. Out of consideration for the supplier statistics, a note should be entered in the inspection record, including cases in which an exemption is granted.

Chapter 7

Plated-Through Board Inspection

7.1 Introduction

This chapter discusses the quality of plated-through holes in detail because their quality greatly affects the board reliability. The inspector should note that some specifications—for example, requirements regarding complete solder filling of plated-through holes—are in fact holdovers from a time when less ductile copper deposits caused greater risks of hole barrel cracking. The cracking could be caused by a thermal shock during mass soldering.

7.1.1 Uniformity of Quality

Like the diameter of drilled holes mentioned in Chapter 6, all through-platings of a batch are very similar. Either all plated-through holes are good or nearly all bad, whereas the intermediate range with some good and some bad plated-through holes almost never occurs.

When the PCB manufacturer uses modern production equipment and has an extensive quality control system, there is no reasonable cause to assume that the plated-through holes will differ much in quality. We can therefore limit the routine inspection of plated-through holes as described in Paragraph 6.8.5a regarding the inspection of hole diameters.

7.1.2 Causes of Defects

It is important that the inspector have some knowledge of the mechanisms which cause defects or reduce the reliability of the through-plating; he can better recognize possible defects and better substantiate his complaints to the PCB manufacturer.

Although the following considerations mostly apply to plated-through boards, some also apply to nonplated-through boards.

7.1.2a Effects of Machining

First, we will study the machining (the drilling operation), as this is of decisive importance to the plating quality. Then we will deal with the plating process, emphasizing the various types of defects.

A short description of the drilling and plating operations should help the inspector to understand fully how the defects arise. The individual processes are shown in the block diagram, Figure 7.1.

After being drilled, the holes are inspected for drilling quality, particularly for burrs. Large burrs (see Paragraph 7.2.4) are a sign of poor drilling, which is possibly accompanied by a rough or gouged hole wall (Paragraph 7.2.5) or by epoxy smear (Paragraph 7.2.6).

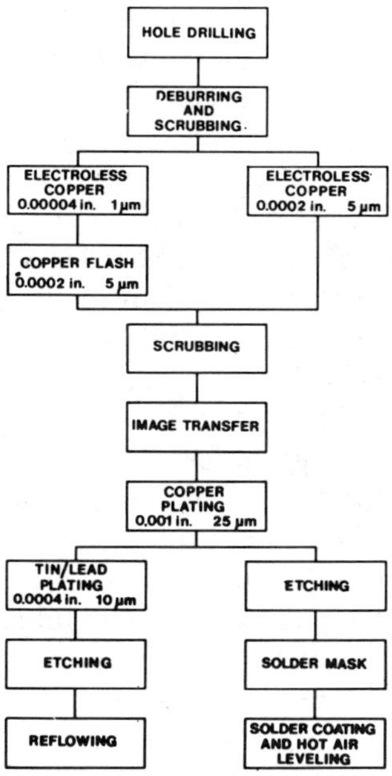

Fig. 7.1. Block diagram of part of the production processes.

When burrs occur within a small local area of the production panel, a light hand sanding with a fine-grit silicon carbide paper is performed. It is very important that the sanding paper be of good quality, with a uniform grit size. Otherwise, it can cause deep scratching which cannot be removed. If the production panel is widely burred it will usually be scrapped. In principle, burring problems should always be solved at the drilling stage and not by sanding.

7.1.2b Effects of Deburring and Scrubbing

Following the inspection of the drilling quality, the board is treated in a deburring/scrubbing machine. The modern machines usually combine the functions of degreasing, rinsing, wet deburring and scrubbing, rinsing, and drying. The boards are scrubbed (brushed) by several pairs of brushes; by combining brush types it is possible to remove small burrs and to clean and polish the surface. In order to avoid streaking the copper surface, the brushes move back and forth across the conveyor width.

Some PCB manufacturers round the hole edge by using special brushes and increasing the brushing pressure, while other PCB manufacturers prefer sharp hole edges. Paragraph 7.2.4d deals with this in detail.

The hole walls of multilayer boards must be very carefully cleaned of possible epoxy smear to ensure a reliable metallic contact between the plated hole wall and the inner pads.

The cleaning can be a mechanical cleaning (vapor honing or vapor blasting), a chemical cleaning by immersion in an acid bath, or plasma etching. In some cases etchback of the base material is performed so that the copper pads protrude slightly from the hole wall. See Paragraph 7.2.6c.

7.1.2c Effects of Electroless Copper Deposition

After the board, especially the holes, has been inspected for surface cleanliness, it is ready for electroless copper deposition. It is absolutely necessary for quality through-plating that the holes be completely clean; that is, with no drill chips, compressed copper dust, dirt, etc., in the holes.

Common practice is to use an electroless copper bath giving a heavy deposit of 0.00012 in. (3 μm), perhaps even 0.00020 in. (5 μm). Earlier electroless copper baths deposited only 0.00002 to 0.00004 in. (0.5 to 1 μm) copper, which required a subsequent electroplating, or copper flash, to build up a 0.00020 in. (5 μm) thick copper deposit able to endure the subsequent handling.

In order to secure a good adhesion of the photo resist, the board is mechanically scrubbed. The scrubbing removes the unavoidable oxide layer which reduces the adhesion of the dry film. This oxide should not be confused with the black oxide which, in the case of dry film solder masks, is applied to the copper for better adhesion.

When scrubbing, it is very important to avoid excessive pressure or the brush will remove some of the copper deposit from the edge of the hole. Defects arising from this treatment are dealt with in Paragraph 7.2.4d.

7.1.2d Effects of Electrolytic Copper Deposition

After transferring the image as a plating resist, electrolytic copper is deposited, with the goal of achieving a ductile (elastic) copper deposit. If subjected to excessive heating—for example, during reflowing and mass soldering—the base material expands drastically in the direction of the Z-axis (Figure 4.3), up to three times as much as the copper wall of the hole. This applies strong tension to the copper wall, and when the copper is too hard or brittle—that is, with too low a coefficient of elasticity—it can cause serious cracking, often at the hole circumference.

Since several different copper plating baths are in use, each with its own advantages and disadvantages, it is important to make sure that the PCB manufacturer uses a good type of copper bath. The inspector who inspects only the finished boards is restricted to looking for possible plating defects. Symptoms of such defects are described in detail in Paragraph 7.3.

7.1.2e Effects of Tin/Lead Deposition

A tin/lead plating follows the copper plating. Some PCB manufacturers, however, have replaced the former by a solder coating with subsequent hot-air leveling. (A preceding etching of the pattern is, of course, implied.) This process is rather new, but it is becoming more common within the PCB industry.

Briefly, the process consists of immersing the board in molten solder. The tin/lead layer is then leveled by hot air immediately after the board is withdrawn from the solder bath. During the leveling, the board is passed between two air knives which perform the leveling by blowing any excess solder off the board and out of the plated-through holes.

A solder mask frequently is applied to the board on top of the copper, for which reason tin/lead will be deposited only on the solder pads and in the solder holes. As Paragraph 9.4 describes in detail, this procedure implies that the solder mask does not wrinkle during mass soldering as no conductors, ground planes, etc., are covered with tin/lead.

The inspector will find no perceptible difference between solder holes that have been tin/lead plated and reflowed in the normal way, and solder holes that have been solder coated and hot-air leveled. See Figure 7.2. The latter method has the advantage of maintaining the tin/lead composition over the entire board.

A tin/lead plated board always shows some variations in the tin/lead alloy composition, caused by a varying field line

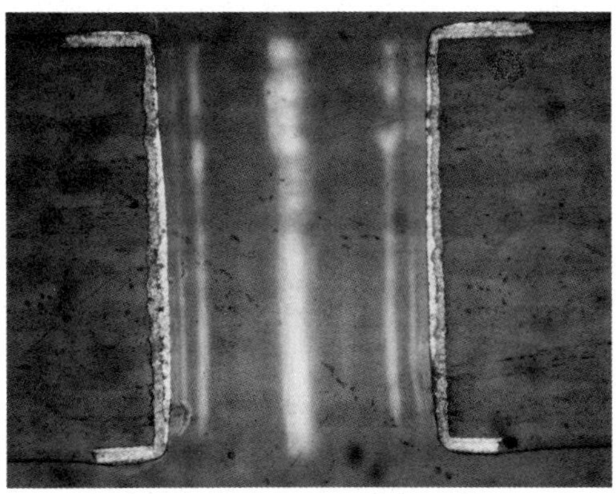

Fig. 7.2A. Microsection of plated-through hole in tin/lead plated and reflowed board.

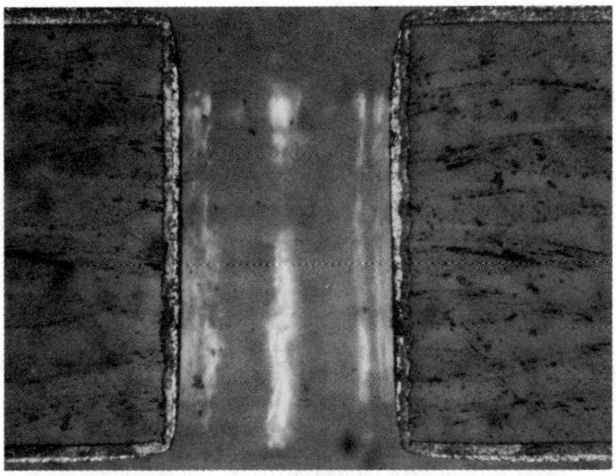

Fig. 7.2B. Microsection of plated-through hole in solder coated and hot-air leveled board.

density in the plating bath. As the tin content increases with the current field line density—greatest in the open areas of the pattern—reflowing as well as soldering problems can be caused by the higher melting point of the tin/lead in such areas. See Paragraph 8.3.3. Here again the PCB designer can eliminate these problems by balancing the board density properly.

7.2 Drilling Quality

Paragraphs 6.2 and 6.8 dealt with hole drilling from the perspective of assembly, specifically with hole position and diameter. The quality of the hole drilling itself is, however, of decisive importance to the finished plated-through hole, and we now consider characteristic defects caused by poor hole drilling:

- Burrs
- Rough and gouged hole walls
- Epoxy smear
- Protruding glass fibers
- Drilling delamination

7.2.1 Inspecting for Drilling Defects

Good hole drilling is characterized by the absence of the aforementioned defects. The quality of the drilling process depends on many drilling parameters, so only the most important ones are discussed here. Most of the defects cannot be observed directly because they are more or less concealed by the tin/lead layer, and it would serve no useful purpose to try inspecting the sample specifically for these defects.

The inspector should look for any perceptible defects at the same time he inspects the hole plating (Paragraph 7.3). The following discussion supplies enough background on these defects so that the inspector can both recognize them and understand their importance to quality.

A thorough investigation can be carried out only on the basis of microsections, which are studied with a microscope. Such investigations should be performed regularly, or at least once or twice a year, and always when approving a new supplier with an unknown quality level.

7.2.2 Inspecting Multilayer Boards

With multilayer boards the considerations are quite different since the consequences of inadequate drilling quality and hole quality are far greater than in the case of ordinary double-sided plated-through boards.

The production panel always contains a test coupon which the PCB manufacturer can use for microsectioning. In case of highly specified boards, such as for military applications, MIL-P-55110C requires the PCB manufacturer to prepare and deliver microsections for each board so that the customer himself can evaluate the quality. (MIL-P-55110C also applies this requirement to ordinary double-sided plated-through boards.)

In other cases—for example, large and complex boards for computer applications—the PCB manufacturer may be required to prepare and deliver a few microsections representing the total batch. Finally, there are other types of multilayer boards for which the customer does not want microsections.

7.2.3 PCB Manufacturer's Use of Microsections

The conscientious PCB manufacturer always takes some microsections of a batch so that he can control the quality. The customer (inspector) can always order microsections (photos) for delivery with the boards.

Obviously, the cost of the microsections are included in the price of the batch. A cheaper solution is to have some test coupons delivered with the batch, so that the inspector can prepare his own microsections.

Many PCB manufacturers currently make, on a routine basis, microsections in order to survey and control the quality of both multilayer and ordinary double-sided plated-through boards. Unfortunately, this does not apply to all PCB manufacturers, for which reason the inspector should check which quality control measures his PCB manufacturer performs in that respect. Because some PCB manufacturers of ordinary double-sided plated-through boards lack the necessary equipment to make microsections, the inspector must realize that quality flaws can occur from time to time.

7.2.4 Burrs

Drilling can cause burrs at the hole edge as irregularly-occurring raised rings. The height of the burrs depends very much on the drilling, but heights from 0.001 to 0.002 in. (25 to 50 μm) are not unusual unless the manufacturer takes steps to avoid burring.

7.2.4a Effects of Stacking

The production panels are normally drilled in stacks of two or three boards (A, Figure 7.3). Multilayer boards, however, are drilled one at a time. The burrs form at the outer sides of the stack where the drill bit breaks through (a) and where it is withdrawn (b).

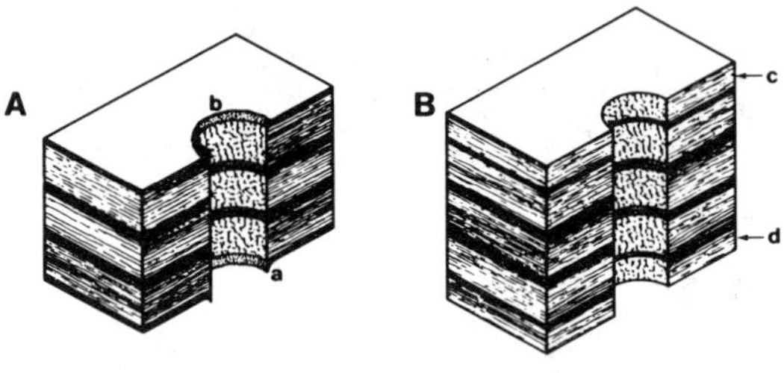

Fig. 7.3. Stacking of boards for hole drilling. A: With no entry and back-up materials, burrs occur at a and b. B: Entry (c) and back-up (d) materials prevent the occurrence of burrs.

To avoid this problem, the PCB manufacturer employs entry and backup materials, which are placed as shown at B, Figure 7.3. In this way, the copper foils at the outer sides of the stack receive the same type of clamping as the internal foils. Some PCB manufacturers use a phenolic paper laminate for the entry and backup materials, while others use an aluminum-clad hardboard.

Stacking before drilling requires scrupulous cleanliness. If a dust particle gets between two boards of the stack, these boards will, over a certain area, be separated enough from each other to allow burring.

7.2.4b Effects of Drill Bits

Some other precautions reduce burring. Only first-class drill bits should be used, and they must have a geometry suitable for drilling in the base laminate. The optimum drill bit geometry is slightly different for phenolic paper laminates and epoxy glass laminates.

It is very important that the flute angle be correct and that the flute itself be very smooth (polished) so that the chips do not clog the drill bit. The drill bit has to be sharp and, as the glass fiber material is quite abrasive, the drill bit should not be allowed to drill more than 5000 holes in a 0.062 in. (1.6 mm) epoxy glass laminate corresponding to a total drilling depth of 300 in. (~8 m). This applies to ordinary doubled-sided plated-through boards. In multilayer boards the limit of utilization is often as low as 1000 holes.

7.2.4c Effects of Drilling Parameters

The drilling parameters shall, of course, be correct, not least the drill speed and the ratio between drill speed and drill feed. In general, drill speeds of 35,000 to 50,000 revolutions per minute are common in the case of a 0.031 in. (0.8 mm) drill bit, together with drill feeds from 0.001 to 0.002 in. (25 to 50 μm) per revolution. If the drill feed is larger, the drilling will resemble punching and the hole wall will become very rough.

On the other hand, the drill feed per revolution should not be too low, as the drill bit remains in the hole too long, which raises the temperature of the drill bit and the hole wall excessively. This can cause epoxy smear as mentioned in Paragraph 7.2.6. Obviously, the drilling parameters are of crucial importance to the quality of the drilled hole and also to the reliability of the plated-through hole.

7.2.4d Undesirability of Burring

Why is burring so undesirable? In non-plated-through boards a large burr at the hole edge can, under certain circumstances, be the starting point of a later crack in the solder fillet. See Figure 7.4.

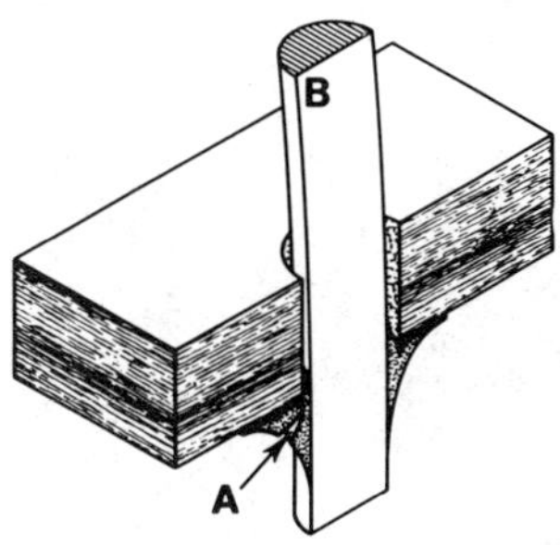

Fig. 7.4. Crack in solder fillet caused by burr. A: Crack. B: Component lead.

In plated-through boards, burrs cause a heavier copper deposit in the area around the burrs because they attract more current field lines during plating. See Figure 7.5. At the same time, the layer thickness of the hole wall becomes too thin. The greatest danger is that the copper deposit has too coarse a structure in areas with excessive current field line densities.

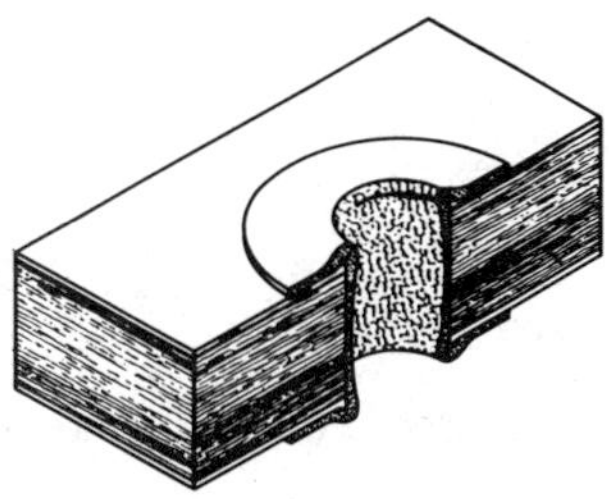

Fig. 7.5. Crack along hole edge caused by burr.

The coarse structure creates internal stress in the copper. In this way we have a serious risk of cracking along the hole edge, which can be triggered during the thermal shocks to which the board is subjected—for example, during reflowing and mass soldering.

The PCB manufacturer exerts considerable effort to avoid forming burrs during drilling, and to remove burrs by sanding and scrubbing operations should they arise in spite of the measures taken. Heavy-handed deburring should always be avoided, as this causes the burrs to be pressed back into the holes as shown in Figure 7.6. If this occurs, we will again face the risk of later cracking along the hole edge, as previously described.

7.2.4e Recognizing Burrs

Due to the fact that reflowing, as well as hot-air leveling, levels out the area around

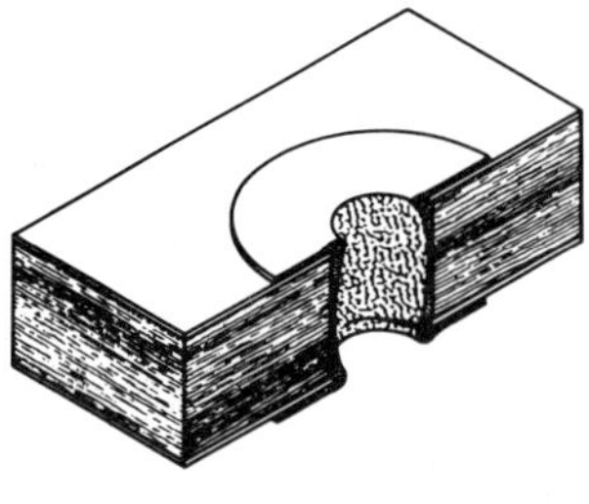

Fig. 7.6. Incorrect deburring technique presses the burrs back with the risk of cracking later on.

a burr, it can be very difficult, or even impossible, for the inspector to recognize burrs. A raised ring around the solder hole can indicate a burr, but reliable identification requires microsectioning.

7.2.4f Board Disposition

7.2.4f1 Nonplated-Through Boards

We reject boards with burrs higher than 0.002 in. (50 μm), and the fault is rated as an A-defect. Burrs with a height from 0.001 to 0.002 in. (25 to 50 μm) are accepted and rated as a B-defect, whereas burrs with a height below 0.001 in. (25 μm) are fully accepted. No remarks are therefore entered in the inspection record.

7.2.4f2 Plated-Through Boards

In consideration of the effects mentioned in Paragraph 7.2.4d, we demand smaller burrs in plated-through boards than in nonplated-through boards. The occurrence of burrs higher than 50 percent of the copper foil thickness with a maximum of 0.001 in. (25 μm) is therefore regarded as cause for rejection and consequently rated as an A-defect. Lower burr heights are acceptable, and no remarks are entered in the inspection record.

7.2.5 Rough and Gouged Hole Wall

Drilling must leave a reasonably smooth inner hole surface, which makes heavy demands on the drilling machine as well as the drill bits used. The bearings of the drilling machine must have no play, as this will cause the drill bit to wobble, especially at high loading, and cause a gouged hole wall.

7.2.5a Effects of Drill Bit Geometry

The drill bit geometry should be adapted to drilling in three different materials with widely different properties: copper, glass fibers, and epoxy or polyimide. If the drill bit has an off-center point or the two flutes are asymmetrical, the drill bit will wobble during the drilling and cause a gouged hole wall, occasionally with such deep grooves that they can be seen, even after reflowing. These grooves should not be confused with the more or less visible projections occurring in the holes of multilayer boards because of heavy etchback.

The heavy wear to which the drill bit is subjected causes rounding and cratering of the cutting edges. This increases the friction and raises the temperature of the drill bit so much that the epoxy chips melt and cake together with the glass particles on the bit surface. Instead of cutting, the drill bit then starts "chunking" the hole, which becomes very rough.

7.2.5b Nail Heading

Another phenomenon which sometimes occurs in multilayer boards is nail heading. Under unfavorable conditions a certain thickening of the internal pads can take place at the hole barrel. A microsection of a hole shows that the internal pads develop a cross section corresponding to a nail with a conical head, which gives it this name. See Figure 7.7.

This phenomenon, analogous to the burring previously described, is caused by a mechanical deformation of the copper resulting from an incomplete cutting action of the drill bit. The reason can be an incorrect drill bit geometry, a worn drill

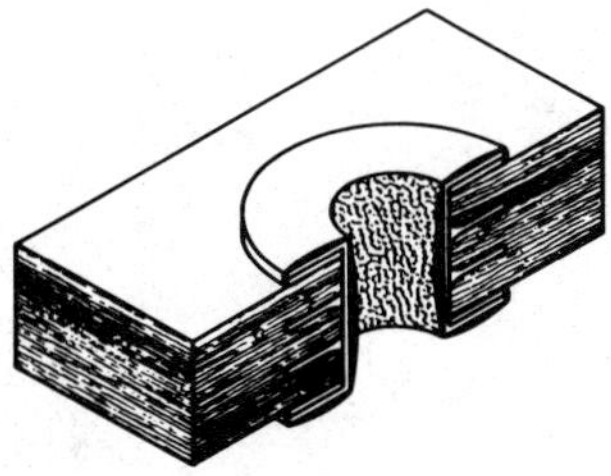

Fig. 7.7 "Nail heading" caused by mechanical deformation during drilling of hole.

bit, or an incorrect drill speed and/or feed. The mechanical deformation causes the copper to harden, with an increased risk of cracking.

Under especially unfavorable conditions, nail heading can increase the internal copper to twice the thickness at the hole edge. Generally, the inspector should not accept an increase larger than 50 percent of the original copper thickness. Exceeding this limit is cause for rejection and rated as an A-defect.

An inspection requires microsectioning the board, but in order to save an expensive multilayer board it is common practice to use a test coupon. The test coupon is placed on the production panel, and therefore manufactured under the same circumstances as the multilayer board.

Epoxy smear is often found internally in the hole in conjunction with nail heading. This is described in Paragraph 7.2.6.

In order to avoid nail heading, the PCB manufacturer uses only new drill bits for drilling multilayer boards, and frequently limits the number of holes to 1000 per bit. The drill bits are thereafter used for drilling ordinary double-sided plated-through boards. After having drilled a total of 5000 holes, the drill bit is resharpened. It can then be used for another 2000 holes before the next resharpening.

A worn and resharpened drill bit is not equal to a new drill bit. The bit material deteriorates in use, and for each subsequent resharpening the allowed number of holes should be reduced. Most PCB manufacturers limit the number of drill bit resharpenings to three. After the third resharpening, the drill bit is used on non-plated-through boards only.

7.2.5c Hole Wall Inspection

Inspection of the drilled hole wall can only be performed by microsection. It can, however, be difficult to assess the drilling quality as it is no small problem to set definite limits on acceptable irregularities in the hole wall. The inspector must compare the microsection with the ones shown in Figure 7.8, which depict various degrees of roughness.

A completely smooth hole wall is not preferred as the plating shows a slightly lower pull-out strength than when the hole wall is slightly rough. This does not apply to polyimide boards as the hole wall will always appear smooth. See Paragraph 5.8 concerning pull-out strength.

The greatest disadvantage of a rough hole is the risk of uncontrolled and unreliable plating. This is dealt with in Paragraph 7.3, which describes various faults caused by a rough hole wall.

7.2.6 Epoxy Smear

Epoxy smear (also called resin smear) is an epoxy contamination in the drilled hole. Under adverse drilling conditions (Paragraphs 7.2.4 and 7.2.5) a vigorous local heating of the drill bit can take place so that the epoxy chips soften and melt. Consequently, a very thin epoxy layer is smeared on the hole wall, including the copper foil. (See A, Figure 7.9). Epoxy smear can also be caused by incomplete curing of multilayer boards.

7.2.6a Effects of Epoxy Smear

After copper plating, ordinary double-sided boards show electrical connection from the solder pad to the copper depos-

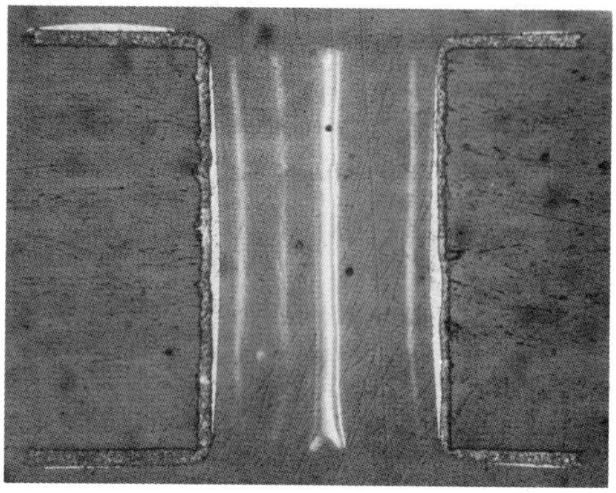

Fig. 7.8A. Microsection showing an acceptable roughness of the drilled hole wall in a plated-through hole.

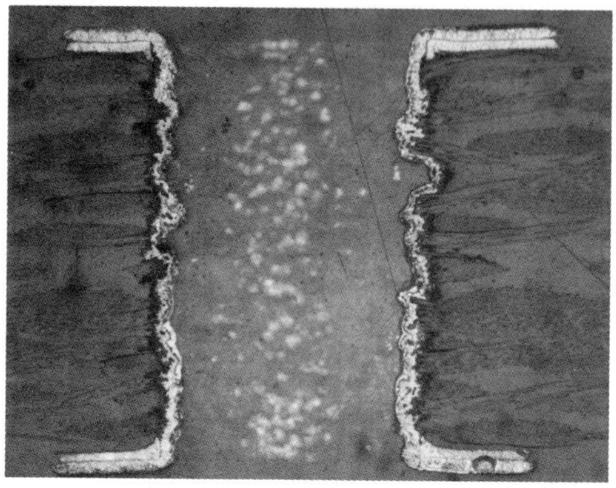

Fig. 7.8C. Microsection showing an unacceptable roughness of the drilled hole wall in a plated-through hole.

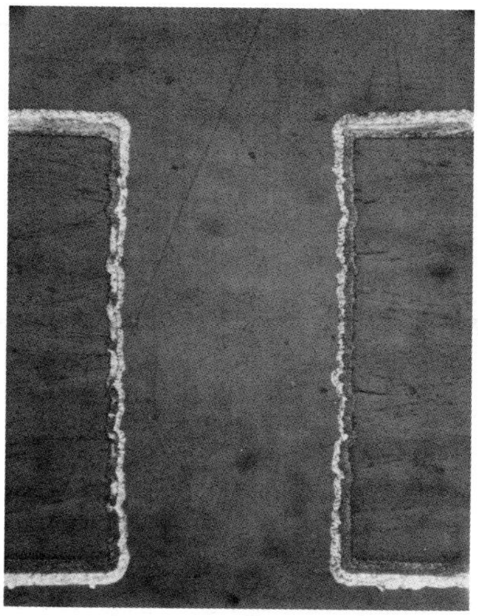

Fig. 7.8B. Microsection showing a just acceptable roughness of the drilled hole wall of a plated-through hole.

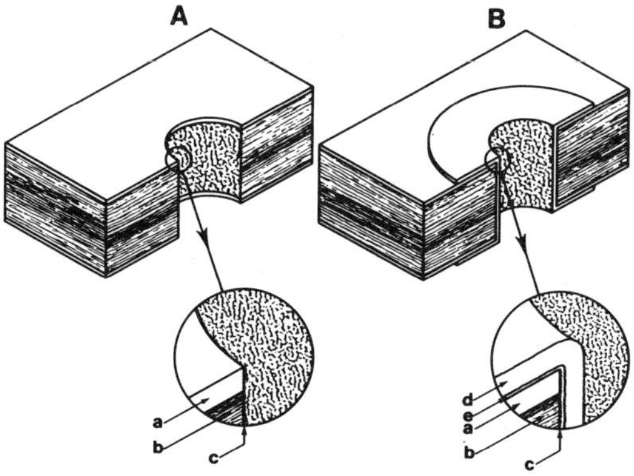

Fig. 7.9. Epoxy smear in solder holes of a double-sided board.
A: Drilling with epoxy smear over the entire hole wall. B: Owing to the epoxy smear, the hole wall does not get a metallic contact with the edge of the copper foil.
a = copper foil. b = base material. c = thin layer of epoxy smear. d = plated copper. e = electroless copper.

ited on the hole wall. It cannot be denied that, owing to the epoxy smear, there is a weak region in the plating accompanied by stressed copper. (See B, Figure 7.9.) Under adverse circumstances, the result is cracking along the hole edge from the thermal effect of mass soldering.

Because the coefficient of thermal expansion of the base material is considerably higher than that of the deposited copper, heat creates a tensile stress in the wall of the plated-through hole. Whether this causes cracking depends on other factors, including the appearance of the interface zone and the ductility of the deposited copper.

An organic contamination of the copper bath or over-replenishment with additives can cause the copper deposit to become brittle and more inclined to cracking.

Usually, PCB manufacturers of ordinary double-sided and plated-through boards are not particularly concerned about epoxy smear since the copper baths normally used deposit a very ductile (soft) copper.

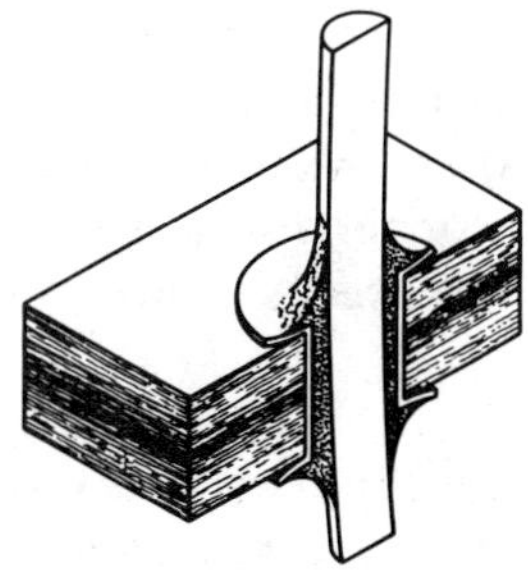

Fig. 7.10. Lifting of solder pad at pad rim.

7.2.6b Inspecting for Epoxy Smear

The inspector again faces a concealed fault which cannot be discovered with visual inspection. When cracking occurs after soldering, it is recommended to make microsections so that the reason for the fault can be found and complaints made.

In this connection, it should be noted that solder holes with epoxy smear along the copper foil edge withstand mass soldering better than hand soldering. This is because of the preheating which takes place as part of the mass soldering process. Furthermore, the soldering temperature and time are very carefully controlled in the mass soldering process. Hand soldering exposes the solder holes to the full thermal shock, frequently with prolonged soldering time and excessive soldering temperature.

That a large expansion of the base material really takes place is demonstrated by the fact that the rim of a solder pad with narrow annular rings sometimes lifts from the base material because of prolonged soldering or excessive soldering temperature. As the base material contracts after the solder has solidified, the solder pad will often remain lifted at the rim as shown in Figure 7.10.

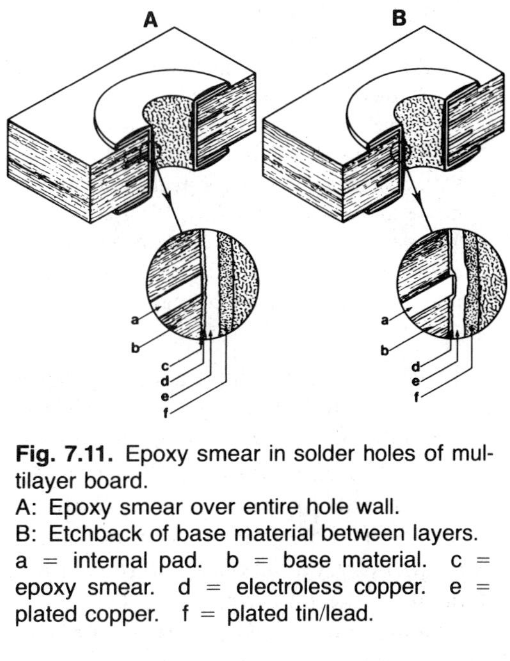

Fig. 7.11. Epoxy smear in solder holes of multilayer board.
A: Epoxy smear over entire hole wall.
B: Etchback of base material between layers.
a = internal pad. b = base material. c = epoxy smear. d = electroless copper. e = plated copper. f = plated tin/lead.

7.2.6c Precautions in Multilayer Boards

In the case of multilayer boards, the PCB manufacturers take stronger precautions against epoxy smear. An epoxy layer which is not removed prior to the electroless deposition can result in the pads of the inner layers not making electrical connection with the copper layer on the hole wall. The epoxy layer acts as an insulating barrier. See A, Figure 7.11.

As mentioned earlier, no more than approximately 1000 holes are drilled in multilayer boards with a single drill bit, after which it is replaced with a new one. Since the entry and backup materials can also contribute to epoxy smear, these materials preferably should be free of epoxy, plasticizers, etc. Materials clad with aluminum or copper foil are desirable because the foil acts as a heat sink for the drill bit.

After the mechanical cleaning of the holes, possible epoxy smear is chemically cleaned. There are various methods:

> The PCB manufacturer can perform a vapor honing or vapor blasting, which means a high-pressure flushing of the holes with microscopic glass balls suspended in water. It is, however, difficult to clean very small and/or very deep holes.
>
> An alternative is a chemical cleaning performed by immersing the board in a bath of concentrated hot sulphuric acid.
>
> The newest method is plasma etching, which uses a microwave technique to remove a thin layer of epoxy. Plasma etching will probably be the preferred method in the future, a major reason being that chemical cleaning can be dangerous to the operator.

In certain cases, the cleaning process is supplemented by an etchback process, which removes a little of the base material between the pads. (See B, Figure 7.11.) The pads will protrude slightly with respect to the hole wall and allow better bonding between the pads and the copper deposit on the hole wall.

The etchback process consists of a bath of concentrated sulfuric acid followed by a bath of concentrated hydrofluoric acid. The former attacks the epoxy, and the latter the exposed glass fibers. Etchback should normally not exceed 0.002 in. (50 μm). There are differing opinions regarding the benefits of etchback, which is frequently not performed at all.

There is some risk that ultrafine channels will be created in the base material. Copper can be deposited in the channels during the electroless copper process and short-circuit the hole wall; for example, to a ground plane in an inner layer.

The PCB manufacturer normally delivers a test coupon with the multilayer board so that the hole quality can be verified by microsectioning.

7.2.7 Protruding Glass Fibers

Because the base material contains glass fibers, the drill bit can catch a fiber and tear it out along part of the hole circumference. The result is a deep groove and, often, a glass fiber protruding into the hole.

During plating, the glass fiber terminates an excessive portion of the current field lines and receives a heavy deposit, leaving insufficient deposits on adjacent areas of the hole wall (Figure 7.12). Single cases are accepted with no remarks in the inspection record. Even a plated fiber cannot impede the insertion of a component lead into a solder hole, and is unlikely to interfere with soldering. The fiber is more likely to break off and leave a void in the hole wall

As mentioned in Paragraph 7.3.3d, we will accept some voids in the hole wall plating. If no more than 1 percent of the solder holes have protruding glass fibers, we will accept the boards and rate the fault as a B-defect. When the 1 percent

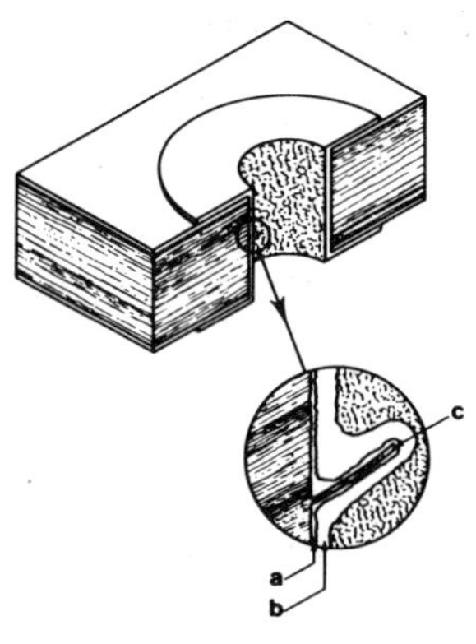

Fig. 7.12. Protruding glass fiber in plated-through hole.
a = electroless copper. b = plated copper. c = protruding glass fiber.

limit is exceeded, the PCB manufacturer has lost control of the process. This fault is cause for rejection and rated as an A-defect.

7.2.8 Drilling Delamination

The drill bit exerts an outward force, which can possibly cause the pad to lift at the hole edge. In addition, the individual glass fiber layers deeper in the hole can delaminate (Figure 7.13). Clamping an entry material against the stack with the drilling machine's pressure foot counteracts this tendency to delaminate.

As mentioned in Paragraph 4.5.4, delamination is cause for rejection because of the risk that cavities created by the delamination fill with moisture or electrolyte. This usually results in blowholes during mass soldering and/or reduced insulation resistance. This fault is an A-defect even though it cannot be detected by visual inspection. It can be suspected when the production department complains of excessive blowhole frequency; the only certain test is microsectioning.

7.3 Plating Quality of Plated-Through Holes

In the previous paragraphs we have addressed, among other things, the preconditions of achieving a good through-plating of the solder holes of the board. We now deal with a number of matters of great importance in the quality assessment of the finished boards:

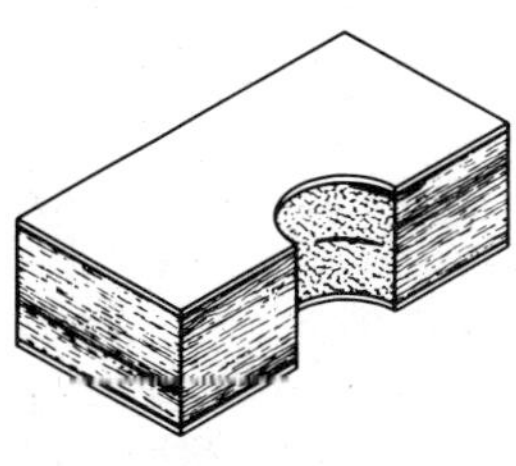

Fig. 7.13. Delamination caused by drilling.

Plating thickness
Voids in the hole wall
Cracks in the hole wall
Uneven hole wall
Plating pockets
Nodules
Burning
Step-plating

7.3.1 Plating Thickness

Overly thin plating in plated-through holes occurs for such reasons as insufficient plating time or low current density. The customer usually specifies an average thickness of 0.001 in. (25 μm), but because of unavoidable variations inherent in the process, many PCB manufacturers prefer a somewhat thicker plating.

7.3.1a Acceptable Thickness

A copper deposit thicker than 0.001 in. (25 μm)—for example, 0.0016 in. (40 μm)—only means that the solder holes are better than specified, when they still meet the diameter tolerances. This is more difficult with tin/lead plating because the reflowing causes such radical changes in the layer thickness that there is no point in dealing with the question here. Paragraph 8.1 deals with reflowing problems.

Because the plated tin/lead layer serves as an etch resist, the PCB manufacturer finds it in his own interest to deposit a layer thick enough to protect the underlying copper. The necessary thickness is about 0.00024 to 0.00028 in. (6 to 7 μm). A PCB specification can therefore require a minimum thickness of 0.00028 in. (7 μm) or 0.0004 in. (10 μm).

Prescribing a maximum value is common practice, as too much tin/lead can cause certain soldering problems associated with tin/lead melting in the holes. Because the plating tolerance can be as much as +100 percent, the maximum value is 0.0006 or 0.0008 in. (15 or 20 μm). The PCB manufacturer will avoid depositing more tin/lead than specified in order to avoid problems meeting the diameter tolerances after reflowing the boards.

The inspector should note that most PCB specifications call for a copper plating thickness of 0.001 in. (25 μm); for example, MIL-STD-275D. This thickness is commonly understood to be the average thickness, because the rough hole wall causes unavoidable variations in the plating thickness.

To guarantee a minimum of 0.001 in. (25 μm) of copper requires a much greater average (nominal) thickness—for example, 0.0016 in. (40 μm)—which, of course, adds to the price. An average thickness of 0.001 in. (25 μm) is normally specified, but supplemented with a specified minimum copper thickness of 0.0006 in. (15 μm).

Isolated areas with copper thickness below 0.0006 in. (15 μm) are regarded as voids in the hole wall. According to Paragraph 7.3.3d, only three voids per hole are allowed, for which reason a certain restriction of such areas is introduced.

7.3.1b Inspecting Plating Thickness

The only way of achieving a reasonably true impression of the plating thickness and evenness is to look at a microsection with a microscope. The inspector determines the average plating thickness by measuring the plating thickness at three or five places at each side of the barrel cross-section (see A, Figure 7.14), after which the average value is calculated. The inspector should avoid isolated thin or thick areas when measuring the thickness. On the other hand, when determining the minimum thickness, he of course should look for isolated thin areas.

In the case of multilayer boards, the determination of the average thickness should be based on thickness measurements opposite the internal pads. See B, Figure 7.14.

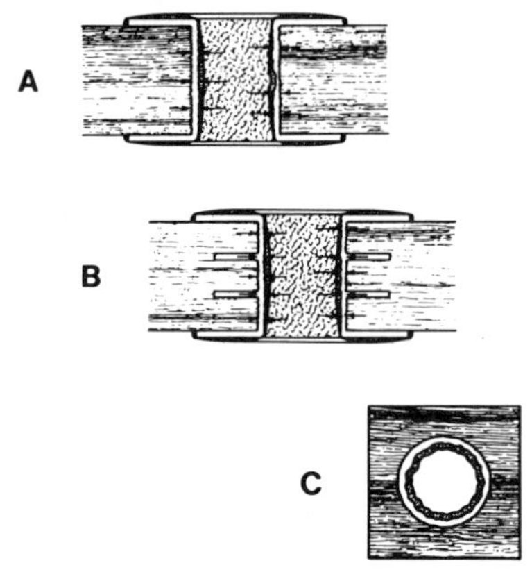

Fig. 7.14. Determination of average plating thickness based on microsectioning the board. A: Double-sided board with the microsection laid through the hole axis. B: Multilayer board with the microsection laid through the hole axis. C: Microsection laid in parallel with the board.

It is frequently preferred that the cross-section be placed horizontally as shown in C, Figure 7.14. The reason for this is the difficulty in locating the exact vertical cross-section through the axis of the hole. See A and B, Figure 7.14. A small displacement will result in a larger thickness measurement than is actually the case.

Microsectioning a board is destructive, and is therefore most expediently carried out on a test coupon manufactured with the batch. It should be noted that the plating thickness varies across the production panel, especially when the pattern is unbalanced.

The largest plating thickness will be found at the board edges, and the smallest at the center of the board. The ratio between the plating thicknesses at these places can frequently be as high as 1.5 to 1. Measuring on test coupons does not imply that the minimum plating thickness is met over all the board. This would require a determination of the ratio between the external and internal plating thicknesses for each type of board. The determination requires the preparation of a plating profile based on a relatively large number of microsections. These should, of course, be based on the same size of holes.

Instead of using test coupons it is practical to arrange that the PCB manufacturer deliver with the batch a reject board which can be used for microsectioning. It is important to inform the PCB manufacturer of the reason for wanting the board, as it should not be one rejected for plating faults. The PCB manufacturer will frequently cut off a corner of the board to provide identification, and to prevent accidental use.

The board is usually microsectioned so that it sections five to six holes representative of the batch.

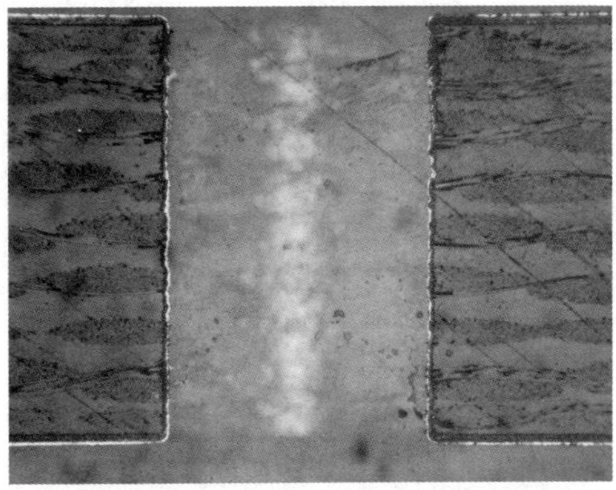

Fig. 7.15. Microsection of plated-through hole with a thin and even copper plating.

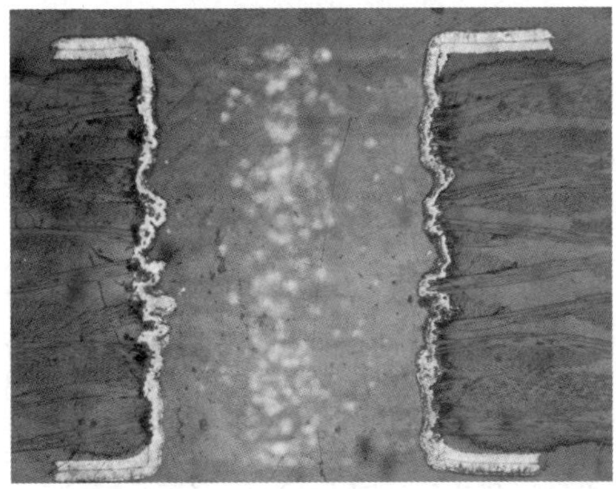

Fig. 7.16. Microsection of plated-through hole with a thick and rough copper plating.

7.3.1c Acceptance Criteria

We can now lay down the following acceptance criteria where t_a represents the average thickness and t_m the minimum thickness:

1. $t_a \geq 0.001$ in. (25 μm). The plating thickness is considered fully satisfactory.
2. $0.0008 < t_a < 0.001$ in. ($20 < t_a < 25$ μm). The plating thickness is accepted as tolerable, but the fault is rated as a B-defect. The PCB manufacturer should be notified that the specification has not been met.
3. $t_a \leq 0.0008$ in. (20 μm). The plating thickness is definitely too small, and the fault is rated as an A-defect. The extent to which the fault can be exempted depends on the actual thickness and the appearance of the hole drilling. In general, an even but thin copper plating of 0.0007 in. (18 μm) is more reliable than a thick but uneven copper plating of 0.0009 in. (22 μm).

   Assuming that the hole drilling is even, as shown in Figure 7.15, we will exempt the boards for plating thicknesses down to 0.0007 in. (18 μm). If the hole drilling is uneven, as shown in Figure 7.16, we will reject the boards for a plating thickness of less than 0.0008 in. (20 μm).
4. $t_m \geq 0.0006$ in. (15 μm). The minimum thickness of the copper plating shall nowhere be less than 0.0006 in. (15 μm). If it is less, and Requirements 1 or 2 are met, or an exemption is granted for Requirement 3, the batch is accepted.
5. $t_m < 0.0006$ in. (15 μm). When isolated areas of the copper plating have a thickness of less than 0.0006

in. (15 μm), such areas have the same status as voids in the hole wall. See Paragraph 7.3.3d with regard to the acceptance criteria. When these are fulfilled together with Requirements 1 or 2 above, or an exemption is granted for Requirement 3, the batch is accepted.

Microsectioning is not a normal routine matter for all incoming inspection departments, and a detailed description of the technique will therefore be found in Chapter 25. As the preparation is somewhat inconvenient, the inspector can, as part of his daily inspection, resort to other methods, which are dealt with below.

The copper plating always determines the strength and reliability of the plated-through holes. A preliminary measurement of the copper plating thickness will therefore give a clear indication of the necessity of preparing microsections.

An investigation of the copper plating thickness of a large number of boards from a wide selection of PCB manufacturers shows that the prescribed tolerances can easily be exceeded. In about 25 percent of the cases, a copper thickness of less than 0.001 in. (25 μm) was found, and in one-third of these cases the copper thickness was less than 0.0006 in. (15 μm). It should perhaps be added that most of the out-of-tolerance conditions occurred in boards from smaller PCB manufacturers, although in some cases out-of-tolerance conditions were found in boards from larger and better-known PCB manufacturers.

It is recommended that the inspector occasionally check the copper thickness in the plated-through holes, and always in the case of deliveries from a new PCB supplier. In practice, two methods are available, a resistance measurement and a beta-backscatter measurement. The latter can be used only in the case of pure copper platings, which means that a tin/lead overplating must not be present.

7.3.1d Resistance Measurement

Because the hole resistance is inversely proportional to the copper plating thickness, the thickness can be ascertained by measuring the resistance of the copper cylinder in the hole. Obviously, we will achieve an average value as the method cannot reveal smaller defects in the hole wall, such as unevenness or voids.

7.3.1d1 Effect of Tin/Lead Layer

We can disregard the effect of the tin/lead layer since the conductivity of tin/lead is about one-tenth that of copper. On the average, the thickness of the reflowed tin/lead layer of a plated-through hole is only about half the thickness of the copper. This means that the measured value of the copper resistance is about 5 percent too low, which in turn means that the determination of the thickness is about 5 percent too high.

7.3.1d2 Multilayer Boards

It should be noted that a resistance measurement is not applicable to multilayer boards as the inner pads cause the measured resistance to be too low. This effect becomes more pronounced with more layers in the board. The PCB designer can introduce some special holes intended for measuring the copper thickness. Obviously, the holes must not contact any internal pads.

7.3.1d3 Measuring Method

In principle, the measurement consists in determining the voltage drop V across the hole when a current I is fed through the hole as shown in Figure 7.17. The resistance R is determined from the expression $R = V/I$, from which the thickness of the copper layer is then determined by means of the curves shown in Figure 7.18.

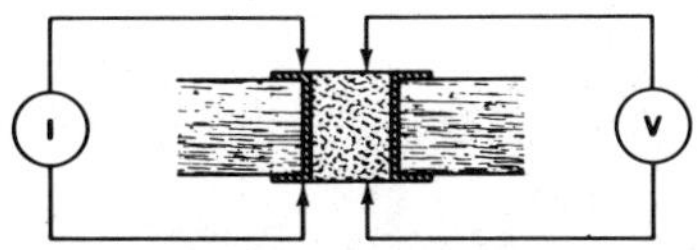

Fig. 7.17. Principle of a 4-terminal resistance measurement. I: Constant current generator. V: Microvoltmeter.

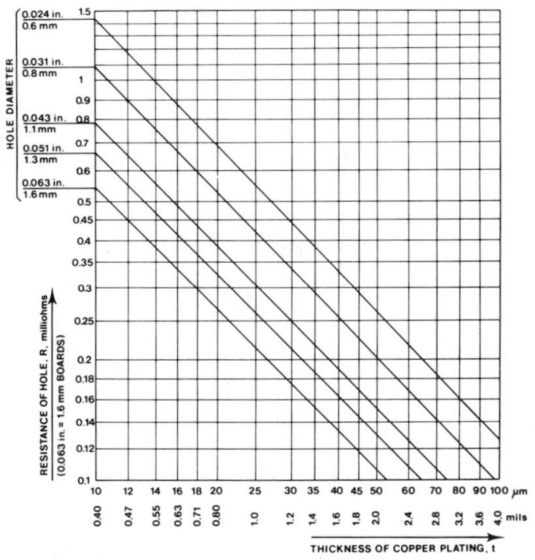

Fig. 7.18. Diagram for determining copper plating thickness as a function of hole size and measured resistance. Curves are valid for a 0.062 in. (1.6 mm) board.

For example, the resistance of a 0.031 in. (0.8 mm) hole in a 0.062 in. (1.6 mm) board is about 0.4 milliohm when the copper thickness is 0.001 in. (25 μm).

Because the contact resistances between the measuring equipment and the pads can easily be larger than the hole resistance, it is necessary to use a four-terminal measuring technique. Figure 7.17 shows that the current circuit as well as the voltage circuit have their own set of contacts. In this way the voltage drop across the current contact does not affect the measurement. Since the voltage is on the order of 0.5 millivolt, it is better to avoid the influence of external fields, by using a dc measuring technique.

The curves shown in Figure 7.18 are based on the expression:

$$R = \propto \frac{T}{\pi t(D + t)}, \text{ where}$$

R = resistance in hole
∝ = specific resistance of copper
T = board thickness
t = copper plating thickness
D = finished hole diameter

When measuring in inches, the expression becomes

$$R = 0.2155 \frac{T}{t(D + t)} \mu\Omega, \text{ and}$$

when measuring in millimeters, it becomes:

$$R = 5.475 \frac{T}{t(D + t)} \mu\Omega$$

The through-plating has the shape of a cylindrical tube as shown in Figure 7.19. A more accurate calculation takes into account that the length of the plated-through tube shall not include the thickness of the copper foils of the base laminate, for which reason it should be reduced by twice the foil thickness.

Furthermore, the effective diameter D should be increased by twice the thickness of the tin/lead layer, if we are to use the specified diameter of the finished hole. Finally, a uniform current distribution in the plated tube is implied. This is not al-

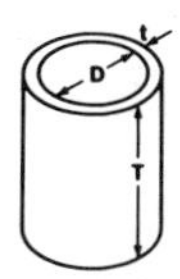

Fig. 7.19. Plated-through hole represented by a cylindrical tube. D = internal diameter of hole. t = thickness of copper plating. T = thickness of board.

ways the case, although the use of conical contact probes improves the current distribution.

The curves shown in Figure 7.18 do not take these matters into account, and they give a copper plating thickness about 5 percent too high. When the previously mentioned 5 percent due to the tin/lead plating is added, the measured thickness will be about 10 percent too high. This applies, however, only for nominal hole sizes. Owing to the fact that the tolerance of the diameter of copper-plated holes frequently is ±10 percent, the measuring error can easily reach 20 percent.

For the above reasons, it is not really correct to reject boards on the basis of a resistance measurement. The method should instead be regarded as a quick, convenient, and nondestructive test which can indicate abnormal plating conditions. This applies not only to the average copper plating thickness as dealt with above, but also to the occurrence of plating cracks which show up as abnormally high resistance values. When abnormal plating conditions are indicated, the only dependable method is to examine microsections under a microscope.

In order to pick up holes offering the largest chance to find the lowest plating thickness, the inspector should select small holes, usually 0.028 or 0.031 in. (0.7 or 0.8 mm), preferably located in dense areas at the center of the board. As previously mentioned, the plating profile of the board shows the lower plating thicknesses in the middle of the board, particularly if the pattern is not balanced. The inspector should note that the hole under investigation should not be connected in parallel with other holes, as this can cause incorrect results.

7.3.1d4 Measuring Equipment

There are several different types of measuring equipment on the market, but where the incoming inspection department does not wish to invest in such equipment, it is not too difficult to make the probes necessary. In conjunction with an ordinary power supply and a microvoltmeter such probes provide an excellent and cheap equipment for measuring microresistance.

The probes are shown in Figure 7.20 and a cross-section of a probe in Figure 7.21. The current probe is conical and rests on the hole edge, so that a fairly uniform current distribution in the hole wall is achieved. The voltage probe consists of a PTFE insulated piano-type wire, or still better, a hard-drawn gold wire, located in an eccentrically drilled channel in the current probe.

Fig. 7.20. Probe device for a 4-terminal measurement.

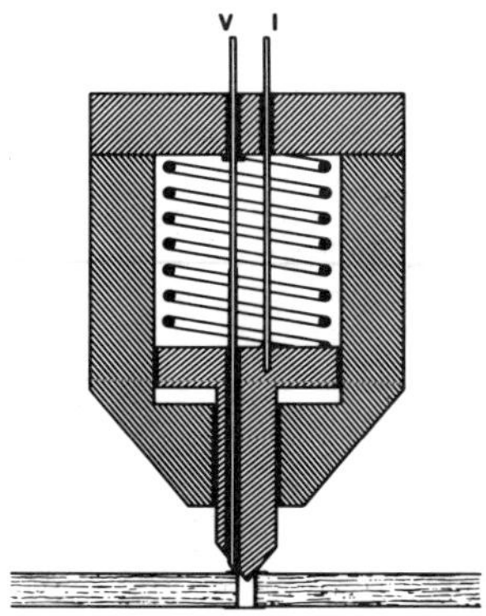

Fig. 7.21. Cross-sectional drawing of one of the probes used in the probe device shown in Fig. 7.20. V: voltage contact. I: current contact.

Since the eccentricity is about 0.020 in. (0.5 mm), the voltage probe will hit about 0.004 in. (0.1 mm) within the edge of the annular ring of a 0.031-in. (0.8 mm) hole. Consequently, larger holes will require other probes with a proportionally larger eccentricity.

An ordinary dc power supply can be converted to a constant-current generator simply by inserting a resistance in the current circuit. This resistance should be large compared with the resistance of the plated-through hole. A practical value is 5 ohms (5 watts); at a voltage of 5 volts, the measuring current becomes 1 ampere, regardless of the resistance of the hole.

If, for example, a voltage of 400 microvolts is measured across the hole, the resulting resistance is 400 microohms or 0.4 milliohm. The curves in Figure 7.18 show a plating thickness of about 0.001 in. (25 μm) for a 0.031 in. (0.8 mm) hole.

As mentioned earlier, the measured result is approximately 10 percent too high. A compensation can be achieved by increasing the current by 10 percent and converting the voltage reading to microohms by using the original value of the current (1 ampere). In this case, we obtain a resistance of 440 microohms, corresponding to a plating thickness of approximately 0.0009 in. (22.5 μm).

If the incoming inspection department does not wish to make its own probes, or it has had poor results with commercially available probes, the PCB designer might be persuaded to introduce in the board one or more special test holes as shown in Figure 7.22, if space permits.

The pads are designed in such a way that it is possible to make a four-terminal measurement. By turning the solder pad on the other side of the board by 180 degrees, symmetrical current and voltage conditions are achieved. The extra cost of drilling the test hole plus four terminal holes is negligible.

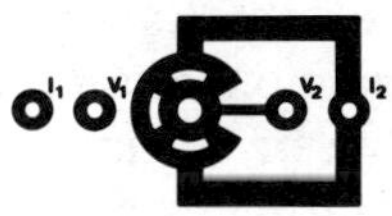

Fig. 7.22. Solder pad configuration designed for a 4-terminal measurement.

7.3.1e Beta-Backscatter Method

The basic element of this method is a low-energy isotope emitting beta rays (fast electrons) and a Geiger-Müller tube that collects the electrons reflected (back-scattered) by the object being measured. The beta rays are emitted in a narrow beam directed at a 45-degree angle to the wall of the hole under investigation. See Figure 7.23. The hole wall reflects part of the beta rays back to the detector, which delivers a signal to a calculating circuit. Here the thickness of the copper plating is calculated and the result displayed.

The thinner the copper layer, the fewer beta rays are reflected, since the greater part of the beta rays penetrate the copper layer and disperse into the base material. The degree of reflection (beta-backscatter) is therefore a measure of the plating thickness. An initial calibration, based on known standards, is easily made.

It should be noted that the equipment uses completely harmless low-energy isotopes such as carbon-14, promethium-147, thallium-204, and strontium-90. As the isotopes emit the beta rays in a relatively random manner, the results of the measurement depend on the measuring time, which is typically between 15 and 30 seconds.

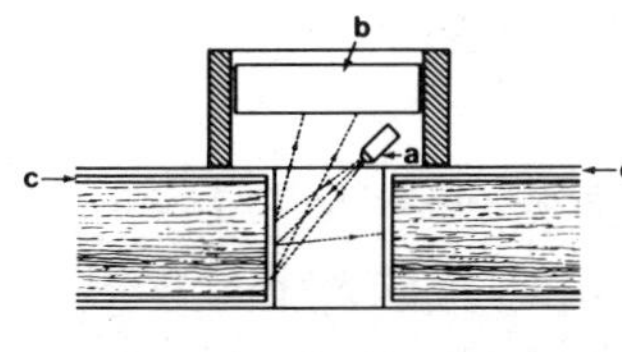

Fig. 7.23. Principle of measuring plating thickness by means of a beta-backscatter equipment. a = isotope emitting beta rays. b = Geiger-Müller tube. c = copper foil. d = plated copper.

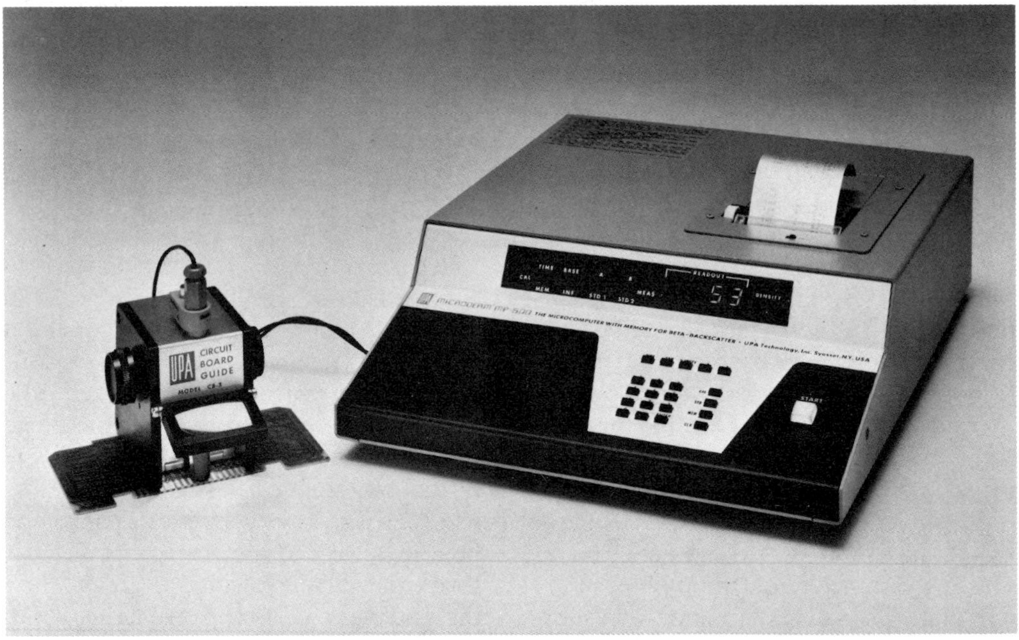

Fig. 7.24. Example of a beta-backscatter meter (courtesy UPA Technology, Inc.).

7.3.1e1 Limitations of the Method

Unfortunately, the method cannot be used for measuring the thickness of the copper plating if it is covered by an overlying tin/lead layer. It is therefore restricted to applications in conjunction with boards having copper plating only, possibly covered with a protective lacquer. The measuring method is therefore not of particular interest to an incoming inspection department, since the vast majority of plated-through boards of today are tin/lead plated. Incidentally, beta-backscatter equipment is fairly expensive.

7.3.1e2 Benefits of the Method

On the other hand, the PCB manufacturer will benefit from the beta-backscatter measuring method because it is possible to measure the copper plating thickness in a nondestructive manner immediately after plating; for example, as a process control. The measurement is carried out in the course of a very few minutes. In this way it is possible to achieve a high degree of certainty that the finished boards will have the right copper plating thickness. Most PCB manufacturers have therefore purchased beta-backscatter measuring equipment. A typical example is shown in Figure 7.24.

7.3.1f Eliminating the Need for Measurement

It is evident from the preceding discussion that the determination of the copper plating thickness can be a little problematic in the incoming inspection department. Instead, the inspector can emphasize preventive measures, the most important one being an inspection visit to the PCB manufacturer. One of the matters which are investigated during such a visit is the control of the plating process, including the necessary measuring equipment.

When the PCB manufacturer's records of plating thickness show satisfactory results, the probability of fulfilling the demands regarding plating thickness is such that it is not necessary to perform fixed routine tests. Random tests carried out by measuring the resistance of plated-through holes serve, therefore, to satisfy the inspector regarding the continued correctness of the plating thickness.

7.3.2 Voids in Hole Wall

Voids in the hole wall of plated-through solder holes are usually caused by one or more faults in the manufacturing process. We will therefore deal with some of the sources of these faults in order that the inspector may achieve a reasonable background to enable him to ask the right questions in conjunction with a factory inspection.

7.3.2a Drilling Technique

When the drill feed and/or speed, or the drill bit geometry, is incorrect or (frequently) a worn drill bit is used, the base material becomes burned by the excessive heating of the hole wall. This implies that certain areas of the electroless copper are deposited on an unstable base, and therefore flake. Electrolytic copper and tin/lead, therefore, are not deposited at such places. See A, Figure 7.25.

The hole wall can also be contaminated; for example, as a result of oil from the compressed air supply. This prevents the surface from accepting the electroless copper where contamination has occurred, and as mentioned above, no electrolytic copper and tin/lead is deposited. See B, Figure 7.25. In both these cases the result will be voids in the wall of the finished hole.

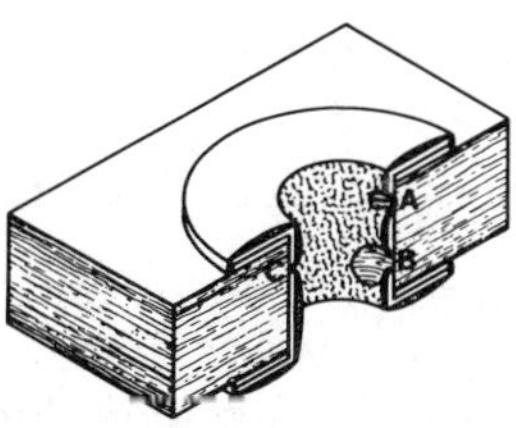

Fig. 7.25. Various types of holes in the hole wall. A: Small void. B: Larger void. C: Pinhole.

7.3.2b Cleaning after Drilling

Drill particles which become lodged in the hole wall are not always removed, frequently because the rinsing procedure is inadequate. Electroless copper is deposited on such particles. However, they may become detached and enter into the electrolytic bath due to the influence of the temperature and the flushing of the holes caused by the mechanical agitation of the board in the bath. Where such drill particles have worked loose, no electrolytic copper and tin/lead is deposited, and we find pinholes in the hole wall. See C, Figure 7.25.

7.3.2c Deposit of Electroless Copper

During the copper reduction process, hydrogen forms, and some hydrogen bubbles can be trapped in the holes and prevent the deposit of the electroless copper. The result is that no electrolytic copper and tin/lead can be deposited at later stages. See B, Figure 7.25. It is therefore common practice to agitate the boards back and forth in the bath to flush the holes so that the bubbles are removed.

Where supervision is negligent—for example, in maintaining temperature—or the bath constituents are not properly maintained, the rate of deposit can decrease so that the deposited layer is very thin and discontinuous. Isolated areas will therefore not receive any deposit of the electrolytic copper and tin/lead; these areas will appear as small holes or as larger nonplated areas. See A and B, Figure 7.26.

Insufficient bath filtration can cause small particles to adhere to the hole wall; the particles subsequently may work loose

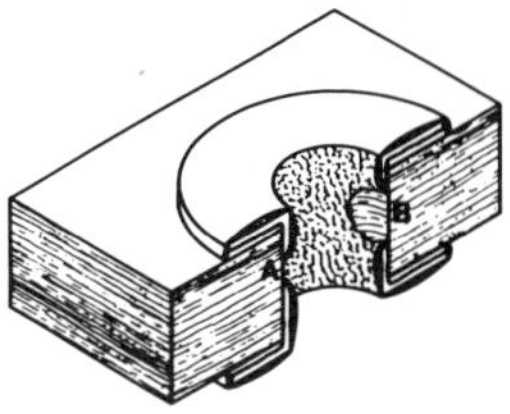

Fig. 7.26. Other types of holes in the hole wall. A: Larger void B: Large nonplated area

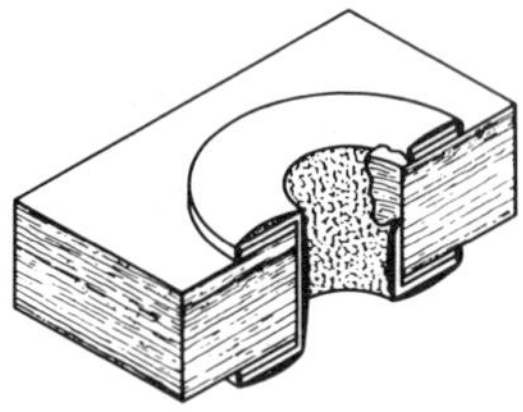

Fig. 7.27. Hole wall with holes caused by remnants of screen printing ink.

and enter the electrolytic copper plating bath. This case is analogous to drill particles in the holes, and the result is voids, or rather, pinholes in the hole wall. See C, Figure 7.25.

Another cause can be the oxidation of the very thin layer of deposited chemical copper, either as a result of insufficient drying of the board, or because of excessive delay before the subsequent plating process. In addition, in this case no deposit of electrolytic copper and tin/lead takes place, and the result will be voids in the hole wall. See B and C, Figure 7.25.

It should be noted that not all electroless copper baths are equally good. In some baths it can be problematic to cover the ends of the glass fibers where they appear in the hole wall. This is not necessarily reflected in voids or pinholes in the hole wall because the subsequent electrolytic copper deposition will build up and close such areas. The reliability of the through plating, however, should be called in question.

7.3.2d Image Transfer

When the image is transferred by screen printing, misprinting can occur in the form of an insufficient side-to-side registration. The screen printer will, in a few cases, try to save such boards by washing off the printing ink. It is difficult to avoid all contamination of the holes with a film of ink. Since the color pigments are frequently washed away, the film is practically invisible and therefore difficult to detect.

The film acts as an insulator and no electrolytic copper and tin/lead will be deposited at such places. See Figure 7.27. It should therefore be taken for granted that washing and subsequent reprinting will not be performed. The necessary adjustment of the screen printing machine, including the regular process control of the printing accuracy, is performed by means of a single board selected for this purpose. After each printing, the ink is washed off and the board reprinted until the machine is correctly adjusted; it is then discarded and the entire batch printed.

7.3.2e Cleaning before Plating

The electroless copper surface must be cleaned before the electrolytic copper bath, particularly if the board has been set aside for such a long time that the electroless copper oxidizes. Incorrect cleaning solutions—for example, an etching solution, or anodic cleaning, where the board acts as an anode and gives off copper—can puncture the thin layer of electroless copper or, in extreme cases, remove it entirely. See B and C, Figure 7.25. The result is again that no deposit of electrolytic copper and tin/lead occurs.

7.3.2f Deposit of Electrolytic Copper

Holes or overly thin areas in the electrolytic copper layer can be due to a low throwing power, which indicates the ratio between the copper thickness in the holes and on the surface (Figure 7.28). A care-

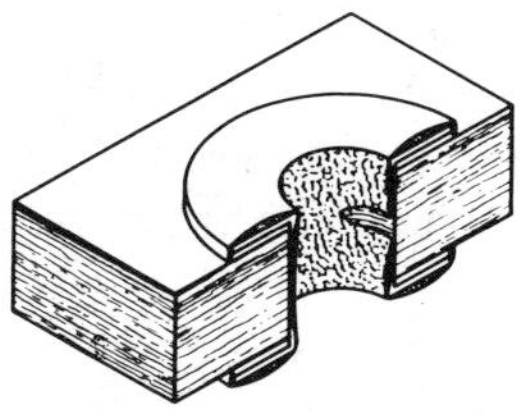

Fig. 7.28. Hole wall with holes or thin areas.

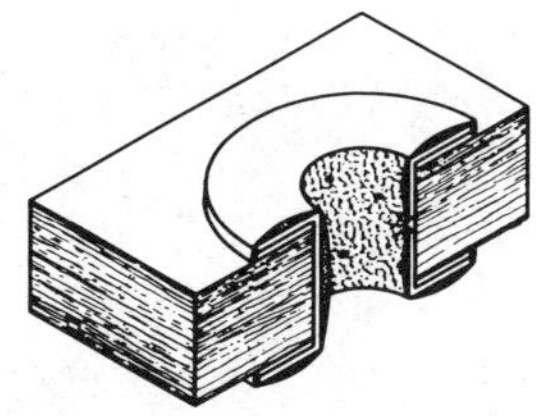

Fig. 7.29. Hole wall with holes caused by overly thin or excessively porous tin/lead plating.

ful analysis and maintenance of the balance of the bath constituents is absolutely necessary. Commonplace causes are incorrect process conditions; that is, incorrect plating time and/or plating current.

Another cause of overly thin copper can be a design error. When the hole diameter is less than 40 to 50 percent of board thickness, too little copper frequently is deposited into such holes. A compensation can be achieved by increasing the plating time, but in turn, too much copper is deposited in larger holes. This causes problems regarding meeting the diameter tolerance unless such holes are drilled in an enhanced oversize.

Obviously, the conditions become fairly incalculable, and almost require individual treatment of the holes in the preproduction stage, especially when the pattern is not balanced. Since an individual treatment is generally impossible, it has to be accepted that the design error results in a somewhat thinner copper layer in very small holes. The inspector should draw the attention of the PCB designer to this problem so that it can be corrected.

7.3.2g Deposit of Electrolytic Tin/Lead

During the subsequent etching process, the tin/lead layer serves as an etch resist. If the layer is too thin, which means less than 0.00028 in. (7 μm), or is too porous, it is very likely that the underlying copper will be attacked during etching. See Figure 7.29.

A frequent cause of an insufficient tin/lead layer is incorrect process conditions; that is, incorrect plating time and/or plating current. More problematic is reduced throwing power, which can be caused by contamination of the bath. One reason could be copper contamination caused by drag-in from the preceding bath. Another reason could be organic contamination in the form of decomposition products of the organic additives of the bath. The reduced throwing power can, as discussed above, also be due to a design error regarding hole sizes.

7.3.3 Inspection of Voids in Hole Wall

It appears from the preceding discussion that great care should be taken during the many processes involved in the manufacture of plated-through boards. It is not particularly reassuring, on an inspection visit, to see that the plating current of an electrolytic bath is monitored by a heavily corroded ammeter, and that the plating time is measured with an ordinary kitchen clock. Or to see that the operator, some time after the alarm of the clock has sounded, pulls the rack out of the bath and rinses it more or less leniently in the rinsing baths before it is moved to the next bath.

In order to manage the human factor, and also to refute charges of uncontrolled processes, many PCB manufacturers now employ automatic processes, using computer-controlled plating lines. See Figure 7.30. In some cases, the plating lines are provided with automatic control systems which analyze the baths and automatically replenish the used chemicals.

Fig. 7.30. Automated plating line (courtesy Printca AS).

7.3.3a Value of Factory Visits

A well-spent control cost is to visit the factory, actually a preventive measure. When the inspector, by doing so, convinces himself that the quality control system is adequate and in good order, he can with good reason expect the delivered boards to meet the demands regarding the quality of the holes. The inspector should always inspect the boards in the incoming inspection department since experience shows that from time to time, strokes of bad luck can happen, even at the most professionally advanced PCB manufacturer.

7.3.3b Level of Inspection

As mentioned in the introduction of this chapter, and further supported by the survey given above, the hole quality will, in general, be quite uniform throughout the total batch. We can therefore use the same limited sample size as stated in Paragraph 6.8.5a regarding the tolerance on diameter. This means that it is sufficient to inspect a few holes.

Owing to the fact that a lacking balancing of the pattern implies a thicker plating than required, there is no reason to increase the inspection level in the case of unbalanced boards. The holes to be investigated are distributed as mentioned in Paragraph 6.8.5a

7.3.3c Inspection Methods

We will now see how to carry out a visual inspection. As the most common hole size is around 0.031 in. (0.8 mm), the inspector has to use optical aids; for example, a microscope with a fish-eye optical system. In this way, it is possible to view the entire hole wall at one time. See Figure 7.31. Another and perhaps more common

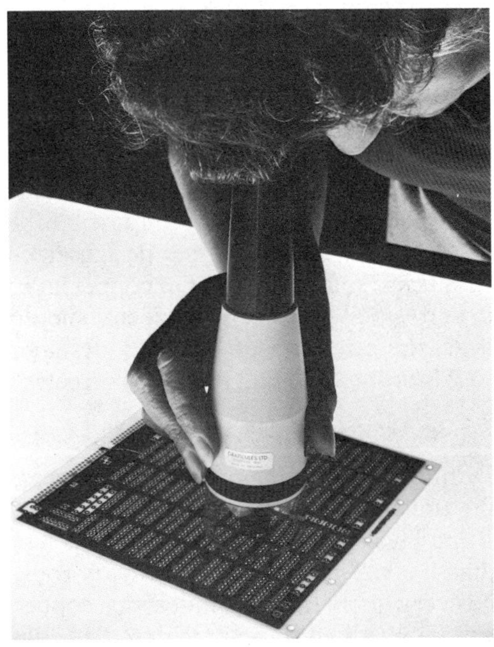

Fig. 7.31A. Inspection of hole wall by means of a microscope with a "fish-eye" optical system (courtesy Graticules Limited).

Fig. 7.31B. Long shot of plated-through hole as seen through the microscope shown in Fig. 7.31A.

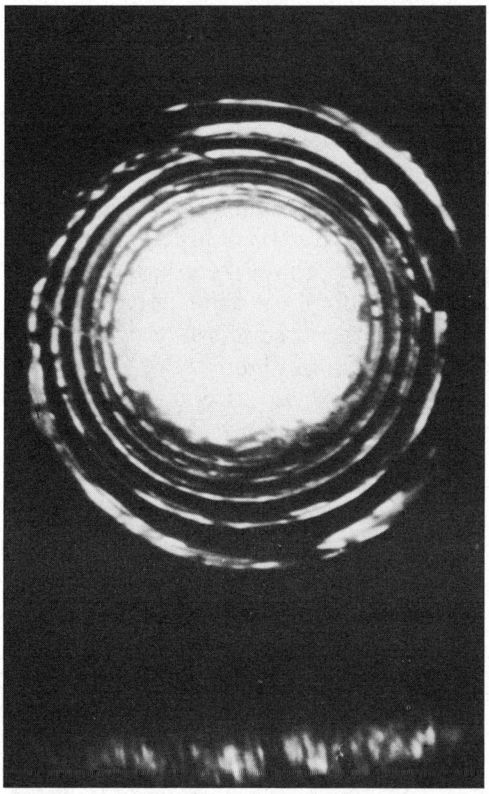

Fig. 7.31C. Close-up of plated-through hole as seen through the microscope shown in Fig. 7.31A.

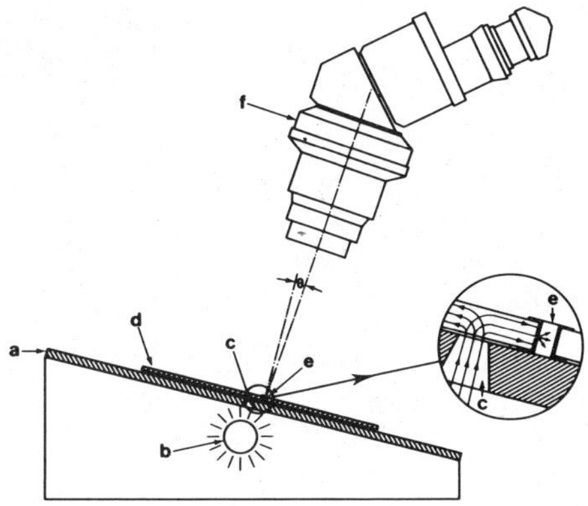

Fig. 7.32. Cross-sectional drawing of test setup for revealing voids in plated-through holes. a = tilted tabletop with a slot. b = 7 W fluorescent tube. c = slot 0.08 x 1 in. (2 x 25 mm). d = board. e = plated-through hole. f = stereo microscope.

aid is a stereo microscope which is characterized by a large range of depth definition.

Whichever type of microscope the inspector uses, the problem is the difficulty of seeing possible voids in the hole wall, not least because of the reflections from the bright metal. This can be solved by means of a test setup as described below.

In all simplicity, it consists of an illumination unit and a stereo microscope as shown in the cross-sectional drawing of Figure 7.32. The illumination unit consists of a tilting tabletop (a) made from 0.06 in. (1.5 mm) aluminum, and with a slot (c) of 0.08 × 1 in. (2 × 25 mm). A 7 watt fluorescent tube (b) is placed just below the slot through which the light passes. When a board (d) is placed on the tabletop, part of the light is transmitted lengthwise through the board as shown by the arrows. The board is now placed so that the hole (e) under investigation is withdrawn a little with respect to the slot to avoid direct light through the hole.

If there are no voids in the hole wall, the hole will appear dark. Any voids will appear quite clearly as bright spots or areas. The optical axis of the microscope

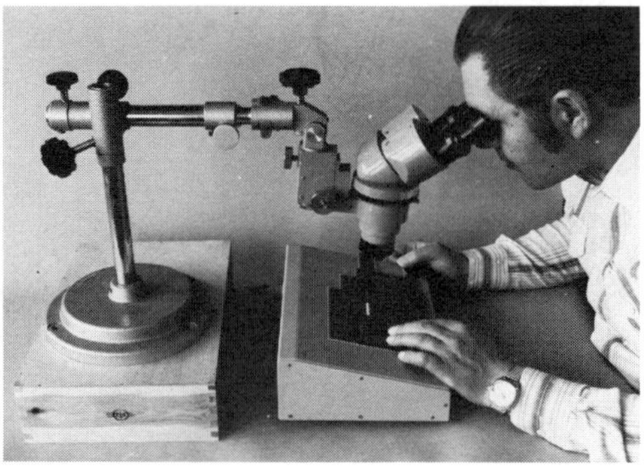

Fig. 7.33. Practical use of the setup shown in Fig. 7.32.

Fig. 7.34. Wall of plated-through hole with void, revealed by means of the test setup shown in Fig. 7.32.

should form a small angle with the axis through the hole under investigation so that it is easier to view the hole wall. The necessary magnification is 10 to 20 times.

In order to avoid a false light transmission, the tabletop can be covered with matte black paper. Furthermore, it is advantageous to place the test setup under subdued light. Figure 7.33 illustrates the practical use of this setup, and Figure 7.34 shows a plated-through hole with voids in the hole wall, viewed through a microscope.

7.3.3d Acceptance Criteria

Voids in the hole wall can be accepted under the following conditions:

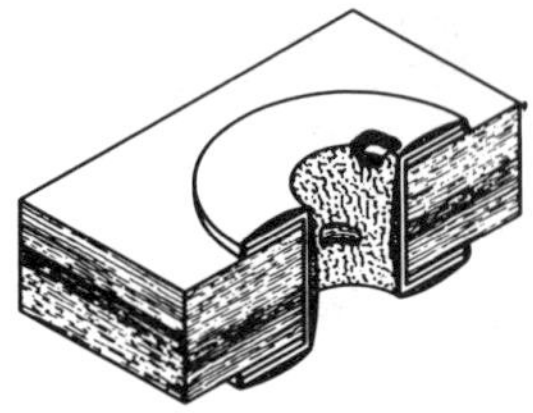

Fig. 7.35. Voids in plated-through hole.

1. No more than three voids per solder hole in no more than 2 percent of the solder holes.
2. The total area of voids per solder hole shall not exceed 5 percent of the area of the hole wall.
3. No voids shall be found at the pad-to-barrel transition.

These faults are illustrated in Figure 7.35. It is impracticable to measure the size of the voids for which reason the area is left to the inspector's personal judgment. In general, this is no problem as there are either no voids or so many that the decision is quite obvious.

Some PCB specifications contain a further limitation regarding the maximum size of the individual voids: The largest dimension in the axial direction shall not exceed 25 percent of the board thickness, and orthogonally it shall not exceed 25 percent of the circumference of the hole. This is shown in Figure 7.36. Obviously, this assessment is extremely difficult to perform, but experience shows that at least one of Requirements 1 to 3 has been violated when voids occur to such a degree.

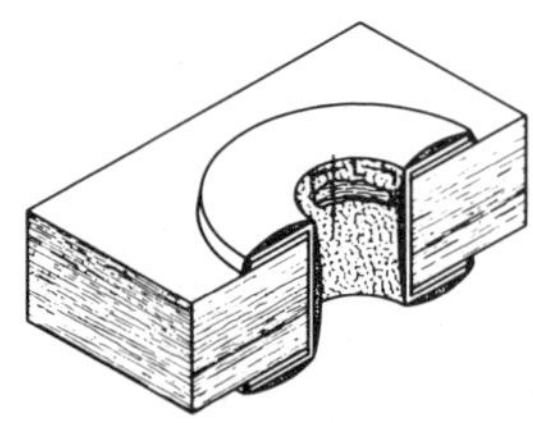

Fig. 7.36. Maximum size of voids in plated-through holes. $t \leqslant 25$ % of thickness of board. $L \leqslant 25$ % of circumference of hole.

When one of Requirements 1 to 3 is exceeded, the fault is rated as an A-defect, and consequently the board is rejected. It is useless to stage a rescue by exempting such holes which could actually have been nonplated-through (solder holes which on the component side are not connected to the circuitry by conductors). Experience shows that there will be so many circuit-active holes that a screening will be a waste of time.

7.3.4 Cracks in the Hole Wall

Concurrently with the above inspection for voids, the hole wall is inspected for cracks. See A, Figure 7.37. Cracks are very dangerous, particularly circumferential cracks, because they can be the cause of intermittent faults in the finished board. This applies in particular to via holes which normally, in the case of hand-soldered boards, are not soldered. A popular name of the phenomenon is "barrel cracking."

7.3.4a Effect of Solder Masks

Another point is that in many densely packed boards there is no room for clearance between the solder mask and the solder pad of the conducting hole. The PCB designer therefore chooses to overprint the solder pad with the solder mask, and the solder holes become more or less closed so that they are not filled with solder during mass soldering.

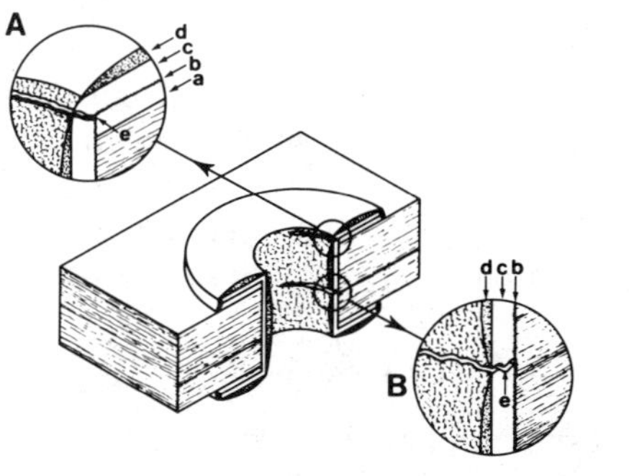

Fig. 7.37. Hole wall with cracks. A. Corner cracking. B: Barrel cracking. a = copper foil. b = electroless copper. c = plated copper. d = plated tin/lead. e = crack.

7.3.4b Board Disposition

Cracks are an indisputable cause for rejection, and the fault is rated as an A-defect.

7.3.4c Detecting Cracks

The resistance measurement mentioned in Paragraph 7.3.1d can prove helpful in detecting cracks, as these can cause resistance values outside the normal range. This should, however, be taken with certain reservations as the increase in resistance caused by a crack can be less than the decrease caused by a thicker plating.

In practice, all flawless plated-through holes of the same size will exhibit nearly the same resistance value. The inspector should therefore create a reference by measuring the resistance of a number of holes. Abnormal deviations with respect to the reference will, in all probability, indicate cracking or larger voids in the hole wall.

7.3.4d Microcracks

Microcracks are virtually invisible, but they are just as dangerous as the visible ones. They occur frequently in the transition area between the solder pad and the plated-through barrel. A popular name of the phenomenon is "corner-cracking." See B, Figure 7.37.

When the solder does not fill the hole completely, circumferential cracks can cause intermittent interruptions. Such interruptions occur when the base material expands in consequence of a temperature increase. Shorter cracks, however, are also dangerous since they can grow.

As a matter of fact, one should distinguish between cracks caused by thermal actions, and cracks caused by improper

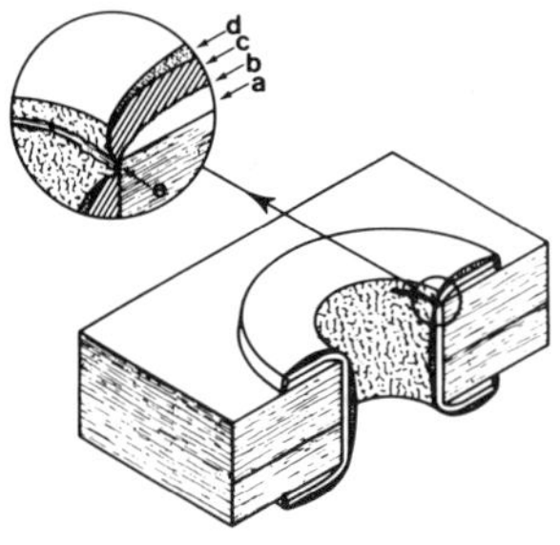

Fig. 7.38. Hole wall with crack caused by excessive scrubbing subsequent to electroless copper stage.
a = copper foil. b = electroless copper. c = plated copper. d = plated tin/lead. e = crack.

manufacturing techniques. Figure 7.38 shows an example of a very heavy sanding of the surface of the copper foil, which resulted in removing the copper foil at the edge of the hole. The crack is due to the absence of electroless copper and not to any temperature variation. In a thermal crack, the two sides of the crack will fit together, which is not the case with other cracks as some material will be missing.

Owing to the fact that it is nearly impossible to find microcracks, the inspector has to make sure in other ways. Knowledge of the type of copper bath can give a certain indication. Some copper baths are more likely than others to deposit the desired ductile (soft) copper in the holes. Therefore the holes do not exhibit any tendency to cracking due to thermal action, to which the board is subjected during reflowing and soldering.

It is not sufficient to know the type of copper bath as the brighteners are often added in order to achieve a brighter deposit or a shorter plating time. Unfortunately, this has some disadvantages, including a somewhat harder copper deposit, which is more susceptible to cracking.

7.3.4e Checking the Tendency to Crack

The simplest method of investigating the tendency to crack is to select a board and pass it through the mass soldering machine. A row of IC holes is then microsectioned and the holes studied with the aid of a microscope. If the tendency to crack, however, is not very pronounced, the inspector will be lucky indeed to find cracked holes with the first trial.

The most reliable knowledge of the tendency to crack is achieved by having some special test coupons made at convenient intervals. The test coupons are produced together with an ordinary order and should be placed with one test coupon per production panel. They will therefore be representative of the boards produced.

When the test coupons, which are subjected to a thermal shock test, are ordered every second or third month, the inspector has good possibilities of tracing possible changes in the copper hardness so that the PCB manufacturer can be contacted.

The test is not a routine test and a detailed description is therefore given in Chapter 29, which also describes the test pattern. It should be mentioned that the test coupon has 50 holes connected in series. By measuring the resistance of the row of holes during the shock test, it is possible to reveal the tendency to crack. Furthermore, a permanent change in the resistance of the holes gives an indication of the hardness of the copper.

7.3.4f Microsections

When a defective hole is localized—for example, by means of the microresistance measuring equipment mentioned in Paragraph 7.3.1d4—it can be microsectioned and studied with the aid of a microscope.

7.3.5 Uneven Hole Wall

During the visual inspection of the solder holes—which best is carried out with a microscope—the inspector should also assess the unevenness of the hole wall. Many degrees of unevenness can occur, and a

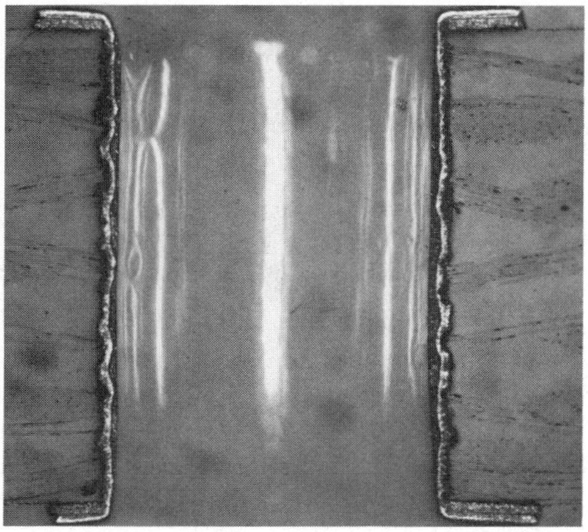

Fig. 7.39. Microsection showing how the drilling grooves outlined in the copper plating are leveled after reflowing.

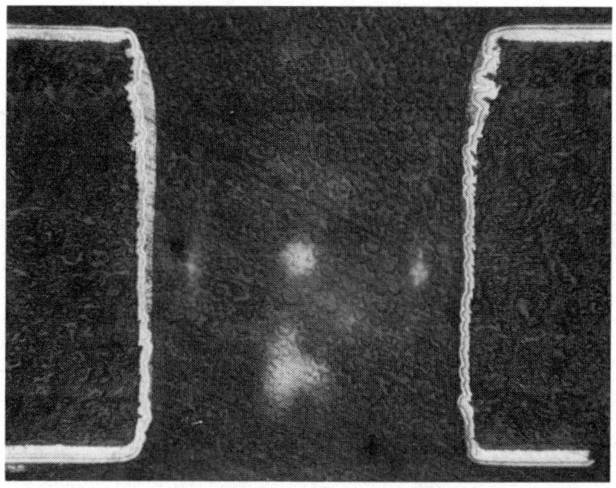

Fig. 7.40. Microsection showing how irregularities in the copper plating are leveled after reflowing. Contrary to Fig. 7.39, the hole drilling is very smooth.

very rough or grooved hole wall is a typical sign of poor drilling (Paragraph 7.2.5).

A certain filling-up of the drilling grooves always takes place during the plating processes, followed by a leveling achieved by reflowing. Even in case of fairly deep drilling grooves, this leveling effect can mislead the inspector (Figure 7.39). The safest method of assessing the hole quality is by microsectioning.

It can be very difficult to ascertain by visual inspection whether the surfaces of the holes are so uneven that the board should be rejected, and generally no criteria for rejection are stated in the PCB specification. Frequently it is demanded that the surface of the holes shall be even, and in such cases the customer is, at any rate, entitled to expect reasonably smooth and even holes.

As shown in Figure 7.40, the leveling effect can veil the real conditions. The reliability of the hole depends on the copper plating and not so much on the tin/lead plating. In order to be absolutely sure of the quality of the hole, it is necessary to study its structure in detail which implies microsectioning.

The PCB specification can also be written so that it specifies a hole drilling which ensures that, on one hand, the pull-out strength of the finished hole is met (Paragraph 5.8.3), and on the other hand, that the board can stand a certain soldering influence without cracking of the plated-through holes taking place.

A very rough drilling with deep grooves in the base material implies such a risk of internal stress in the deposited copper that there is a latent danger of thermal cracking. This fault is therefore rated as an A-defect, and the board is rejected.

If the inspector, at the visual inspection, observes a very clear outline of grooves in the drilled wall, he should sacrifice one of the boards for microsectioning so that the drilling quality can be assessed in a more objective manner.

In multilayer boards a rough hole wall can also be caused by excessive etchback. The purpose of etchback is, as mentioned earlier, to remove all epoxy smear and expose a little of the internal pads in order to achieve a better interface with the through-plating. When the etchback process lasts too long, the surface becomes very uneven with deep pits in the glass fiber weaving.

The plating will be very thin at the bottom of the pits, although reflowing of the tin/lead layer will show a certain leveling

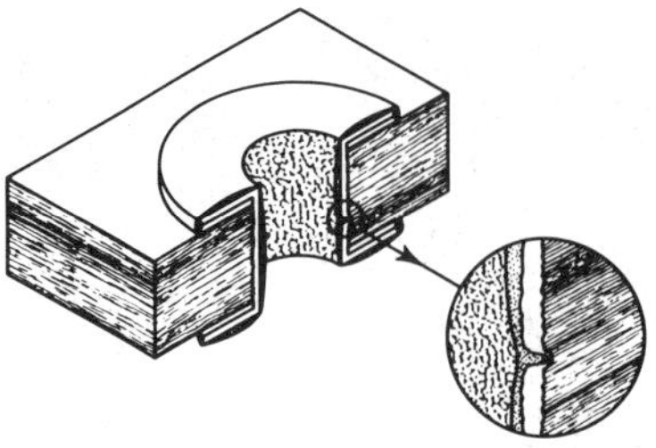

Fig. 7.41. Example of the leveling effect of the reflowed tin/lead layer.

effect. See Figure 7.41. The fault is rated as an A-defect and the board should be rejected.

7.3.6 Plating Pockets

An uneven hole drilling can, during the copper plating process, cause electrolyte or moisture to be encapsulated at the bottom of one or more drilling grooves or other deep uneven areas, such as those caused by glass fibers torn out of the hole walls. Plating pockets can be closed or open as shown at A and B, Figure 7.42.

As a result of the violent heating of the board which takes place during mass soldering, the vapor pressure in the plating pockets can increase so much that the vaporized electrolyte or moisture makes its way out. In the case of closed plating pockets, the hole wall will usually rupture which does not happen if the pockets are open.

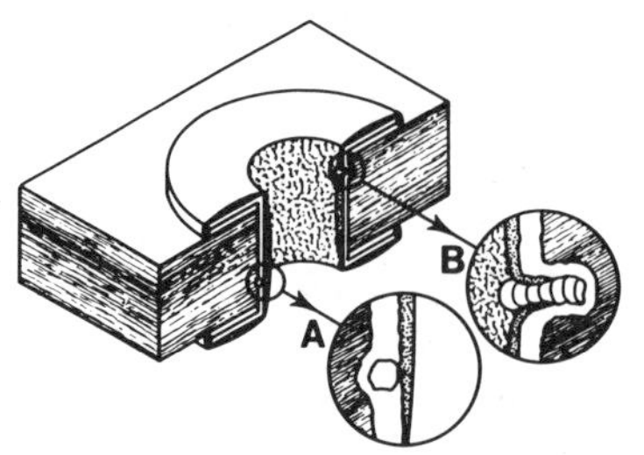

Fig. 7.42. Plating pockets in plated-through holes.
A. Closed pocket. B: Open pocket.

The gasses force their way through the solder as long as it is not solidified, which creates the blowholes in the solder fillets. This is described in detail in Paragraph 10.7.4.

Plating pockets are rated as an A-defect, and the board is rejected. It is, however, not possible for the inspector, during a visual inspection, to tell whether a void or a pinhole covers an open plating pocket. For very good reasons, the inspector cannot see closed plating pockets at the visual inspection.

It is not practicable to prepare microsections from each batch being delivered. The inspector has to base his decision on the experience of the PCB manufacturer in question, and await information from the assembly line concerning any unusual occurrence of blowholes in the soldered boards. If the case is suspect, he can prepare microsections, and perhaps find some plating pockets.

The inspector should be aware of the fact that blowholes are not necessarily a sign of plating pockets as there can be other causes; for example, poor control of the fluxing preceding the mass soldering. See therefore Paragraph 10.7.4.

7.3.7 Nodules

During visual inspection, the inspector may frequently see some small, irregular nodules. See Figure 7.43. There are two main causes of the formation of nodules: poor hole drilling and contaminated plating baths.

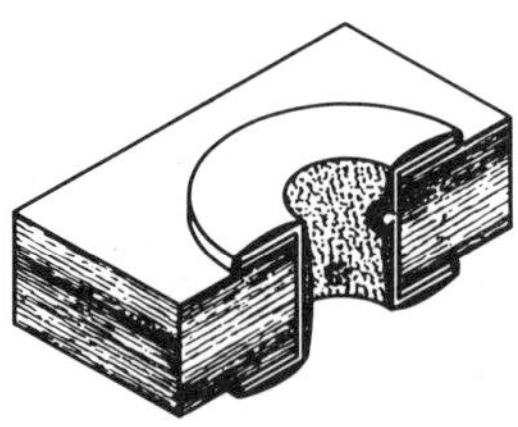

Fig. 7.43. Nodules on hole wall of plated-through holes.

In the case of poor hole drilling, the hole wall of the drilled hole exhibits small protrusions of frayed glass fibers. Drill particles lodged in the hole wall, possibly in conjunction with epoxy smear, can also occur. A good drilling technique is, as mentioned in Paragraph 7.2.5, of decisive importance to the evenness of the surface.

Contamination of the copper bath is often due to organic material, frequently in the form of decomposition products of the additives. Other causes can be particles of the plating resist which have worked loose in the plating bath. Finally, the contamination can be due to insufficient cleaning of the boards which results in drill dust or drill chips being washed off into the plating bath.

The bath should be kept clean by continuous filtration so that all foreign particles are filtered out. It is quite obvious that a careful cleaning of the boards before plating is very important.

Large nodules in plated-through holes can impede the component insertion when the tolerance on diameter is exceeded. Generally, however, nodules will not prevent the insertion as they will break off. Unfortunately, the remaining nonplated areas can be the starting point of barrel cracking, for which reason large nodules are cause for rejection and consequently rated as an A-defect. Figure 7.44 serves as a guide for the inspector.

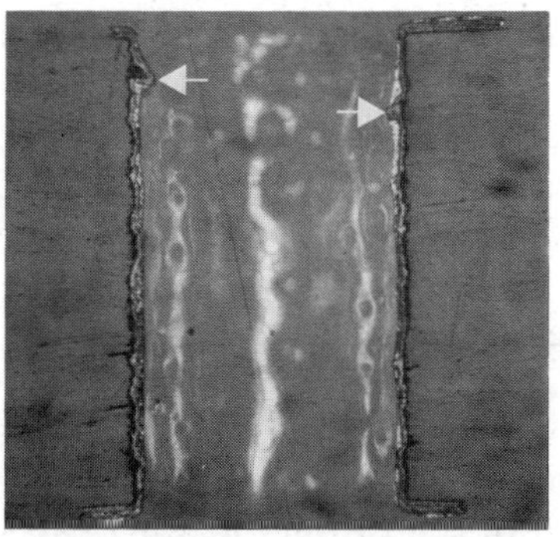

Fig. 7.44. Microsection showing the maximum acceptable size of nodules.

Another unfortunate circumstance is that the nodules, which mostly consist of copper, work loose in the solder pot or solder wave. The copper forms intermetallic compounds with the tin, and when the copper content of the solder exceeds about 0.2 percent, the solder becomes sluggish and the solder fillets become gritty.

7.3.8 Burning

Another form of nodules arises by burning, which can take place during the plating process if the plating current is too high. Since the current field lines concentrate around sharp edges, the nodules will tend to develop at the transition from the hole wall to the solder pad, along the outer rim of the solder pads or along the conductor edges. See Figure 7.45.

The causes of burning can be incorrect bath conditions, including inadequate copper concentration, excessive acid concentration, and low temperature so that burning occurs at a lower current density than specified for the bath in question.

Because plating baths are usually very carefully analyzed and maintained, they are rarely the cause of burning. The most obvious reason is an incorrect, or even completely missing, determination of the area to be plated, so that the maximum current density for pattern plating in the copper bath in question—for example, 20 amp/sq. ft. (2.2 amp/dm^2)—is exceeded.

Burning can also be caused by bad de-

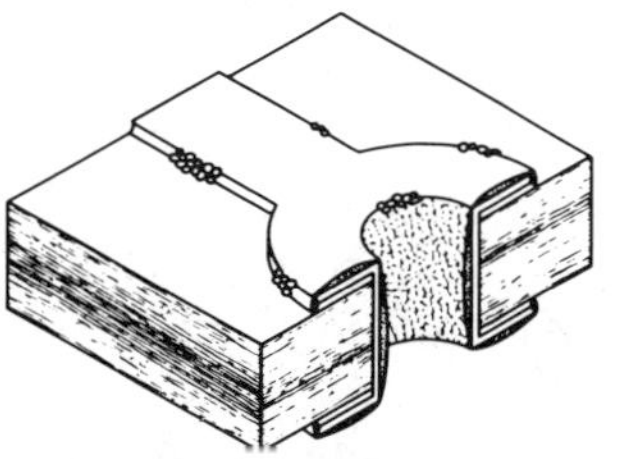

Fig. 7.45. Most common occurrence of nodules caused by burning.

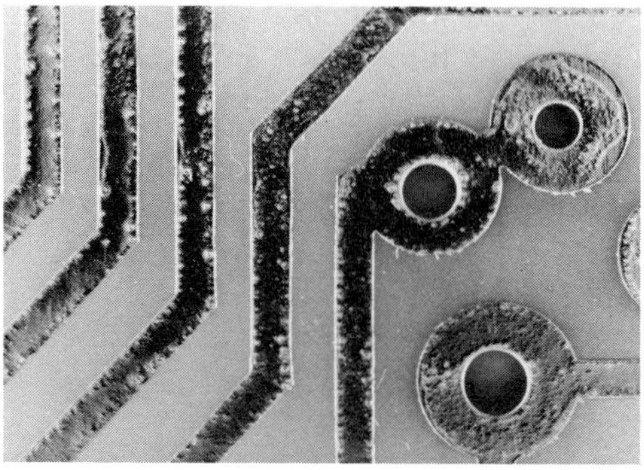

Fig. 7.46A. Board with widespread occurrence of nodules caused by burning during plating.

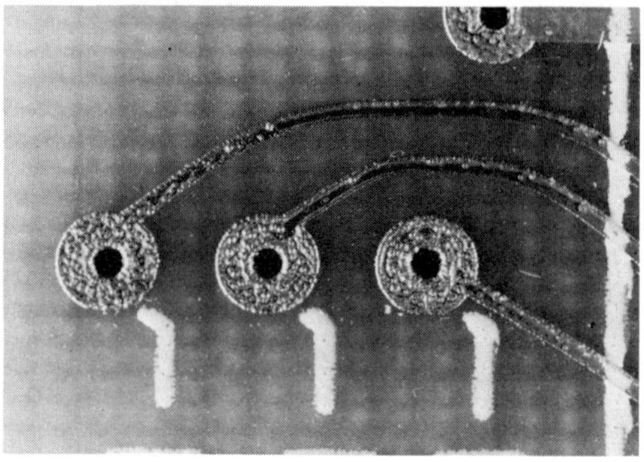

Fig. 7.46B. Board with widespread and heavy occurrence of nodules caused by burning during plating.

sign which leaves isolated pads to attract excessive currents in the plating bath.

It is quite easy to plate tin/lead on top of such nodules but the reflowing process does not succeed very well and leaves areas with poor wetting. However, this gives the inspector a very good indication of the solderability properties of the boards during the soldering process later. In extreme cases of widespread occurrence of nodules, the inspector should reject the board and rate the fault as an A-defect. See the examples shown in Figure 7.46.

7.3.9 Step-Plating

This phenomenon, which fortunately is extremely rare, is also called skip-plating or tear drops. It appears on the solder pads as differences in level, visible to the naked eye. The differences in level can be rather sharp, and frequently the dividing lines run in circumferential ways around the solder holes. See Figure 7.47.

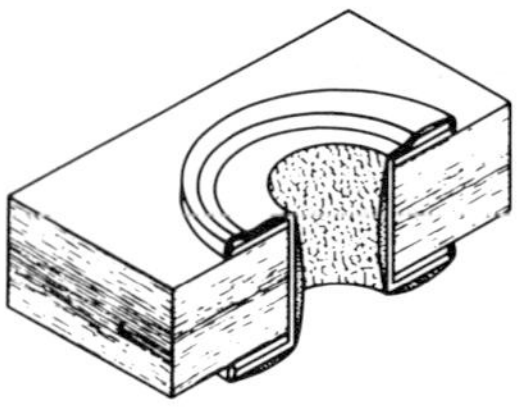

Fig. 7.47. Example of step-plating.

The cause is often a contaminated copper plating bath which contains overconcentrated organic additives. This implies that the copper deposit gets out of control and results in sudden variations in the layer thickness.

Whether the fault should be regarded as cause for rejection depends on the copper thickness at the place in question. A PCB specification usually requires a plating thickness of 0.001 in. (25 μm) copper in the holes and on the surface.

The latter point can be made the subject of a more particular interpretation, namely that the total copper thickness of the solder pads as well as the conductors shall not be less than 0.001 in. (25 μm). When the base material is clad with a ½ oz. = 0.0007 in. (17.5 μm) copper foil or thicker, the inspector will probably find at least 0.0008 to 0.001 in. (20 to 25 μm) copper on the areas in question. The matter is a little more problematic when the copper foil of the base material is ¼ or ⅛ oz. (9 or 5 μm). In all probability, the total copper will be too thin. The largest problem, however, is how much copper to find in the through plating itself. Unfortunately, the deposit on the hole wall will most likely be very thin.

An exemption implies microsectioning and subsequent measurement of the total copper thickness in the step-plated areas. If the copper thickness is below 0.0008 in. (20 μm), the board is rejected and the fault is rated as an A-defect. If the copper thickness is above 0.0008 in. (20 μm), the board is accepted and the fault rated as a B-defect.

7.4 Contaminated Holes

The solder holes of the board must not be contaminated because the soldering quality will be degraded, not least owing to outgassing of the impurities during soldering. On the whole, the inspection is carried out in the same way and to the same extent as described in Paragraph 7.3.3, which deals with inspection for voids in the hole wall. Surface illumination is an advantage, because it makes it easier to study the hole. We shall now deal with the following types of faults:

- Dust in holes
- Brown and black crusts
- Screen printing ink in holes

7.4.1 Dust in Holes

The holes are inspected for dust and larger foreign particles such as chips originating from contouring. Incomplete cleaning is rated as a B-defect, and the holes are blown clean with compressed air. It is important that the compressed air does not contain oil, which may come from the compressor.

7.4.2 Brown and Black Crusts

The holes and the pads are also inspected for brown or black crusts. There are several reasons for their occurrence: remnants of the flux used during the reflowing, and remnants of the plating resist. Flux remnants are due to an incomplete cleaning of the board after reflowing, and remnants of plating resist originate from contaminated cleaning baths for the plating resist (Paragraph 3.6, Reference A12 of the Reading List) when, for example the distilling equipment is slightly overloaded.

The crusts will normally flow up through the solder holes together with the solder when the board is soldered. They do not reduce the solderability of the board, but as the flux remnants usually are active, their occurrence should be taken quite seriously. More or less invisible flux remnants will presumably lie between the conductors, for which reason there is a risk of collecting humidity at such places. This can result in corrosion and reduced insulation resistance.

These considerations imply that crusts should be cause for rejection and the fault rated as an A-defect.

7.4.3 Screen Printing Ink in Holes

Except for via holes, the solder holes must not contain screen printing ink originating from the solder mask or the component notation. Ink in the holes can possibly outgas during the soldering process and cause blowholes and, as a consequence, reduced solderability of the solder fillets. The fault is rated as an A-defect, and the board is rejected.

Another result of printing ink in the holes can be, in the case of mass soldering, that the solder does not rise through the plated-through hole but stops at the ink.

Stains of ink in a few holes—for example, 5 percent of the total number of solder holes—should not result in a board rejection. The ink will float away during the soldering, and even if it, at worst, should result in some blowholes during mass soldering, the reliability of the solder fillets should not cause any rejection.

The PCB manufacturer, however, is not always directly to blame when screen printing ink is found in the holes. The fault can in many cases be due to the fact that the PCB designer may not understand the limits to printing accuracy involved in screen printing. Therefore, the artwork can have been prepared in such a way that it is quite unavoidable to have the ink running down into the holes. Unless the PCB manufacturer has expressed a reservation, he is, however, considered responsible for the fault. See also Paragraphs 11.5 and 13.2.2.

A rare, but at the same time treacherous fault is the presence of a very thin, transparent epoxy film which prevents wetting of the solder spot. This fault is due to an erroneous screen printing of, for example, a solder mask. The ink is washed off by means of a solvent and the board is reprinted.

The pigments are removed, but unfortunately a very thin and invisible epoxy film can remain in the holes and on the solder pads, preventing complete wetting of the board during soldering. The purchase specification should therefore state that misprints shall not be washed off; that is, the board must be rejected at the printing stage.

If this is the case, the fault will be extremely rare, and if it occurs as an exception, only a very limited number of boards will be involved. The microresistance measurement mentioned in Paragraph 7.3.1d3 will, in many cases, reveal the fault because of unstable contact conditions.

Generally it cannot be economically justified to inspect a batch of boards especially for epoxy films in the holes. Instead, the inspector should wait for feedback from the assembly line. Obviously, the fault is an A-defect and the board rejected.

7.5 Landless Plated-Through Holes

Landless plated-through holes are plated-through holes with no solder pads on one side of the board. Such solder holes are used very rarely today although they can occur in special applications, such as UHF boards with almost no room for solder pads.

Landless holes were once common because they were used in single-sided boards wtih plated-through holes. The advantage was that a real component side with a circuit pattern was not needed so that no registration problems occurred.

Because of process problems, many PCB manufacturers are against producing such boards. Following the electroless copper deposition, the whole component side must be covered by a plating resist. When screen printing the resist there is a certain risk that some printing ink flows down into the holes, and when using dry film, the holes are tented; that is, closed to one side.

Even a very careful rinsing means the risk of transferring liquid from one bath to the next which could be very detrimental for the baths. Furthermore, the deposits of copper and tin/lead will be fairly limited at the closed end of the holes, or even miss.

For the above reasons, several PCB manufacturers offer, out of consideration to their baths, to place solder pads on the landless side. Such solder pads will in many cases be only 0.004 in. (0.1 mm) larger than the drilled holes so that, in reality, the annular rings are broken. Such cases must necessarily be disregarded at the incoming inspection.

The two pattern imaging methods (screen printing and photoprinting) can result in through-plating problems as shown at A and B, Figure 7.48. When the imaging is done by screen printing, there is a tendency to achieve a protruding collar in continuation of the plating within the hole. However, the collar can fall off after board assembly and cause short circuits. When the imaging is performed by photoprinting—that is, via a dry film—

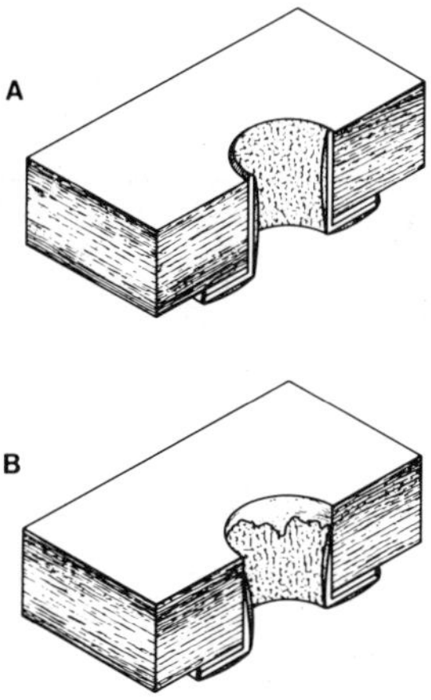

Fig. 7.48. Cross-sectional drawing of landless plated-through holes. A: Protruding collar on the landless side of the board. B: Missing plating at the landless end of the hole.

the through plating can be very thin towards the component side, or even missing.

Excepting the conditions mentioned below, all requirements regarding the quality of the through-plating of landless plated-through holes are, in principle, the same as valid for ordinary plated-through holes.

7.5.1 Plating Depth

In consequence of the above conditions, it is accepted that the plating covers 70 percent of the hole depth only. When the plating covers the total depth of the hole, the protruding collar on the landless side must not work loose and fall off.

A test is most simply performed by means of the scalpel method mentioned in Paragraph 5.8.4 and illustrated in Figure 5.35. The inspector checks whether the protruding collar is proof against a reasonable standard influence, for which the inspector should develop a certain experience. Alternatively, the collar shall be removed by a suitable mechanical treatment so that the hole plating is flush with the surface of the board.

Failure to meet these requirements is cause for rejection, and the fault is rated as an A-defect.

7.5.2 Plating Thickness

The plating thickness specified for ordinary plated-through holes must be achieved in at least 50 percent of the hole depth, after which the thickness may fall to zero.

Failure to meet this requirement is cause for rejection and the fault is an A-defect.

7.5.3 Miscellaneous Conditions

For obvious reasons, the quality requirements valid for ordinary plated-through holes—that is, drilling (Paragraph 6.8), hole quality (Paragraphs 7.2 to 7.4), and reflowing (particularly Paragraph 8.7.1)—apply only for 50 percent of the hole depth. For the pull-out strength only half the normal value applies. (See Paragraph 5.8.3.)

With the above limitations of the specifications, assessment criteria valid for ordinary plated-through holes can also be used for landless plated-through holes.

Chapter 8

Reflowing Inspection

8.1 Introduction

Reflowing melts the plated tin/lead layer and metallurgically restructures the tin/lead by changing it from a relatively porous deposit to a homogeneous and nonporous alloy so that the storage time of the board is extended.

8.1.1 Metallurgical Effects of Reflowing

Electrolytically deposited tin/lead contains tin-rich regions with discrete pockets of lead; after reflowing, the two metals form a more uniform alloy. Furthermore, an intermetallic compound between the tin and the underlying copper is created, and this is a very important prerequisite for a good wetting of the copper surface during soldering.

8.1.2 Reflowing Quality and Solderability

One of the most important properties of a printed circuit board is that it can be soldered, and the quality of the reflowing can serve as an indication of the solderability. An unsolderable board will not reflow satisfactorily, and the inspector therefore has in the reflowing a convenient means of assessing the solderability of the board. For this reason we deal in detail with reflowing in order that the inspector can achieve a broad background and better assess the board quality.

8.1.3 Reflowing Methods

Reflowing was once performed by dipping the board in hot oil. This was a relatively slow process as it had to be done manually. Another disadvantage was that the oil was used dangerously close to the flash point. Most PCB manufacturers today use infrared reflow equipment, which makes automatic handling possible. A conveyor belt carries the board at an adjustable speed through the necessary processes: fluxing, preheating, reflowing, cooling, and cleaning.

8.1.4 Reflowing Multilayer Boards

Many multilayer boards, however, are still reflowed in hot oil, because it is far easier to control the thermal influence on the board than by infrared reflowing. As the bond between the plated-through holes and the internal pads can be damaged by thermal shocks, a higher degree of security is achieved by hot-oil reflowing.

It must be added that with infrared reflowing it can be very difficult to achieve full melting in the solder holes if the multilayer board is relatively thick; for example, 0.125 in. (3.2 mm). By adequate preheating in an oil bath, the multilayer board can reach a reasonably even temperature throughout the total board thickness so that the subsequent hot-oil reflow gives a better result throughout the solder holes.

8.1.5 Reasons for Reflowing

Unless otherwise stated on the master drawing, all tin/lead plated boards should be reflowed, as substantial advantages are achieved: removal of the dangerous tin/lead overhang along the conductor edges, and a considerable lengthening of the permitted storage time. See Paragraphs 3.5.5, 4.5.6, and 4.13.4 of Reference A12, Reading List.

8.1.6 Level of Inspection

As a direct consequence of the automated infrared reflowing process a very uniform result may be expected, so that we can reduce the inspection level considerably. The reflow quality should be inspected concurrently with other inspections of the board—for example, of the pattern—and not as a separate inspection. The sample size is the same as indicated earlier.

8.1.7 Faults Associated with Reflowing

There are disadvantages and direct faults associated with reflowing. We will, therefore, describe various cases as listed below:

- Base material
  - Warp and twist
  - Discoloration
  - Layer movement
  - Surface resistance
- Appearance of the tin/lead surface
  - General requirements
  - Freeze lines
  - Alloy composition
  - Surface faults
- Overhang
- Tin/lead coverage of pattern
  - Coverage of pattern edges
  - Coverage of hole edges
- Bridging and hole blocking
  - Incomplete reflowing
  - Wetting after reflowing

8.2 Base Material

Whether reflowing is performed by dipping the board in hot oil or by infrared heating, the base material is subjected to large thermal stresses. This can very well result in permanent damage to the board as described below.

8.2.1 Warp and Twist

As mentioned earlier, the base material can have a certain internal mechanical stress which is relieved by the vigorous heating. In ordinary single-sided and double-sided boards the stress originates from the lamination of the base material, and in multilayer boards, from the board lamination process, whether it is performed by the PCB manufacturer or by the base material manufacturer. In the latter case the process is called mass lamination.

A major contributor to warp and twist is a large difference between the patterns of the solder and component sides of the board, and in the case of multilayer boards, also because of a possible unsymmetrical buildup.

When a large difference exists between the pattern density of the component and solder sides of the board—for example, an unrelieved ground plane on one side and a very open pattern on the other side—the finished board will inevitably warp and twist. The PCB manufacturer usually takes the precaution of expressing reservations in such cases. He can, however, make some improvement by using a better (and presumably more expensive) base material.

The problem is worst in very large boards, meaning boards with dimensions of 16 in. (400 mm) or more. The inspector should call the PCB designer's attention to the technical problems which an incorrect PCB design can cause.

Warp and twist are addressed in Paragraph 4.4, which describes the inspection in detail.

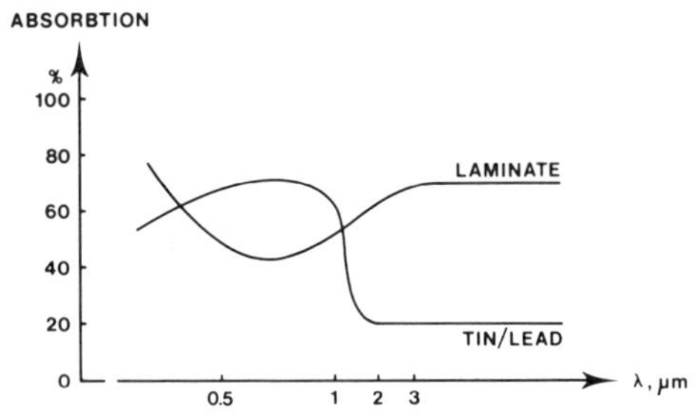

Fig. 8.1. Absorption of infrared radiation as a function of wave-length λ of the radiation, shown for tin/lead plating and for laminate.

8.2.2 Discoloration

Where infrared reflowing is adopted, it can be difficult to control the temperature of the board precisely, one reason being unavoidable variations in the pattern density. Consideration for the PCB designer merits mentioning some of the problems met when the reflow machine has to be adjusted for changing pattern densities.

The wavelength λ of the infrared light depends upon the temperature of the heating lamps, and therefore upon the power supplied to the lamps. As shown in Figure 8.1, the absorption of the infrared radiation depends on the wavelength λ of the radiation and of the material (tin/lead and base material). This implies that a change in the temperature of the heating lamps, and thereby in the wavelength of the radiation, can reduce the coefficient of absorption for tin/lead, and increase it for the base material.

Other important absorption factors are a possible oxidation of the tin/lead layer and the color of the base material. Since very large variations in pattern density can occur on the same side of the board, as well as from side to side, it is no simple matter to adjust the reflow machine to a universal and reasonable optimum setting, so that a broad spectrum of boards can be accepted.

An incorrect setting of the reflow machine can result in a discoloration of the base material and, in worst-case conditions, the start of blistering. If the PCB manufacturer, after auditing the documentation package, does not express any reservations regarding possible faults arising from the reflowing, the customer is fully entitled to expect satisfactory results.

A faint local discoloration is rated as a B-defect and the board is accepted. A widespread discoloration, beginning as light brown, is cause for rejection and rated as an A-defect.

In order to achieve the best possible results, the PCB designer should design the board with the most even pattern density possible. Large ground planes should be relieved as they are usually difficult to reflow without damaging the board, and an uneven pattern should be balanced.

8.2.3 Layer Movement

When reflowing very large boards—for example, computer boards 20 × 20 in. (500 × 500 mm) or larger—a small movement of the individual layers forming the multilayer board can sometimes occur. This is due to an uneven thermal influence on the two sides of the board, usually because of an uneven side-to-side pattern density, but possibly also because of uneven distribution of the infrared radiation.

Usually the layer movement will only be noticeable in the outermost areas of the board—that is, along the edges. Here the holes can be slightly inclined—for example, 2 degrees. This corresponds to a side-to-side movement of approximately 0.002 in. (0.05 mm) in the case of a 0.063 in. (1.6 mm) board.

The fault can be very difficult to detect. A tightly fitting gauge inserted in the hole will not necessarily give a reliable indication because the hole might not be absolutely cylindrical after reflowing. The best way is to prepare a microsection of the hole and measure the movement on a suitably enlarged photo of the microsection.

Generally, the inspector should not inspect the boards for possible layer movements. The effect is seldom important and is addressed here only for the sake of completeness. See also Figure 4.23 of Paragraph 4.5.8.

8.2.4 Surface Resistance

Although the surface resistance has been addressed in Paragraph 4.6.5b, there are reasons to comment on the reduction of the surface resistance caused by reflowing. This particularly applies to infrared reflowing where the surface resistance can be reduced by a factor of 1000 or more. Also, hot-oil reflowing can lead to a decrease in surface resistance, but to a lesser extent. As infrared reflowing now prevails, the inspector should have a certain knowledge of these matters.

Fluxing is always performed prior to reflowing, whether an infrared reflow or a hot-oil reflow system is used. Frequently, flux types in general use can cause a slight softening of the surface. With infrared reflowing, the base material is so highly influenced by temperature effects that the flux encrusts a little on the surface.

In practice, it is very difficult to remove all traces of flux residues after reflowing, and permanent damage to the board surface results. Owing to the fact that flux residues are polar (dissociated into ions) the surface resistance drops, especially after moisture penetration.

Corrosion problems can also arise, and a thorough cleaning should always be performed immediately after reflowing in order to remove as much of the flux residues as possible. Hot-oil reflowing allows a better control of the temperature, for which reason flux encrusting is far less pronounced.

Nearly all PCB manufacturers use infrared reflowing today, and the reduction in surface resistance is generally accepted. It should be noted that where identical boards are supplied by different PCB manufacturers, there can be extremely large differences between the surface resistances. They may vary by as much as a factor of 1000 or more, and the variation is usually caused by different reflowing techniques. The degree of activation of the flux used is fairly decisive and, together with the cleaning technique, plays an important part.

The inspector should occasionally perform comparative surface resistance measurements on boards supplied by different PCB manufacturers. The inspection criteria are detailed in Paragraph 4.6.5b and not repeated here.

8.3 Appearance of the Tin/Lead Surface

In effect, the surface of the reflowed board is also a conclusive indication of the solderability. The inspector can therefore use the appearance as a direct indication of quality. An efficient assessment, however, presumes knowledge of a number of different types of surface defects, as described below.

8.3.1 General Requirements

The reflowed tin/lead layer should appear even and shiny in the same way as a correctly mass-soldered board. The surface must be smooth with a fine-grained and homogeneous structure and should not exhibit a dark or a bluish tinge, which is usually due to unfavorable process conditions.

It should be noted that a shiny tin/lead surface can cause a somewhat reduced adhesion of the solder and insulation masks. Some PCB manufacturers solve the problem by brushing the reflowed tin/lead surface before applying the masks. On such boards, the tin/lead layer will exhibit a greyish color, frequently with traces after brushing. Brushing can lower the surface resistance of the board because metal particles can be worked into the surface of the base material.

Incidentally, the adhesion to the tin/lead surface is of importance on the component side only, as the conditions on the solder side change completely because of the restructuring which takes place in the tin/lead layer when mass soldering the board.

Within the limitations stated below, certain nonfunctional surface defects are accepted. The faults are only permitted on conductors and ground/voltage planes, and not on solder pads or in solder holes.

Gritty surface: 5 percent
Grey surface: 10 percent

The percentages apply to the area of the defective conductor or ground/voltage plane. Such faults are rated as B-defects and do not imply a rejection of the board. In a larger extent—for example, large and unbroken areas with a gritty appearance—this fault is rated as an A-defect and the board is rejected.

On one side of the board small marks of, say 0.05 × 0.10 in. (approximately 1 × 2 mm) or less, are sometimes spread over the tin/lead surface. These unimportant marks originate from the conveyor of the reflow machine.

8.3.2 Freeze Lines

Freeze lines occur frequently and give the tin/lead surface an appearance which, to a certain degree, resembles that of an alligator skin. Owing to the unavoidable locally occurring variations in the alloy composition, some points start to freeze.

As the tin/lead contracts a little during the freezing process, material must necessarily be provided from the surrounding areas still molten. Where such locally freezing areas meet, boundary lines, usually called freeze lines, are created in the form of more or less deep and broad grooves in an irregular pattern. See Figure 8.2.

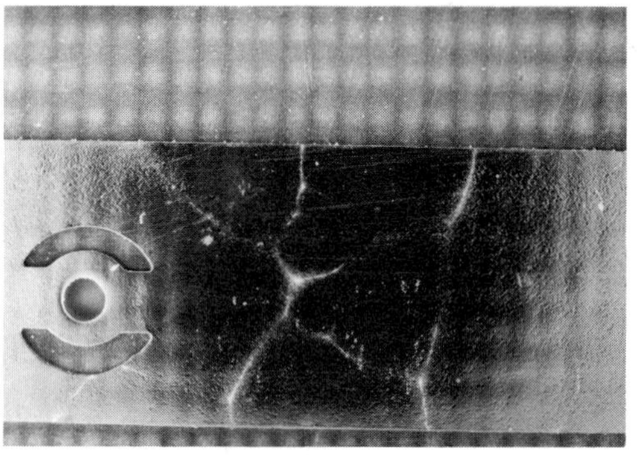

Fig. 8.2A. Board with freeze lines after reflowing.

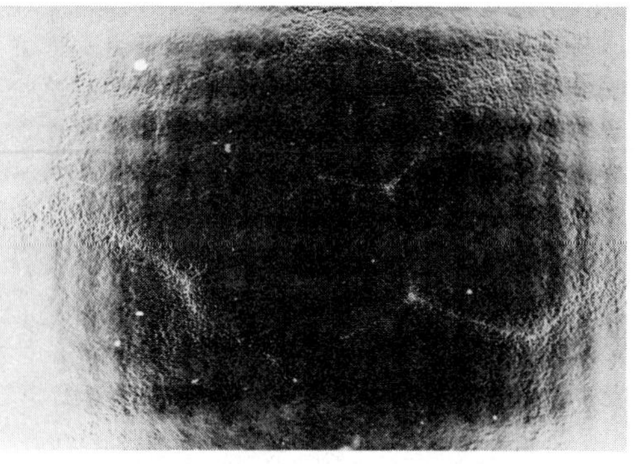

Fig. 8.2B. Close-up of board with freeze lines after reflowing.

Freeze lines are not faults, but merely the natural result of the freezing process in the tin/lead surface after reflowing. This effect is most pronounced on large ground/voltage planes and increases with the layer thickness.

8.3.3 Alloy Composition

Plating of tin/lead is a rather touchy process because of the concurrent deposition of two different materials. One of the greatest problems is to deposit the desired alloy composition all over the board. Most PCB specifications demand an alloy composition of nominally 60 percent tin and 40 percent lead with an accepted variation of the tin content from 50 to 70 percent along the board surface. In some specifications, the accepted variation is only from 55 to 70 percent.

As mentioned earlier, the density of the current field lines in the plating bath strongly affects the alloy composition, and the tin content increases with increasing current density. Generally, the highest tin content occurs at the board edges and the highest lead content at the center of the board.

Modern tin/lead baths with a high throwing power will reduce the variation mentioned above. A good balancing of the pattern is always very advisable to achieve a uniform alloy composition.

The appearance of the tin/lead surface can give the inspector a certain basis for a visual assessment of the alloy composition:

1. The tin content is very close to 60 percent: the surface is very bright with a very fine-grained structure. The freeze lines are very thin and nearly invisible.
2. The tin content is above 60 percent: the surface is still very bright with a very fine-grained structure. The structure changes at a tin content of 75 percent and above, where randomly oriented feather-shaped crystals form.
3. The tin content is below 60 percent: the surface becomes more coarse-grained, and the freeze lines broaden and become grey. The adjoining areas are light grey. At a very low tin content (below 25 percent) the whole surface becomes grey.

The solderability of a tin/lead plated surface does not, however, depend upon the tin/lead composition, whether it tends towards the tin or the lead side. As shown in the tin/lead phase diagram of Figure 8.3, the melting temperature increases with deviation from the eutectic alloy

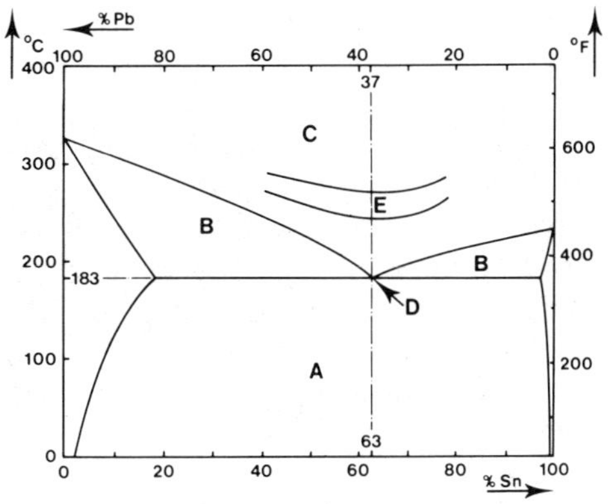

Fig. 8.3. Tin/lead phase diagram.
A: Solid range. B: Pasty range. C: Liquid range. D: Eutectic point. E: Recommended interval of soldering temperature. Sn: Tin Pb: Lead

composition of 63 percent tin and 37 percent lead.

The eutectic point is characterized by the merging of the melting lines and the freezing lines at a temperature of 361°F (183°C). The temperature of the mass soldering machine has to be somewhat above the melting curve, which separates the pasty range and the range of completely molten tin/lead (Figure 8.3). The temperature is frequently kept at 480°F (250°C).

When the tin/lead alloy composition moves outside the range from 50 (or 55) to 70 percent tin, it is necessary to increase the temperature of the mass soldering machine, which is not very practical. This also applies, of course, to the temperature of the reflow machine.

Owing to the broader pasty range, which the solder fillets have to pass through before they freeze, there is an increased risk of cold solder joints and internal stress in the solder fillets. The tin/lead alloy should therefore always lie within the specified range of 50 (or 55) to 70 percent tin.

The PCB manufacturer usually can meet the alloy specification; the above assessment criteria preferably are used when the appearance of the tin/lead surface deviates significantly from the customary appearance.

A chemical analysis of the alloy composition—for example, based on a little tin/lead scraped from a conductor—is normally unnecessary. The experienced operator of the mass soldering machine will soon discover a faulty alloy composition and report it to the incoming inspection department.

8.3.4 Surface Faults

The reflowing process, as well as the preceding plating processes, can result in a variety of faults on the surface of the finished reflowed board. First, we will address some specific types of faults. It is, however, somewhat difficult to define conclusively the effects of the faults on the soldering process, as the importance of the faults naturally depends on the degree to which they occur.

In principle, the tin/lead surface should be bright and smooth, and deviations regarded as abnormalities. The survey given below serves primarily to call the attention of the inspector to certain types of deviations. When in doubt, the inspector can make a solderability test by passing a bare board through the soldering machine and assess the result. See Paragraph 8.7.2.

A dark surface is often due to a contamination of the tin/lead with a foreign metal, frequently copper. The contamination can originate from the tin/lead bath, but it can also be due to the mechanical cleaning of the tin/lead plated surface prior to the reflowing. For example, using the same brushing machine for cleaning copper and tin/lead surfaces can transfer copper to the tin/lead surface. Finally, certain cleaning agents used for cleaning the tin/lead plated surface can react with the flux used in reflowing, and produce darkly stained areas. The inspector should realize that the fault can reduce solderability.

A dark surface can also be due to tin oxides caused by adverse cleaning and neutralization processes. This implies an inclination of the tin/lead surface to oxidize prior to the reflowing. The oxides and the tin/lead can form intermetallic compounds which reduce the effect of reflowing and cause a dark grey surface, possibly with a certain granular appearance and possibly combined with some dewetting (Paragraph 8.7.2).

Large crystal formations of a matte appearance, which makes the surface resemble a galvanized surface, also indicate contamination, frequently of iron. The cause can be a direct contamination of the tin/lead bath or an improper mechanical cleaning; for example, using steel wool.

As mentioned earlier, hot-oil reflow is used for certain applications, and when it is performed manually, there is a risk that the operator will quench the hot board in the cleaning solution so that the tin/lead freezes immediately. This causes a surface structure with a combination of fine crystal formations and internal stress in the solder holes.

A gritty surface appearing after reflowing is frequently due to excessive use of organic additives in the tin/lead plating bath. The additives can be occluded in the tin/lead plated layer, but during reflowing they will gasify and leave tiny craters in the surface, giving it a gritty appearance.

A gritty appearance can also be due to the cleaning of the tin/lead surface prior to reflowing. When the cleaning is performed with abrasives, nonmetallic particles can become embedded in the surface and cause a gritty surface after reflowing. If the abrasive particles occur extensively, dewetting of the surface can occur. See the detailed discussion of dewetting in Paragraph 8.7.2.

8.4 Overhang

Most PCB specifications require that all tin/lead plated boards based on epoxy glass

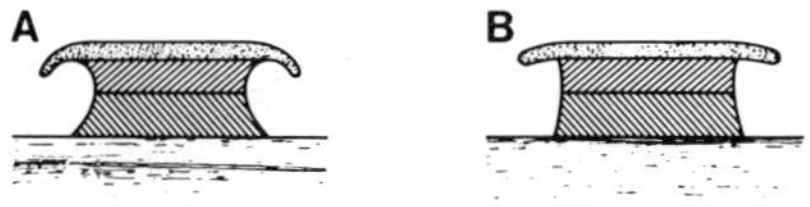

Fig. 8.4. Cross-sectional drawing of conductor after etching. A: Pattern plating. B: Panel plating.

be reflowed to eliminate the tin/lead overhang along the conductor edges or pad rims. An incomplete reflowing caused by an incorrect tin/lead alloy composition or an overly thick tin/lead layer introduces the risk that the overhang will break off, producing slivers.

There are two fundamentally different methods of plating: pattern plating and panel plating. With the first method, only the pattern is plated, first with copper and then with tin/lead. In the second method, the whole panel is plated with copper whereas only the pattern is plated with tin/lead.

During etching, where the tin/lead acts as an etch resist, the conductors ideally develop a cross-section as shown in Figure 8.4. See Paragraphs 3.4.3 and 3.4.4 of Reference A12. Both methods produce an overhang of tin/lead.

The protruding edges, which can be 0.0004 to 0.0008 in. (10 to 20 μm) thick and from 0.001 to 0.002 in. (25 to 50 μm) wide, break loose very easily under the slightest mechanical influence. See Paragraph 5.10 for further details.

Also, when the board is provided with solder and insulation masks, it is very important that the reflowing is performed satisfactorily so that no slivers are embedded in the masks.

8.4.1 Inspecting for Overhang

It is not necessary to perform a routine test, as a reflowing which meets the requirements regarding a bright and homogeneous surface will also melt down any overhang. Only in cases of deviating alloy composition (Paragraph 8.3.3) or surface faults (Paragraph 8.3.4) which cause incomplete reflowing, should the inspector check whether all overhang has been eliminated during reflowing. If this is the case, the edges of all conductors and solder pads should feel smooth and round.

The assessment can be difficult when the board is covered with solder and insulation masks. Where the solder pad rim is not covered by the mask, the inspector can perform a tape test on a number of IC solder pads (Paragraph 5.10) or he can use the scalpel method (Paragraph 5.8.3 and Figure 5.35). A nonreflowed overhang will quickly reveal itself in the form of tiny slivers, which break off along the solder pad rims.

8.4.2 Board Disposition

Nonremoval or unsatisfactory removal of the overhang by reflowing is cause for rejection and consequently rated as an A-defect. A remedy, in the form of a repeated reflowing, is not applicable in the case of boards with solder and insulation masks. The masks will wrinkle or flake off; furthermore, the component identification will flow and become distorted where it is printed across unmasked tin/lead areas.

8.5 Tin/Lead Coverage of Pattern

A very controversial issue is the tin/lead coverage of the pattern after reflowing. This especially applies to the pattern edges and the internal edges of the solder holes—that is, the transition zone between pad and hole. We deal with this matter in more detail below.

8.5.1 Coverage of Pattern Edges

Many specifications for printed circuit boards for military and aerospace applications demand full coverage of the pattern edges after reflowing. Many electronics companies also demand this in their PCB specifications although it makes heavy demands on the PCB manufacturer, affects the surface resistance, and is perhaps completely unnecessary.

It should be noted that in MIL-STD-275 a notable change from the C issue of 1970 to the D issue of 1978 has taken place. It is no longer required that the edges be covered with tin/lead after reflowing (MIL-STD-275D, Paragraph 5.6.4.5).

We will now consider the problems through the eyes of the ordinary customer. On the face of it, one would expect that the overhanging tin/lead, during reflowing, flows down and covers the edges. The first problem is that during etching all edges appear in copper, which quickly oxidizes, and secondly, they can become contaminated with salt residues.

8.5.1a Removing Contamination and Oxides

The first condition for full coverage is that the oxides and the contamination be removed; this requires the use of a fairly activated flux. Unfortunately, as mentioned in Paragraph 8.2.4, the flux residues can become encrusted into the board surface with the risk of decreasing the surface resistance of the finished boards.

Another method is to treat the board with chemical tin (immersion tin) immediately before reflowing. The disadvantage is that tin salts can be left on the surface of the base material, decreasing the surface resistance.

8.5.1b Effects of Profile

The next condition of full coverage of the edges is an acceptable profile; that is, a profile which does not make it too difficult for the melted tin/lead to flow down along

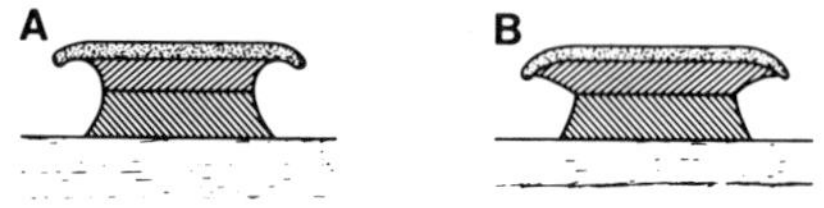

Fig. 8.5. Cross-sectional drawing of pattern plated conductor after etching. A: Favorable profile for edge coverage. B: Unfavorable profile for edge coverage.

the edges. The profile depends on many factors; two factors are the manufacturing method (either pattern plating or panel plating, see Figure 8.4) and the etching technique, which has a large influence on the actual profile.

Regarding the first factor, it is difficult to achieve full coverage of the edges when the board is pattern plated, which most boards are today. The profile should be vertical or slanting outwards, so that the conductors are widest at the base. The inspector should be aware of the difference between the two profiles shown in A and B, Figure 8.5. Profile A implies a greater probability of full coverage than profile B, which also has less tin/lead available for the coverage.

In practice, it can be difficult to continuously achieve edge profiles which are easily covered with reflowed tin/lead. Today, the use of base materials with thinner copper foils—for example, ½ oz. = 0.0007 in. (17.5 μm) or less—has made full coverage somewhat easier to achieve.

8.5.1c Drawbacks of Full Edge Coverage Demands

When the PCB specification explicitly requires full coverage, the PCB manufacturer must add an inspection stage and furthermore, be prepared for a lower yield than normal. This means a price increase, for which reason the customer should carefully consider the need for full coverage. In this connection, it should also be considered that there will always be some coverage of the edges, perhaps 75 percent or more. The problem faced by the PCB manufacturer is to ensure full coverage.

8.5.1d Reasons for Full Edge Coverage

There are two reasons for the full-coverage requirement. It can have been taken, without serious reflection, from a PCB specification for nonplated-through boards with roller tinning or boards with reflowed tin/lead surface plating. (Additional details are presented in Paragraphs 3.9.4 and 4.12.2 of Reference A12.) With these types of boards, the etched edges usually have a profile which ensures full tin/lead coverage. Unfortunately, the difficulties of achieving the same results with plated-through boards have not been considered when preparing the PCB specification.

The other reason can be the desire for full assurance against board corrosion, and in this respect bare unprotected copper is very vulnerable. Most boards, however, are currently provided with solder and insulation masks, and although the masks can leave a certain clearance around the solder pads, all conductors will, on the whole, be covered by the masks. Short circuits created by possible corrosion products will therefore almost never be found.

It should be noted that the most difficult points are where one or two conductors are running between a pair of IC solder pads. As mentioned in Paragraphs 11.3 and 11.4.1, there is a certain risk of a misregistered screen printed mask so that one of the conductor edges is not covered.

For the greatest possible protection against external environmental influences, and also a better assurance of a good electrical insulation between solder pads and conductors, some electronics companies seal the assembled boards with conformal coating. The protective layer can be epoxy, polyurethane, silicone, or polystyrene, to mention a few types, and the layer thickness is frequently from 0.0004 to 0.002 in. (10 to 50 μm).

There are good reasons to bring the problem of mealing to light. This is the designation for a local occurrence of white spots under the solder and/or insulation masks. Moisture in the form of water vapor can always penetrate a mask, either through pinholes or directly (but somewhat slower) by diffusion.

When the mask covers dissociated (polar) flux residues originating from a strongly activated flux, moisture can dissolve the flux residues which can ingress into small cavities beneath the mask (A, Figure 5.22), where they form the white spots mentioned above.

The fault will, in all probability, not be visible at the time of inspection, but in the course of time there is a risk of an insulation problem as mentioned in Paragraph 8.2.4. It should therefore be considered whether a higher degree of reliability, in spite of everything, is achieved by renouncing the requirement of full coverage of the pattern edges, as milder flux types can be used.

The easiest inspection method is to look at one or two boards at a very low angle, with the surface well lighted. Uncovered edges will show themselves by the reddish tinge of the copper. When the board is masked, the inspector will have to do without the conductors and look at the solder pads.

8.5.1e Board Disposition

Whether incomplete tin/lead coverage of the edges should be considered as an A-defect depends on the PCB specification. The board is rejected in the case of an A-defect.

8.5.2 Coverage of Hole Edges

Coverage of the internal edge of the solder hole with a reasonably thick tin/lead

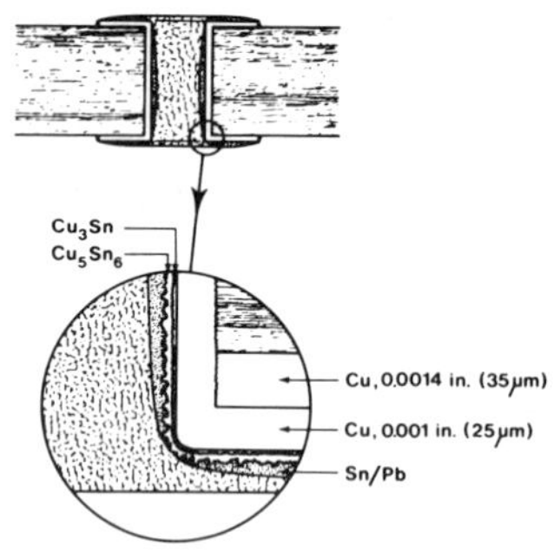

Fig. 8.6. Cross-sectional drawing of plated-through hole. Enlarged section shows conditions at the transition between pad and hole.

layer is a severe problem connected with reflowing; the problem affects all PCB manufacturers regardless of the reflow method and the reflow equipment used. The tin/lead which melts during reflowing will, due to the surface tension, tend to pull away from the edges.

A plated-through hole has the cross-sectional appearance shown in Figure 8.6. The tin/lead gathers in a range in the middle of the solder hole and on the middle of the annular ring, whereas it becomes very thin at the edges.

8.5.2a Problems of Hole Edge Coverage

The most controversial problem is the coverage of the hole edge—that is, the transition zone between pad and hole. Here the thickness of the reflowed tin/lead becomes only 0.00002 to 0.00004 in. (0.5 to 1 μm). Many electronics companies have adopted PCB specifications with requirements for a certain thickness at the hole edge—for example, 0.00004 to 0.00008 in. (1 to 2 μm)—from specifications for boards used in military or aerospace applications.

It should be noted that MIL-STD-275D no longer requires a distinct thickness of the tin/lead layer at the hole edge. In the European Space Agency (ESA) specification PSS-50 (QRM-17E), issue 1, a minimum thickness of 0.00004 to 0.00008 in. (1 to 2 μm) is tolerated as a minor fault.

It turns out to be very difficult for a PCB manufacturer to achieve and guarantee a thickness of, for example, 0.00004 in. (1 μm) at the hole edge with certainty. The PCB manufacturer will, in all probability, perform the reflowing at the lowest possible temperature so that the tin/lead layer just melts with little runoff. This technique implies the risk that local areas with a deviating tin/lead alloy composition (and a higher melting point), do not reflow properly. There is, furthermore, a risk that the full crystalline restructuring does not occur.

8.5.2b Intermetallic Compounds

Actually there is a certain insincerity in specifying the tin/lead thickness at the hole edge, since the thickness of the intermetallic tin/copper compounds is supposed to be included in the total thickness. By diffusion, the intermetallic compounds Cu_3Sn and Cu_5Sn_6 are formed in the transition zone between the tin/lead and the copper with a layer thickness of 0.00002 to 0.00004 in. (0.5 to 1 μm) or more, depending upon the reflow temperature. See Figure 8.6.

These layers are of decisive importance to good solderability, since wetting requires the intermetallic compounds. The problem, however, is that the Cu_5Sn_6 crystals exhibit a very fast crystal growth. Under certain conditions such as prolonged storage at a high temperature, the crystals can penetrate the thin tin/lead layer which characteristically appears at the hole edge. With oxidation of the developed Cu_5Sn_6 crystals, the solderability is substantially reduced.

8.5.2c Effects of Tin Diffusion

Another contribution to the reduction of solderability is that the tin/lead surface

Fig. 8.7. Filling of mass soldered plated-through hole. Solder filling is stopped at the upper hole edge.

becomes tin-depleted because of tin diffusion, and therefore becomes relatively more lead rich. The result is sometimes that, during mass soldering, the solder cannot pass fully up through the hole, but is stopped at the upper hole edge owing to the reduction of the solderability (Figure 8.7).

8.5.2d Effects of Design on Single-Layer Boards

The inspector should note that the reason for not filling the hole completely could also be due to so large a heat transfer from the solder pad on the component side that the temperature through the solder hole drops so much that the solder does not flow. The heat transfer, for example, can be due to an unfortunate design practice, namely to connect very wide conductors to the solder pads without tapering. The PCB designer should be made aware of this problem.

8.5.2e Effects of Design on Multilayer Boards

In multilayer boards, the heat transfer can be due to another type of undesirable design, which allows the plated-through hole to have a good thermal contact all the way around to a ground or voltage plane. The design should be changed so that the hole has contact to a pad which is thermally isolated from the plane but electrically connected to the plane through one or two straps. See Figure 91e of Reference A12.

8.5.2f Realistic Specifications

It should be noted that MIL-STD-275D no longer requires that the solder wet the hole wall entirely and extend outwards completely around the hole, wetting a minimum of 0.005 in. (0.13 mm) outwards from the hole.

For cost reasons, it is generally not advisable to specify any particular layer thickness of tin/lead at the solder hole edge. However, in order to have a certain grasp of the situation, it can be specified that the tin/lead thickness shall be sufficient to ensure full solderability after storage of a specified period, usually 12 months for reflowed boards. In the author's experience, the storage period can, in practice, be considerably longer, and several examples of boards with satisfactory solderability have been observed even after some 30 months of storage.

8.5.2g Inspection Approach

The inspector should not undertake any routine inspection of the degree of tin/lead coverage at the hole edge, as this requires microsectioning. This is a tedious and destructive test, and instead it is suggested that microsectioning is performed in conjunction with the routine examination of test coupons as mentioned earlier.

If, in routine assembly work, serious soldering faults occur, these faults may be traced back to an unsatisfactory coverage of the hole edges. Here the inspector should contact the PCB manufacturer, who may possibly be able to optimize the reflowing process.

8.6 Bridging and Hole Blocking

In Paragraph 5.2.1 we dealt with those types of short circuits in the pattern which are caused by the manufacturing processes preceding reflowing, particularly the image transfer process. In very rare cases the reflowing can cause bridging between very closely spaced parts of the pattern, usually the conductors. The bridging means, however, that the tin/lead has been deposited in an exceptionally thick layer, usually above 0.0008 to 0.001 in. (20 to 25 μm).

In principle, the fault is rated as an A-defect and consequently cause for rejection. On the other hand, it occurs so rarely that it is, in fact, easier to remove a single short circuit than to return the board. The easiest way of removing the bridge is with a soldering iron and a solder sucker or solder wick. When the board is provided with a solder and/or insulation mask, it is easier to use an engraver as shown in Figure 4.36. Subsequently, the mask is repaired by applying a one-pack ink with a brush.

A more common occurrence is the blocking of a few solder holes. The cause is, as mentioned above, an exceptionally thick deposit of tin/lead, frequently in conjunction with a large tin/lead overhang. Owing to the large amount of molten tin/lead, the tin/lead tends to flow down into a solder hole, which thereafter becomes more or less blocked. For further details, see Paragraph 6.8.5, which also describes inspection methods.

8.7 Solderability of Reflowed Boards

Reflowing, which is the equivalent of melting the deposited tin/lead layer, provides, on the whole, the same sort of indication of solderability that will be achieved when mass soldering later. Here we deal with incomplete reflowing and wetting after reflowing.

8.7.1 Incomplete Reflowing

The word reflowing can be interpreted in two ways: First, whether the melting of the tin/lead layer has been uniform over the whole board surface and in all solder holes, and second, whether the melting has reached the full depth of the tin/lead layer itself. We will now study these matters separately.

It is quite simple to inspect the boards for melting over the whole surface, as the appearance must be bright and smooth. In this way, the appearance will contrast with the grey and matte appearance of a nonreflowed board.

Let us now assume that the reflow machine has been adjusted in such a way that, in general, a correct reflowing can be expected. The board receives sufficient heat to melt the tin/lead layer, and furthermore, the margin is sufficient to overcome reasonable variations in the pattern density.

8.7.1a Causes of Incomplete Reflowing

Cases of incomplete melting over the whole surface can only be ascribed to a special cause. The margin is possibly not large enough to accommodate large variations in pattern density, but an experienced operator usually can perform a fairly correct assessment of the pattern density, and adjust the reflow machine accordingly.

The most frequent cause is the presence of an incorrect tin/lead alloy composition in local areas, for which reason a considerably larger amount of heat is necessary. As shown in Figure 8.3, a relatively small deviation of the tin/lead alloy composition can increase the melting temperature so much that such areas do not melt.

The incorrect alloy composition is often caused by an unbalanced pattern or by the occurrence of large unrelieved voltage and ground planes, and in a way, the fault is therefore self-inflicted.

An adverse alloy composition can also be found in the center region of the solder holes although the newer tin/lead baths have been substantially improved in this respect. The alloy composition can be the reason that no reflowing takes place in a

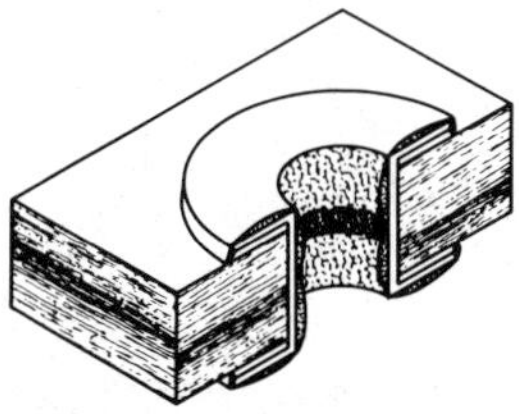

Fig. 8.8. Lack of reflowing in center area of plated-through hole shows up as a ring-shaped area with a diverging structure. The phenomenon is often called a "black hole."

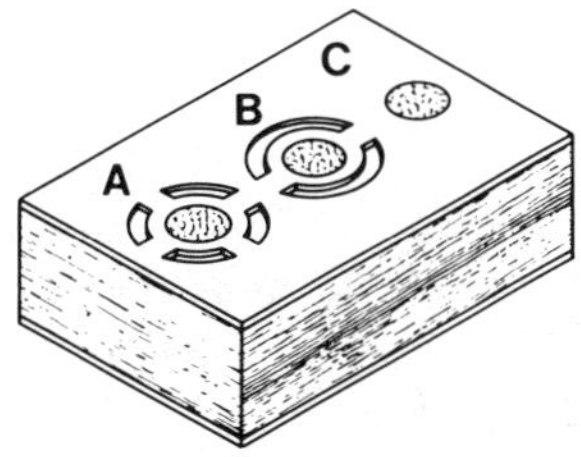

Fig. 8.9. Solder holes in ground or voltage plane. A and B: Solder pads and holes are thermally isolated from plane. C: Solder hole is in thermal contact with plane.

ring-shaped area in some of the solder holes. See Figure 8.8. Such holes are frequently called "black holes."

8.7.1b Board Disposition

Since incomplete reflowing is an unmistakable warning of problems to come when the board is mass soldered at a later stage, the fault is rated as an A-defect and the board rejected. In support of the rejection is the risk that the tin/lead overhang along the edges of the pattern has not been melted and removed from the areas in question.

When the problem is the lack of reflowing of smaller internal areas of a larger voltage or ground plane, the reason is probably a marginal adjustment of the reflow machine. It is a precondition that solder pads and holes which are to be connected to such planes be designed as shown in A or B, Figure 8.9, where the solder pads are thermally isolated from the plane. Complete reflowing internally in the solder holes as well as on the solder pads is to be anticipated, even if the voltage or ground plane suffers from incomplete reflowing.

The fault will in all probability not cause any soldering problems, for which reason it is rated as a B-defect. An exemption is not valid when the solder holes are designed as shown in C, Figure 8.9, as no thermal isolation from the voltage or ground plane occurs.

8.7.1c Inspection Methods

For obvious reasons, it is impossible to inspect visually to what degree the reflowing has been performed in depth of the tin/lead layer. Normally it is not necessary that the inspector performs a special investigation by preparing microsections. It is more expedient to let the examinations of test coupons comprise an assessment of how deep the reflowing has gone, in order to see whether intermetallic tin/copper compounds have formed.

8.7.2 Wetting after Reflowing

Wetting of the reflowed surface is the ability of the molten tin/lead to flow out onto the underlying copper, which is said to be wetted. Since a good wetting is tantamount to a good soldering, it is possible for the inspector to perform a relatively reliable assessment of the solderability just by looking at the reflowed surface.

Good solderability of the board is characterized by a smooth and unbroken surface of the reflowed tin/lead. Owing to the surface tension of the molten tin/lead, a cross section of a conductor or a pad shows a slight convex shape of the tin/lead surface.

8.7.2a Significance of Contact Angle

Where there is a reduced solderability, the surface will appear uneven, and a more

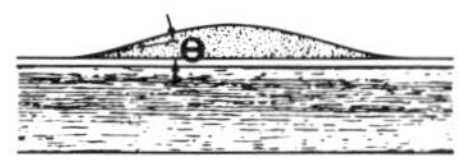

Fig. 8.10. Definition of contact angle θ between solder pad and solder.

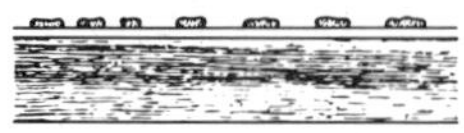

Fig. 8.11. Cross-sectional drawing of area with nonwetting.

or less sharp boundary line between the tin/lead and the copper will be found. We will therefore introduce the term "contact angle θ" which means the angle between the tin/lead and the copper as shown in Figure 8.10.

The smaller the contact angle, the better the tin/lead has flowed and wetted the copper. Satisfactory wetting (or soldering) is now characterized by contact angles between 0 and 45 degrees, whereas poor wetting is characterized by contact angles greater than 45 degrees.

8.7.2b Effect of Surface Contamination

In practice, it is possible to deposit a tin/lead layer on top of a contaminated copper surface. The contamination can be either ordinary dirt, oil, or grease, or it can be due to improper cleaning of the copper surface with abrasives, because abrasive particles easily become embedded in the soft copper.

When the tin/lead layer melts during reflowing—corresponding to the later soldering—the tin/lead cannot wet the copper surface at places where it is contaminated. This implies that formation of the intermetallic tin/copper compounds necessary for the metallurgical bonding does not occur. We now consider the following conditions:

1. Nonwetting: The tin/lead layer exhibits very sharp boundary lines around the contaminated areas and the contact angles will be very large (Figure 8.11). During reflowing, the tin/lead pulls back from the contaminated areas because there is no metallurgical bonding to the copper, which therefore is visible.

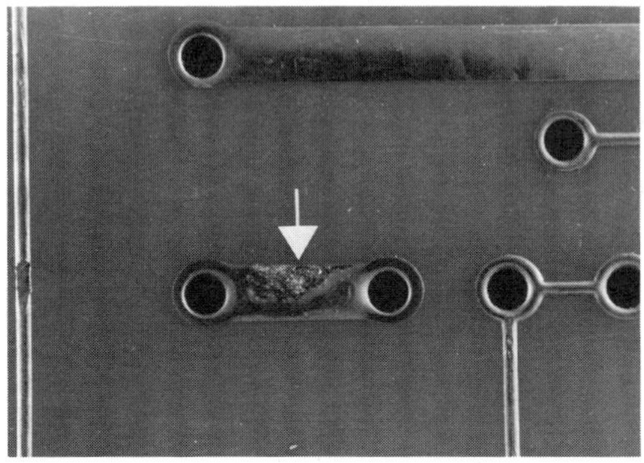

Fig. 8.12. Board with nonwetting after reflowing.

   Nonwetting is rated as an A-defect and the board is rejected. Figure 8.12 shows an example of a board exhibiting nonwetting.

2. Dewetting: Although the tin/lead has wetted the copper surface during reflowing, it has pulled back in the form of round or longish accumulations with large contact angles. See Figure 8.13. At places where the tin/lead has pulled back, a very thin layer of tin/lead is left which covers the copper. Soldering such areas, however, is very difficult.

   Dewetting is not allowed on solder pads or inside the solder holes. Occurrence of dewetting at such places is rated as an A-defect and is a certain cause for rejection.

   Dewetting amounting to 25 percent of the area of a conductor or a voltage/ground plane is accepted

Fig. 8.13. Cross-sectional drawing of area with dewetting.

Fig. 8.14. Board with dewetting after reflowing.

Fig. 8.15. Measuring the thickness of reflowed tin/lead by means of a height gauge.

provided that no soldering is to be performed on the dewetted areas. The fault is rated as a B-defect. Dewetting exceeding the 25-percent figure is an A-defect and the board is rejected. Figure 8.14 shows an example of a board exhibiting dewetting.

8.7.2c Inspection Methods

In the introduction to this chapter it was mentioned that the inspector can perform a relatively simple assessment of the board's solderability on the basis of the appearance of the reflowed surface. There are reasons in favor of mentioning a single exception which is caused by a long-term effect so that it cannot be revealed at the visual inspection.

This problem is the diffusion which takes place in solid metals, also called "solid state diffusion." During prolonged storage under the influence of temperatures above normal room temperatures, intermetallic tin/copper compounds form.

Since the tin diffuses into the copper, the surface becomes relatively lead rich. This is especially pronounced the thinner the layer. To this must be added that the Cu_5Sn_6 crystals can grow out through the thin layer, after which the crystal surfaces oxidize. The outcome is a reduced solderability with the possible occurrence of nonwetting or dewetting.

The inspector cannot be blamed if boards with these solderability problems pass undetected through inspection since two factors are missing at that time: the long-term effect of diffusion and a raised temperature. If the inspector suspects that the reflowed tin/lead layer is extremely thin, he can check his findings by performing a thickness measurement. The easiest way is to scrape the tin/lead layer away until the underlying copper appears. The thickness is then determined with a height gauge which measures the height difference between the bare copper and the untouched tin/lead surface (Figure 8.15). The measurement is not particularly accurate, perhaps within only 10 to 20 percent, but adequate. When the tin/lead layer is only 0.00004 to 0.00012 in. (1 to 3 μm) thick, there is a high risk that solid-state diffusion is present, and the inspector should contact the PCB manufacturer about the matter.

The fact that a very thin tin/lead layer has the effect described above is also known from reflowing by the hydro squeegee method. In this system the tin/lead is melted with hot oil which is sprayed over the board under high pressure. The oil also removes all superfluous tin/lead from the pattern and the solder holes. The result is that the tin/lead layer of the finished board is very thin, perhaps below 0.00004 in. (1 μm). Because such boards suffer from poor solderability after storage, the method is seldom used now.

It may seem somewhat illogical to bring these matters to light and call the solderability of the solder holes in doubt. In particular, solder holes are characterized by having, after reflowing, a very thin tin/lead layer at the hole edge. The author is convinced that many cases of the failure to fill a solder hole up to, or beyond, the upper edge of a plated-through hole is actually due to the effect described above. However, owing to the fact that the remaining part of the solder hole shows a good wetting, the PCB user has to accept an absence of total filling of the solder holes as a blemish of no functional importance.

Chapter 9

Solderability Preservation Inspection

9.1 Introduction

Although pure copper has excellent solderability, oxidation quickly reduces the solderability to the point where the board becomes unacceptable. A strongly activated (aggressive) flux can remove the oxide layer if it is not too thick or if the soldering time is long enough. Unfortunately, such fluxes can cause a long-term degradation of the surface resistance, even after a very careful cleaning. Paragraph 8.2.4 discusses surface resistance.

While milder fluxes are generally preferred, they require the PCB manufacturer to suitably coat the board surface to prevent—or at least delay—environmental effects on board solderability. This chapter deals with various types of surface protection:

- Tin/lead plating with reflowing
- Selective tin/lead plating with reflowing
- Solder coating with hot-air leveling
- Selective solder coating with hot-air leveling
- Roller tinning
- Organic coatings
- Miscellaneous protective coatings

9.2 Tin/Lead Plating with Reflowing

The most common surface treatment of plated-through boards is tin/lead plating. The subsequent reflow process serves, among other things, to remove overhang and to allow a longer storage time due to the restructuring of the tin/lead layer. Chapter 8 dealt with this topic.

Frequently, a distinction is made between thick and thin reflowing (or fusing) with the transition thickness at 0.0003 in. (7.5 μm). The predominant methods of thick reflowing—when the tin/lead plating thickness allows—are infrared heating and hot-oil immersion; thin reflowing is achieved with the hydrosqueegee method described in Paragraph 8.7.2c.

The latter method is seldom used now because the resulting layer is extremely thin, frequently less than 0.00004 in. (1 μm). The thinness makes it difficult for the inspector to assess the solderability of the board, particularly the tendency to dewetting. The entire tin/lead layer is so thin that it is impossible for the inspector to see any variations in the tin/lead thickness, meaning that he cannot detect any dewetted areas.

Chapter 8 describes the inspection criteria for thick reflowed boards.

9.3 Selective Tin/Lead Plating with Reflowing

A severe problem with ordinary tin/lead plated boards is the presence of tin/lead on all conductors, which implies that the overlying solder mask will wrinkle during mass soldering. The PCB manufacturer can solve this problem by changing the manufacturing process so that tin/lead is plated only on the solder pads and in the solder holes.

9.3.1 Selective Plating Process

The first steps of the process are exactly the same as used when manufacturing ordinary tin/lead plated boards; that is, drilling of holes, electroless deposition of copper, and application of plating resist with subsequent deposition of electrolytic copper on the pattern including the solder holes (pattern plating). After this stage the process changes a little as a 0.0002 to 0.0004 in. (5 to 10 μm) thick tin/nickel layer is applied. The alloy composition is 65 percent tin and 35 percent nickel.

A further plating resist is now applied, leaving the solder pads open but covering the conductors and other areas such as ground planes. Tin/lead is then deposited in the normal manner, after which the board follows the normal manufacturing processes: removal of plating resist, etching, reflowing, and application of solder and insulation masks. By using an alkaline ammonia etchant, not much overhang should occur, which is very fortunate as a tin/nickel overhang cannot be removed by reflowing.

9.3.2 Advantages of Selective Plating

The additional processing makes boards with a selective tin/lead plating more expensive than ordinary tin/lead plated boards. In return, they offer the following advantages:

1. The solder mask does not wrinkle or flake off as the mass soldering process does not affect the tin/nickel surface.
2. The solder pads can better resist heavy scratching because of the relatively hard tin/nickel layer.
3. The plated-through holes are strengthened because of the tin/nickel deposit which also opposes possible tendencies to create blowholes in the solder fillets.
4. The board is not subjected to thermal shocks as when solder-coated and hot-air leveled (Paragraph 9.4).

Relevant criteria for inspection of ordinary tin/lead plated and reflowed boards are also valid for boards selectively plated with tin/lead and a hard tin/nickel intermediate layer.

9.4 Solder Coating with Hot-Air Leveling

This method has gained a strong foothold over the last few years. The board does not need to be tin/lead plated, but may very well be a bare copper board, as the solder-coating machine has its own solder pot with eutectic tin/lead, which means an alloy composition of 63 percent tin and 37 percent lead. See Figure 9.1.

9.4.1 The Leveling Process

After pickling (light etching), the board enters the machine for fluxing and pre-

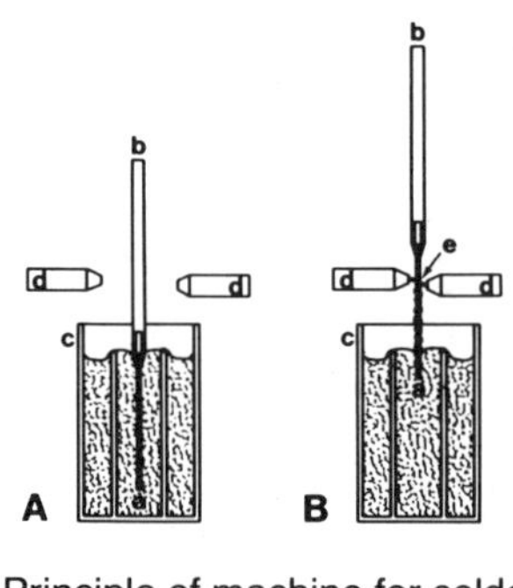

Fig. 9.1. Principle of machine for solder-coating and hot-air leveling. A: Board immersed in solder pot. B: Board withdrawn and leveled. a = board. b = clamp fixture. c = solder pot. d = air knife. e = hot air jet.

heating. The subsequent few seconds of immersion in the solder pot completely wets the copper, including the pattern edges. This occurs because the coverage, unlike that on ordinary reflowed boards, does not depend on the amount of tin/lead on the surface, and because the edges are in direct contact with the molten tin/lead.

It should be noted that the D issue of MIL-STD-275 states that the conductors of solder-coated boards shall be completely wetted with tin/lead; this presumably includes the edges.

During withdrawal from the solder pot the board passes between a pair of hot air jets called air-knives. The air-knives level the tin/lead on both sides of the board by removing all excess tin/lead.

Some machines reorient the board so that it cools in a horizontal position. This method prevents the variation in tin/lead distribution (also called solder sagging) which always occurs across a solder hole when the board cools in a vertical position, letting gravity pull the still-molten tin/lead downwards. This phenomenon also occurs during hot-oil reflowing. On the other hand, when the board cools horizontally, the solder sags from top to bottom in the hole. This causes a somewhat heavier deposit at the bottom of the solder hole and the associated pad.

The diameter tolerance of the insertion holes plays a decisive part in boards for automatic component insertion; see Paragraph 10.3.6 of Reference A12 of the Reading List. An advantage when the component side is turned upwards during cooling is that the diameter of the insertion hole will be slightly larger on the component side and will facilitate the insertion of the component lead.

Incidentally, a similar condition occurs in the case of infrared reflowing. The PCB designer should therefore state on the master drawing that the component side should be turned upwards during the cooling and indicate the component side, so that the operator can identify it easily.

After cooling, the board is automatically passed to a cleaning station where the flux residues are cleaned off.

When the machine is properly adjusted, there is no perceptible difference between solder holes which have been infrared reflowed or hot-air leveled. See Figure 7.2. The inspection criteria are therefore similar to those for reflowed boards.

9.4.2 Hot-Air Leveling Problems

Some PCB manufacturers claim that the solder coating method yields a greater hole-to-hole uniformity than does the ordinary tin/lead plating with reflowing. This applies, however, only to the tin/lead alloy composition which naturally will be the alloy composition of the solder pot.

Otherwise, the statement should be viewed skeptically because both adjustment and maintenance of the machine are fairly critical. There are many adjustments, and the following discussion calls the inspector's attention to possible faults.

During his inspection visits to the PCB manufacturer, the inspector should satisfy himself that ongoing maintenance routines for the machine have been established.

9.4.2a Dwell Time in Solder Pot

The dwell time shall be appropriate for the board characteristics. If it is too long, blistering can occur (Paragraph 4.5.3) and if it is too short, dewetting can take place because the flux has insufficient time to remove possible oxides.

The optimum dwell time depends to a certain degree on the nature of the board—among other things, the ratio between the copper area and the board area, and also the board thickness. In addition, the dwell time depends on whether the board is an ordinary plated-through board or a multilayer board.

The inspector should realize that even if preheated, the board receives a certain thermal shock when immersed in the solder pot. If the copper in the through-platings is not sufficiently ductile, barrel cracking or corner cracking can occur.

9.4.2b Speed of Withdrawal

After solder coating, the board passes between air-knives where the speed of withdrawal plays an important part. Insufficient speed causes excessive removal of tin/lead, and under worst-case conditions only a very thin coating, similar to that of hydrosqueegee-treated boards, is left. As a matter of fact, this implies the formation of intermetallic tin/copper compounds which can create solderability problems later. Excessive speed leaves too much tin/lead on the board with a risk of blocked holes.

A manufacturer of solder coating and hot-air leveling equipment warrants that no more than 0.1 percent of the holes will be blocked as long as the ratio between the board thickness and the hole diameter is kept less than 2. This warranty assumes that the machine is adjusted correctly.

9.4.2c Adjustment of Air-Knives

All the air-knife adjustments have to be correct. This applies particularly to the air pressure, the clearance between the air-knives and the board, and the angle of the air-knives with respect to the board. Incorrect adjustments can cause the following faults:

A thin tin/lead layer with the risk of solderability problems due to the formation of intermetallic tin/copper compounds.

A thick tin/lead layer with some solder holes too small, or even totally closed. This can occur in bands across the board, specifically when one of the air-knives becomes partially blocked by tin/lead splatter.

Front-to-back variations in tin/lead thickness of the board.

9.4.2d Webbing

When the solder pot level is too low, flux can accumulate on the surface of the molten tin/lead. The flux becomes tar-like from evaporation of the solvent and threatens to contaminate the board surface. The flux can become intermixed with the tin/lead which is circulated just as in a wave soldering machine, or it can stick to the surface when the board is withdrawn from the solder pot. Small tin/lead drops can form during the leveling, and when these drops are trapped by the sticky, tar-like flux, they can possibly solder to the pattern and create short circuits.

9.4.2e Orientation of Production Panel

As mentioned earlier, the molten tin/lead tends to run downwards. If the board has a tin/lead coated edge connector, the edge connector should not turn downwards in the machine because solder sagging would increase the thickness of the tin/lead layer so that insertion of the board in the receptacle is hampered or even impossible.

In such cases where the edge connector is provided for testing, the excess amount of tin/lead is scraped off and possibly deposited in the receptacle, which can cause short circuits later.

9.5 Selective Solder-Coating with Hot-Air Leveling

A solder mask applied over tin/lead plated areas will wrinkle somewhat after mass soldering. The reason is that the tin/lead melts under the solder mask and flows away, and as the solder mask becomes plastic from the heat, it tends to follow the solidified tin/lead surface. Chapters 11 and 12 deal with solder and insulation masks.

In order to avoid wrinkling of the solder mask, many PCB users prefer to buy boards without tin/lead on conductors or larger areas such as ground and voltage planes.

Wrinkling or flaking of the mask on wide conductors can also be avoided by omitting the mask on such conductors. This allows tin/lead to build up during mass soldering, and since wide conductors are usually provided to carry a high current, this buildup is an advantage.

The selective solder coating is based on covering the pattern—except the solder pads—with solder and insulation masks which are applied directly on the copper surface, either as screen-printed masks or as dry or liquid film masks. Subsequently the board is solder coated and hot-air leveled as described in Paragraph 9.4.

No tin/lead shall be found on the plated-through board prior to application of the masks. The problem is therefore to manufacture plated-through boards with copper plating only, since etching of the pattern is not allowed to attack the copper plating in the solder holes. Fortunately, there are several methods with which to accomplish this. The methods are described below, so that the inspector can acquire a reasonably broad knowledge.

There can be many different causes of defective hole platings, and the inspector should take care when inspecting selectively solder-coated boards. The inspection criteria for tin/lead plated and reflowed boards also apply in the case of solder-coated and hot-air leveled boards.

9.5.1 Tenting Technique

The first step after copper plating is to apply an etch resist in the form of a dry film which protects the pattern against etching. The dry film tents all solder holes and via holes, which means that they are sealed against intrusion of etchant and therefore protected against etching. See Figure 9.2.

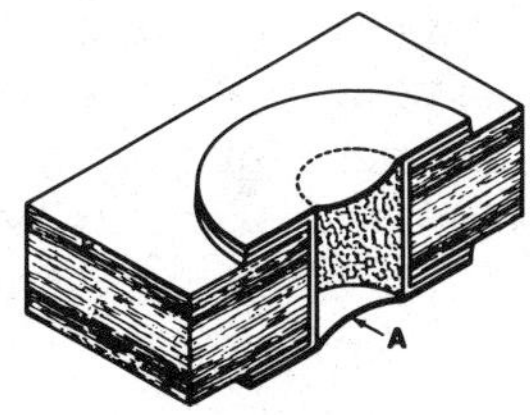

Fig. 9.2. Copper-plated hole tented with dry film. A: Dry film.

After all unwanted copper has been etched away, the dry film is removed. The board now has the pattern and the plated-through holes in bare copper, and is ready for the application of solder and insulation masks, followed by selective solder coating and hot-air leveling of solder holes and solder pads.

As the dry film has to seal the holes against etchant intrusion, it has to adhere extremely well and be absolutely tight. The first condition requires that the annular rings be reasonably wide. The PCB designer should determine the external diameter D of the solder pad relative to the nominal diameter d of the solder hole according to the expression

$$D \geqslant 1.4\,(d + 0.004)\text{ in.}$$
$$(D \geqslant 1.4\,(d + 0.1)\text{ mm})$$

This can be converted into the following expression which indicates the necessary width w of the annular ring:

$$w \geqslant 0.2d + 0.0028\text{ in.}$$
$$(w \geqslant 0.2d + 0.07\text{ mm})$$

In case of a 0.031 in. (0.8 mm) solder hole, the minimum width of the annular ring is 0.009 in. (0.23 mm), which is less than the normal design value of 0.01 in. (0.25 mm) for photoplotted filmwork, and 0.012 in. (0.30 mm) for manual artwork. Incidentally, oblong or square pads favor adhesion of the tenting dry film to the pads.

The holes to be tented should not be too large. The usual type of dry film can tent holes with diameters up to 0.16 in. (4 mm), but a somewhat thicker and more expensive dry film can tent holes with diameters up to approximately 0.25 in. (6.3 mm).

The tenting dry film shall be tight, with no holes through which etchant can be forced into the solder hole and attack the copper plating. It is therefore assumed that there are no pores and that the imaging is perfect.

Possible center holes of the solder pads on the master film drawing shall be closed. An overlooked center hole means a hole in the tenting dry film and therefore a defective through-plating. Finally, the hole drilling shall leave no burrs, and the registration of the pattern relative to the hole drilling shall be precise enough to prevent any broken annular rings.

9.5.2 Temporary Tin/Lead Resist Technique

Another method is to manufacture a tin/lead plated board, but with a much thinner tin/lead layer than usual, frequently only 0.00012 in. (3 μm) thick. With a special etching technique and a nonporous tin/lead deposit it is possible to etch the pattern without attacking the copper located beneath the tin/lead layer. When the tin/lead layer is stripped, the pattern and the plated-through holes appear in bare copper.

It is extremely important that the tin/lead etch resist be nonporous and that the etching machine have a short etching time. Otherwise voids in the copper plating of the solder holes can occur. A combination of solder and insulation masks and lacquering (Paragraph 9.7) can protect the copper from oxidation.

9.5.3 Differential Etching

Differential etching starts with a base material clad with ultrathin copper foil; for example ⅛ oz. = 0.00018 in. (5 μm) thick. Following a quite normal copper plating of holes and pattern, a differential etching is performed, in which 0.00024 in. (6 μm) of copper is etched away in a very carefully controlled process. The pattern appears after all unwanted copper has been etched away.

The pattern and plated-through holes also lose 0.00024 in. (6 μm) copper, but this is compensated for during the copper plating. The result is a board with the pattern and the plated-through holes showing in bare copper.

Many PCB manufacturers do not care to use a base material clad with ultrathin copper foil, and prefer the two previous methods mainly because of problems with the adhesion of the copper foil. Added to the fear of reduced adhesion is a fear that the ultrathin copper foil is not always completely impermeable, so that epoxy can bleed through and become deposited on the surface during the laminating process.

Finally, the etching process has to be controlled extremely carefully in order not to remove too much copper. Otherwise, an undesirable undercut of conductors can result in heavily reduced adhesion of very narrow conductors.

If the etching is inadequate, copper traces can be found on the surface so that the insulation resistance is reduced (Paragraph 4.6.3). However, it can take quite a lot of over-etching to guarantee removal of all such minute traces of copper. A differential etching is therefore something of a balancing act and perhaps not so attractive as it might appear, thus the inspector should be aware of the possibility of the defect types mentioned above.

9.6 Roller Tinning

One way of protecting the surface of non-plated-through bare-copper boards is roller tinning. Another is board lacquering, described in Paragraph 9.7.

The roller tinning machine is relatively simple in its construction. In principle, it consists of two rollers, the lower one running in a bath of molten solder, usually with a tin/lead alloy composition of 63/37 or 60/40. The fluxed board is fed in between the two rollers, and the lower roller deposits a layer of tin/lead on the copper areas of the board. Although this process seems quite simple, the inspector should be aware that it can cause some problems.

Succeeding paragraphs address the following problems:

- Surface appearance
- Layer thickness
- Formation of drops
- Insertion holes
- Coverage of pattern areas
- Lifted copper
- Copper contamination
- Dross
- Solderability

9.6.1 Surface Appearance

Many requirements regarding the appearance of the reflowed board can, as a matter of course, be applied on roller tinned boards. The reader is therefore referred to those specifications stated in Paragraphs 8.3.1, 8.3.2, and 8.3.4 which do not imply a tin/lead plating and the involved processes.

The alloy composition of the deposited layer will naturally be the same over the whole board surface, and equals the alloy composition of 63/37 or 60/40 of the tin/lead bath in the roller tinning machine.

9.6.2 Layer Thickness

Here we distinguish between thin and thick roller tinning. Formerly it was only possible to achieve a very limited layer thickness of 0.00004 to 0.0001 in (1 to 2.5 μm), which was inadequate to ensure good long-term solderability. Newer roller tinning machines can apply a substantially thicker layer of 0.0006 to 0.001 in. (15 to 25 μm) because the lower roller is no longer smooth, but provided with grooves.

The layer thickness, however, does not depend entirely on the surface and pressure of the rollers, the speed of travel, and the temperature, but also on the pattern itself. In general, large areas such as ground and voltage planes will develop a rather limited thickness, whereas small areas such as solder pads and narrow conductors will achieve a somewhat greater thickness. It seems fairly difficult to indicate values of the layer thickness as these vary considerably from one PCB manufacturer to the next, depending on the machine type being used. The survey given below should therefore be taken with some reservations.

Thin Roller Tinning

Solder pads and narrow conductors:	0.00016 in. (4 μm)
Wide conductors:	0.00008 in. (2 μm)
Ground and voltage planes:	0.00002 in. to 0.00004 in. (0.5 to 1 μm)

Thick Roller Tinning

Solder pads and narrow conductors:	0.0006 to 0.001 in. (15 to 25 μm)
Wide conductors:	0.0002 to 0.0004 in. (5 to 10 μm)
Ground and voltage planes:	0.00012to0.0002in. (3 to 5 μm)

In principle, the inspector should check whether the layer thickness stated in the PCB specification is met. Performing an accurate measurement can be rather tedious, but a useful measurement can be made by scraping the tin/lead layer off until the underlying copper just appears. The thickness is now determined by measuring the height difference between the bare copper and the untouched tin/lead surface. It is most practical to use a height gauge as shown in Figure 8.15. The measuring accuracy is about 10 to 20 percent, which is quite adequate.

In practice, it is usually unnecessary to inspect the layer thickness. The decisive factor is the solderability of the board, and in that respect the inspector can rely on prompt feedback from the assembly line if there are problems with mass soldering. See the remarks at the end of Paragraph 8.7.2 regarding solid state diffusion in conjunction with very thin tin/lead layers.

Incorrect layer thickness is, in principle, an A-defect, but practical experience regarding the solderability of the board should control the decision to return the boards for a repeated roller tinning.

9.6.3 Formation of Drops

It is more or less unavoidable that small, flat tin/lead drops form at the trailing edges of the pattern as the board leaves the rollers. See Figure 9.3. The height of these drops should not exceed 0.020 in. (0.5 mm), and exceeding this limit should be rated as a B-defect. The measurement can be made by means of a height gauge as shown in Figure 8.15, or by means of a micrometer gauge.

When an edge connector needs only a tin/lead coating—for example, when used for connecting test equipment—no drops should form on the connector tabs. Drops could not only cause the contacts in the receptacle to be overstrained, but introduce the risk of tin/lead scrapings entering the receptacle housing.

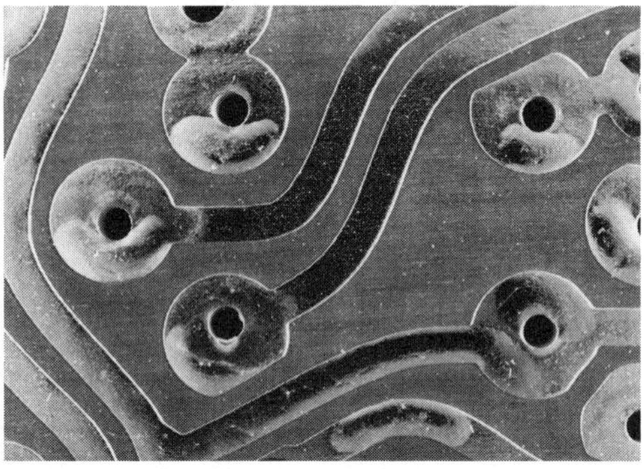

Fig. 9.3 Roller-tinned board with solder drops.

Drop formations on the connector tabs can be avoided if the edge connector passes through the rollers first. In this way, possible drops will form outside the effective contact area of the tabs. Violating this requirement is rated as an A-defect, and the board is returned for a repeated roller tinning.

9.6.4 Insertion Holes

Depending on the manufacturing method, one of the following problems may show up on the board:

1. Some insertion holes (solder holes with no through-plating) have been closed during roller tinning by a thin layer of tin/lead being drawn across the hole in the solder pads.
2. The solder pads are not fully solderable at the internal hole edge.

In the first case, all holes were drilled prior to roller tinning, introducing a certain risk of closing a number of holes, perhaps 5 percent, during the roller tinning. Closed holes should be opened, either by using an awl or by desoldering with a soldering iron.

Opening closed holes is a laborious and expensive process which also implies the risk of missing some of the holes. Therefore some PCB manufacturers place a piece of alcohol-wetted foam rubber on the side of the board not to be roller tinned, and pass both through the roller tinning machine. The vapor pressure arising from the alcohol opposes hole closure.

Overlooked closed holes are accepted in a proportion of 2 holes per 16 sq. in. (100 cm^2), and the fault is rated as a B-defect as the holes are easily opened. A larger number of closed holes is rated as an A-defect, and the board is repaired or returned depending on the extent and immediate need.

In the second case all holes are drilled after roller tinning. This implies certain drilling problems, partly because the boards are not quite plane because of the drop formation mentioned in Paragraph 9.6.3, and partly because the drill flutes can become clogged with tin/lead chips.

The bare copper ring which appears where the copper foil is drilled through is undesirable. During storage of the board, these rings can oxidize so heavily that a satisfactory mass soldering is possible only with a relatively strong flux. As mentioned earlier, this can reduce the surface resistance and introduces a risk of later corrosion problems.

When the purchase specification requires all holes to be drilled prior to roller tinning, the inspector shall verify that this has been done. The inspection is best carried out by means of a 7X or 10X magnifier, and the criterion is that tin/lead shall be present on the internal edge of the solder pad.

In principle, failure to meet the above requirement is cause for rejection, and consequently the fault is rated as an A-defect. Actual practice, however, is more lenient.

Provided no problems concerning solderability, surface resistance and corrosion have been found in cases where the boards were drilled after roller tinning, it is to be assumed that the fluxing is strong enough to clean the copper, but on the other hand sufficiently mild so as not to create corrosion problems. In such cases it seems irrational to reject fully usable boards.

When in doubt, the inspector can arrange a trial soldering by passing one of the boards through the mass soldering machine. In order to accelerate the oxidizing process he can bake the board for 1 or 2 hours at 230°F (110°C) before soldering.

Over a longer period, it may happen that boards delivered from certain PCB manufacturers have not been manufactured according to the purchase specification regarding hole drilling before roller tinning. If such boards can be mass soldered with satisfactory results, the logical response would be to eliminate the requirement from the purchase specification, at least for the PCB manufacturer in question.

9.6.5 Coverage of Pattern Edges

Although the pattern edges retain a favorable profile during etching, it is not certain that roller tinning will fully cover the edges. One reason is that the amount of tin/lead available is rather limited, particularly in a thin roller-tinning machine, and another reason is that the flux should not be too strongly activated.

As a matter of fact, we find the same conditions as described in Paragraph 8.5.1 for infrared reflowing. PCB specifications frequently do not require coverage of the pattern edges; therefore, the inspector does not have to inspect the edges.

It should be noted that the above also applies to the internal hole edge of the solder pads. These edges are usually reasonably well-covered, but the PCB manufacturer cannot guarantee full coverage.

9.6.6 Lifted Copper

In Paragraph 5.8 we dealt with lifted solder pads and conductors as well as the corresponding inspection. In the case of roller tinned boards, however, the inspector should particularly be on the alert for lifted copper. Not only is roller tinning a rather harsh process, but the adhesion of the copper to the base material is reduced as long as the board is hot.

If the board is handled carelessly when it leaves the roller tinning machine, there is a risk that narrow conductors will lift. This increases to an alarming degree if the roller tinning did not succeed fully the first time, and is repeated on the still-hot board. Figure 5.30 also applies to roller tinning. Lifted conductors and solder pads are an indisputable cause for rejection and the fault is rated as an A-defect.

9.6.7 Copper Contamination

If the tin/lead in the roller tinning machine is contaminated with more than 0.5 percent copper, the tin/copper eutecticum can form and the inspector can expect solderability problems with the finished boards.

The cause of this contamination is a dissolving of copper from the copper surface of the boards. Such contamination is characterized by a darkened appearance of the finished board. In such cases a solderability test should be performed. This is most easily done by passing a board through the mass soldering machine. Reduced solderability is cause for rejection, and the fault is rated as an A-defect.

9.6.8 Dross

Dross—the tin and lead oxides occurring on the surface of a bath of molten solder—also forms in the roller tinning machine. Unless an antidross agent is added, there is a risk that dross will be deposited on the roller-tinned board. This is an unquestioned cause for rejection and rated as an A-defect.

9.6.9 Solderability

The surface of roller-tinned boards must be fully solderable; the reader is referred to the inspection criteria stated in Paragraph 8.7.2. The inspector should be aware of the fact that in many cases a tendency to dewetting does not turn up until the roller-tinned board has been soldered. Dewetting is therefore not so easy to demonstrate as in the case of reflowed boards.

A contaminated copper surface should normally be suspected as the cause of dewetting; it frequently results from inadequate cleaning by sanding before roller tinning. The failure mechanism is exactly as described in Paragraph 8.7.2 for reflowed boards. The difference is that in case of reflowed boards the PCB manufacturer will notice possible faults as soon as the board is reflowed, whereas in the case of roller-tinned boards, he cannot see any fault.

The mechanical action during sanding can cause particles to become embedded in the soft copper and covered by a very thin layer of copper. In time, copper forms an intermetallic compound with tin, and during soldering the tin/lead will pull back from the particles since there is no longer any copper to wet.

9.7 Organic Coatings

A common means of surface protection for bare-copper boards is lacquering. This can also be used on plated-through boards when the manufacturer does not want a metallic protective coating such as a tin/lead plating or a solder coating of the copper in the plated-through holes. The purpose of this is to ensure a tighter tolerance on diameter, which can become important for satisfactory automatic component insertion.

The inspector can only inspect the surface protection by performing a solderaability test, most easily done by passing the board through the soldering machine. Usually this is not necessary in the case of deliveries from a known and approved PCB manufacturer.

In conjunction with the preservation of the good natural solderability of copper there are some problems which the inspector should be familiar with. The following paragraphs address these problems.

9.7.1 Cleaning

The copper surface must be totally cleaned of impurities and oxides before the protective lacquer is applied.

Although many PCB manufacturers take care to clean the surface before applying the protective lacquer, the copper surface can oxidize slightly again. The reason is frequently that the process includes a short drying period in which the wet boards are dried with hot air. Added to this is the possibility of an extra waiting period, if there is a long wait at the application station.

9.7.2 Solubility

The protective lacquer shall be soluble during the soldering process, particularly in the case of mass soldering where the soldering time is short, such as 3 seconds. Therefore it is necessary that the flux, aided by board preheating, can dissolve and remove the protective lacquer so that the flux has sufficient time to clean the copper surface within the soldering time.

A slow oxidation will inevitably take place even through the best protective lacquer, thus the storage time of lacquer-coated boards is rather short, frequently only 6 to 12 weeks, although a storage time of 6 months has been reported.

9.7.3 Types of Coating

A fairly common type of lacquer is the rosin-based lacquer which is applied by spraying, dipping, or rolling. In the case of mass soldering it is extremely important that the coating can be dissolved and removed by the flux in conjunction with adequate preheating of the board. Only in this way can the flux contact the copper surface and clean it so that a good wetting can take place during soldering.

To avoid the above drying period and the attendant risk of oxidation, some PCB manufacturers use a water-dip lacquer, which is actually a water-displacing lacquer. Unfortunately, this type of lacquer becomes so hard after storage for some time that it is very difficult to dissolve during mass soldering and therefore causes unsatisfactory wetting.

The cause is an unavoidable polymerization of the lacquer—transformation of part of the lacquer to long-chained molecules—which shows up as a white adhering powder after soldering.

In order to avoid such problems the PCB specification can specify a combined process consisting of dipping in a water-dip lacquer to displace the water film, followed by dipping in a compatible rosin-based lacquer. It is of fundamental importance that the board be transferred to the vat with the water-dip lacquer immediately after pickling (light etching) and rinsing. It is just as important that the board remain there until it can be transferred to the vat with the rosin-based lacquer.

9.8 Miscellaneous Types of Surface Protection

The preceding methods of surface protection of the board are commonly used within the electronics industry. There are, however, some other methods, but due to the fact that they are only used to a very limited extent, we deal very briefly with them. No inspection criteria are given here.

9.8.1 Tin Plating

Boards are seldom tin plated today. The main objection is that the boards cannot be reflowed. This is because the melting temperature is so high that the base material can be damaged by reflowing, which implies discoloration, cracking in the plated-through holes, and warp and twist.

The greatest disadvantage of the missing reflowing is the risk of broken-off overhang as described in Paragraph 8.4, especially in areas where the plating has been carried out with a high density of current field lines in the plating bath. This implies a more brittle deposit with a greater risk of breaking off the overhang.

On the other hand, the metallurgical restructuring is not so necessary as the tin oxide formed on the surface protects against corrosion and retains the solderability. Yet another disadvantage is the formation of whiskers, which is dealt with in Paragraph 5.2.2.

Formerly, tin plating was very popular in Germany and to a certain degree in France. However, there recently has been a very marked tendency towards tin/lead plating, which is the preferred type of coating in English-speaking countries.

9.8.2 Gold Plating

Commercial boards are very seldom completely gold plated, and with the current gold prices its use will probably become even rarer. In the case of boards for very special applications—for example, space technology—the entire pattern can be gold plated, but in such cases the inspection criteria will be stated in the purchase specification.

Gold is most frequently used for selective plating of edge connectors. The reader is referred to the detailed discussion in Chapter 14. Another common application is gold plating of strip-line boards requiring special precautions to achieve demanding high-frequency properties.

It is appropriate to caution against soldering on gold-plated areas such as solder pads and plated-through holes if the plating thickness exceeds 0.00008 in. (2 μm). The solder will very quickly dissolve the gold and form an intermetallic gold/tin compound, AuSn. The solder fillet will be weakened or lose all strength at a gold concentration of approximately 20 percent. Since soldering is done quickly, there is not time enough for the gold to diffuse right into the solder fillet, causing the dangerous gold concentration to remain in a transition zone.

In some cases a very thin gold plating with a thickness of 0.000004 to 0.00002 in. (0.1 to 0.5 μm) is specified. Unless the plating is carried out very carefully, the very thin gold layer will be so porous that the underlying copper oxidizes. During soldering the thin layer of gold is dissolved quickly, but the flux has no chance to clean the underlying copper surface, and the result is poor wetting. Obviously a very thin layer of gold implies the risk of reduced solderability.

9.8.3 Tin/Nickel Plating

A tin/nickel plating is sometimes specified, and it is deposited as an intermetallic compound consisting of tin/nickel in the ratio of 65/35. Unfortunately, the solderability of tin/nickel deteriorates rapidly over a few days, and even the use of a highly activated flux, with all the problems this implies, helps very little.

The problem can be remedied by applying a gold flash of about 0.00001 in. (0.25 μm) on top of the tin/nickel layer. Because of the barrier action of the tin/nickel, no corrosion of the copper can take place through pores in the gold flash, which retains the good solderability of the fresh deposited tin/nickel.

9.8.4 Immersion Tin

Immersion tin is an electroless deposit of tin on the surface of nonplated-through boards as well as plated-through bare-copper boards. In this way a certain surface protection is achieved, but unfortunately it is not possible to deposit more than approximately 0.00002 in. (0.5 μm) tin, a thickness quite inadequate to ensure solderability over an extended period.

As mentioned earlier, tin acts with copper to form intermetallic tin/copper compounds. Because the tin layer is very thin, its entire depth rapidly converts to intermetallic compounds which oxidize and greatly reduce the solderability.

Immersion tin is very seldom used, but it can be justified on plated-through boards for automatic component insertion, as a narrower tolerance on the hole diameters than valid for tin/lead plated and reflowed boards can be achieved. However, it precludes long-term storage.

When inspecting boards with an immersion-tin surface protection, the inspector is advised to follow the directions given in Chapter 7 concerning copper plating. Since the tin layer is very thin, it cannot conceal possible faults in the copper plating. A solderability test is best performed by passing a board through the mass soldering machine. If the solderability proves insufficient within a specified period, the fault is rated as an A-defect, and the boards are returned.

Chapter 10

Solderability Inspection

10.1 Introduction

When the production department assesses the solderability of a board, the criterion is usually the appearance of the solder side of the populated and soldered boards. If the solder fillets look like those illustrated in Figure 10.1, the solderability is good. If they look like those in Figure 10.2, the solderability is poor. A good solder fillet (solder joint) is characterized by:

1. Full wetting, which means that the solder pad, including any plated-through hole, as well as the component lead, are fully wetted by the solder. Wetting, including the term contact angle, has been dealt with in Paragraph 8.7.2.
2. A concave conical shape and a very small contact angle. This means a smooth transition between the solder and the pad as well as between the solder and the component lead.
3. A smooth, bright surface.

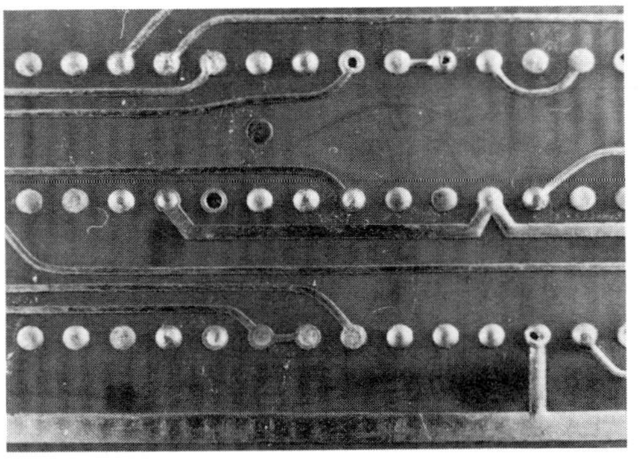

Fig. 10.1. Board with good solderability.

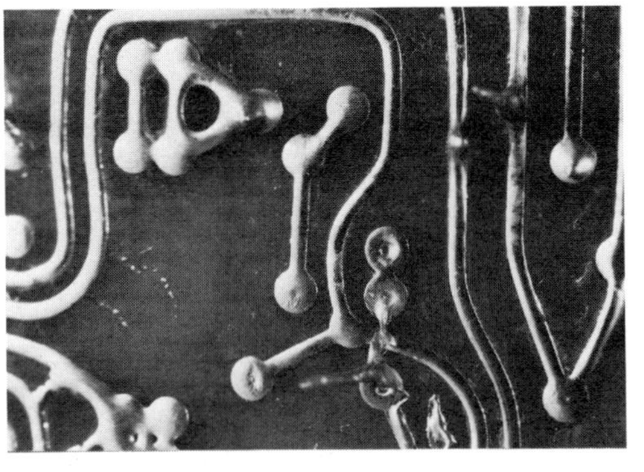

Fig. 10.2. Board with poor solderability.

Board soldering means the formation of a metallic connection from the solder pad, including any plated-through hole, to the component lead as shown in Figure 10.3.

During soldering, an alloying process takes place as mentioned earlier. This process creates intermetallic tin/copper compounds in a relatively thin transition zone between the solder and the parts being joined.

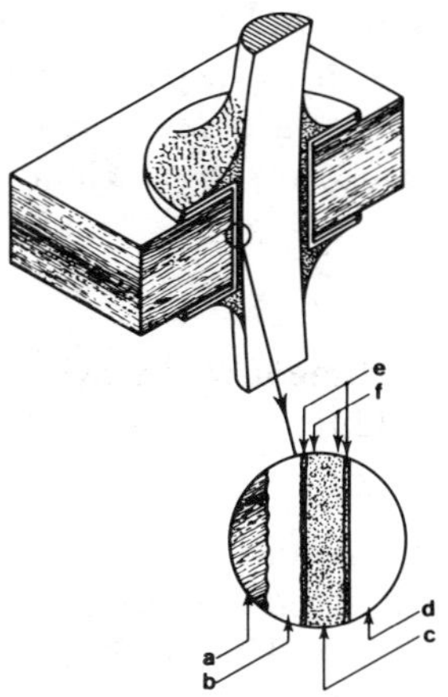

Fig. 10.3. Cross-sectional drawing of a solder joint.
a = base material. b = hole wall (plated copper). c = solder (tin/lead). d = component lead (copper). e = intermetallic layer (Sn_3Cu and Sn_5Cu_6). f = tin depleted zone.

10.2 Factors Affecting the Soldering Quality

The PCB manufacturer frequently is blamed for poor soldering results when the operator complains of poor or reduced board solderability, especially in the case of mass soldering.

In reality, the soldering quality depends on a number of factors, some of which are the responsibility of the PCB manufacturer and some of the PCB user and/or the PCB designer. It is therefore no simple matter to define the solderability, still less to assign the responsibility for an unsatisfactory soldering result. This section discusses these matters.

10.2.1 Solderability of the Board Itself

Paragraph 8.7.2, which described the tin/lead plated layer's wetting of the copper surface after reflowing, used the surface appearance (the wetting) as a measure of the board's solderability.

Since the same conditions are valid in the case of soldering—regardless of the type of surface protection (Chapter 9)—we can use the criteria listed in Paragraph 8.7.2 when assessing the soldering results. These criteria, however, will be influenced by the conditions described in the following paragraphs.

10.2.2 Design of the Board

The pattern can strongly influence the quality of the finished solder fillets, which should be considered when assessing the quality. The size of the solder hole should properly match the diameter of the component lead, so that the solder can satisfactorily fill the hole by capillary action. See Paragraph 6.5.4a of Reference A12.

The pads shall be shaped to minimize unnecessary heat loss; wide conductors shall also be connected to the pads so as to minimize heat loss. In addition, the designer should try to keep the pattern as symmetrical as possible. See Paragraph 6.5.7 of Reference A12 concerning design rules.

10.2.3 Solderability of the Component Lead

It is important that the component lead be solderable. The lead is frequently made from tinned, hard-drawn copper wire, but if the tinning is thin and/or porous, the underlying copper can oxidize and cause serious problems.

In other cases, the component lead is made from an iron/nickel alloy, such as kovar. Such alloys are not particularly solderable and, in any case, require the use of a highly activated flux. When these alloys are used, the component leads are frequently gold plated, unfortunately to a thickness of 0.00001 in. (0.25 μm) or less. During soldering, this thin layer can dissolve in the solder, which can no longer wet the component lead as shown in Figure 10.4.

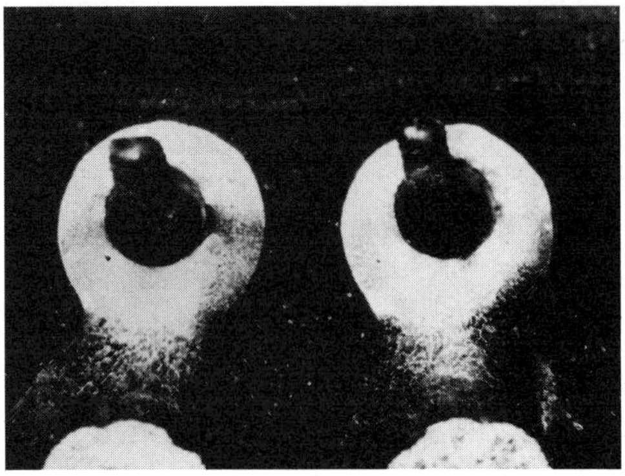

Fig. 10.4. Incomplete wetting of gold plated component lead. (Courtesy Tin Research Institute.)

Many problems encountered in mass soldering are caused by poor solderability of the component leads rather than of the boards themselves.

10.2.4 Fluxing

The applied flux serves both to clean the surface of the board and the component leads of oxides, and also to prevent reoxidation while the boards are being heated to soldering temperature.

The nonactivated flux designated R (for rosin) is so mild that it is almost incapable of removing oxides; therefore, it is rarely used. An activated flux is obtained by adding activators such as amine hydrochlorides. With the mildly activated flux designated RMA (rosin mildly activated), the amount of activator is so small that the flux, after soldering, is inert and noncorrosive. For this reason, RMA flux is widely used within the electronics industry.

With additional amounts of activator, the flux becomes highly activated and has the designation RA (rosin activated). The flux is very active, and is used only where it can be removed with certainty after soldering; otherwise the board will corrode heavily.

Water-soluble fluxes can also be used when soldering boards. They are extremely active and can ensure soldering of boards with very poor solderability. However, they are extremely corrosive and require thorough cleaning after soldering.

10.2.5 Adjustment of the Soldering Machine

Whether a wave soldering machine or a drag soldering machine is used, good soldering always requires that the machine be correctly set. The machine could conceivably operate marginally so that boards from one PCB manufacturer are soldered satisfactorily while boards from another PCB manufacturer produce unsatisfactory results. Several factors can affect the soldering.

10.2.5a The Flux Station

The flux station of the soldering machine must operate correctly; this means, among other things, that the flux must have the correct viscosity. As the solvent in the flux evaporates fairly quickly, it must be replenished several times a day.

10.2.5b Preheating

The board should be preheated at a suitable temperature over a suitable length of time so that the acidic activators in the flux can be released before the soldering starts. In this way, the flux can clean the solder spots of oxides within the soldering time, which can be as short as 3 seconds.

The preheating removes excessive solvents from the flux in order to prevent sputtering during soldering, which could cause blowholes in the solder joints. Finally, the preheating reduces the thermal shock to which the board is subjected during soldering. Under adverse conditions, the shock could induce cracking of the plated-through holes.

10.2.5c Soldering Temperature and Time

The soldering temperature and time must, of course, be correct. The soldering temperature frequently is 480°F (250°C), and the soldering time varies from 3 to 5 seconds, depending on the pattern.

10.2.5d Soldering Machine Dynamics

Also, the dynamic conditions of the soldering machine must be adjusted correctly. In the case of a wave soldering machine they include the height, shape, and velocity profile of the wave, as well as the inclination of the board. In the case of a drag soldering machine, it is a question of the conditions under which the board leaves the solder bath, primarily speed and inclination.

10.2.5e Impurities

The solder should be analyzed regularly for impurities, such as copper and/or iron, which in the course of time have been shed from the boards and component leads. If such impurities are excessive, the quality of the solder joints will decrease, showing itself in the form of solder bridging, icicles, large contact angles, and solder joints of a grey, gritty, and uneven character.

10.2.5f Solderability Assessment

This survey indicates that the soldering process comprises many parameters, which can independently and unfavorably affect the quality of the soldering. It would be advantageous if the PCB manufacturer and the customer determine the solderability on the same unambiguous, objective basis of assessment.

A number of measuring methods have been developed, but the lack of generally accepted, standard measuring methods is a considerable hindrance. The reason is not only that many institutions, organizations, and larger electronics companies have already established internal standards of solderability, but also that there exists a number of measuring instruments based on fundamentally different principles.

Incidentally, very few electronics companies other than the largest ones have found reason to purchase measuring equipment for solderability determination. The soldering machine instead is used for this test, whether it is a wave or a drag soldering machine.

It is important to the PCB user that the solderability determination be made under normal working conditions. A trial soldering of an unpopulated board makes it possible to predict, with a high degree of certainty, whether the boards of a particular batch will solder satisfactorily.

In order to perform a similar solderability determination, several PCB manufacturers have purchased a small soldering machine. There are, however, reasons to oppose this practice. Satisfactory operation of a soldering machine depends, to a large degree, on the operator's practical experience regarding the optimum adjustment of the machine. The required experience will not be gained by sporadic operation.

The author believes that many PCB batches have been rejected by PCB users on the grounds of poor solderability. In many cases, however, the inspector should suspect substandard operation and maintenance of the soldering machine. For the PCB manufacturer, it would perhaps be better to arrange with a neighboring PCB customer to have samples passed through his soldering machine. This will give a far better and more reliabie assessment.

10.3 IPC Solderability Test

The IPC (Institute for Interconnecting and Packaging Electronic Circuits) suggests a "Wave Solder Test" as a solderability test.

This is described in IPC-S-803, "Solderability Test for Wave Soldered Printed Wiring Boards," to which the reader is referred for additional details.

In short, the test consists of soldering a test board in a soldering machine, the type of which is agreed on between the PCB manufacturer and the PCB user. The IPC test prescribes primarily a wave soldering machine, but there should be no objection to using a drag soldering machine.

The solder is specified as Sn 60 or Sn 63—a solder with an alloy composition of 60/40 or 63/37, respectively. The flux shall be a nonactivated rosin flux having a nominal composition of 25 percent (by weight) of water-white gum rosin in a solvent of a 99 percent isopropyl alcohol. The specific gravity of the flux shall be 0.843 ±0.005 at 77°F (25°C).

The board which is to be soldered shall be used as received, and handled carefully in order to avoid surface contamination; for example, by fingerprints. After soldering, the board is cleaned of flux and inspected.

According to the IPC, the criterion for acceptable solderability is a relatively uniform and adhering coating of solder covering at least 95 percent of the pattern being tested. The remaining 5 percent of the pattern surface may contain small pinholes, nonwetted, dewetted, and rough areas only. These defects shall not occur in a single area. Examples of nonwetting and dewetting are shown in Figures 8.12 and 8.14.

10.4 IEC Solderability Test

The IEC (International Electrotechnical Commission) has described a solderability test in its Publication 68-2-20, Fourth Issue, of 1979. In short, the test measures the soldering time necessary for satisfactory wetting, and it is carried out with the test equipment shown in Figure 10.5

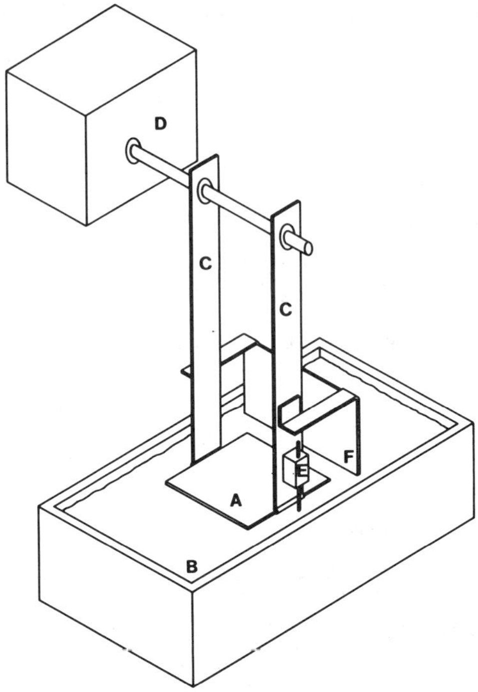

Fig. 10.5. Equipment according to the IEC for solderability testing. A = sample. B = solder bath. C = arm. D = motor. E = contact for timer. F = oxide scraper.

Briefly, it consists of a solder bath and a motor-driven arm to which the sample is attached. The size of the sample is 1.2 × 1.2 in. (30 × 30 mm) with an effective soldering area of 1 × 1 in. (25 × 25 mm). The arm length is 4 in. (100 mm), and in its lower position the sample is parallel with the surface of the solder bath, and in contact with the solder. In this position, the sample shall not be immersed in the solder bath by more than its thickness, and care should be taken, when designing the sample holder, that the solder does not flow over the upper side.

The solder bath shall have a minimum depth of 1.6 in. (40 mm), and a rectangular surface of 3 × 4 in. (75 × 100 mm), or a circular surface with a diameter of 4.8 in. (120 mm). The composition of the solder alloy shall be as follows:

Tin:	60 ±1 percent
Antimony:	Maximum 0.5 percent
Copper:	Maximum 0.1 percent
Arsenic:	Maximum 0.05 percent
Iron:	Maximum 0.02 percent
Lead:	Remainder

An oxide scraper is attached to the arm so that the surface of the solder bath can be cleaned of oxides immediately before the sample contacts the solder. With a contact needle, which is mounted on the arm and connected to an electronic timer, the contact time of the sample in the solder bath is determined. Adjusting the motor speed varies the contact time within a range of 1 to 10 seconds.

Before the test, the sample is fluxed by dipping, and the excess flux drained off by keeping the sample vertical for 5 minutes. The flux can be one of the following types, which are all composed of 25 percent (by weight) rosin and 75 percent isopropyl alcohol:

Type A: A nonactivated flux composed as stated above.

Type B: A flux like type A, but mildly activated with enough diethyl ammonium chloride added to ensure a chloride content of 0.2 percent relative to the rosin content.

Type C: A flux like type A, but with enough diethyl ammonium chloride added to bring the chloride content to 0.5 percent.

Since the type C flux is usually used, the formula is given below:

Rosin	1 oz. (25 g)
Isopropyl alcohol	3 oz. (75 g)
Diethyl ammonium chloride	0.016 oz (0.39 g)

The fluxed and dried sample is now attached to the arm, and the solderability test carried out in one of the following ways: either as a determination of the degree of wetting achieved at a fixed soldering time—for example, 3 seconds—or as a determination of the shortest soldering time for satisfactory wetting.

The last method requires the testing of a number of samples at various soldering times; for example, 1, 2, 3, 5, and 8 seconds. Both methods require a solder bath temperature of 455°F (235°C) with a tolerance of +9 and −0°F (+5 and −0°C). The temperature is measured with a calibrated thermocouple placed 0.75 to 1 in. (19 to 25 mm) beneath the surface of the solder bath.

After all flux residues are cleaned off with isopropyl alcohol, the samples are inspected. A satisfactory wetting requires that at least 95 percent of the pattern surface is bright and smooth with no defects. The remaining 5 percent of the pattern surface may contain small pinholes, nonwetted, dewetted, and rough areas only. These defects shall not occur in one area. See Figure 10.6, which shows typical examples. Also, the plated-through holes are inspected for solderability (wetting). Figure 10.7 shows examples for acceptable and unacceptable forms.

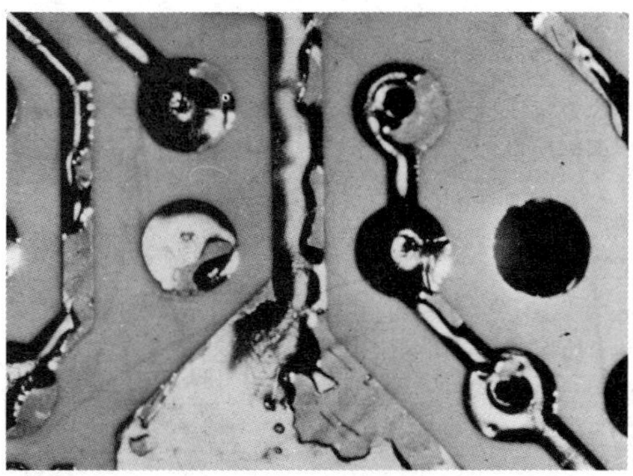

Fig. 10.6. Nonwetting ascertained by solderability testing. (Courtesy Tin Research Institute).

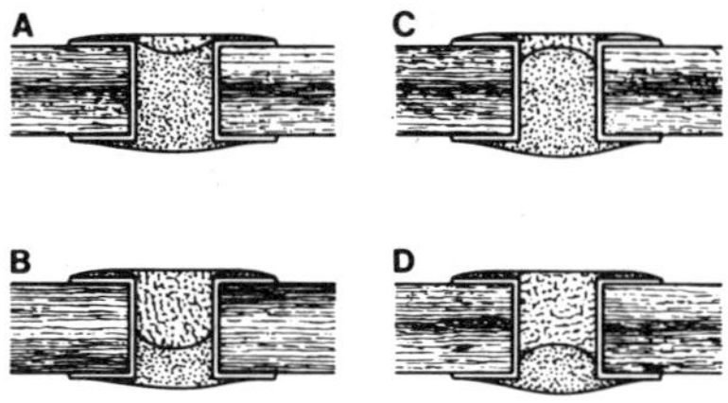

Fig. 10.7. Various degrees of wetting of plated-through holes. A and B: Acceptable wetting. C and D: Unacceptable wetting.

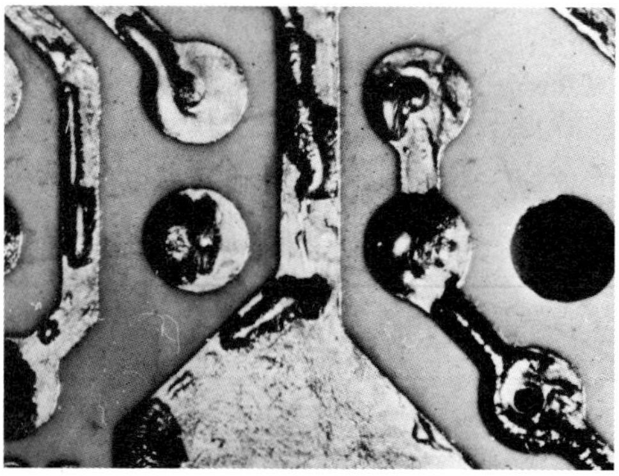

Fig. 10.8. Dewetting ascertained by dewetting test. (Courtesy Tin Research Institute).

A dewetting test is performed in very much the same way as a solderability test, except that the soldering time is 5 seconds, and the soldering temperature is 500°F (260°C) with a tolerance of +9 and −0°F (+5 and −0°C). If the solder pulls back and leaves a very thin coating of solder, the sample has a tendency to dewetting. Figure 10.8 shows typical examples.

10.5 Resistance to Soldering Heat

Many PCB specifications require that the finished board withstand a certain soldering heat when mass soldered. Typical conditions are:

Soldering temperature:	500 ± 9°F (260 ±5°C)
Soldering time:	10 ±1 s

Note: According to MIL-P-55110C the temperature is 550 ±10°F (288 ±5°C)

These soldering conditions shall not cause delamination or blistering as mentioned in Paragraphs 4.5.3 and 4.5.4, and the copper shall not lift. Finally, the plated-through holes shall not crack.

In case of deliveries from approved PCB manufacturers, it is unnecessary to investigate whether the boards can withstand the specified requirements concerning soldering heat. It is somewhat different in the case of a new PCB manufacturer, whose product quality is more or less unknown, or alternatively, if the production department complains of the aforementioned defects.

In such cases it will be an advantage to use the test equipment described in Paragraph 10.4 for assessing the resistance to soldering heat. A practical solution is the introduction of a microswitch which automatically stops the arm in its lower position, and a start button, by means of which the operator, after a suitable time, can start the movement of the arm away from the solder bath. The exact soldering time is measured by means of the timer. As a reference the inspector must, of course, use the actually specified acceptable soldering conditions which can vary considerably according to the base material; for example, FR2, FR3, or FR4.

After the soldering test as described above, the sample is inspected visually. If the inspector finds delamination, blistering, or lifted copper, such faults are rated as A-defects. Cracking in plated-through holes is also rated as an A-defect, but ascertaining this requires microsectioning and microscopic inspection.

If the sample cannot withstand the required soldering heat, the test should be repeated on another sample. If this sample also develops A-defects, the batch should be considered doubtful, for which reason it is returnable.

The test equipment can also be used to determine the duration of the soldering heat which the sample can withstand. It is most practical to know the margin relative to the actual soldering conditions followed by the customer.

The temperature of the above solder bath is therefore lowered to the temperature of the soldering machine, usually 480°F (250°C), from which the longest soldering time that does not affect the sample is determined. The difference between this time and the normally used soldering time is the soldering time margin, which should not be less than 5 seconds. If this is the case, the boards can be exempted. Naturally, the PCB manufacturer should be advised not to use inferior base material for the next deliveries.

10.6 Solderability after Storage

Even after prolonged storage the boards should be solderable. In the case of hand soldering, the problem is not particularly pronounced because the operator, within reasonable limits, can extend the soldering time until the solder flows. This implies that the flux has had sufficient time to clean the surface being soldered.

It is quite different when the boards are mass soldered, as this leaves only a rather short time for the flux to clean the surface. It is therefore imperative that the board be fully solderable, even after it has been stored for some time.

For this reason most PCB specifications contain requirements concerning full solderability after storage for a specified period of time. It is necessary to distinguish between various possible methods of surface protection when specifying storage periods:

Tin/lead plated boards:
- Nonreflowed: minimum 6 months
- Reflowed: minimum 12 months

Solder-coated/hot-air leveled boards: minimum 12 months

Bare-copper boards:
- Lacquered: minimum 3 months
- Immersion-tinned: minimum 2 weeks

The storage conditions shall, of course, be reasonably good; the following conditions are often specified:

Temperature:	59 to 95°F (15 to 35°C)
Relative humidity:	45 to 75 percent
Packing:	Polyethylene bags

It is difficult for the inspector to test the boards for storage durability since this requires an accelerated aging which, in turn, implies an accurate knowledge of the equivalent time of aging. According to a report "Solderability Testing of Printed Circuit Boards," prepared by the Swedish Institute of Engineering Research (P.O. Box 5506, S-114 85 Stockholm, Sweden) the IEC test Ca—damp heat, steady state, for 10 days—corresponds to an aging of 6 months with tinned material.

The above test is described in the IEC Publication 68-2-3 (1969), "Basic Environmental Testing Procedures." The aging is performed by placing the board for 10 days in a test chamber with a temperature of 104°F (40°C) and a relative humidity of 90 to 95 percent. After removal from the test chamber the board is conditioned for 2 hours under normal ambient conditions, depending somewhat on the actual surface protection.

The effect of aging seems a little uncertain since there is no unambiguous reference. The inspector should obtain from stock a number of obsolete boards so that he can start storage durability tests of his own. Every third month a board is taken for a solderability test, either in the mass soldering machine or by means of the test equipment described in Paragraph 10.4. In the course of 1 to 2 years the inspector will gain valuable experience at a relatively low cost.

10.7 Miscellaneous Soldering Faults

A number of faults can occur during mass soldering of printed circuit boards. These faults can be due both to improper manufacturing of the boards and to unsuccessful soldering techniques. Another reason could be improper design.

Sometimes the production department tends to blame the board, causing the incoming inspection department to become involved. It is important that at least the incoming inspector has a reasonably broad knowledge of the causes of such complaints, so as to avoid unfounded complaints to the PCB manufacturer. We will therefore address the following faults:

- Solder webbing
- Solder bridging
- Icicling
- Blowholes
- Apparent nonwetting

10.7.1 Solder Webbing

Solder webbing, or just webbing, is an adhesion of solder to the base material outside the pattern. Sometimes the appearance can resemble a cobweb, from which the name is derived. The solder can form thin threads which short-circuit parts of the pattern as shown in Figure 10.9.

This can be due to an incompletely cured base material or solder mask (either a screen-printed solder mask or a photopolymer solder mask) or to incorrect flux viscosity. In the case of a bare-copper board, the cause could be incompatibility between the flux and the surface protection.

In order to find the cause, the curing can be tested with the pencil test described in Paragraph 4.5.8, while the flux viscosity is tested in the usual manner.

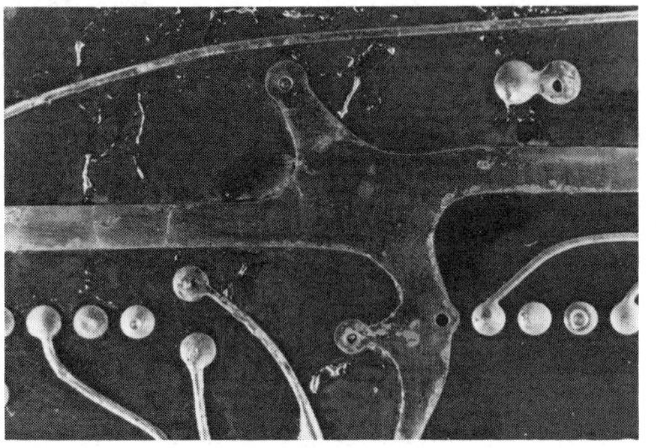

Fig. 10.9. Board with webbing.

Incompatibility between an organic protective lacquer and the flux is best tested by fluxing with another type of flux and then soldering the board. It is unnecessary to change the soldering machine's flux, as the alternative flux can be applied to the board before it goes into the soldering machine; i.e., by brushing or dipping.

10.7.2 Solder Bridging

Solder bridging is the formation of short circuits between various parts of the pattern; for example, between conductors and solder pads. See Figure 10.10. Solder bridging can be caused by incorrect adjustment of the soldering machine; for example, by too low a soldering temperature, incorrect board inclination, or solder bath contamination.

Fig. 10.10. Board with solder bridging.

The reason can also be a poor PCB design with adjacent parallel conductors combined with an undesirable angle between the direction of the conductors and the board's direction of travel in the soldering machine. See Paragraph 6.5.3 of Reference A12.

It should be noted that bridging from a solder pad to an adjacent conductor is a function of the pad diameter and the projection of the component lead beyond the face of the board. This type of bridging can often be eliminated by shortening th component leads.

In all the above cases the PCB manufacturer cannot be blamed for possible soldering faults. Incidentally, a solder mask will normally reduce the tendency to solder bridging by a considerable degree.

10.7.3 Icicling

Icicling, as shown in Figure 10.11, can occur frequently and can be caused by incorrect adjustment of the soldering machine, such as the soldering temperature, the speed of travel, and the board inclination.

Basically, the cause is that the solder has insufficient time to pull back to the solder bath before the solder side of the board is withdrawn from the solder bath or the solder wave. Icicling is a soldering fault for which the PCB manufacturer cannot be blamed.

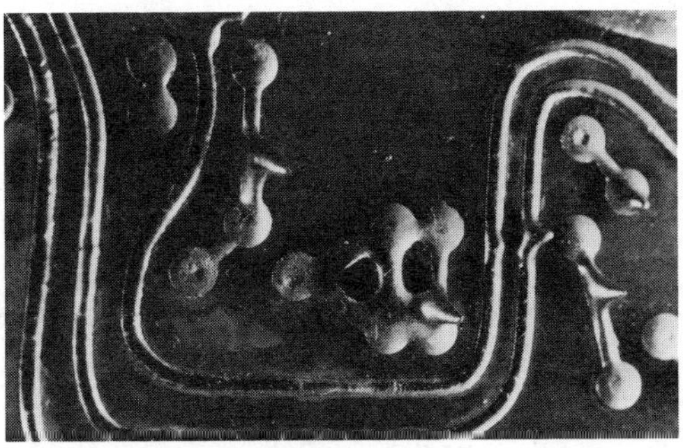

Fig. 10.11. Board with icicles.

10.7.4 Blowholes

After mass soldering it is sometimes possible to find blowholes in the solder joints of plated-through holes. See Figure 10.12. A blowhole frequently covers a larger cavity in the solder joint as illustrated at A, Figure 10.13. The blowhole is produced by steam and/or gasses which form during the soldering process, and make their way through the still-liquid solder.

Blowholes are usually present on the solder side because the solder solidifies last on this side. Concealed and closed pockets, which cannot be detected by visual inspection, can also form. See B, Figure 10.13. The reliability of the solder

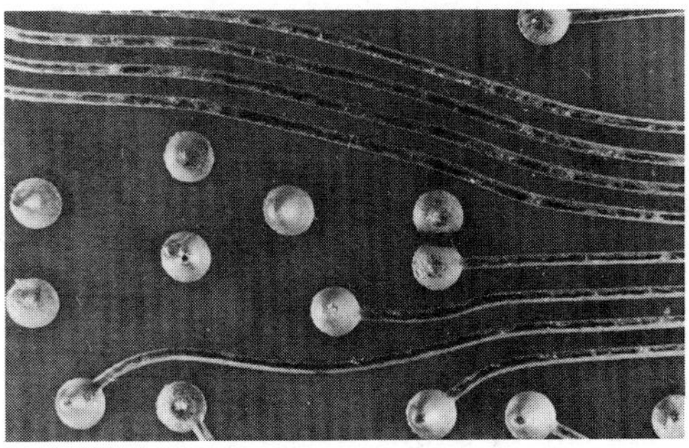

Fig. 10.12. Board with blowholes.

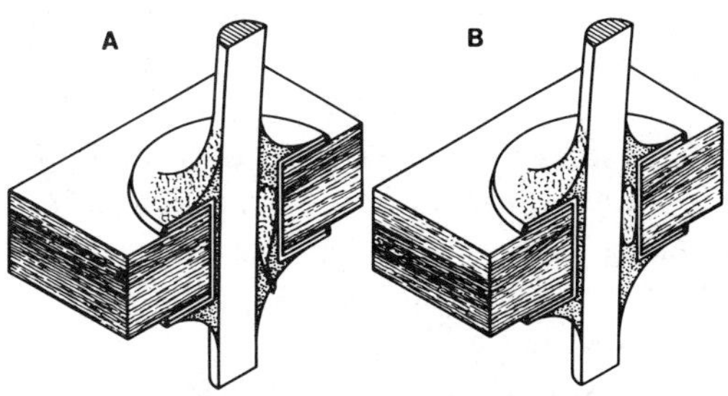

Fig. 10.13. Cross-sectional drawing of solder fillets with blowholes. A: Blowhole leading to large cavity in solder fillet. B: Concealed and closed cavity in solder fillet.

joints are suspect as the contact area between the component lead and the solder can be greatly reduced.

10.7.4a Causes of Blowholes

Blowholes can occur for many reasons, some of which are caused by the PCB manufacturer and others by the PCB user. Where the PCB specification makes reservations regarding blowholes, the inspector should, in fairness, distinguish between the various causes described below.

A pronounced manufacturing fault is the occurrence of plating pockets, which are dealt with in Paragraph 7.3.6. Also drilling delamination can, as mentioned in Paragraph 7.2.8, cause blowholes. Yet another cause can be occluded organic additives which produce gasses in the tin/lead plating when heated. An incompletely cured base material reacts similarly. Finally, it should be mentioned that the 20 percent bromination of the epoxy which makes the base material flame retardant can sometimes develop gasses when heated during soldering.

Blowholes can also result from a poor mass soldering technique, primarily incorrect fluxing, but also because of dross in the solder bath.

10.7.4b Moisture Removal

During storage, and possibly also during shipment to the PCB user, the boards can absorb moisture, which vaporizes during mass soldering and causes blowholes. It is always prudent to dry the boards by baking them in an oven immediately before assembly and soldering; i.e., 1 to 2 hours at 230°F (110°C).

Some manufacturers of base material suggest baking at 260°F (125°C), corresponding to the glass transition temperature; that is, the temperature at which the epoxy softens. The reputed advantage is that the water molecules are better able to withdraw from the regions around the drilled holes into the base material. As the concentration of water in the base material adjacent to the hole platings is lowered, the risk of producing blowholes is reduced correspondingly.

In return, there is a risk of a permanent deformation of the boards, particularly when they are placed in a rack during baking. It is therefore recommended that the temperature be limited to 230°F (110°C).

10.7.4c Evaluating the Problem

When the assembly department encounters problems regarding the formation of blowholes, the first step is to determine the extent: Is it a general problem or does it occur only in particular deliveries?

In the first case, the operation of the soldering machine should be checked, especially the fluxing station. When there is no established routine for baking the boards, they can be experimentally baked for 1 to 2 hours at 230°F (110°C). If this measure removes the blowholes, the PCB manufacturer is absolved of blame.

On certain assembly lines, where each operator inserts a few—for example, 5—components, the turnaround time may be as much as 2 days, or 4 days over weekends. When the relative humidity is high, the boards can absorb so much moisture over this period that baking is a wasted effort.

One should not ignore the fact that the PCB manufacturer can be responsible for an unusually high water content in the base material, either because of process conditions or, more often, inadequate packing.

If the blowholes do not disappear, or if the blowholes appear in certain batches, the PCB manufacturer is probably to blame. The base material is frequently at fault, and changing the base material supplier might correct the condition.

10.7.4d Board Disposition

It is very debatable whether the occurrence of blowholes after mass soldering should be regarded as an A-defect or a B-defect. The author knows of no real guidance in the technical literature. According to his experience, a small number of blowholes are acceptable, particularly when the solder joints on the component side are good. He therefore ventures to set up the following assessment criteria:

1. If the blowholes are less than 0.01 in. (0.25 mm) at the crater, and affect less than 2 percent of the total number of solder pads, the board is accepted with no remarks in the inspection record. This decision implies evidence of good joints on the component side.
2. When the blowholes are as described in Requirement 1, but affect 2 to 5 percent of the solder pads, the fault is rated as a B-defect. The fault is exempted and the board is accepted.
3. When the blowholes as described in Requirement 1 affect more than 5 percent of the solder pads or the blowholes are larger than 0.01 in. (0.25 mm), the fault is rated as an A-defect and the board is rejected.

Assembled and soldered boards can be salvaged by hand soldering, but this requires extra work, especially if the blowholes are widespread.

10.7.5 Apparent Nonwetting

If gas pockets between the board and the solder bath are created during mass soldering, especially during drag soldering, the solder cannot wet the pattern at the pocket locations. This could lead to the misconception that the problem is dewetting, especially in the case of lacquered bare-copper boards.

The cause is frequently that the solvent vehicle has insufficiently dried in the preheating station, for which reason the excess volatiles create gas pockets. The fault should not be mistaken for common nonwetting which is always cause for rejection.

The fault is easily verified by resoldering, preferably after the fluxing and preheating stations have been overhauled. If the fault disappears after resoldering, it was apparent nonwetting, for which the PCB manufacturer is blameless.

Chapter 11

Screen Printed Solder and Insulation Mask Inspection

11.1 Introduction

11.1.1 Conditions Preventing Solder Bridging

The steadily increasing pattern density of modern printed circuit boards presents an increased risk of solder bridging during mass soldering, which has also gained widespread use in recent years. In principle, the following two main conditions will prevent solder bridging:

1. The solder in the mass soldering machine is absolutely pure—that is, not contaminated by copper, gold, cadmium, or other materials originating from boards and components passing through the machine.
2. All conductors on the solder side are running in the same direction as the board travel when passed through the mass soldering machine.

Under these conditions, there is no opportunity for sidewise solder bridging, even with the minimum spacing of .004 in. (0.1 mm). In practice, it is quite different. A low contamination of the solder bath always exists as a result of daily use; a small amount of copper and other metals dissolves from every board passing through the mass soldering machine.

One has to live with a slowly increasing contamination until it reaches the limit, at which time the solder bath is exchanged (cleaned).

It is not possible to completely fulfill the design requirement in normal practice. Consistent separation between longitudinal and transverse conductors on the two sides of the board usually will be hopelessly inexpedient because, among other things, this requires a considerably higher number of via holes which, in turn, require more board area. Design practice therefore is a compromise.

Even when the many design problems have been considered, and the solder bath kept reasonably pure—that is, with no particular contamination—solder bridges can still form.

11.1.2 Benefits and Problems of Solder Masks

Removing solder bridges is a rather demanding process which both degrades the appearance of the board somewhat and implies a risk of overlooking some solder bridges. It is therefore not surprising that solder masks have recently become very common as they can prevent solder bridging. Another very important advantage is that the PCB designer regains his design freedom, although it is always prudent to design with a view towards avoiding solder bridging.

A parallel tendency is the growing use of applying an insulation mask to the component side. Since insulation masks present more or less the same problems as solder masks, we will limit the discussion to solder masks.

Until now, screen printing has been the most common method of applying solder masks, but the demand for photopolymer masks is steadily increasing, especially for use with solder-coated and hot-air leveled boards. This chapter deals only with screen-printed masks. The following chapter addresses photopolymer masks.

Unavoidable inaccuracies connected with even a very carefully performed screen printing hinder the accurate placement of the mask with respect to the pattern. Oddly enough, this seems to be one of the major technical antagonisms between the PCB manufacturer and the PCB user.

Unfortunately, many PCB manufacturers have yet to realize how important an accurate screen printing of the solder mask is for a well-soldered board with virtually no solder bridging.

The problem is compounded by the fact that the PCB designer is frequently ignorant of the technical problems associated with screen printing. Moreover, the pattern density is often so high that, in practice, it is impossible for the PCB designer to design a solder mask artwork fulfilling the screen-printing demands. See Paragraph 6.9 of Reference A12, Reading List.

The trend towards increased board dimensions means that board sizes of 16 × 16 in. (400 × 400 mm) or more are no longer rare. Obviously, registration problems are severe in the case of such large boards.

For these reasons, the inspector should be able to fairly assess the solder mask; that is, to determine whether a possible fault is due to inattentive screen printing or to a solder mask artwork which ignores the tolerances associated with screen printing. The following paragraphs thoroughly describe solder mask problems so that the inspector can gain a wide background for assessing possible faults.

11.2 Artwork Problems

It has been mentioned earlier that the artwork must be free of built-in production traps. This applies particularly to solder and insulation masks which, unfortunately, are not always designed with sufficient care.

The PCB manufacturer should always express any reservations regarding inadequate clearance and/or poor registration between the solder pads and the solder mask apertures. Even when the PCB manufacturer has made no objections, it can be difficult to reject boards because of solder mask faults if the reason is poor artwork. The result can never be better than the documentation. We will now discuss the following artwork problems:

Clearance around solder pads
Parallel conductors
Intentional overprinting of solder pads

11.2.1 Clearance around Solder Pads

The primary purpose of a solder mask is to prevent solder bridging. This implies that all parts of the pattern which are not to be soldered are covered by the solder mask. One of the most difficult places to achieve this is where a conductor runs between two IC pads as shown in Figure 11.1. A displacement of the solder mask of only 0.006 in. (0.15 mm) means a serious risk of the solder mask not covering one of the conductor edges.

As stated above, it is not always possible to achieve the necessary accuracy by

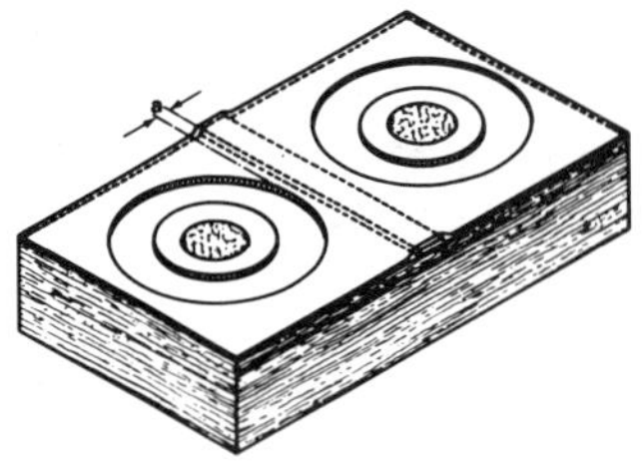

Fig. 11.1. Conductor laid between two IC solder pads.
a = margin of solder mask displacement is often only 0.005 to 0.006 in. (0.12 to 0.15 mm).

screen printing and we face a conflict between demand and execution, which invalidates a rejection. The inspector should therefore be able to ascertain whether a displaced solder mask is caused by artwork shortcomings or by a less-attentive screen printing. Paragraph 11.3 gives further details.

The PCB designer has a major impact on the success of a screen-printed solder mask. As mentioned in Paragraph 8.8.6 of Reference A12, manual taping should be organized in such a way that the solder mask artwork is prepared before the solder side artwork. This implies that, when taping the latter, it is more natural to show consideration for the necessary clearance between conductors and solder pads. When a CAD system is used, an automatic routine should be introduced to ensure a minimum clearance between the conductors and the solder mask apertures, particularly where a conductor passes very close to a solder pad.

11.2.2 Parallel Conductors

Adjacent parallel conductors can also present a problem, since it can be difficult to apply the printing ink to the bottom of the space between such conductors, especially when the direction of printing is perpendicular to the conductors.

As shown earlier in Figure 5.22, the printing ink will probably cover such conductors so that no solder bridging occurs, but the remaining cavities create a risk of corrosion and reduced insulation.

When the pattern thickness is about 0.004 in. (0.1 mm), the lower limit of the spacing—to allow total filling of the space between parallel conductors—is 0.012 in. (0.3 mm). This varies a little from one PCB manufacturer to another. The pattern thickness is derived as follows: 1 oz. = 0.0014 in. (35 μm) copper foil, 0.001 in. (25 μm) plated copper + 50 percent, 0.0004 to 0.0008 in. (10 to 20 μm) plated and reflowed tin/lead.

Where the conductors are spaced even closer, the PCB designer should, for etching, as well as for screen printing the solder mask, prescribe a base material with a ½ oz. = 0.0007 in. (17.5 μm) copper foil instead of 1 oz. = 0.0014 in. (35 μm).

When the pattern thickness is increased to 0.006 in. (0.15 mm), the minimum spacing for complete filling increases to 0.02 in. (0.5 mm).

11.2.3 Intentional Overprinting of Solder Pads

The PCB designer can choose to have some pads overprinted, usually only at critical locations. In each case he should assess the risk that the printing ink, at a 0.01 in. (0.25 mm) displacement of the solder mask, can flow down into the solder holes. See Paragraph 6.9.4 of Reference A12.

The PCB designer should also consider that the hole drilling can be so eccentric that hole breakout with respect to the solder pad may occur. See Paragraph 5.9, which deals with the width of the annular ring.

Displacement of the solder mask can cause printing ink to flow into the solder holes. Since displacements are inherently unpredictable, the PCB designer should avoid overprinting solder holes and use other types of solder masks, such as the dry or liquid film masks described in Chapter 12.

We have ignored here a total overprinting of via holes, which is sometimes necessary, and which can also increase the reliability of the finished board. There are several examples where solder has flowed up through non-overprinted via holes, and damaged the insulation lacquer of resistors mounted over the holes. This resulted in unstable resistors because they were center tapped.

The inspector should ascertain whether an overprinting is intentional or a screen printing defect. The decision is based on an investigation of the registration between the solder side and the solder mask. The inspector registers a positive solder mask film (opaque with transparent apertures) with respect to a positive solder side film (transparent with an opaque pattern), using the registration marks of the films. This is facilitated by placing the film set on a light table.

If no luminous ring appears around the solder pads, we have an intentional overprinting, at least from the screen printer's point of view. The width of the luminous ring indicates the size of the clearance, which should preferably be greater than 0.010 to 0.012 in. (0.25 to 0.30 mm). See Figure 11.2.

There is every reason to draw attention to the risk connected with a general permission to overprinting solder pads. The inspector will, in all probability, find that the extent of overprinting increases with time, until it comes alarmingly close to the edge of the solder hole with the increased risk of finding printing ink in the hole. This results in some loss of solderability in such holes.

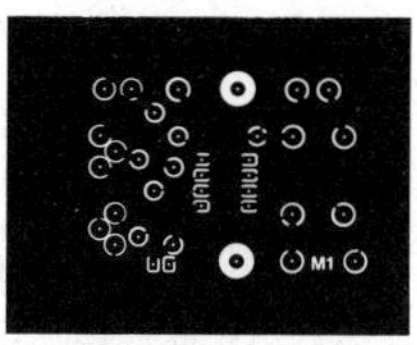

Fig. 11.2. Checking the films for intentional overprinting of solder pads.

11.3 Screen Printing Faults

In the case of screen-printed solder masks, the subsequent manufacturing processes do not put any demands on the quality of the solder mask. Actually, the screen printing of solder masks is one of the last processes performed. It is quite common that the screen printer, as well as the inspector of the final inspection department, may not appreciate the importance of a well-printed solder mask. This frequently contributes to inadequate quality of the solder mask.

Screen printing of solder masks is indeed an art owing to the vast number of parameters which come into play. The inspector should therefore appreciate the natural limits he can expect of a solder mask. Below we deal with some typical screen printing faults so that the inspector can better discuss quality problems with the PCB manufacturer:

- Insufficient registration
- Displacement across the board

11.3.1 Insufficient Registration

During the normal image transfer preceding the plating process, achieving a good registration between the hole pattern and the PCB pattern is a strict necessity. Otherwise, there is a risk of broken annular rings. The screen printer must therefore always exercise care.

The printing process itself is also somewhat easier as the surface at this time is completely even. In contrast, the solder mask is printed on top of the finished pattern, which means a relatively uneven surface.

11.3.1a Effect of Positioning System

In the case of solder masks, an unsatisfactory registration sometimes occurs. One reason can be an inferior positioning system. The board is frequently located by means of a guiding frame, which simply consists of a few pieces of base material taped to the printing table. In this way the board is located relative to its contour, which implies that the screen printing is already affected by the tolerance of the contour. The tolerance depends upon the chosen method of contouring, as stated below:

Shearing	±0.012 to ±0.020 in.
Sawing (with no sawing jig)	(±0.3 to ±0.5 mm)
Sawing (with a saw jig)	±0.008 in.
Pin routing	(±0.2 mm)
Blanking	±0.004 in.
NC routing	(±0.1 mm)

Add to this a possible initial adjustment error of the screen printing machine. This inaccuracy occurs when the screen printing frame is adjusted in the X and Y directions until a suitable registration between the solder pads and the solder mask apertures is achieved.

11.3.1b Benefit of Tooling Holes

A better method of guiding the board is achieved when the board or the production panel is provided with internal tooling holes to be used when fixing the board or panel to the table of the screen printing machine. This allows a far better reproducibility because the possible misregistration between the PCB pattern and the solder mask may normally be expected to be less than ±0.002 to ±0.004 in. (±0.05 to ±0.10 mm).

11.3.1c Permanent Dimensional Changes

Quite another reason for poor registration can be the permanent change of the board dimensions mentioned in Paragraph 6.4.3. The change can, under worst-case conditions, amount to ±0.05 percent, or ±0.004 in. (±0.10 mm) over a length of 8 in. (200 mm).

Owing to the weave structure of the glass layers of the laminate, or to the instability of the paper layers of phenolic paper laminates, as well as to differences in the pattern density of the board's two sides, or layers in the case of multilayer boards, such changes will often be different in the X and Y directions of the board. Furthermore, they frequently will be nonlinear.

The differences in X and Y dimensional changes make it difficult to reliably compensate for these changes by adjusting the size of the solder mask film. This could be done when photographing the solder mask artwork, by using an empirical correction of the reduction ratio.

Another possibility would be to adjust the tension of the screen after transferring the solder mask image to the screen. As shown in the next paragraph, incorrect stretching of the screen has some undesirable effects.

11.3.2 Displacement across the Board

A common fault is a more or less variable displacement between the solder mask and the pattern. This is caused during the printing process by the stretching of the screen, which increases with larger board sizes.

11.3.2a Inherent Displacements

A rather small displacement is inherent in the screen printing principle as it is necessary to keep a certain small off-contact distance between the screen and the board surface, frequently 0.12 in. (3 mm). Dur-

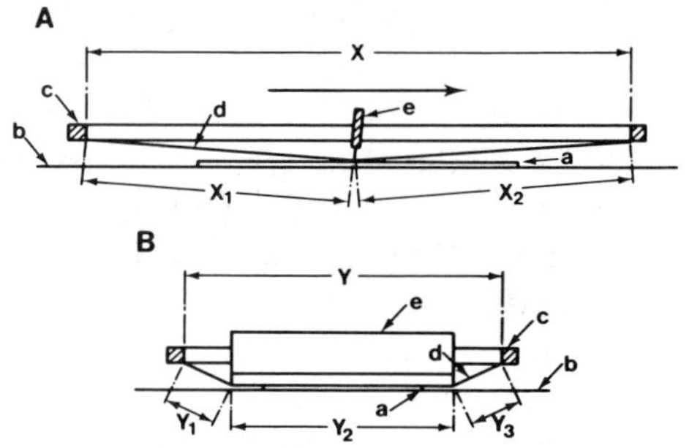

Fig. 11.3. Stretching of screen printing screen depends upon the distance between frame and board.
A: Cross-section corresponding to direction of printing.
B: Cross-section transversally to direction of printing.
Lengthwise: $\Delta X = X - (X_1 + X_2)$
Crosswise: $\Delta Y = Y - (Y_1 + Y_2 + Y_3)$
Typical values: ΔX = 0.0012 in. (0.03 mm), ΔY = 0.004 in. (0.10 mm) for the following values: X = 24 in. (600 mm), Y = 16 in. (400 mm), and width of squeegee = 9.5 in. (240 mm).
a = board. b = printing table. c = frame. d = screen. e = squeegee.

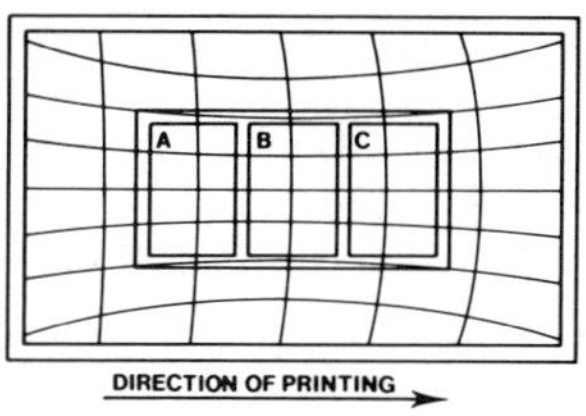

Fig. 11.4. Illustration of nonlinear displacement lengthwise and crosswise to direction of printing. A–B–C: Production panel with three boards.

ing printing, the screen is pressed down against the board surface by the squeegee, which causes a certain stretching of the screen. This is shown in Figure 11.3, which also illustrates that the largest displacement will occur transversely to the printing direction. This varying displacement is shown in Figure 11.4.

11.3.2b Incorrect Screen Stretching

Another reason for displacement is incorrect stretching of the screen, which

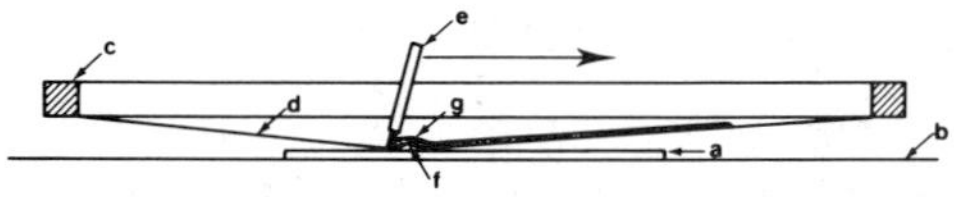

Fig. 11.5. A slack screen causes a rolling over the board surface. a = board. b = printing table. c = frame. d = printing screen. e = squeegee. f = rolling of screen. g = printing ink.

means that the screen is not mounted correctly on the screen frame. During the travel of the squeegee (see Reference A12, Paragraph 3.2.1) a slack screen almost rolls over the printing surface as shown in Figure 11.5. In order to touch the printing surface, the screen has to stretch a little as shown in Figure 11.3, but if the elastic limit (for permanent elongation) of the threads of the gauze is exceeded, the screen will be drawn out further for each print run. Subsequent stretching will only worsen the situation.

11.3.2c Stainless Steel Screens

Some PCB manufacturers prefer, in expectation of a higher printing stability, stainless steel screens. It should be considered that, contrary to the image transfer preceding the plating processes, we have a board with a fairly uneven surface. This requires a screen with a higher elasticity than can be obtained with a stainless steel screen.

After a fairly limited number of printing runs a steel screen will show a permanent dimensional change, causing a reduced printing accuracy. Therefore most PCB manufacturers prefer to use a polyester screen, metallized in some cases. The metallization encapsulates all the weaving intersections and provides better stability and easier ink passage.

When a displacement of 0.004 in. (0.10 mm) over a length of 10 in. (250 mm) occurs with a polyester screen, the displacement could be as much as 0.01 in. (0.25 mm) in the case of a stainless steel screen.

11.3.2d Additional Factors

In addition to incorrect screen stretching, there are other factors affecting the printing accuracy:

1. The frame should be reasonably large with respect to the printed circuit board in order to avoid overstretching the screen.
2. The width of the squeegee must be adapted to the width of the board. Too wide a squeegee, particularly in conjunction with too small a frame, will considerably overstretch the screen which will result in crosswise displacements. See Figure 11.3.
3. The hardness and profile of the squeegee as well as the pressure of the squeegee against the screen must be optimized, and the speed of its travel must be correct.
4. The viscosity of the printing ink must be suitable for the task.
5. Changes in temperature or humidity affect the emulsion layer of the screen.

The above statements indicate clearly that screen printing is a difficult art, setting high demands of skill and experience on the screen printer. The PCB designer, as well as the inspector, must accept the natural limitations to the quality of screen printed solder masks.

The registration of the solder mask can be improved a little by the following measures:

6. During trial printing, the screen printer can displace the frame a little with respect to the pattern so that the finished solder mask registers exactly at the center of the board. In this way the extent of misregistration is halved where it was formerly at its worst. In return, a similar misregistration, but in the opposite direction, occurs where there previously was full registration.
7. The solder mask can be applied in two operations, but with intermediate curing and printing in opposite directions. In this way, there is a greater chance of covering both edges of a conductor running between two IC solder pads.
8. Very large boards can, with a certain overlapping, be screen printed in sections, so that the effective extent of the displacement is reduced.

It is strongly recommended that the above precautions be discussed with the PCB manufacturer when the quality of the solder mask is unsatisfactory. A combination of Items 6 and 7, possibly including Item 8, will improve the registration, but it will, of course, also increase the cost.

11.4 Efficiency of the Solder Mask

The primary purpose of the solder mask is, as stated earlier, to prevent solder bridging during mass soldering. Therefore, the inspector tries to assess the efficiency of the solder mask in this respect. The board is inspected visually, preferably along diagonals as mentioned in Paragraph 6.8. First and foremost, the inspector checks whether the solder mask covers the conductor edges adjacent to solder pads.

11.4.1 Aperture Spacing

When the minimum spacing between a conductor and a mask aperture is 0.024 in. (0.6 mm), the solder mask must cover all conductor edges. The spacing is composed of a pad clearance of 0.012 in. (0.3 mm) demanded by the PCB manufacturer, a possible displacement of the solder mask of 0.010 in. (0.25 mm) and a

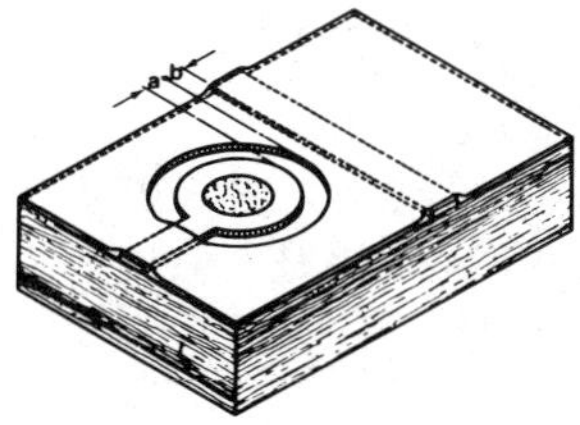

Fig. 11.6 Design requirements on the solder mask for achieving full coverage of conductor adjacent to a solder pad.
a = necessary clearance of 0.012 in. (0.3 mm) between solder pad and mask aperture. b = distance from mask aperture to edge of conductor should be minimum 0.012 in. (0.3 mm).

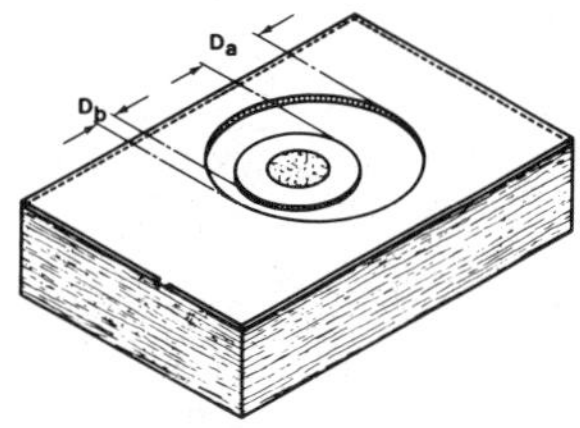

Fig. 11.7. Determination of solder mask displacement.

small addition to ensure that the printing ink can really cover the conductor edges. See Figure 11.6.

According to the above discussion of screen printing faults (Paragraphs 11.3.1 and 11.3.2), the inspector has to accept that a conductor running between two IC solder pads is not always completely covered by the solder mask. This applies also to such cases where screen printing in opposite direction has been carried out; see Item 7 above.

When the PCB designer has designed the artwork with intentional overprinting of solder pads at critical places (a conductor between two IC pads), the inspector needs the films to determine whether full coverage of a conductor can be expected.

11.4.2 Assessment Criteria

It is extremely difficult to state unambiguous criteria for rejection, since many degrees of displaced screen printing occur. It should be noticed that a slightly displaced solder mask in general will not increase the risk of solder bridging very much, provided the solderability of the board is satisfactory and the soldering machine is adjusted correctly.

As a matter of fact, assessment of the efficiency of the solder mask is left to the subjective judgment of the inspector, although some guidelines can be given. Under all circumstances, it will be very helpful if the inspector gains personal experience of the efficiency of the solder mask relative to its various deficiencies. It is strongly recommended that from time to time the inspector follow the soldering process to get a possibility of comparing actual solder bridging with shortcomings of the solder mask.

When the displacement of the solder mask exceeds 0.010 in. (0.25 mm), so that the nominal clearance of 0.012 in. (0.3 mm) is nearly or totally used, we have a printing error, provided the registration of the artwork is correct. The displacement is determined as shown in Figure 11.7. By measuring D_a and D_b with a 7X or 10X magnifier with graticule, the displacement ΔD is determined as $\Delta D = \frac{1}{2}(D_a - D_b)$, independent of any possible change of the aperture diameter with respect to the nominal diameter.

11.4.3 Measuring Problems

As the apertures of the solder mask at times can be somewhat distorted, it can be difficult to determine the displacement accurately. The distortion is usually caused by the squeegee which cannot press the screen gauze right down to the base material immediately around the solder pads, so that the printing ink cannot be deposited sufficiently close to the solder pads.

The problem is most pronounced in the case of a very high (thick) pattern in conjunction with small clearances, particularly when the screen gauze is made of stainless steel and therefore rather inelastic. In such cases, the most accurate way of measuring the distortion is to choose a solder hole in a ground plane. If necessary, the PCB designer can introduce a measuring spot at each corner of the board; that is, a solder hole with a fairly wide annular ring intended for partial overprinting. As no difference in height is present here, screen printing will not result in an unwanted distortion of the aperture in the solder mask.

11.4.4 Board Disposition

Displacements up to 0.010 in. (0.25 mm) are accepted with no remarks entered in the inspection record. Displacements up to 0.016 in. (0.40 mm) are rated as a B-defect because this fault will not necessarily result in a dramatic increase of the number of solder bridges after mass soldering. It is, however, reasonable to contact the PCB manufacturer and enjoin him to increase the printing accuracy.

Large displacements—above 0.016 in. (0.40 mm)—often mean that many conductor edges are not covered by the mask where the conductors pass very near solder pads. Furthermore, so large an overprinting of the pads implies a large risk of printing ink in the solder holes. See Paragraph 11.8. In such cases, the fault is rated as an A-defect, and the boards are returned for remake.

In cases of extreme need, the boards can possibly be used, but this implies a very careful inspection after soldering. It is likely that a certain amount of repair work (removal of solder bridges and manual post-soldering of unsatisfactorily soldered joints as well as joints exhibiting blowholes) will be necessary. Add to this a certain number of faults, perhaps 10 percent, which will pass inspection.

11.5 Unintentional Overprinting of Solder Pads

In Paragraph 11.2.3 we discussed intentional overprinting of solder pads, which for one reason or another is required by the PCB designer. We now deal with unintentional overprinting caused mainly by improper screen printing.

11.5.1 Registration versus Overprinting

The inspector should understand the difference between inaccurate registration of the solder mask and overprinting of solder pads. A small registration error does not necessarily cause the solder pads to be overprinted, but it can easily result in conductors not being completely covered by the solder mask. On the contrary, large registration faults (displacements) can cause overprinting of the solder pads.

11.5.2 Waivers and Exemptions

The purchase specification may be written so that it is a little less demanding on the PCB manufacturer. It can, for example, waive overprinting of 25 percent of the solder pads, but only to the middle of the annular rings affected.

Since it is not possible to know beforehand in which direction an eccentric hole drilling will drift, the aforementioned exemption should not be perceived as acceptance of a general displacement of the solder mask. It should rather be viewed as a possibility of avoiding rejection of boards where some solder pads, for one reason or another, have been slightly overprinted.

Random overprinting can also be caused by bleeding of printing ink from the screen, particularly when the rim of the mask aperture is a little too close to the solder pad. In this way the screen will deposit too much ink so that it overflows the solder pad.

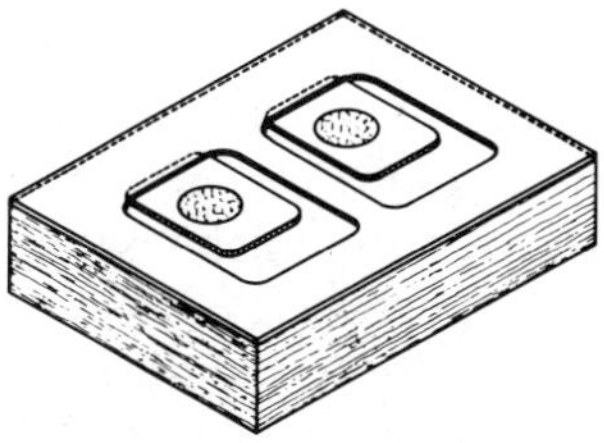

Fig. 11.8. Oblong solder pads set lesser requirements to the printing accuracy.

A general exemption for overprinting all solder pads to the middle of the annular ring is of little use to the PCB designer as nobody knows the result of the drilling operation in advance.

11.5.3 Effect of Solder Pad Design

Some solder pad designs are more favorable with regard to further exemptions. In many cases the PCB designer uses oblong solder pads, and on digital boards they are often oriented in the same direction. When the solder mask is printed in the longitudinal direction of the solder pads, a general overprinting does not imply soldering problems. See Figure 11.8.

11.5.4 Board Disposition

Although it is easier for the inspector to require no overprinting of solder pads, he should assume a more functional attitude and only reject boards in those cases where overprinting will adversely affect the soldering process. The inspector should, on the other hand, not go too far when exempting boards as the PCB manufacturer may then possibly reduce his own standards regarding printing quality.

If overprinting towards the middle of the annular ring of the solder pad has taken place, and the number of defects is less than 25 percent of the total number of solder pads, we will exempt the fault, which is rated as a B-defect. A further exemption can be given according to the above discussion if the solder pads are oblong and oriented in the same direction. Overprinting right up to the rim of the solder holes is cause for rejection and consequently rated as an A-defect.

11.5.5 Epoxy Film

A treacherous, but fortunately rare, type of fault is a thin and transparent film of epoxy covering the pad and perhaps also the hole wall. There can be two causes of this fault. In the case of overprinting, it can happen that a small amount of epoxy has segregated from the color pigment and either fully or partially overflows the solder pad as a transparent film.

The other cause is misprinting discovered by the screen printer immediately after printing. It is completely useless for the screen printer to wash the board off under the delusion that it can be reprinted. The washing always leaves a very thin film of transparent epoxy on the pads and in the solder holes.

A measure is to introduce a clause in the purchase specification prohibiting the washing off of misprinted solder masks. It is most unlikely that such a fault will be found in the receiving inspection department because the epoxy film is usually invisible, but it appears later as soldering problems during mass soldering. Naturally, the fault is cause for rejection and rated as an A-defect.

11.5.6 Fully Overprinted Solder Pads

Total overprinting of a few solder pads in a certain part of the batch is often due to an adhesion fault of the emulsion layer on the screen gauze. This fault can cause some of the pads in the emulsion layer to fall off during screen printing so that the screen gauze deposits ink on top of the corresponding solder pads on the board.

Such faults can easily pass through the PCB manufacturer's final inspection as well as the user's incoming inspection because of both the rarity of the problem and its probable restriction to a small portion of the batch. The fault is cause for rejection and rated as an A-defect.

11.6 Thickness of the Solder Mask

The thickness of the solder mask vitally affects its resistance to wrinkling where it is applied over tin/lead plated conductors and voltage/ground planes. Several different factors determine the thickness of the mask, including the relative printing area (open area of the gauze) and the thickness of the gauze itself.

11.6.1 Screen Characteristics

The thickness of the screen-printed solder mask is normally around 0.006 to 0.008 in. (15 to 20 μm) per printing run. Nevertheless, by optimum choice of the screen (among other things, the mesh count per inch and the thread diameter) and the type of stencil (for example, made by the direct/indirect method), a thickness of 0.0012 to 0.0014 in. (30 to 35 μm) can be obtained per printing run. The stencil is the pattern of the printing screen, consisting of open and closed areas.

11.6.2 Printing Characteristics

The thickness of the finished solder mask also depends on the type of ink used. In the case of a one-pack ink, the dry matter of the ink amounts to 40 to 50 percent. This means that a printing thickness of 0.008 in. (20 μm) is reduced to 0.003 to 0.004 in. (8 to 10 μm) when the ink dries. In the case of an UV ink or a two-pack ink the shrinkage is considerably less, perhaps only 5 to 10 percent when the mask is cured.

11.6.3 Measuring the Thickness

It is seldom necessary to measure the solder mask thickness when the inspector has gained experience with the screen-printing capability of the PCB manufacturer in question. However, if the inspector wishes to check the thickness, he can use a very low inspection level, as the printing process seems to be highly reproducible with respect to the thickness. A very limited number of boards is selected—for example, five boards—and the thickness measured at two different places of each board of the sample. The average of all 10 measurements is then calculated.

Unfortunately, it is impossible to measure the thickness of the solder mask with a micrometer. The resolution is no better than 0.0005 in. (in the metric system, 10 μm) and therefore quite inadequate. Instead, the inspector can use a height gauge as described in Paragraph 9.6.2 and illustrated in Figure 8.15. The thickness of the mask is the height difference between an area covered with the mask and an adjacent uncovered area.

In order to facilitate the inspection, the PCB designer should introduce measuring apertures at suitable places in the solder and insulation masks. A consequence is that the thickness measurement should not necessarily be made at the board edges, from which the masks usually are slightly withdrawn.

The solder mask thickness can also be measured by means of an electronic measuring instrument based on the eddy current principle. It is, however, necessary to perform the measurement at places having an underlying metal area at least the size of the probe—for example, 0.4 to 0.6 in. (10 to 15 mm) in diameter.

It is not possible to measure on top of conductors, so the measurement is limited to voltage/ground areas or to specific measuring areas introduced on the board. The greatest objection against the method is, in the author's experience, a relatively low reproducibility, meaning that repeated measurements at the same spot yield inconsistent results.

The solder mask thickness can also be determined by microsectioning the board so that the thickness can be measured directly. Microsectioning gives a good indication of the thickness variations across a conductor. Obviously, this measurement is destructive, and there is a certain risk of erroneous measurements; for example, if the solder mask is affected by the molding and machining of the sample.

A light-sectioning microscope provides very high measuring accuracy. The measurement is carried out along an edge, provided the mask as well as the base laminate are visible over a length of 0.04 in. (1 mm). Since the measurement is rather special, Chapter 26 gives a detailed description.

The last three methods are seldom used, partly because they can be difficult to perform and partly because they require relatively expensive equipment. The simple measuring method based on the height gauge is therefore most commonly used.

11.6.4 Assessment Criteria

There are different opinions regarding the necessary thickness of a solder mask when it is applied on a tin/lead plated board and a reasonable resistance to wrinkling caused by soldering is required. The problem is further complicated by the fact that various brands of printing ink differ slightly in their resistance to wrinkling.

There is a choice between one- and two-pack inks, and between oven-curing and ultraviolet-curing (UV-curing) inks. The choice should be made in collaboration with the PCB manufacturer. As an example, it is useless to prescribe an UV-curing ink if the PCB manufacturer lacks the necessary equipment.

The inspector is recommended to obtain practical experience regarding the response of the type of ink, including the actual thickness of the solder mask, to wrinkling during mass soldering. The problem is that the underlying tin/lead layer melts during mass soldering. As the solder mask softens because of the temperature and flows a little, the structure of the solder mask surface will change when the board cools and the tin/lead solidifies.

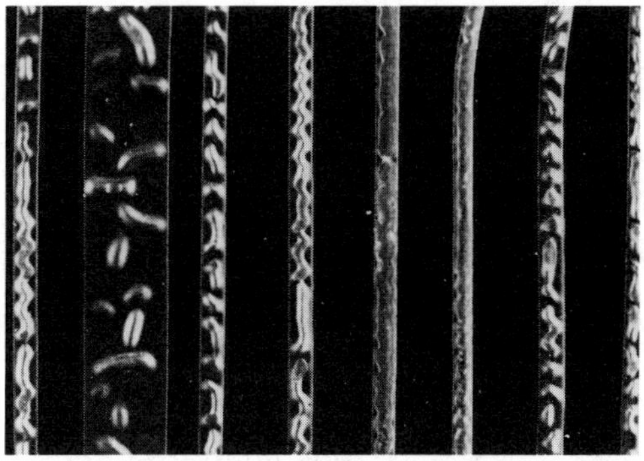

Fig. 11.9. Board with various degrees of solder mask wrinkling caused by mass soldering.

At most, a slight wrinkling at narrow conductors up to 0.02 in. (0.5 mm) and a little more at wider conductors up to 0.08 in. (2 mm) should occur. In the case of still wider conductors it is not possible to talk of wrinkling, but rather of a heavily undulated surface with peaks and valleys. See Figure 11.9, which shows different degrees of wrinkling.

Some flaking of the solder mask from very wide conductors and large voltage/ground planes is almost unavoidable after mass soldering. The most radical solution of this problem is to relieve the solder mask artwork so that the solder mask is not printed on top of such areas.

11.6.5 Board Disposition

A general specification can state an average solder mask thickness of at least 0.006 in. (15 μm) when measured at an area of 2 × 2 in. (50 × 50 mm) maximum. In the author's experience, the thickness should be at least 0.001 in. (25 μm), which requires one or two printing sequences, depending on the technique of the PCB manufacturer and the type of ink. Thicknesses below the specified limit are rated as A-defects, and the boards are returned to have an extra layer screen-printed.

When the solder mask is applied on bare copper there is no tendency to wrinkling, and the thickness is not particularly critical. Thicknesses down to 0.0004 in. (10 μm) are therefore accepted, whereas thicknesses below this limit are rated as an A-defect and the board is returned for reprinting.

In the case of insulation masks, the matter is a little different. The insulation mask is not affected by mass soldering, but as the name implies, there are certain requirements regarding the electrical insulation properties. For example, it may be desirable to achieve a certain assurance against voltage flash-over from components touching conductors or voltage/ground planes on the component side.

Although the dielectric strength of most two-pack printing inks is of an order of magnitude of 1 kV per 0.001 in. (40 kV per mm or 40 V per μm), a thickness of at least 0.0004 in. (10 μm) will normally be specified as a precaution. Thicknesses below 0.0004 in. (10 μm) are rated as A-defects, and the board is returned for reprinting.

An insulation mask can also give a better background contrast for component notations and also a more level surface on which to print the component notation.

11.7 Electrical Insulation Properties

The applied mask (solder mask as well as insulation mask) must not reduce the insulation properties of the finished board perceptibly, and this will frequently be stated in the general purchase specification. The electrical insulation data for the ink can be obtained from the PCB manufacturer, but they will normally be of the following order of magnitude:

Surface resistance: 10^{12} to 10^{13} Ω
Volume resistance: 10^{14} to 10^{15} Ωcm

These values correspond, on the whole, to those specified for the board itself, and a measurement is usually unnecessary.

The inspector should beware, however, of certain black printing inks which are used where a light-tight mask is specified. If the pigmentation is based on carbon black, the insulation resistance may drop by a factor between 100 and 1000, or even more. In certain cases, this is extremely detrimental, not least because the leakage resistance is very unstable. The PCB designer should therefore contact the PCB manufacturer in advance and agree on a suitable printing ink.

For this reason, the inspector should test the insulation resistance of all board deliveries having a black light-tight mask. As the reduction of the insulation resistance depends only on the printing ink, the sample can be a single board.

Insulation measurements are most conveniently carried out by means of a megohmmeter as shown in figure 11.10. The distance between the two measuring points could, for example, be 1 in. (25 mm), and the measuring result is then compared with earlier results or with a reference measurement made on an ordinary board with a normal solder mask.

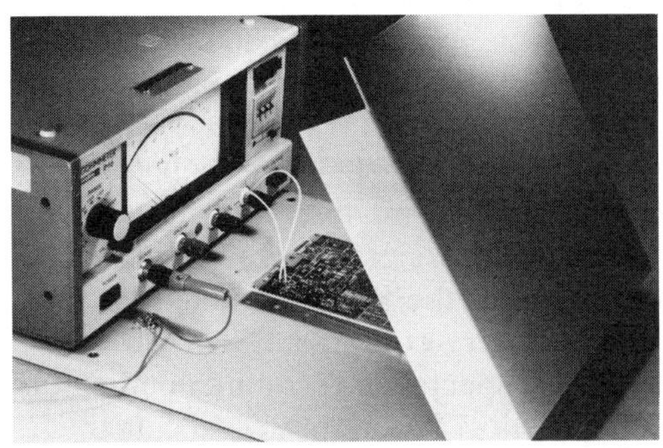

Fig. 11.10. Measuring insulation resistance of board provided with a screen printed black light-proof mask.

If the resistance value found drops to one-tenth or less of the normal readings, the board should be considered defective. The sample size is increased according to General Inspection Level II, and if the same reduction is still found, the batch should be rejected. Where the impedance level of the circuit is low and no significant offset currents occur, it is possible, in collaboration with the PCB designer, to exempt the batch.

The PCB designer should realize that his position towards the PCB manufacturer is rather weak when he specifies only a black mask with no indication of the insulation resistance required.

11.8 General Quality Requirements

Below we address the general quality requirements such as:

Appearance
Adhesion
Curing
Chemical resistance
Thermal stability

11.8.1 Appearance

All edges and apertures of the solder mask must have a good definition with no smear, which means that the apertures shall have the intended shape and size. See Figure 11.11. It is not unusual for a PCB manufacturer, on his own initiative, to touch up the artwork or filmwork in order to increase the apertures and in order to facilitate the screen printing process. In turn, the solder mask is no longer efficient against solder bridging. The purchase specification should state that the PCB manufacturer must not change the artwork without permission from the customer.

Fig. 11.11. Board with a satisfactory quality of solder mask.

11.8.1a Extent of Inspection

Since all boards will presumably show the same quality of printing, the inspection is carried out on only a very few boards of the sample. If the requirements are not met, the inspection should be extended so that a possible rejection of the batch is based on the full sample size. The boards are returned for reprinting unless they are urgently needed, so that a certain risk of solder bridging has to be accepted.

11.8.1b Embedded Metal Particles

Metal particles embedded in the solder mask can be accepted without any remarks entered in the inspection record if they occur in a limited number—for example, one defect per 16 sq. in. (100 cm^2) maximum—provided they are not so close to the pattern that the insulation distance is unacceptable.

Particles occurring more frequently than this or too close to the pattern cannot be accepted and are consequently rated as an A-defect. As the metal particles occur at random, the inspector should use the full sample size according to General Inspection Level II.

In general, the metal particles will be copper chips from the drilling operation. In the case of plated-through boards, metal chips cannot adhere throughout all the subsequent processes (image transfer, plating, and etching), so they presumably originate from the drilling of unplated holes. Such holes are usually drilled in a second drilling operation shortly before the screen printing of the solder mask. Inadequate cleaning of the board allows the chips to reach the screen printing stage and become encapsulated.

The PCB manufacturer should not necessarily be assessed as delinquent with respect to board cleaning. The problem is that the chips can become so hot that some of them cake to the epoxy surface and become very difficult to remove. This is another reason to design the board in such a way that a second drilling operation is avoided whenever possible.

11.8.1c Surface Quality

The surface of the solder mask must be even, and the following requirements must be met:

1. Pinholes numbering up to 10 per 16 sq. in. (100 cm^2) can be accepted without any remarks entered in the inspection record.
2. If the pinholes exceed Requirement 1, the board is returned for mask reprinting, and the fault is rated as a B-defect.
3. Larger, unintentionally skipped areas can be accepted up to 2 per 16 sq. in. (100 cm^2) provided they do not introduce any risk of solder bridging. No remarks are entered in the inspection record.
4. Larger, unintentionally skipped areas exceeding Requirement 3 are rated as a B-defect and the board is returned for reprinting.

If skipped areas occur at the same locations, this is most likely caused by a stencil defect. Random occurrences can be due to contamination—for example, silicone—causing the printing ink to pull back, almost corresponding to solder dewetting. If so, there is a general reduction of the adhesion of the mask, for which reason a tape test should be performed as described in Paragraph 11.8.2a.

Reduced adhesion is an indisputable cause for rejection, and the fault is consequently rated as an A-defect. The inspection should be based on a full sample size according to General Inspection Level II, as pinholes and larger skipped areas occur at random.

It is acceptable for the PCB manufacturer to repair pinholes and skipped areas by touching up (applying ink with a brush), if the workmanship is good.

11.8.2 Adhesion

Solder and insulation masks usually adhere to the board so well that the adhesion need not be tested. Insufficient adhesion can be due to insufficient curing of the printing ink after screen printing (see Paragraph 11.8.3) or to contamination of the board surface or the printing ink—for example, by silicone.

The printing ink can also fail to adhere in limited areas. Such areas are frequently characterized by a lighter color; they can alternatively be greyish because of the air underneath the mask. The fault is assessed in the same manner as unintentional holes and skipped areas mentioned in Paragraph 11.8.1.

11.8.2a Adhesion Test

When the inspector suspects that the mask has reduced adhesion, he should inspect the adhesion. The easiest way is to perform a tape test on a single board, very similar to the tape test for tin/lead overhang described in Paragraph 5.10. The solder mask is cross-hatched in squares with a cutter—for example, 5 × 5 squares, with an edge length of 0.1 in. (2.5 mm). The tape test is carried out on the cross-hatched area, and the number of squares torn off is used for assessing the adhesion.

11.8.2b Acceptance Criteria

In the IPC standard IPC-SM-840 "Qualification and Performance of Permanent Polymer Coating (Solder Mask) for Printed Boards," certain criteria of acceptance are based on a quality division into three classes. Within the individual classes the percentages of lifted squares stated below should not be exceeded:

Underlying Material	Class 1	Class 2	Class 3
Bare copper	25%	25%	5%
Gold or nickel	50%	50%	10%
Tin/lead plating	50%	50%	10%
Reflowed tin/lead	50%	50%	25%
Base laminate	5%	5%	5%

These values apply to tape tests performed before as well as after mass soldering, except Class 1, which only applies to a tape test before mass soldering.

Incidentally, the results of a tape test preceding mass soldering can be compared with the results of earlier tape tests to obtain a relative assessment of the adhesion of the solder mask.

11.8.2c Board Disposition

Insufficient adhesion is cause for rejection and consequently rated as an A-defect. The inspector should contact the PCB manufacturer about returning the boards on the basis of the existing test results.

11.8.3 Curing

The solder mask must be fully cured or mass soldering can cause webbing. See Figure 10.9. Another effect is an abnormally low insulation resistance. Incomplete curing is an indisputable cause for rejection and rated as an A-defect.

Normally, it is not necessary to check whether the printing ink is fully cured, but when the inspector suspects the curing—for example, if the surface has a slightly tacky or soft feel—he should perform a pencil test. The test is based on a set of pencils of various hardnesses; the inspector uses trial and error to find the pencil hardness that does not scratch the surface when the point is moved across the solder mask with a constant pressure and at a 45 degree angle.

Obviously, the result is uncertain and can be so operator-dependent that it can only be compared with earlier results obtained by the same operator. If such results are supplemented with a soldering test in order to determine the propensity for webbing, the inspector nevertheless has a reasonably useful indication of the quality of the solder mask.

There can be other causes of webbing: incorrect adjustment of the soldering machine (preheating, soldering temperature, speed), incorrect fluxing (insufficient solvent or inadequate drying after fluxing), and contaminated solder. See Paragraph 10.7.1.

In cases where the pencil test indicates an incomplete curing of the solder mask, a practical test can be performed. As this test is destructive, it should only be made on a single board. Webbing is cause for rejection and rated as an A-defect. The inspector should contact the PCB manufacturer about returning the boards on the basis of the existing test results.

11.8.4 Chemical Resistance

The solder mask must resist solvents normally used in assembly and soldering—for example, chlorothene, isopropyl alcohol, Freon, and not least, flux. Usually it is unnecessary to test the resistance of the solder mask to chemicals. If the inspector for any reason—for example, feedback from the assembly line—suspects inadequate chemical resistance, he should make a test.

A single board is selected and a few drops of the solvent or flux used are applied to the solder mask. After about one minute the solder mask is vigorously rubbed with a white rag at the places under test.

If the surface is resistant, no color contamination of the rag will be found, and the surface of the solder mask cannot be scratched by the fingernail at the places tested. Otherwise the chemical resistance of the solder mask is inadequate. This is cause for rejection and consequently rated as an A-defect.

As all the remaining boards will almost certainly behave in the same manner, it is not necessary to extend the inspection. The inspector should contact the PCB manufacturer about returning the boards on the basis of the existing test results.

11.8.5 Thermal Stability

The thermal stability of solder masks was superficially dealt with in the discussion of wrinkling appearing where the solder mask covers tin/lead plated conductors and voltage/ground areas. The solder mask must, at all other places—that is, where applied on the base material or on bare copper—withstand the heat absorbed during mass soldering.

Under the soldering conditions dealt with in Paragraph 10.5 (500°F (260°C) for 10 seconds), the solder mask must not show any sign of damage such as burning, color change, blistering, or delamination. Such faults are necessarily considered cause for rejection and consequently rated as A-defects. These faults, however, are so rare as to be mainly hypothetical, for which reason special tests are unnecessary. Screen printing inks for solder masking are usually based on the same materials and behave quite similarly, regardless of where the board is manufactured.

Chapter 12

Photopolymer Solder Mask Inspection

12.1 Introduction

From the preceding chapter, it appears that screen-printed solder masks will in general exhibit a limited registration accuracy, so that one cannot give modern densely packed boards a 100 percent protection against solder bridging caused by mass soldering.

12.1.1 Automatic Solder Bridge Removal

Some newer types of mass soldering machines remove solder bridges by means of an air-knife. The principle resembles hot-air leveling of solder-coated boards (Paragraph 9.4).

There is, however, a slight difference between the two principles. In the case of hot-air leveling, most of the solder remaining on the board after solder coating is removed. In the case of soldering it is desirable to remove only the amount of solder which can cause solder bridging and icicling. This requires an extremely careful and stable adjustment of the various parameters of the air-knife.

12.1.2 Principles of Photopolymer Masks

To satisfy increasing demands for solder masks with both very accurate registration and a very small clearance relative to the solder pads, new types of solder masks have been developed in recent years. These masks are based on a photopolymer material which is applied to the board, either in a solid form as a dry-film solder mask or in a liquid form as a liquid-film solder mask.

The solder mask pattern is transferred photographically by a contact copying process using ultraviolet (UV) light. A negative film of the solder mask is used—that is, the film is transparent and the apertures of the mask are opaque. This causes the exposed areas to polymerize so that they become insoluble when developed.

Nonpolymerized areas (all the apertures) are dissolved and washed off by the developer, after which the mask appears. The last process is a post-curing, which gives the mask a higher resistance to the conditions it will be subjected to by the PCB user, primarily mass soldering and the solvents used in soldering.

12.1.3 Accuracy of Photopolymer Masks

The photographic transfer of the solder mask image makes it possible to obtain a much better accuracy than by screen printing, so that the registration accuracy is no longer 0.01 in. (0.25 mm) but 0.002 in. (0.05 mm). The photographic transfer makes the registration between the solder

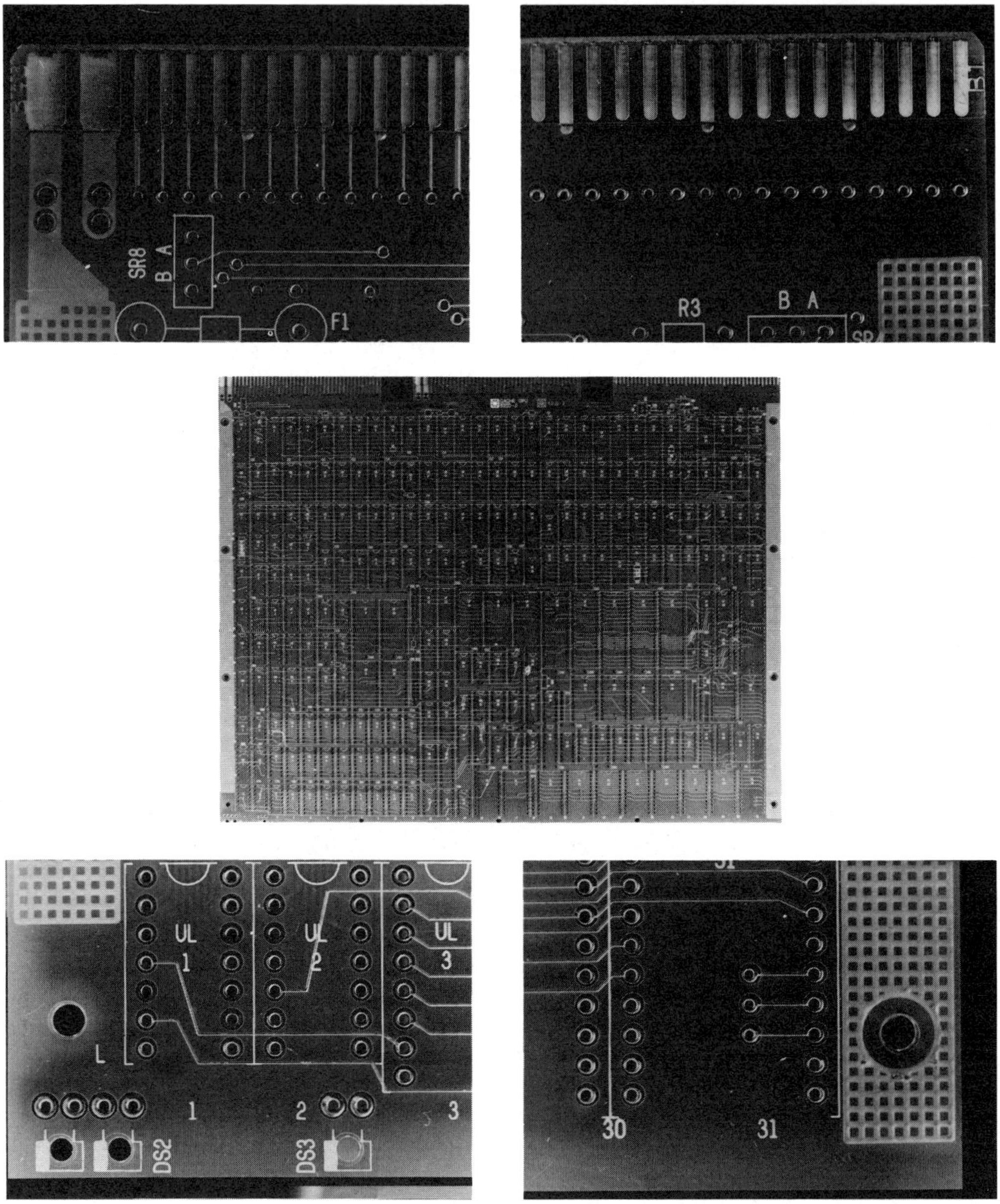

Fig. 12.1 Corner sections of a 12 x 15 in. (300 x 380 mm) multilayer board with dry-film solder mask. The sections show a solder mask register of 0.002 in. (0.05 mm) all over the board.

pads and the solder mask apertures, on the whole, uniform all over the board. This is apparent from Figure 12.1, which shows sections taken from each corner of a 12 × 15 in. (300 × 380 mm) board. The misregistration is about 0.002 in. (0.05 mm), when measured as described in Paragraph 11.4.2. The better uniformity is primarily due to the fact that, contrary to screen printing, no stretching of the image-carrying medium (the solder mask working film) occurs. A contributing factor is the use of an accurate registration system based on guide pins and holes.

12.1.4 Additional Advantages

Another advantage of photopolymer solder masks is that the clearance between a solder pad and the corresponding aperture can be made much smaller in the case of screen-printed solder masks. In some cases the clearance can even be zero. This is shown in Figure 12.2.

The greatest advantage is that it is now possible to achieve a very high certainty of full coverage of the edge of a conductor adjacent to a pad. The edge coverage in itself also greatly improves, as shown in Figure 12.3.

Fig. 12.2. The clearance between solder pad and solder mask aperture can be made very small.

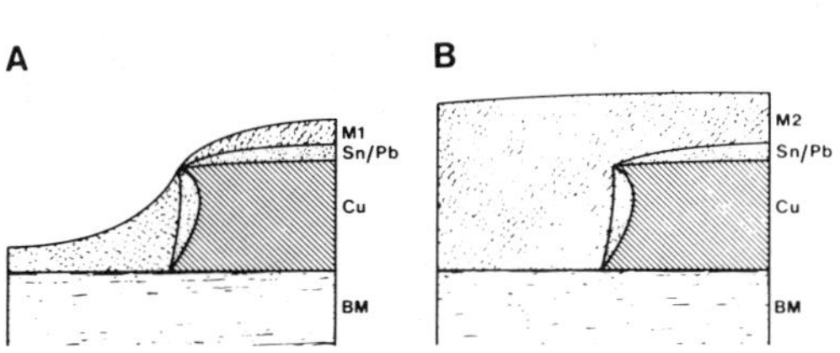

Fig. 12.3. Edge coverage of conductor. A: Screen-printed solder mask. B: Dry film solder mask.
M = solder mask. Sn/Pb = tin/lead layer on conductor.
Cu = copper of conductor. BM = base material.

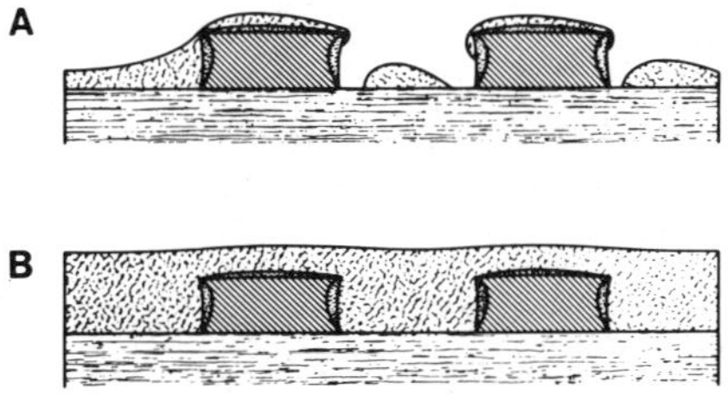

Fig. 12.4. Filling of the space between two parallel conductors. A: screen-printed solder mask. B: dry-film solder mask.

Moreover, photopolymer solder masks offer the great advantage of filling the space between two adjacent parallel conductors completely with the solder mask, which is not always the case with screen-printed solder masks. See the cross-sections shown in Figure 12.4, which show the difference between the screen-printed solder mask and a photopolymer solder mask.

12.2 Photopolymer Solder Masks

Photopolymer solder masks are not nearly so common as screen-printed solder masks. It is, therefore, necessary to deal more closely with photopolymer solder masks, to ensure that the inspector can fully understand why they are, in many respects, superior to screen-printed solder masks. In any case, a knowledge of photopolymer solder masks will be vital to the result of an inspection of the PCB manufacturer's processes. The following paragraphs distinguish between dry-film solder masks and liquid-film solder masks.

12.2.1 Dry-Film Solder Masks

The dry films are delivered in rolls and consist of a thermostable photopolymer film sandwiched between a 0.001 in. (25 μm) protective polyolefin foil and a 0.001 in. (0.25 μm) polyester base film. The photopolymer film can be obtained in various thicknesses—for example 0.0021 in.

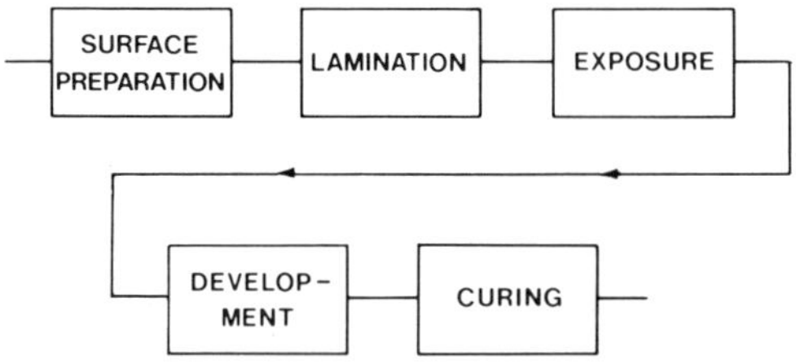

Fig. 12.5. Process stages when manufacturing boards with dry-film solder masks.

(53 μm), 0.003 in. (75 μm), and 0.004 in. (100 μm). The PCB manufacturer's production processes comprise the stages shown in Figure 12.5 and described in the following subparagraphs.

12.2.1a Surface Treatment

A satisfactory adhesion requires a pretreatment of the board surface. Most faults concerning a reduced adhesion, air pockets, etc., can be traced back to inadequate pretreatment and conditioning of the board. Degreasing and brushing are to be performed and in the case of bare-copper boards, also a black oxidation of the copper. New types of dry films, however, do not require the oxidation process.

To remove all moisture after cleaning, the boards are baked at 250°F (120°C) for 15 to 30 minutes.

12.2.1b Application

The application of a dry film, also called lamination, is somewhat more difficult to perform when it is a solder mask and not an ordinary plating resist. In the first case, the board surface is very uneven, with height differences between the base material and the pattern surface of 0.002 to 0.004 in. (50 to 100 μm) or more, whereas in the second case the surface is very even.

When all spaces between parallel conductors are to be completely filled, leaving no air pockets, the ordinary laminator, which applies the dry film under pressure and heat, is inadequate; a vacuum laminator is needed.

In principle, the vacuum laminator consists of a heated chamber, which can be opened, closed, and evacuated, plus an automatic feeding system for the dry film. The vacuum laminator can apply dry film on both sides of the board at the same time. The protective foil is removed immediately before the dry film is applied to the board and automatically reeled up.

12.2.1c Exposure

Photopolymer dry films for solder masks are normally negative-acting; this means that a negative film of the solder mask must be used—that is, a transparent film with opaque solder mask apertures. When exposed to light, the dry film polymerizes and becomes the solder mask itself. Under the opaque areas the dry film remains unchanged and will therefore be removed (dissolved) during the subsequent development.

The working film is registered relative to the board by means of indexing pins and holes, whereupon the board with the working film attached is placed in the exposure machine. The polyester base film remains on the photopolymer film and protects it against scratching, contamination, and oxidation.

Furthermore, it prevents the working film from sticking to the photopolymer film. A disadvantage is that the photographic resolution is somewhat reduced by the light diffusion caused by the 0.001 in. (25 μm) separation between the emulsion layer of the film and the photopolymer film.

Ultraviolet light with a wavelength between 320 and 400 nanometers is used for the exposure. The intensity must be sufficiently high to ensure complete polymerization so that the solder mask becomes thermostable at a soldering temperature of 500°F (260°C).

12.2.1d Development

After exposure, the base film is peeled off and the solder mask developed. All non-polymerized areas, preferably the solder pads, are washed off by the developer, which is an organic solvent. The solder mask is now sufficiently hard to allow inspection. If defects are found, it is still possible to strip the solder mask by means of a solvent.

12.2.1e Post-Curing

A post-curing is necessary in order to fully develop the desired properties of the solder mask. Post-curing can be carried out in two ways. When optimum adhesion as well as high resistance to solvents is required, it is necessary to use a heat treatment combined with UV curing. This combined treatment is particularly important in the case of solder-coated and hot-air leveled boards. For other applications UV curing alone suffices.

Lack of, or insufficient, post-curing often shows up in the form of webbing after mass soldering, because the solder has a tendency to adhere to surfaces not fully cured.

12.2.2 Liquid-Film Solder Masks

Until now, liquid-film solder masks have found only a relatively limited use, but some PCB manufacturers expect that this technique will gain a fairly wide acceptance, for which reason we will deal with it briefly. Similar to dry-film solder masks, liquid-film solder masks utilize a photopolymer material based on epoxy, and apart from the application itself, which also includes a drying stage, the processes are, on the whole, the same. The reader is therefore referred to Figure 12.5.

12.2.2a Surface Treatment

The surface treatment is performed in exactly the same way as described in Paragraph 12.2.1a, possibly supplemented by ultrasonic cleaning.

12.2.2b Application

As the name implies, the film is applied in liquid form. The special method of application is the curtain coating technique, in which the board at high speed—about 6 ft/sec (2 m/sec)—passes through a falling curtain of lacquer, the thickness of which is 0.012 to 0.020 in. (0.3 to 0.5 mm).

An additive increases the surface tension of the lacquer so much that it does not run down at all into holes with sizes up to approximately 0.12 in. (3 mm). In this way it is possible to cover the base material as well as the pattern with a very uniform layer of lacquer without filling the solder holes.

By adjusting the viscosity of the lacquer it is possible to fill the space between two parallel conductors, even when the spacing is as small as 0.004 in. (0.1 mm). After drying, the board is ready for exposure.

12.2.2c Exposure

Exposure is performed by means of UV light in exactly the same way as for dry-film solder masks. As there is no base film, a higher photographic resolution can be expected, and not least, a very minor light diffusion.

12.2.2d Development

Development is performed in exactly the same way as for dry-film solder masks—that is, with an organic solvent.

12.2.2e Post-Curing

Just as in the case of dry-film solder masks, a post-curing is necessary. This is performed as a heat treatment, during which the epoxy attains the full mechanical, chemical, and electrical properties required.

12.3 Artwork Problems

As mentioned earlier, photopolymer solder masks distinguish themselves by a totally different accuracy than found in the case of screen-printed solder masks. Therefore, there are reasons for assessing the way in which the artwork problems discussed in Paragraph 11.2 will affect photopolymer solder masks. The following discussion refers to dry-film solder masks only.

12.3.1 Clearance around Solder Pads

During exposure of the solder mask, a light diffusion, also called coving, takes place, partly due to the thickness of the protective polyester film. This reduces the diameter of the mask apertures in the developed solder mask by 0.002 to 0.004 in. (50 to 100 μm).

When the PCB designer uses the same size of mask apertures and solder pads, which seems very practical for some methods of preparing manual artwork, the rim of the solder pads will be overlapped by the finished solder mask by 0.001 to 0.002 in. (25 to 50 μm). Normally this has no effect on the solderability of the board.

Some PCB manufacturers, however, prefer to avoid overlapping and use instead a solder mask film with mask apertures oversized by 0.004 to 0.008 in. (0.1 to 0.2 mm). This is achieved by reversing a positive solder mask film (opaque with transparent mask apertures) into a negative solder mask film (transparent with opaque mask apertures), through a transparent film of suitable thickness laid between the films. In this way light diffusion gives the mask apertures (the opaque pads) the desired oversize.

The result is that when the solder mask image is transferred to the photopolymer solder mask film, the finished solder mask fits closely around the rim of the solder pads. See Figure 12.6.

The PCB designer is advised not to design with a larger clearance than 0.004 in. (0.1 mm) in dense areas. Otherwise he will lose one of the advantages of the photopolymer solder mask: the certainty of covering the edge of a conductor placed very close to a solder pad. It is clearly apparent from Figure 12.7 that there is not much to give away when two conductors are placed between two IC solder pads, particularly when a small general misregistration between the solder mask and the pattern as well as a permanent dimensional change of the board is taken into consideration.

12.3.2 Parallel Conductors

In the case of photopolymer solder masks, adjacent conductors with spacings as small as 0.008 in. (0.2 mm) do not present any problems regarding complete filling of the space. See Figure 12.8, which shows a cross-section of parallel conductors on a dry-film solder-masked board. It should be noted that there is a perfect edge coverage and a reasonable thickness on top of the conductors. Since the dry film is applied under heat, it will soften enough to fit closely to any possible undercut along the edges of the pattern.

12.3.3 Thickness of the Solder Mask

Obviously, the thickness of the dry film is crucial to the thickness of the finished solder mask, but the density and thickness (height) of the pattern also play an important part. Consequently, there will be variations across the board surface, and therefore it is necessary to define the minimum thickness of the solder mask as

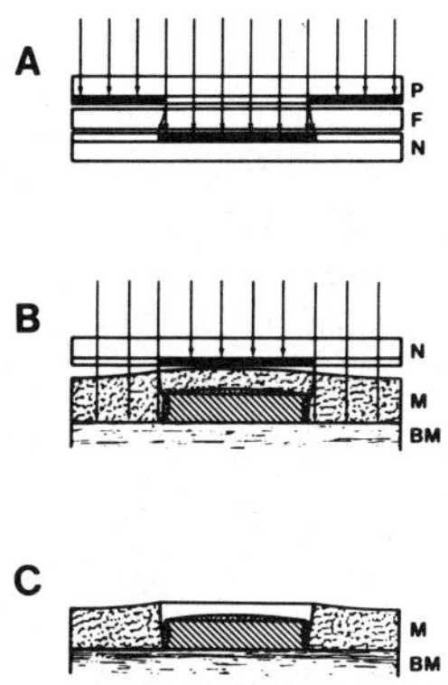

Fig. 12.6. Increasing the aperture size on the solder mask in order to avoid overlapping of the solder pads.
A: Reversing a positive solder mask film (P) to a negative solder mask film (N) with a transparent film (F) placed between (P) and (N).
B: By exposing the dry film mask (M), the solder mask aperture size is reduced because of light creeping. BM = base material.
C: The finished dry-film solder mask fits tightly around the solder pad.

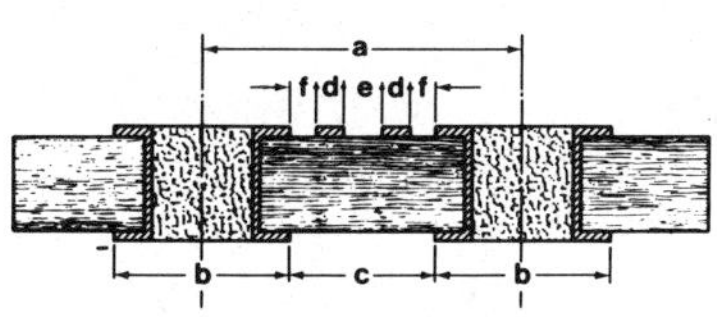

Fig. 12.7. Example of the limited space conditions when laying two conductors between two IC solder pads.

a: Modular distance	=	0.100 in. (2.54 mm)	
b: Diameter of solder pad	=	0.055 in. (1.40 mm)	
c: Distance between solder pads	=	0.045 in. (1.14 mm)	
		Example No. 1	*Example No. 2*
d: Width of conductor	=	0.0080 in. (0.20 mm)	0.0098 in. (0.25 mm)
e: Width of space	=	0.009in. (0.24 mm)	0.0094 in. (0.24 mm)
f: Clearance to pad	=	0.0098 in. (0.25 mm)	0.0080 in. (0.20 mm)

Fig. 12.8. Microsection of parallel conductors on board provided with a dry-film solder mask.

measured at the middle of a conductor. According to IPC-SM-840, "Qualification and Performance of Permanent Polymer Coating (Solder Mask) for Printed Boards," the minimum thickness can be divided into the following three classes:

Class 1: Visible coverage (unspecified thickness)

Class 2: Thickness 0.005 in. (13 μm)

Class 3: Thickness 0.010 in. (25 μm)

General guidelines for selecting the dry-film thickness appear in Figure 12.9, which states the minimum thickness achievable

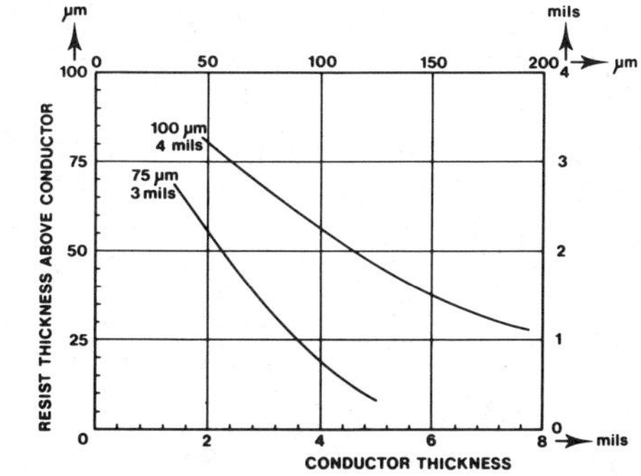

Fig. 12.9 Thickness of dry-film solder mask above a 0.012 in. (0.3 mm) wide conductor as a function of the conductor thickness (height). The curves correspond to two commercially available dry-film types.

on the finished board as a function of the thickness (height) of the pattern, valid for the basic dry-film thickness of 0.003 in. (75 μm) and 0.004 in. (100 μm). The minimum thickness is measured at the middle of a 0.012 in. (0.3 mm) wide conductor.

If the PCB designer is designing for Class 3 and the thickness of the pattern is 0.0033 in. (84 μm), a 0.003 in. (75 μm) thick dry film material is sufficient. The thickness of the pattern is derived as follows:

1 oz. copper foil	0.0014 in. (36 μm)
+ plated copper	0.0010 in. (25 μm)
+ 50% tolerance	0.0005 in. (13 μm)
+ reflowed tin/lead	0.0004 in. (10 μm)
Total	0.0033 in. (84 μm)

12.3.4 Intentional Coverage of Solder Pads

A great advantage of photopolymer solder masks is that the PCB designer does not have to design with intentional coverage of solder pads as it is possible to obtain a close fit of the solder mask around the pads all over the board.

When the solder mask artwork is based on a pad master from which the solder and component side artwork is derived, possibly with an enlargement of the mask apertures as mentioned in Paragraph 12.3.1, it is easier to let all via holes be uncovered by the solder mask. Otherwise, it is necessary to remove all via hole mask apertures (pads) from the solder mask artwork. This can also be achieved by touching up the negative solder mask film (opaque with transparent mask apertures), provided it is possible to indicate which apertures are to be closed.

12.4 Application Faults

When imaging the solder mask pattern, the faults mentioned in Paragraph 11.3, specifically insufficient registration and displacement across the board, are avoided to a very high degree. The imaging is based on a photographic process, and as the film does not stretch in the same way as the screen of the screen printing machine, the total registration fault is normally less than 0.002 in. (50 μm). This is in addition to any possible permanent dimensional change of the board itself, usually limited to ±0.05 percent. See Paragraphs 6.4.3 and 11.3.1c.

For these reasons, inspection of dry- and liquid-film solder masks proceeds substantially faster than of screen-printed solder masks. The inspector can suffice with a very limited sample size when checking the registration at the board corners. A satisfactory registration is characterized by having all mask apertures located fairly concentrically with respect to the solder pads as shown in Figure 12.1.

12.5 Efficiency of the Solder Mask

As mentioned earlier, the primary function of a solder mask is to prevent solder bridging during mass soldering. With the higher accuracy of registration and the greatly reduced incidence of faults, it is easier to inspect dry- and liquid-film solder masks.

Due to the photographic imaging, the size of the mask apertures will be under full control. With a clearance of 0.004 in. (0.1 mm), as mentioned in Paragraph 12.3.1, insufficient coverage of conductor edges adjacent to solder pads requires a rather large and therefore visible misregistration of the solder mask.

If the inspector finds one or more conductors with uncovered edges, the most likely reason is a fault in the working film used for the exposure of the photopolymer solder mask. The defect will therefore repeat itself on boards of the same batch.

12.6 Unintentional Coverage of Solder Pads

Even when the PCB designer shows reasonable consideration for the fact that the solder mask apertures shrink a bit during imaging, the dry-film solder mask can overlap the solder pads a little. When no overlapping is allowed at all, the inspector should, with his knowledge of the design and manufacturing conditions, be able to decide whether the fault is effectively self-inflicted.

12.6.1 Causes of Unintentional Coverage

There are several causes of failure, and under worst-case conditions they all act in the same direction. There can be an outward edge displacement, as mentioned in Paragraph 5.4, combined with registration faults and permanent dimensional changes as dealt with in Paragraph 12.4.

12.6.2 Preventing Unintentional Coverage

In order to eliminate any possibility of unintentional coverage of the rim of the solder pads, it is necessary to design the solder mask apertures with a large clearance. This is determined as at least 0.002 in. (50 μm) for shrinking effects, 0.002 in. (50 μm) for outward edge displacements, 0.002 in. (50 μm) for misregistration and approximately 0.05 percent of the board length to cover a possible permanent dimensional change.

In the case of an 8 × 8 in. (200 × 200 mm) board, the necessary design clearance is about 0.010 in. (0.25 mm). On the finished board, the clearance becomes 0.008 in. (0.20 mm) due to shrinkage of the solder mask apertures. Such a large clearance can, however, be very undesirable and it is not recommended for the following reasons:

1. The photographic method described in Paragraph 12.3.1 for enlarging the solder mask apertures (opaque pads on the film) cannot be used to the necessary extent without drastically reducing the edge resolution. It is always necessary to perform the process in two or more steps.
2. It is impossible to be sure of complete coverage of the outside edges of two conductors placed between two IC pads. This appears from Figure 12.7. When only one conductor is placed between two IC pads, there is a margin of 0.007 in. (0.17 mm), which is sufficient.

Fortunately, actual results very seldom turn out as bad as the calculations for a worst-case condition. Many PCB designers therefore prefer to use the smaller clearance of 0.004 in. (0.1 mm) and risk that the solder mask will overlap the rim of the solder pads, which does not affect the solderability.

12.6.3 Board Disposition

The inspector should accept overlaps to the middle of the actual annular rings with no remarks entered in the inspection record.

12.7 Thickness of the Solder Mask

Compared to screen-printed solder masks, dry-film solder masks are characterized by a better-defined thickness, corresponding to the thickness of the dry-film material used, modified by the thickness (height) of the PCB pattern itself. It is therefore physically impossible for the solder mask thickness, on top of a conductor, to correspond to the thickness of the dry film material used.

12.7.1 Checking the Thickness

In practice, the inspector can only check whether a dry film material of the specified thickness has been used. The measurement is performed as described in Paragraph 11.6. In addition, the inspector can measure the thickness of the solder mask on top of a conductor and compare the result with the value indicated by one of the curves in Figure 12.9.

After mass soldering, there is no pronounced difference in wrinkling between photopolymer solder masks and screen-printed solder masks (Paragraph 11.6). Incidentally, it seems that photopolymer solder masks are frequently used in conjunction with solder-coated and hot-air leveled boards. In such cases, the solder mask will be applied to a bare-copper board, and consequently no wrinkling will occur. It should be noted that the solder mask need not necessarily be a photopolymer mask. An ordinary screen-printed solder mask can also be used in conjunction with this technique.

12.7.2 Board Disposition

Overly thin dry-film material can be cause for rejection, for which reason the fault is rated as an A-defect. Before the boards are returned, the inspector should consider the possibility of exempting them. For example, the boards could be overspecified, which means unnecessarily stringent demands.

It can also happen that the PCB designer is not familiar with the curves shown in Figure 12.9, so that the solder mask has in fact sufficient thickness above the conductors. With a pattern thickness of 0.004 in. (100 μm), the thickness of the solder mask measured on top of a conductor is 0.001 in. (25 μm). As the dielectric strength is 2.5 kV/0.001 in. (100 V/μm), the solder mask has, at the spot in question, a dielectric strength of approximately 2500 volts. This is a much higher dielectric strength than can be obtained with a screen-printed solder mask (Paragraph 11.6). Based on these functional considerations, a rejection seems exorbitant.

Liquid-film solder masks should be assessed on a par with screen-printed solder masks concerning the thickness. The reason is that the mask thickness depends upon the adjustment of the curtain coating machine, meaning that there are many parameters to monitor:

Viscosity of the lacquer
Solid matter of the lacquer
Thickness of the curtain
Speed of travel

The reader is therefore referred to Paragraph 11.6, but he should note that the dielectric strength is of the same order of magnitude as specified for dry-film solder masks: 2.5 kV/0.001 in. (100 V/μm).

12.8 Electrical Insulation Properties

On the whole, the electrical insulation properties of both dry- and liquid-film solder masks are of the same order of magnitude as those of screen-printed solder masks, perhaps a little better. The reader is therefore referred to Paragraph 11.7.

12.9 General Quality Requirements

By and large, we expect a photopolymer solder mask to meet the same quality requirements as a screen-printed solder mask. We therefore address the following requirements:

- Appearance
- Adhesion
- Curing
- Chemical resistance
- Thermal stability

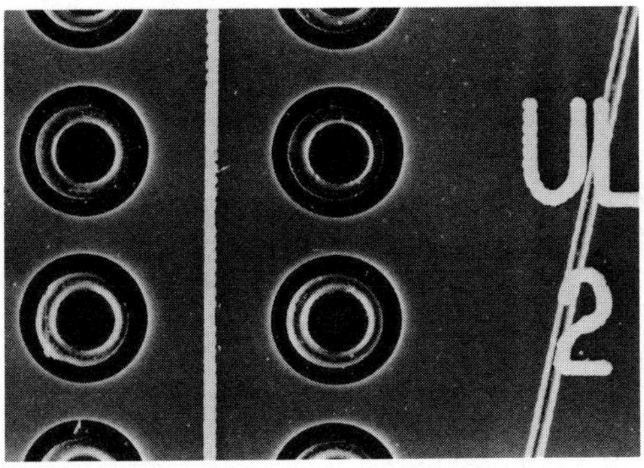

Fig. 12.10 Satisfactory appearance of the aperture rims of a dry-film solder mask.

12.9.1 Appearance

All edges and apertures of the solder mask must be well-defined, specifically that all apertures must have the intended shapes and sizes. See Figure 12.10. It is, however, practical to include a clause in the purchase specification stating that the PCB manufacturer can photographically increase the clearance from 0 to 0.004 in. (0.1 mm) in order to avoid overlapping of the solder pads.

12.9.1a Extent of Inspection

The high reproducibility of the solder mask means that the inspection is performed on only one board of the sample. If the requirements are not met, the inspection should be extended so that a possible rejection of the total batch is determined on the full sample size.

12.9.1b Manual Touch-Up

If rejected, the boards are returned to the PCB manufacturer who can possibly use a one-pack epoxy ink to perform a local manual touch-up. Although dry-film masks can be dissolved with methylene chloride, this is risky because the epoxy base material will soften and swell. This can result in reduced adhesion between copper and base material as well as reduced reliability of plated-through holes. On the contrary, liquid-film solder masks cannot be stripped when cured, so this problem does not arise at all.

12.9.1c Frayed Edges

Although the solder masks can withstand machining of the board, it can happen that a solder mask applied right out to the edges of the board frays after profiling. If this is unacceptable, the PCB designer should mask the solder mask on the artwork or on the filmwork—for example, with red lithotape—so that the finished solder mask is kept 0.04 in. (1 mm) from the board edges.

12.9.1d Holes and Embedded Particles

Encapsulated metal particles together with pinholes and larger unintentional apertures are dealt with as described in Paragraph 11.8.1. The boards are returned to the PCB manufacturer for rework or remake. It is acceptable for the PCB manufacturer to repair pinholes and larger holes by touching up with a one- or two-pack screen-printing ink.

12.9.1e Radial Wrinkles

Around larger holes, radial wrinkles are sometimes seen in the solder mask. Figure 12.11 shows an example of such wrinkles. The occurrence is caused by mechanical tension in the dry film and cannot always be avoided unless the PCB designer relieves the solder mask around such holes. This is of no practical importance and is no cause for entering remarks in the inspection record.

12.9.1f Cracks

The dry-film solder mask must not crack—for example, where it lies across conduc-

Fig. 12.11 Example of a radially occurring wrinkling of a dry-film solder mask around larger holes.

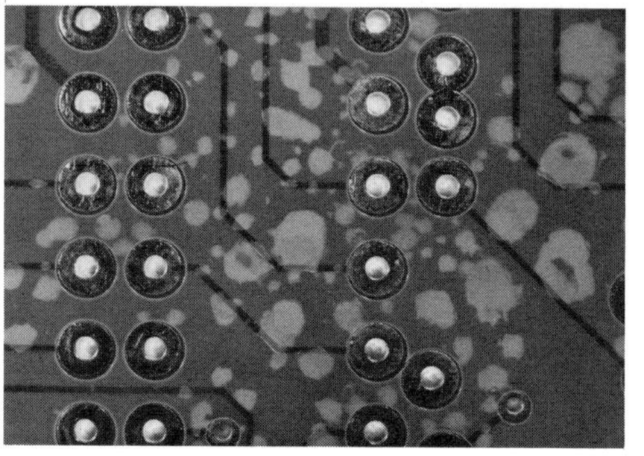

Fig. 12.12. Example of an unsatisfactory adhesion of a dry-film solder mask to the substructure.

tors. The reason is usually insufficient thickness of the dry-film material relative to the thickness of the pattern; see the curves in Figure 12.9. Cracking is an obvious cause for rejection and is rated as an A-defect.

12.9.1g Dry-Film Gloss

Dry-film solder and insulation masks have a substantially glossier surface than screen-printed solder masks. This can be an advantage as well as a disadvantage during assembly. Improper ceiling illumination can cause unpleasant light reflections from the insulation masks, but assembly tables with spotlight indication of the insertion points give a clearer indication. Dry-film solder masks seem more advantageous than screen-printed solder masks in the case of mass soldering as less flux adheres after soldering.

12.9.2 Adhesion

The adhesion of photopolymer solder masks to the boards should be dealt with in exactly the same way as for screen-printed solder masks, for which reason the reader is referred to Paragraph 11.8.2. Figure 12.12 shows a section of a solder mask with poor adhesion The problem areas have a lighter color due to the lifting of the mask from the substructure. The reason is usually board contamination at the affected locations, and the board is rejected. The fault is consequently rated as an A-defect.

12.9.3 Curing

A major requirement of a fully cured solder mask is a post-curing after the development of the solder mask (see Paragraphs 12.2.1e and 12.2.2e). The inspection is analogous to the inspection described in Paragraph 11.8.3, to which the reader is referred.

12.9.4 Chemical Resistance

Since the same conditions as described for screen-printed solder masks are also valid here, the reader is referred to Paragraph 11.8.4. It should be noted, however, that dry-film solder masks cannot withstand certain types of Freon.

12.9.5 Thermal Stability

Here again we have the same conditions as described for screen-printed solder masks. The reader is therefore referred to Paragraph 11.8.5.

Chapter 13

Component Notation Inspection

13.1 Introduction

Component notations, which are always screen printed, exhibit the same type of screen-printing faults as solder masks (Paragraph 11.3). Since they do not need the same registration accuracy, component notations are seldom subject to the same registration tolerances as solder masks. The inspector should perform a functional assessment of the component notation, concentrating on whether it accomplishes the primary purpose of identifying the components on the board.

A small registration error is of minor importance as long as the printing ink does not get into the solder holes. Although inspection of component notations is usually quite easy, the inspector should know the functional requirements as well as the design conditions.

13.2 Functional Requirements

The following paragraphs deal with some of the printing and design problems which crucially influence the functional requirements made on the component notation:

- Legibility
- Registration
- Distortion

13.2.1 Legibility

The printing must be well-defined and unsmeared as shown in Figure 13.1. This

Fig. 13.1 Example of a satisfactory appearance of a component notation.

requirement cannot be completely fulfilled where the notation is printed across the conductors. The high packaging density of many modern boards leaves the PCB designer with no alternative to placing the notation across some of the conductors. The inspector must, therefore, understand that a slightly blurred printing is basically self-inflicted, so that the PCB manufacturer is blameless.

It is assumed here that the component notation size permits a reasonably good printing quality. This means that the type height should be at least 0.060 in. (1.5 mm) and the line thickness at least 0.012 in. (0.3 mm). Smaller type sizes limit the PCB manufacturer's odds of achieving an acceptable printing quality and good legibility.

The printing can, on the other hand, be so badly done that the component notations are somewhat blurred. This presents the risk of assembly errors because certain figures resemble each other, particularly the following pairs of figures: 3 and 8, 5 and 6, and 7 and 9. The inspector can assess legibility by simply checking whether these figures are clearly distinguishable.

Reduced legibility of component notations generally does not mean rejection if the boards are otherwise fully functional. Assembly is still possible even though manual assembly is somewhat hindered by the difficulty in identifying correct component insertion locations. The fault is rated as a B-defect, but it is only reasonable to call the PCB manufacturer's attention to the problem or to investigate whether the artwork was prepared improperly.

Reduced legibility is not rated higher than a B-defect because it is becoming much more common to use semiautomatic or fully automatic component insertion techniques. Insertion hole indication is optical for semiautomatic insertion. In the case of fully automatic insertion, the boards are positioned correctly with respect to the insertion unit. The component notation is therefore of no particular operational importance.

During manual component insertion on an assembly line, when the board has no visual insertion aids, the individual operator usually has a limited number of components to insert. The operator soon learns where to insert the components, making the component notation of limited importance here also.

In field servicing, the component notation is also usually of minor importance. The component notations are frequently placed beneath the component bodies so that they cannot be read. If the repairman has no other means of component identification, the notation of course becomes more important.

These considerations should not be interpreted as opposing component notations, but rather to show the inspector that component notations are usually of no real functional importance. A rejection based on imperfect screen printing therefore seems unjustified.

13.2.2 Registration

Like a solder mask, the component notation can under some circumstances be misregistered. The error shows up as a displacement of up to 0.01 in. (0.25 mm). In order to prevent printing ink from flowing into solder holes, the PCB designer should keep a minimum clearance of 0.012 in. (0.3 mm) between the solder holes and component notations or leader lines. As this does not prevent overprinting the solder pads, the clearance should be made with respect to the rim of the solder pads. Overprinting is undesirable because the subsequent soldering damages the component notations.

If the inspector finds solder holes contaminated with printing ink, he should investigate whether the component notation was prepared with sufficient clearance so that the responsibility can be placed. The investigation is performed by laying a positive component-notation film over a positive component-side film, and check whether all component notations, including leader lines, are adequately separated from the solder pads and solder holes.

The risk of reduced solderability makes printing ink in the solder holes unacceptable. The fault is rated as an A-defect, and the boards are returned. This may sound very demanding, but in practice the registration is rarely that bad.

Normally, only a very few places will have clearances insufficient to absorb larger screen-printing displacements, so there are very few recurring faults. Since the defective spots are defined, it is possible to ensure satisfactory soldering quality by a normal post-soldering operation.

Consequently, it is not necessary to return the boards.

The inspector should persuade the PCB designer to increase inadequate clearances. The fault is rated as a B-defect because the PCB manufacturer, during his audit of the documentation package, should have realized that under adverse conditions the printing ink could flow into the solder holes.

13.2.3 Distortion

Screen printing of component notations must not be performed until after the boards have been reflowed. The tin/lead layer of larger ground areas and wide conductors is likely to flow a little while molten, particularly if the board is reflowed in a vertical position in hot oil.

When the tin/lead surface flows, an overlying component notation follows the flow. This can be clearly seen if there are distorted leader lines or contour lines. An insulation mask, however, tends to stabilize the component notation.

For these reasons, a PCB manufacturer working to high professional standards always reflows the boards before screen printing the solder and insulation masks as well as the component notation. If an error occurs in production and the boards pass through the PCB manufacturer's final inspection (which the author has experienced occasionally) the inspector should assume a functional attitude towards the error as described in Paragraph 13.2.2.

13.3 General Requirements

As stated in the following paragraphs, certain fundamental quality demands are made on the component notation:

- Contrast
- Adhesion
- Curing
- Insulation resistance
- Chemical resistance
- Thermal stability

13.3.1 Contrast

It is of great advantage to the operator in the production department that the notation printing ink contrasts well with the base material or the insulation mask.

When no particular color is specified in the purchase specification, the component notation can be in various colors, usually determined by the color of ink already in use when the boards are screen printed.

The most common colors are black and white, but when the board is viewed at a certain angle with glaring reflections from the ceiling lights, the component designations can be difficult to read.

In the author's experience, yellow is a far better color, as it contrasts sharply with most base materials whether they are green (epoxy glass) or brown (phenolic paper and polyimide) or with green insulation masks. Also, the yellow color does not disappear under adverse viewing angles. The PCB designer is recommended to prescribe a definite color, preferably yellow, in the purchase specification.

If the component notation has been accidentally printed in the wrong color, the inspector should exempt the boards as they are fully functional. The fault is rated as a B-defect, and the PCB manufacturer is enjoined to use the color specified.

In some cases a component notation may occur on the solder side. If a solder mask is to be applied here, the component notation should be printed over the solder mask in order to preserve contrast.

13.3.2 Adhesion

The factors which determine the adhesion of the component notation to the substructure are the same as those for solder and insulation masks (Paragraph 11.8.2). The tape test mentioned there can, as a matter of fact, be performed directly on the component notations, and it should not be possible to remove them with tape. Scraping across the component notations with a slightly dull knife should have no effect on them.

It can be difficult to obtain sufficient adhesion to a dry-film mask if the notation is screen printed at the wrong moment—that is, just after the UV curing. A sensible measure therefore is a routine tape test for adhesion of the component notation to dry-film insulation masks. A single-board test is sufficient.

The values listed in Paragraph 11.8.2 provide a rough indication of the adhesion according to the degree of tear-off, based on the number of torn-off figures and letters relative to the total number included in the tape test.

Insufficient adhesion is assessed the same as reduced legibility, which means that the fault is rated as a B-defect.

13.3.3 Curing

In order to achieve a high resistance to mechanical effects, the component notation must be fully cured. Incomplete curing is demonstrated by scraping on one of the leader lines by a slightly dull knife. In this way it can be easily determined whether the printing ink is brittle or soft. Unless the component notation can be removed with a fingernail, incomplete curing is relatively noncritical and should not cause rejection. The fault is rated as a B-defect, and the PCB manufacturer should be informed of the fault.

13.3.4 Insulation Resistance

Since the notation printing ink is normally based on the same type of ink as used for the solder mask, it causes no problems regarding insulation resistance. There are, however, some objections to using black printing ink, as certain types are slightly conductive because they contain carbon black. Under certain circumstances and with no insulation mask, the component notations and leader lines can create leakage paths between high-resistance circuits. See Paragraph 11.7.

13.3.5 Chemical Resistance

As the chemical resistance requirements are exactly the same as for solder and insulation masks, the reader is referred to Paragraph 11.8.4.

13.3.6 Thermal Stability

Thermal stability of the component notation becomes important only when the notation is screen printed on the solder side. As the ink is similar to that used for solder masks, there are normally no problems with thermal stability and no need for the inspector to test for it. See Paragraph 11.8.5 for a discussion of thermal stability as it applies to solder masks.

Chapter 14

Gold Plated Edge Connector Inspection

14.1 Introduction

Despite steadily increasing gold prices, many boards are still designed with gold plated edge connectors. See Figure 14.1. A major reason is the fact that many military specifications continue to prescribe gold plating on the edge connectors. It should be noted that in the United States, newer military projects require two-part connectors.

Many functional deficiencies can be traced back to poorly applied gold plating. This chapter deals in some detail with the quality assessment of edge connectors, especially because many of the assessments are subjective. For this reason, a detailed treatment seems fully justified.

This chapter does not address machining—for example, machining of the slot and chamfering of the connector edge, which are dealt with in Paragraphs 15.7 and 15.8.

Unless stated otherwise, the inspection level is Special Inspection Level S3.

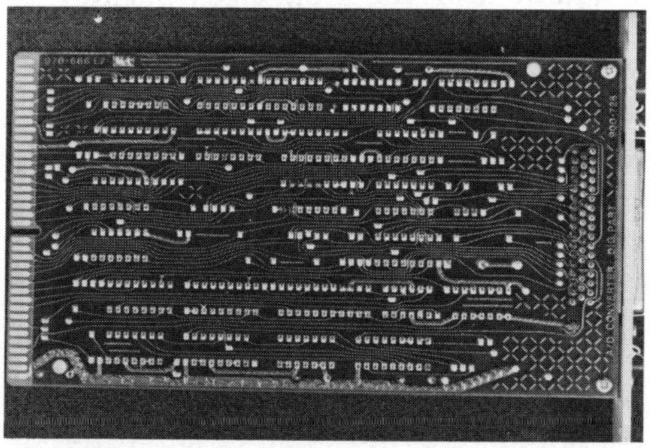

Fig. 14.1 Example of board with a gold plated edge connector.

14.2 Appearance

The initial inspection is performed visually because the appearance of the connector surface reveals much about the quality of the edge connector. Succeeding paragraphs address the following subjects:

- Surface evenness
- Surface brightness
- Discoloration and stains
- Incomplete coverage
- Scratches
- Depressions

14.2.1 Surface Evenness

The evenness of the surface depends to a high degree on the condition of the substructure. A smooth and plain substructure is a precondition of an even, impervious, corrosion-resistant gold layer. See Paragraphs 3.7.2 and 3.7.4 of Reference A12, Reading List. The inspector has, therefore, to assess whether the underlying copper surface has been scratched—for example, by a heavy-handed cleaning process.

14.2.1a Scratches

Although an intermediate layer of nickel smooths out the surface and increases the corrosion resistance, any scratches on the copper structure should not be so large that they can be seen with the naked eye. Otherwise, the board is unacceptable and the fault is rated as an A-defect.

14.2.1b Plated-Over Scratches and Pinholes

When deep scratches and pinholes are plated over, the deposit follows the surface structure and eventually closes the bottom of scratches and pinholes with the risk of encapsulating electrolyte and impurities. Such spots can easily corrode.

14.2.1c Surface Preparation

Although some few PCB manufacturers clean the copper surface with steel wool, this method is extremely undesirable. It unavoidably leaves tiny steel splinters embedded in the base material. The splinters later can rust and cause short circuits between the connector tabs. Steel splinters embedded in the copper surface cause a discontinuous deposit and can become the starting points of later corrosion. Upon learning that steel wool is being used to clean boards—during an inspection visit, for example—the inspector should stop delivery of the boards.

Obviously, the surface to be gold plated must be meticulously prepared to achieve good results. It is quite common for nine out of ten faults in a gold plated edge connector to be caused by inadequate surface cleaning and not by a poorly executed plating process or improper maintenance of the plating bath.

14.2.1d Roughness Criteria

In general, the roughness of the gold plated surface should be less than 0.00002 to 0.00004 in. (0.5 to 1 μm). The measurement is performed with a roughness tester. Roughness exceeding 0.00008 in. (2 μm) is cause for rejection and consequently rated as an A-defect.

14.2.2 Surface Brightness

Gold plated edge connectors must be at least as bright as matte gold but not excessively bright. A very bright surface is undesirable because this usually indicates a brittle plating, which could mean a tendency to crack.

14.2.3 Discoloration and Stains

Discolored and stained contact tabs indicate incorrect plating conditions—for example, excessive current density or a contaminated plating bath. The discoloration is usually of a yellowish brown to a reddish brown color, either over the whole connector tab or in spots.

14.2.3a Discoloration

The worst aspect of discoloration is the fact that it is frequently accompanied by reduced wear resistance and increased risk of corrosion. The reason is usually the burning caused by excessive current density, either generally or locally. The PCB manufacturer can treat light burning by burnishing, but this is very undesirable as the treatment neither reduces the potential porosity nor improves the reduced wear resistance associated with the burned surface.

14.2.3b Stains

Stains can have various causes, including contaminated rinse water, impurities in the plating bath (inorganic impurities or ion contamination), or inadequate precleaning. The inspector should take care not to mistake stains for corrosion products growing up through a porous surface. He must examine such deposits through a microscope in order to determine whether they are stains or corrosion products. Stains differ from corrosion products by showing no vertical growth.

14.2.3c Board Disposition

Boards with discolored and/or stained connector tabs should be rejected, and the fault is therefore rated as an A-defect.

14.2.4 Incomplete Coverage

Incomplete coverage of the underlying metal (either nickel or copper) with gold is an indisputable cause for rejection, even when it occurs in spots, and the fault is rated as an A-defect. The cause of incomplete coverage is usually contamination of the underlying metal surface, which was inadequately cleaned. When the gold is plated on top of copper foil, the reason can be local areas with mold release left over from laminating the base material; basically, the responsibility for leftover mold release belongs to the laminate fabricator, not to the PCB manufacturer.

14.2.5 Scratches

Scratches can only be assessed in a subjective manner by the inspector, who should assume a more critical attitude toward scratches on gold plated contact tabs than on a tin/lead plated surface. Scratches in the gold layer can be dangerous as they can easily allow corrosion in the contact area and cause functional failures. Only very light scratches are acceptable. Figure 14.2 shows various degrees of scratching. Heavy scratching is always cause for rejection and consequently rated as an A-defect.

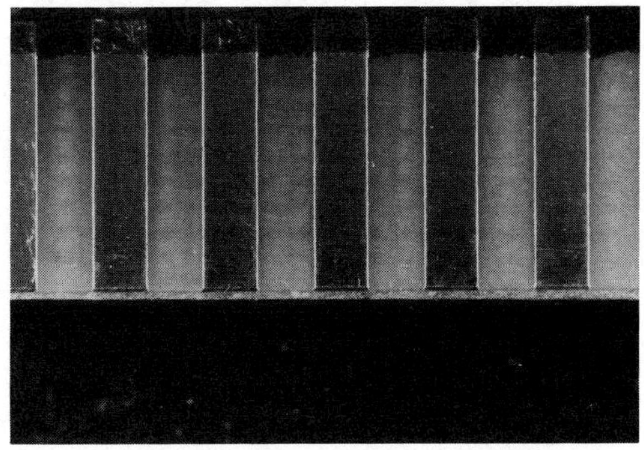

Fig. 14.2A. Gold plated edge connector with acceptable degree of superficial scratching.

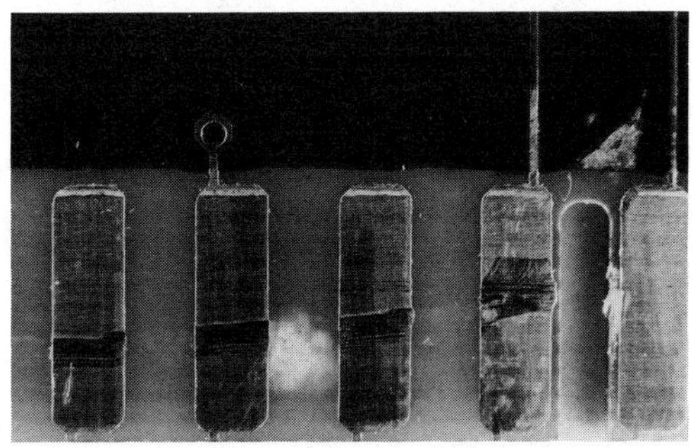

Fig. 14.2B. Gold plated edge connector with a heavy and unacceptable scratching.

14.2.6 Depressions

Like scratches, depressions or dents can only be assessed in a subjective manner. Depressions may exist in the copper foil before plating, but they can also occur after plating because of careless handling of the boards. Sharp-edged depressions can be dangerous because of the stress present at the edges. This can lead to contact failure—for example, from delamination of the gold layer. Sharp-edged depressions are therefore cause for rejection and rated as an A-defect. Rounded depressions as shown in Figure 14.3 can be accepted provided they occur to a limited extent—for example, no more than three depressions per contact. The fault is rated as a B-defect.

Fig. 14.3. Example of rounded depressions in the contact tabs of a gold plated edge connector.

14.3 Plating Thickness

14.3.1 Measuring the Thickness

It can be quite difficult to determine the plating thickness unless the inspector has special measuring equipment based on the beta-backscatter method (Paragraph 7.3.1e). Since this equipment is very expensive, it is available only in a minority of electronics companies, primarily those producing military equipment or other highly rated types of equipment. In such applications, the plating thickness must be measured.

One solution is to make microsections, but measuring layer thicknesses of 0.00004 to 0.00008 in. (1 to 2 μm) requires a good metallurgical microscope fitted with a camera. Measurements made on the photo indicate the thickness.

It is rare indeed to find a metallurgical microscope available in nonmilitary electronics companies. Preparation of microsections is obviously destructive as well as slow and difficult, especially when the operator is inexperienced.

It is desirable to know the plating thickness, because insufficient plating thickness can cause contact failures later. It is therefore recommended that the plating thickness be checked from time to time—for example, once or twice a year for each PCB manufacturer. One possible solution is to send samples to an independent institution or laboratory which has the necessary measuring equipment—for example, a beta-backscatter meter.

Such an investigation is not very expensive and yields valuable information for assessing the capability of the PCB manufacturer. During an inspection of the PCB manufacturer's factory, it is usually possible to borrow his beta-backscatter meter for a short time to check plating thicknesses, which in itself has an excellent preventive effect.

14.3.2 Nonuniform Thickness

In many plating baths, it is not possible to obtain a uniform layer thickness on all the contact tabs of a long edge connector. The result will frequently be a distribution of thicknesses as indicated in Figure 14.4, with the outermost contact tabs having perhaps twice as much deposited gold as the center tabs. This is caused by a higher current density at the outermost contact tabs.

The distribution curve shown in Figure 14.4 is tantamount to a direct loss for the PCB manufacturer as he must meet the specified minimum thickness at the center region.

In the long run, of course, the customer pays for the superfluous gold, so it is of interest to know the distribution curve. With a special plating bath ar-

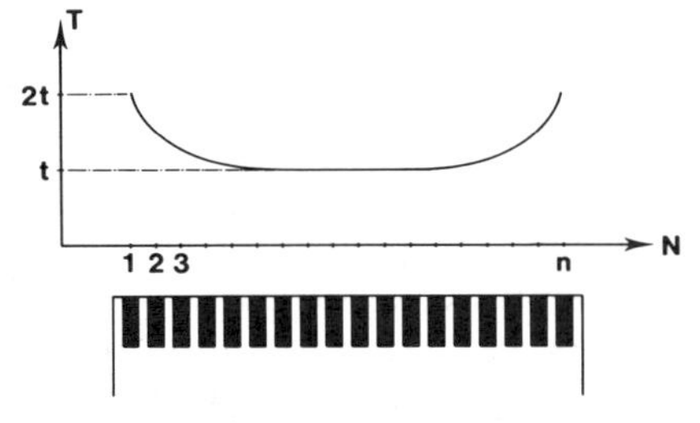

Fig. 14.4. Variation of thickness of gold layer along an edge connector. T = thickness of gold layer. t = nominal thickness of gold layer. N = contact tab number.

rangement—for example, by introducing a special screening of the anodes—the manufacturer can achieve a more uniform plating thickness.

14.3.3 Thickness Requirements

Many different requirements regarding plating thickness can be set up, and the inspector should distinguish between processes where the gold is plated directly onto the copper, and where an intermediate nickel layer is deposited. The plating thickness can, for example, be as follows:

Gold alone:
Minimum 0.0002 in. (5 μm) hard gold

Gold on nickel:
Minimum 0.0002 in. (5 μm) nickel (low stress)
Minimum 0.00006 in. (1.5 μm) hard gold

14.3.4 Board Disposition

If the requirements concerning the plating thicknesses are not met, the whole batch is returned to the PCB manufacturer for screening and/or remake. It seems reasonable to require that the PCB manufacturer include a record of thickness measurements for the replacement batch.

14.3.5 Result of Different Measuring Methods

It seems appropriate to discuss in depth the two methods of measuring the plating thickness of gold: the beta-backscatter method and the microsectioning method. As the two methods give different results, serious disagreements regarding specification compliance can arise between the PCB manufacturer and the customer.

Although unusual, it is to the great advantage of both parties for the purchase specification to indicate the method of measuring the plating thickness. The following discussion shows that it can be very sensible to indicate the measuring method.

In practice, the operator will achieve a lower plating thickness when measuring with the beta-backscatter method than with the microsectioning method. The difference can be as much as 20 to 25 percent.

The reason for the discrepancy is that a plated gold layer exhibits a lower density (specific gravity) than a pure unplated gold layer (i.e., solid gold) because of translocations (defects) in the crystal lattice. This is partly because plated gold is alloyed—for example, with cobalt—in order to achieve the desired hardness, and partly because organic additives are added to the gold bath. The structure of the gold layer becomes more open, although one should not think of pores in this connection. The open structure means that fewer electrons return to the Geiger-Müller tube (Paragraph 7.3.1e), indicating an apparently thinner plating.

The PCB manufacturer will of course introduce a fixed correction factor, but because of several variable factors such as the current density in the plating bath or the condition of the bath, the density of the plated gold can vary a good deal.

In the microsectioning method the linear plating thickness is measured independently of the gold plating density. The inspector should realize that there is a lower measuring limit of about 0.00004 in. (1 μm). This limit requires a well-executed microsection with unrounded edges and a good metallurgical microscope, preferably equipped with a camera.

The difference between the two measuring methods can cause the following problems:

1. The PCB manufacturer measures with a beta-backscatter meter. The customer prepares microsections and finds that the gold layer is approximately 20 percent too thick. The specification is met but the cus-

tomer feels that he pays too much for the gold plating.

2. The PCB manufacturer prepares microsections and measures the linear plating thickness. The customer has a beta-backscatter meter at his disposal and finds that the gold layer is approximately 20 percent too thin and therefore outside the specification. He complains to the manufacturer and requires replacement boards made with more gold on the contact tabs.

These examples show how important it is that the PCB manufacturer and customer agree on how to assess the thickness. The method chosen is a matter of taste. It seems most reasonable to choose the beta-backscatter method as it more accurately indicates the effective plating thickness. Also, it is nondestructive.

14.4 Delamination

Many PCB manufacturers oppose plating gold on nickel because they have experienced delamination or peeling. This means that the gold layer flakes off or can be peeled off, either locally or generally, exposing the nickel layer. The usual cause is incomplete activation of the nickel surface, which again can be caused by several other faults. In order to prepare the inspector for discussions with the PCB manufacturer concerning gold-on-nickel plating techniques, we will discuss the problems of delamination and peeling below.

14.4.1 Additives

With no organic additives, nickel plating has a greyish appearance and there is a risk of pinholes occurring in the deposit. In order to counteract this, the PCB manufacturer always uses organic additives in the form of brighteners to make the surface brighter and more impervious, and wetting agents to inhibit the formation of pinholes.

If too much of these additives is added to the plating bath, the deposit will be subjected to large internal stresses, which make it difficult to obtain good adhesion between the nickel layer and the gold layer. After nickel plating, these additives remain on the surface and are extremely difficult to remove. The board must therefore undergo several cleaning processes and an activating process, which in principle consists of pickling (light etching) with hydrochloric acid.

Some PCB manufacturers assert that the cleaning problem is insignificant because they use additives to such limited extent that it is only necessary to rinse the board in water before the gold plating. It should be noted, however, that if the water is too cool, it will not remove the additives; this inevitably causes adhesion problems.

The organic additives inhibit crystal growth. As shown in A of Figure 14.5, each plating run starts with a few crystal nuclei on the surface. With no additives these nuclei grow larger and cause a fairly coarse-grained crystal structure.

The picture is quite different when organic additives are added to the plating bath. As shown in B, Figure 14.5, the

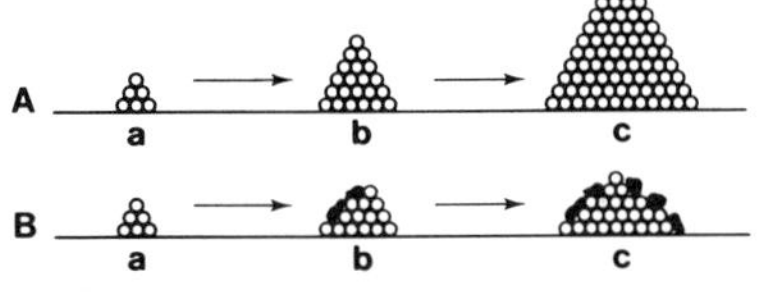

Fig. 14.5. Effect of organic additives in plating baths. (Additives appear in black).
A: Bath without organic additives. a = crystal nuclei. b = crystal is growing. c = final size of crystal.
B: Bath with organic additives. a = crystal nuclei. b = additives adhere to crystal and restrain its growth. c = growth of the crystal is heavily restrained.

organic additives are deposited on the crystals and restrain further growth. New crystal nuclei are created continuously on the surface which is to be plated, and these nuclei also grow to only a limited extent.

The result is an impervious deposit of finer grain, but organic material remains on the plating surface where it acts as an inhibitor. It is therefore necessary to remove the organic material so that the gold plating can adhere to the surface.

14.4.2 Nickel Oxidation

Another cause of delamination can be an oxidized nickel surface. This can easily happen if the board dries between the activating process and the gold plating and is therefore exposed to the air. The board should be kept in a water bath until the gold plating process starts.

14.4.3 Low-Stress Nickel

Many PCB specifications demand low-stress nickel. This requires a special nickel plating bath—for example, a modified Watts nickel plating bath to which a stress-reducing agent is added. The plated nickel has a low internal stress factor, is ductile, and can be activated relatively easily.

14.4.4 Nickel Thickness

The PCB specifications also frequently state that the nickel layer must not be too thick. The reason is that a thick nickel layer in itself contributes to internal stress in the layer. The limit is often 0.0006 in. (15 μm)—that is, a nominal thickness of 0.0003 in. (7.5 μm) + a 100 percent plating tolerance.

14.4.5 Nickel Delamination

The nickel layer with the superjacent gold layer can also delaminate. This is usually caused by insufficient cleaning of the copper or by a contaminated cleaning bath. For example, it is not uncommon that a nearly invisible oil film is present on the surface of the cleaning bath, caused by using the bath over a very long period of time. Another reason can be oxidation of the copper surfaces immediately before starting the nickel plating. The boards should therefore always remain in a waiting bath to prevent drying and subsequent surface oxidation.

14.4.6 Tape Test

The inspector should perform a routine tape test on the edge connector to ensure that no tendency to delamination occurs. The test should be performed as a sampling test. If it is clear that the plated layers are lifted from the substructure, the tape test is, of course, unnecessary. Paragraph 5.10 describes the procedure.

If plated material comes off on the tape, the board is rejected without further inspection and the fault is rated as an A-defect. The sampling test is then extended to a 100 percent test, and all boards with delamination are rejected and returned to the PCB manufacturer. Fresh tape must be used for each test. Figure 14.6 shows some examples of delamination of the gold/nickel layer.

The inspector should note that most

Fig. 14.6. Delamination of nickel/gold layer of an edge connector.

delamination problems occur at the transition zone between the conductor and the connector tab. This especially applies to boards with a butt joint between the tin/lead layer and the nickel/gold layer. See Paragraph 14.6.3.

14.4.7 Burnish Test

The adhesion can also be assessed on the basis of a burnish test, which corresponds to cold-working the surface. The surface is smoothed with a tool provided with a smooth hemispherical end. Such a tool is easiest made by attaching a ¼ in. steel ball to a metal rod of suitable dimensions with a metal-to-metal adhesive.

The burnishing is done by rubbing the contact area rapidly and firmly for 15 seconds. The pressure must be sufficient to leave visible traces of the burnishing. The gold layer must not delaminate or blister as a result of the test. If this happens, the board is rejected and the fault is rated as an A-defect.

14.4.8 Problem Identification

By studying the underside of the delaminated layer sticking to the tape, the inspector can obtain a fair indication of the source of the trouble. Figure 14.7 shows the build-up of the layers. The causes of peeling are discussed below.

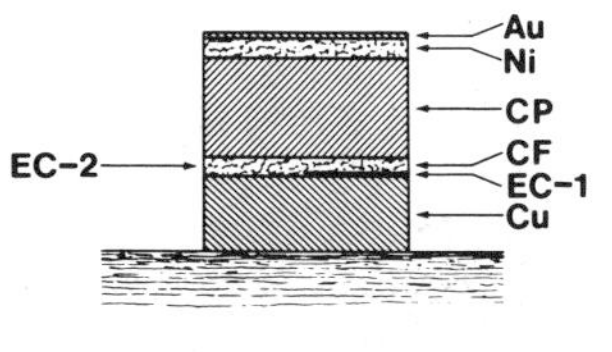

Fig. 14.7. Layer build-up of edge connector. Au = gold, 0.00006 in. (1.5 μm). Ni = nickel, 0.00020 in. (5 μm). CP = copper plating, 0.001 in. (25 μm). CF = copper flash, 0.00016 in. (4 μm). EC-1 = electroless copper, 0.00004 in. (1 μm). Cu = copper foil, ½ oz. = 0.0007 in. (17.5 μm) or 1 oz. = 0.0014 in. (35 μm). EC-2 = alternative method: electroless copper, 0.00020 in. (5 μm).

Nickel and gold peelings are easily identified by the color. However, a complication occurs if the nickel layer itself delaminates, so that the peeling as well as the material remaining on the contact tab have the appearance of nickel. This type of delamination is usually caused by an interruption of the current during the nickel plating process.

If the underside of the peeling as well as the material left on the contact tab show copper, the problem is a separation between the plated copper layer and the copper foil. It is possible to go still further and ascertain where in the copper plating process something went wrong.

An etchant such as ammonium persulfate is used, a drop being deposited on the peeling as well as on the remaining material. Where the copper turns black we have the catalyst which was used in the electroless copper stage. If the peeling becomes black, the fault lies in the treatment preceding the electroless copper stage, and if the remaining material turns black, the fault occurred after the electroless stage.

It can also happen that neither of the surfaces turns black, but retains the copper appearance. This indicates that the separation took place between the actual copper plating of 0.001 in. (25 μm) and the 0.00012 to 0.0002 in. (3 to 5 μm) thick copper layer deposited before the image transfer (Figure 7.1).

In summary, the inspector can use a very simple method to ascertain the weak points of the production process. When inspecting the PCB manufacturer's facility, he can put this knowledge to good use in identifying possible problem areas.

14.5 Edge Coverage of Contact Tabs

Many PCB specifications require that the edges of the contact tabs be covered with gold (and nickel). This requirement eliminates the risk of corrosive attacks from the edge into the copper, which could happen under adverse environmental conditions. The corrosion products, which tend to creep over the surface, could create short circuits between the contact tabs. The machined edges of the contact tabs are excepted for very good reasons.

As all boards of a batch are normally alike with respect to edge coverage of the contact tabs, the inspector need inspect only a single board. Lack of edge coverage along the sides of the contact tabs is cause for rejection and the fault is rated as an A-defect.

14.5.1 Process Variations

A few PCB manufacturers use other methods which deviate somewhat from those described in Paragraph 14.6. Some of the methods leave the edges uncovered; an example is the selective tin/lead plating described in Paragraph 9.3.

With suitable masking, the contact tabs can be gold plated immediately after the tin/nickel plating—that is, before etching the board. This implies, of course, that the edges of the contact tabs will not be covered with gold or tin/nickel.

It should be mentioned that it is also possible to perform the gold plating according to the traditional method after etching the board, so that the edges are covered with gold, but of course not with tin/nickel.

Incidentally, the aforementioned traditional method produces the ideal transition zone between a contact tab and the attached conductor. This is due to the tin/nickel layer which is plated all over the pattern, so that bare copper cannot appear in the transition zone except at the edges.

14.5.2 Board Disposition

Whether a lack of edge coverage is cause for rejection depends entirely on the PCB specification. Without an edge-coverage requirement, the inspector cannot just reject boards when copper shows at the edges of the contact tabs. But when edge coverage is specified, failure to meet this requirement is cause for rejection, and consequently the fault is rated as an A-defect.

14.5.3 Design Considerations

The contact tabs should be designed with rounded corners as shown in Figure 14.8. The connecting strap from a contact tab to the plating bar should be narrower than the contact tab and preferably located asymmetrically. This protects the part of the connector strap left after machining from lifting when the edge connector is inserted into the receptacle; otherwise, the lifting could extend to the contact tab itself.

14.6 Transition Zone between Contact Tab and Conductor

A critical point is the transition zone between the contact tabs and the attached conductors. In some cases the PCB specification describes exactly how to make the transition. This can be very undesirable because the PCB manufacturer may

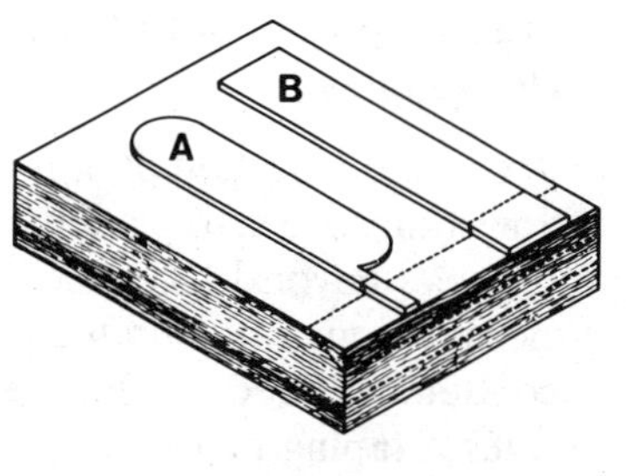

Fig. 14.8 Design of contact tabs.
A: Most desirable design. Bare copper occurs only where the connecting strap to the plating bar is cut when chamfering the edge (shown in dotted lines).
B: Undesirable design. Chamfering causes bare copper over the entire width of contact tab.

be forced to follow procedures with which he is not fully conversant.

Although the PCB manufacturer always assumes full responsibility for the end result, the customer assumes, at least morally, a joint responsibility.

In other cases the PCB specification prescribes no particular manufacturing method, but simply requires that the transition zone be made in a way which allows no corrosion, implying that the transition zone must have no exposed copper. This kind of specification is more attractive to the PCB manufacturer, as he is completely free to follow established manufacturing techniques.

The inspector should know enough of the different manufacturing techniques to ascertain whether the transition zone was produced correctly and meets the PCB specification, or whether it can corrode. The following paragraphs describe three common methods applicable to tin/lead plated boards, and one method for bare-copper boards, including solder-coated and hot-air leveled boards:

- Transition zone with gold overlapping
- Transition zone with epoxy sealing
- Transition zone with butt joint
- Transition zone on bare-copper boards

14.6.1 Transition Zone with Gold Overlapping

The most common method of making this type of transition zone is to plate the nickel and gold layers beyond the contact tabs, so that the layers cover the tin/lead plating of the conductors. See A, Figure 14.9. This implies a taping process.

The first stage is to apply the tape as shown in B, Figure 14.9, after which the tin/lead is stripped from the contact tabs. This means that the tin/lead is dissolved by means of an etchant which does not attack the underlying copper. The exposed copper is now cleaned and dried, after which a second taping is performed.

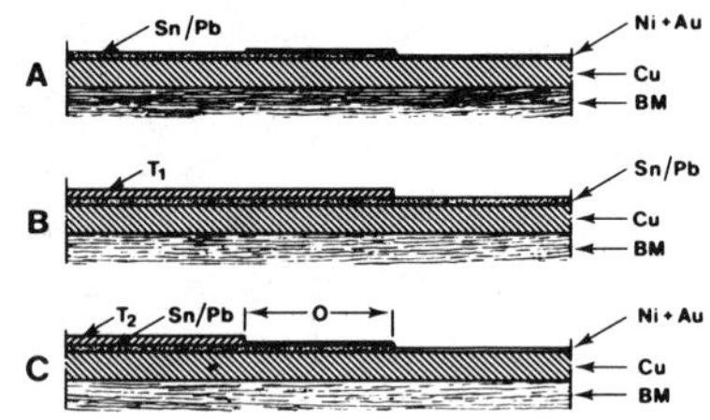

Fig. 14.9. Cross section through edge connector with gold overlapping in the transition zone.
A: Lengthwise cross section through finished contact tab. Nickel and gold overlap the tin/lead layer of the conductor.
B: The first tape is applied, and the board is ready for stripping of the tin/lead.
C: The tin/lead layer is stripped, and the second tape is applied with the displacement "O" with respect to the first tape. Nickel and gold are plated onto the uncovered area, i.e., the contact tab and the conductor end. Sn/Pb = tin/lead. Ni + Au = nickel and gold. Cu = copper. BM = base material. T_1 = first tape. T_2 = second tape. O = zone of overlapping.

As shown in C, Figure 14.9, the tape is displaced about 0.02 to 0.04 in. (0.5 to 1 mm) relative to the first tape in a direction away from the edge. A renewed cleaning, as well as a pickling of the copper, is now performed to prepare the contact tabs for the nickel and gold platings.

It is crucial to a good result that the tape fit tightly around the conductors so that the stripping solution cannot seep along the conductor edges under the tape and attack the tin/lead at unwanted places. See Figure 14.10. The tape should therefore be suitable for this application and applied with adequate pressure to assure a tight fit.

A simple method for avoiding seepage of the stripping solution is to dip the connector edge in distilled water so that any gaps fill with water instead of the stripping solution.

When no nickel layer is used (gold alone, see Paragraph 14.3.3), the method adds the risk that a little lead from the

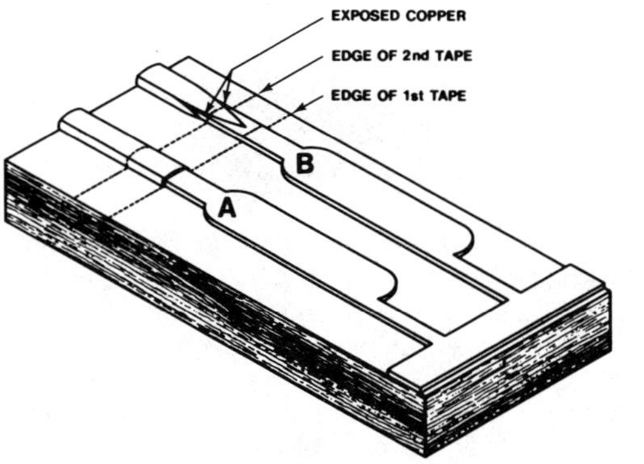

Fig. 14.10 Stripping solution seeping along contact tab edges beneath the tape can result in exposed copper.
A: The tape has shown a good fitting along the contact tab edges so that no seepage of the stripping solution has taken place.
B: Here the tape has adhered inadequately so that the tin/lead has been dissolved along the conductor edges. The second tape was applied a little too far downwards and covers exposed copper which therefore cannot be nickel and gold plated.

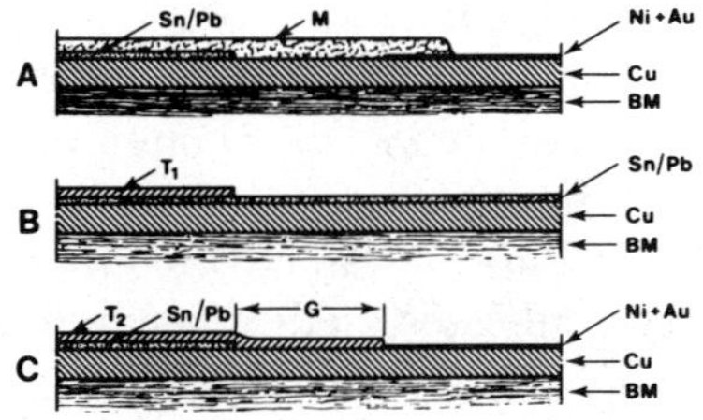

Fig. 14.11 Cross section through edge connector with epoxy sealing of the transition zone.
A: Lengthwise cross section through finished contact tab. The transition zone is sealed by the solder or insulation mask.
B: The first tape is applied and the board is ready for stripping of the tin/lead.
C: The tin/lead layer is stripped, and the second tape is applied with the displacement "G" with respect to the first tape, but in the opposite direction of displacement "O" in Fig. 14.9. Nickel and gold are plated onto the uncovered area, i.e., the contact tab, but not the conductor end as shown in Fig. 14.9.
Sn/Pb = tin/lead. Ni + Au = nickel and gold. Cu = copper. BM = base material. T_1 = first tape. T_2 = second tape. G = gap between contact tab and conductor. M = solder and insulation mask.

tin/lead plated area, which is exposed after the second taping step, is dissolved in the gold plating bath. Unfortunately, gold plating baths are very sensitive to lead contamination, and as little as 30 ppm (parts per million) lead in the gold plating bath will affect the quality of the gold plating, which may take on a brownish appearance. Because of the risk of damaging the gold plating bath, a number of PCB manufacturers oppose this method unless they are allowed to plate "nickel under gold." The nickel protects the gold bath against lead contamination.

14.6.2 Transition Zone with Epoxy Sealing

To avoid the aforementioned problems, many PCB manufacturers prefer to modify the second taping stage so that the tape is displaced to the opposite side. As shown in Figure 14.11, this implies the presence of a gap between the nickel/gold layers and the tin/lead layer.

Bare copper appears in the gap, and the necessary sealing of the gap is accomplished by applying the solder and insulation masks with a slight overlapping of the contact area. If there is no insulation mask, sealing can be accomplished by means of a slight modification of the component notation. The artwork is modified to include a narrow stripe, which is printed across all the gaps, sealing them from corrosion.

When the artwork/filmwork has not been so prepared, the PCB manufacturer can introduce a modification of the masks. As this introduces the risk that the sealing may cover so much of the contact tabs that the effective contact area is affected, the modification should be made only with the customer's consent.

It is more expedient for the PCB designer always to design the solder and insulation masks, or the component notation, to cover the transition zone. Because of possible issue changes, the customer should not leave this to the PCB manufacturer, who will sooner or later forget about it.

In addition, the effective contact area should be indicated on the master drawing (Paragraph 14.7 and Figure 14.14). This has the advantage that the inspector can immediately check whether the effective contact area has been affected by the sealing.

14.6.3 Transition Zone with Butt Joint

Although the current price of gold is relatively high, the processing still forms a large part of the cost of the finished edge connector. In order to keep the price as low as possible, some PCB manufacturers prefer to avoid the second taping stage. See Figure 14.12. This implies that the plated nickel/gold layers exhibit a butt joint with the tin/lead layer.

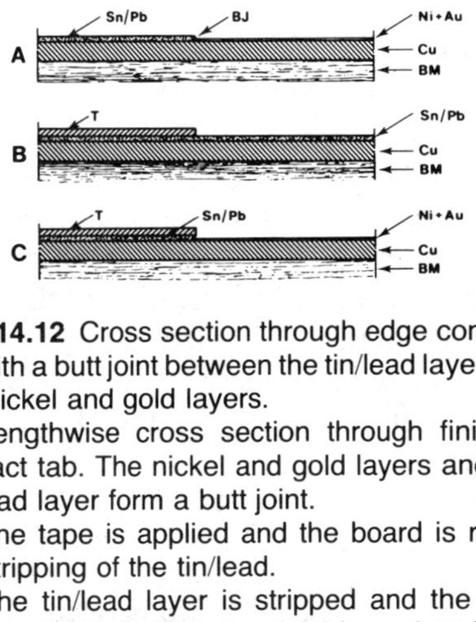

Fig. 14.12 Cross section through edge connector with a butt joint between the tin/lead layer and the nickel and gold layers.
A: Lengthwise cross section through finished contact tab. The nickel and gold layers and the tin/lead layer form a butt joint.
B: The tape is applied and the board is ready for stripping of the tin/lead.
C: The tin/lead layer is stripped and the tape remains in place. Nickel and gold are plated onto the uncovered area and up against the tin/lead face.
Stripping solution seeping along the conductor edges causes exposed copper (as shown in Fig. 14.10) which will not be nickel and gold plated.
Sn/Pb = tin/lead. Ni + Au = nickel and gold. Cu = copper. BM = base material. T = tape. BJ = butt joint.

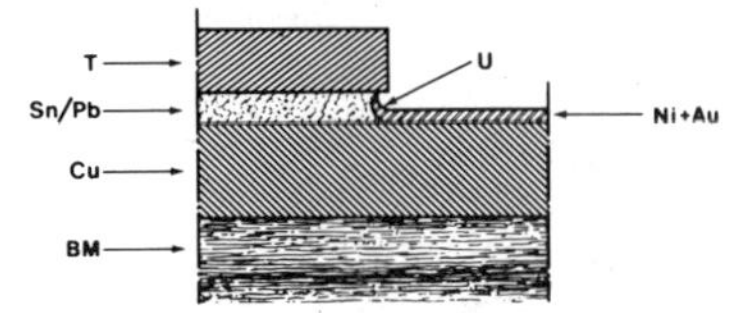

Fig. 14.13 Enlarged cross-sectional drawing of the butt joint shown in Fig. 14.12. Because of the undercut and the shielding effect of the tape there is a risk of overly thin nickel and gold depositions at the tin/lead face.
Sn/Pb = tin/lead. Ni + Au = nickel and gold. Cu = copper. BM = base material. U = undercut. T = tape.

It is imperative to avoid any contamination of the tin/lead edge which is to be joined with the nickel/gold layers. If not, a microscopic crack can form and become the starting point for later corrosion. The contamination can, for example, be due to adhesive from the tape creeping down the tin/lead edge created after stripping the tin/lead from the contact tabs.

Furthermore, undercutting can occur, which means that the tin/lead layer is stripped a little under the tape as shown in Figure 14.13. During the subsequent plating of nickel and gold it may happen that the layer thicknesses become too low under the tape due to the shading effect towards the current lines in the bath.

From time to time, perhaps once or twice a year, the inspector should prepare microsections lengthwise through a contact tab to check the quality of the butt joint.

Because of the aforementioned problems, the sealing mentioned in Paragraph 14.6.2 also applies as there is always some risk that a tiny crack will form in the butt joint.

14.6.4 Transition Zone on Bare-Copper Boards

In the case of bare-copper boards, and also in the case of solder-coated and hot-air leveled boards, the conditions are considerably simpler as it is unnecessary to strip a tin/lead layer from the contact tabs.

The area to be nickel and gold plated is demarcated with tape, after which the processes are carried out in the usual way.

If the bare-copper board is to be lacquered (Paragraph 9.7), it is important not to apply the lacquer to the contacts, as this can cause serious contact problems later on. The PCB manufacturer must very carefully mask the edge connector, and take care to ensure that the tape fits tightly along the edges of the contact tabs so that no lacquer can creep along the edges.

Lacquer on the contact tabs is rated as a B-defect because the lacquer can be removed by careful cleaning with a suitable solvent, such as alcohol.

Some PCB manufacturers prefer to deliver the boards with lacquer all over the surface including the edge connector area as this allows them to avoid the masking process. Since this approach leaves the cleaning of the contact tabs to the customer, the PCB specification should clearly state the method used in applying the lacquer.

In the case of solder-coated and roller-tinned boards, the solder and insulation masks should cover the uppermost part of the contact tabs in order to prevent exposed copper in the transition zone. The problem is that it is necessary to mask the edge connector with tape to avoid coating it with solder. The tape will then overlap the solder and insulation masks. If these masks do not cover the uppermost end of the contact tabs, there is a risk of exposed copper when the tape is removed.

14.6.5 Acceptance Criteria of Transition Zones

Having dealt with the manufacturing methods, we will now determine the acceptance criteria of the transition zone. If the PCB specification demands a special method, it must, of course, be followed. If the PCB manufacturer erroneously uses another method, and the end-product specification does not state any particular requirements for the edge connector plating process, the inspector can waive the error.

A waiver naturally implies that the transition zone has been properly manufactured. The fault is rated as a B-defect and the PCB manufacturer should be contacted to avoid recurrence. If the end-product specification requires a specific edge connector plating process, and the inspector cannot obtain a waiver from the end user or it is undesirable to apply for one, this fault is cause for rejection and consequently rated as an A-defect.

The finished transition zone must be satisfactory, whether a specific manufacturing method has been prescribed or the PCB manufacturer has been allowed a free hand. This means that the underlying copper is not exposed to corrosive attacks.

In the case of overlapping (Paragraph 14.6.1), the nickel/gold layers must completely encapsulate the underlying copper and, in the case of sealing (Paragraphs 14.6.2 and 14.6.4), the transition zone must be covered by the solder and insulation masks, or with a stripe printed at the same time as the component notation.

14.6.6 Inspection Methods for Transition Zones

The inspection is performed visually—for example, with a 7X magnifier—and the inspector should check that no copper is exposed. The inspection is somewhat more difficult in the case of a butt joint (Paragraph 14.6.3), as there may be a microscopic crack between the two butt faces. The inspector should, therefore, occasionally perform a corrosion test, which also tests for pores and pinholes in the contact area. See Paragraph 14.7.10.

14.7 Contact Tab Defects

In the same way as defects occur on the conductors, defects can also occur on the contact tabs as stated below:

Short circuits
Discontinuous plating
Poor edge definition
Width tolerance
Lifted contacts
Projections
Indentations
Porosity
Registration faults

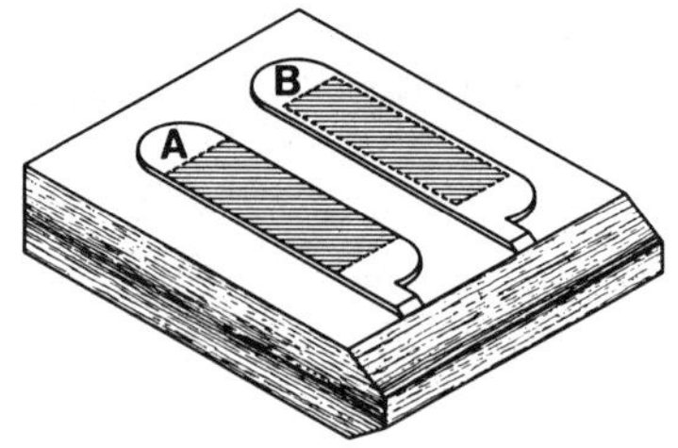

Fig. 14.14. Definition of effective contact area (shown cross-hatched).
A: The effective contact area corresponds to the total width of the contact tab.
B: Here the effective contact area is a little narrower than the contact tab.

14.7.1 Special Inspections during Manufacturing

Some of the demands stated below may appear somewhat extreme, but it should be remembered that an edge connector is a limited area very distinct from the rest of the pattern. The edge connector can easily be made the object of a special inspection during the manufacturing process—for example, in the form of a more severe inspection after the image transfer. This gives an immediate possibility of touching up possible faults in the pattern.

14.7.2 Effective Contact Area

Some PCB specifications indicate an effective contact area which must be flawless. See Figure 14.14. Defects outside this area can be accepted. Before the inspection, the inspector should investigate whether the PCB specification makes any concessions in this respect.

There are reasons in favor of warning against the above tendency. First, many edge connectors have very narrow contact tabs which leave a very limited margin for possible defects, and second, possible corrosion products could grow from pinholes in permitted areas to the effective contact areas, causing functional problems. The following paragraphs ignore defects in permitted areas and treat the contact areas as a whole.

14.7.3 Short Circuits

Short-circuited contact tabs are an indisputable cause for rejection, and the fault is rated as an A-defect. Rework is not acceptable, because there is a considerable risk of delamination, and furthermore, because the edges of the contact tabs will have exposed copper where the short circuits were removed.

14.7.4 Discontinuous Plating

Discontinuity of the gold/nickel layers of the contact tabs—for example, caused by fine transverse cracks—is also an indisputable cause for rejection, and the fault is rated as an A-defect.

There is some disagreement in the electronics industry over the advisability of repairing such discontinuities—for example, by selective plating. In the author's experience, such repairs normally should not be accepted as it is considered impossible, or at least extremely difficult, to achieve a sufficiently smooth transition at the fault location.

14.7.5 Poor Edge Definition

The edge definition of the contact tabs corresponds fairly closely to the edge definition of the conductors (Paragraph 5.5). The only difference is that unevenness greater than 0.004 in. (0.1 mm) should be regarded as cause for rejection, and consequently rated as an A-defect.

14.7.6 Width Tolerance

The tolerance of the contact tab width is very closely related to the general edge displacement of the pattern (Paragraph 5.4). However, it seems more reasonable to use the expression "width tolerance" since the effective width of the contact tabs is important to the functioning of the contacts, particularly if the nominal contact width is 0.04 in. (1 mm) or less.

We will accept here, for example, a width tolerance of ±0.004 in. (±0.10 mm), which corresponds to Requirement 2 of Paragraph 5.4.2. This tolerance should be regarded as an extremely reasonable value by all PCB manufacturers. Exceeding the tolerance is therefore regarded as cause for rejection and rated as an A-defect.

14.7.7 Lifted Contact Tabs

Lifted contact tabs can be caused by the machining of the board, partly by the general contouring and partly by the chamfering of the connector edge (Paragraph 15.8). If a contact tab is lifted or shows burrs at the edge as shown in Figure 14.15, there is a considerable risk that the contact tab will lift more and more with repeated insertions of the board into the receptacle.

Contacts which are lifted or burred are an indisputable cause for rejection, and the fault is rated as an A-defect.

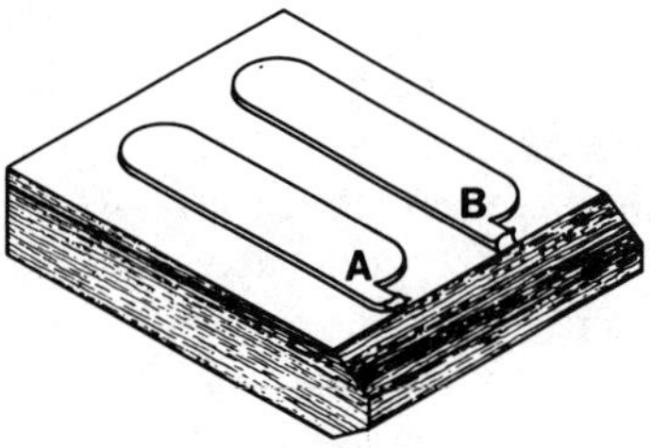

Fig. 14.15. Edge connector defects caused by machining.
A: Lifted connecting strap.
B: Burr at connecting strap.

14.7.8 Projections

Isolated projections along the edges of the contact tabs can be just as dangerous as on the rest of the pattern.

14.7.8a Large Tab Pitch

In the case of edge connectors with contact tabs having a reasonably large pitch—for example, 0.1 in. (2.54 mm, and in the metric system, 2.50 mm) or more—we can formulate the following requirements:

1. At least 75 percent of the nominal insulation distance between two adjacent contact tabs must remain. This requirement must be met at the most unfavorable combination of projections, unevenness along the contact tabs, and possible variations in width. Unwanted metal particles located between two adjacent tabs are not acceptable (Requirement 4, Paragraph 5.6.2).
2. The maximum size of a projection is limited to 0.008 in. (0.2 mm), and the length of the baseline shall be less than 0.02 in. (5 mm). However, Requirement 1 shall always be met (see also, Requirement 1, Paragraph 5.6.2).
3. No more than one projection per contact tab is acceptable.

The measurement is usually performed with a 7X or 10X magnifier fitted with a graticule. If these requirements—considered most reasonable by any professional PCB manufacturer—are not met, the board is rejected and the fault rated as an A-defect. The inspection is extended to a 100 percent inspection and rejected boards are returned to the PCB manufacturer for remake.

14.7.8b Small Tab Pitch

In the case of edge connectors with a very small pitch of 0.05 in. (1.27 mm) or less, a very careful calculation of the entire tolerance chain is necessary as the conditions can become very critical. The following factors should be considered:

1. The possible displacement of the board with respect to the fixed receptacle. The displacement depends on the mechanical tolerances of the guiding system—that is, the slot and the plug, alternatively the width of the connector tongue and the internal width of the fixed receptacle.
2. The length tolerance of the fixed receptacle and the play of the individual contact springs.
3. The tolerance of the pitch and width of the contact tabs of the edge connector.

The varying parameters make it impossible to state fixed and universal criteria for acceptance. The PCB designer should therefore calculate the tolerance chain, starting with the fixed receptacle, and indicate the necessary modifications of Requirements 1 and 2. Not surprisingly, this implies a tightening of these requirements.

Failure to meet the requirements is cause to reject the board, and the fault is rated as an A-defect. A 100 percent inspection is performed, and rejected boards are returned for remake.

14.7.9 Indentations

Isolated indentations, in the same way as projections, degrade the general quality of the finished board and also undesirably reduce the effective contact area. The inspector should realize that the edge connector region attracts larger attention than the remaining part of the pattern, because it represents the electrical boundary between the board and its surroundings and, furthermore, because it is quite regular and gold plated.

On the whole, somewhat stricter requirements should be applied to the edge connector region than to the remaining part of the pattern, as described in the following requirements. They apply under the conditions stated in Paragraph 14.7.8a (a pitch larger than 0.1 in. (2.54 mm, and in the metric system 2.50 mm). See also Requirements 1 to 3 in Paragraph 14.7.8b regarding edge connectors with a smaller pitch.

1. Indentations must not affect the effective contact area, if such an area is specified. See Figure 14.14. When no effective contact area has been specified, the indentation should not have a depth of more than 10 percent of the nominal width of the contact tab. The length of the base line shall be less than 0.02 in. (5 mm). See also Requirements 1 and 5 in Paragraph 5.6.2).
2. No more than one indentation per contact is acceptable.

The inspection is performed in the same way as described in Paragraph 14.7.8a.

14.7.10 Porosity

Another important point in connection with edge connectors is the porosity of the

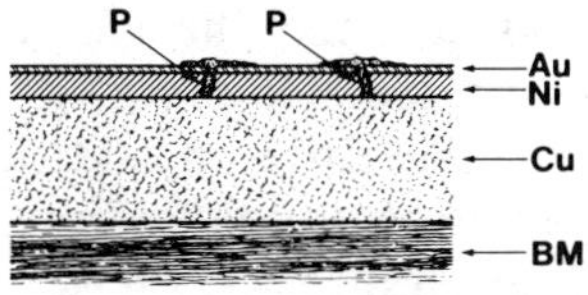

Fig. 14.16 Cross-sectional drawing of contact tab with pinholes/pores, through which corrosion products grow and spread over the surface. P = small pinhole or pore. Au = gold. Ni = nickel. Cu = copper. BM = base material.

plating, meaning the pores and pinholes which can develop in the plating because of poor plating processes. When the board is subjected to a corrosive atmosphere, the underlying copper will be attacked because it is not a noble metal.

14.7.10a Effects of Corrosion

Corrosion products will grow up through pores and pinholes and spread over the surface as sketched in Figure 14.16. Since various contact problems can be traced back to such corrosive effects, the plating should be reasonably impervious.

On the other hand, the situation should not be feared excessively, as millions of boards with gold plated edge connectors are in daily use with no contact problems, and it is quite inconceivable that all these edge connectors have neither pores nor pinholes whatsoever.

14.7.10b Importance of Layer Thickness

As mentioned earlier, it is extremely important that sufficiently thick nickel and gold layers be specified (Paragraph 14.3), particularly a relatively thick nickel layer, as this forms an efficient barrier against corrosion.

Just as important, however, is that the specified thicknesses be met, and also that the plating processes be carried out very carefully. If so, it can be expected that no pinholes will occur and at the most, a very few small pores.

The result is that possible corrosion products growing up to the surfaces as shown in Figure 14.16, achieve a very small extent and thickness. They are therefore removed by the wiping action of the contacts of the receptacle when the board is inserted. If the receptacle is also of reasonable quality, the contact springs will create gas-tight connections which will normally prevent corrosion at the points of contact.

14.7.10c Porosity Specifications

Unfortunately, very few PCB specifications indicate the extent to which porosity in the edge connector tabs is accepted. The IEC-326 recommendation states that porosity tests shold be applied only if explicitly agreed between the PCB manufacturer and the customer. The reason for this is that the confidence level of conclusions drawn from the test results is very limited; consequently, no porosity requirements are stated. In MIL-P-55110C the porosity problem is not mentioned at all.

In BS 9762 (Printed circuits of assessed quality: sectional specification for double-sided printed circuits with plated-through holes), issued by the British Standards Institute, a porosity specification states: Up to 2 pores per contact tab on 10 percent of the contact tabs, and no more than 1 pore on each of the remaining contact tabs.

14.7.10d Inspection Method

The inspection is carried out by inspecting the edge connector tabs of the sample for pinholes and pores with a 7X or 10X magnifier fitted with a graticule. In this way, the inspector is able to find pinholes larger than 0.002 to 0.004 in. (0.05 to 0.1 mm), whereas it is impossible to detect pores. Disregarding pores, we can lay down the following guidelines:

1. No contact tab may contain pinholes within the effective contact area when inspected with a 7X or 10X magnifier.
2. One pinhole, up to 0.004 in. (0.1 mm) in diameter, is accepted per contact, provided it occurs outside the effective contact area or on unused contact tabs.

Exceeding these requirements is cause for rejection, and the fault is rated as an A-defect.

The inspection does not yield any real information about the existence of pores. Since a thorough investigation is relatively complicated and also destructive, it is not routinely done.

However, it is reasonable from time to time—for example, once or twice a year—to perform an investigation so that the plating quality with respect to pinholes and pores can be assessed more objectively. Chapter 27 describes test methods.

Incidentally, some caution should be exercised when comparing the results of an accelerated and thorough laboratory test for corrosion with the much slower corrosive action of even a mildly agressive atmosphere.

14.7.11 Registration Faults

Most boards with edge connectors have contact tabs on both sides of the board, and the inspector should therefore check that the two sets of contact tabs are located correctly with respect to each other. Usually, the registration accuracy does not cause any problems, as the edge connector is part of the pattern.

14.7.11a Detecting Registration Errors

Any large registration error will show up in the form of broken annular rings. Consequently, a hole drilling concentric with the solder pads on both sides of the board reliably indicates that the contact tabs are correctly located. However, if there are broken annular rings on one side of the board, the inspector should measure the misregistration directly.

14.7.11b Registration Requirements

In order to assess a possible registration failure, the inspector should obtain information regarding registration requirements. In some cases the master drawing specifies the registration requirements of the contact sets, especially if the tabs are very narrow—for example, 0.04 in. (1 mm) or less.

In other cases, requirements regarding registration can be stated in the general PCB specification. If no specific requirements are stated, the inspector can avail himself of the following requirement which is so broad that it can be fulfilled by every PCB manufacturer:

1. The misregistration—displacement between the contact tab sets on the two sides of the board—shall be less than 0.008 in. (0.2 mm). See Figure 14.17.
2. In the case of a pitch of 0.05 in. (1.27 mm), the misregistration must be less than 0.004 in. (0.1 mm).

14.7.11c Measuring the Misregistration

The measurement is most easily made from a fixed point which should be conveniently located with respect to the edge connector so that a caliper can be used.

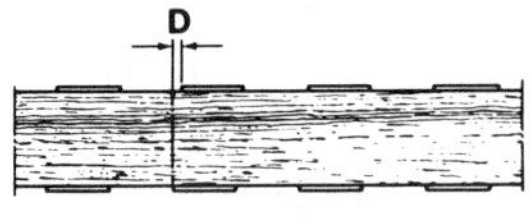

Fig. 14.17. Incorrect side-to-side register between contact tabs of double-sided board. D = displacement between pairs of contact tabs.

It is therefore an advantage to use the edge of the board.

When a polarizing or keying slot is used, the inspector can instead measure the location of the connector tabs adjacent to the slot, and determine any front-to-back registration errors. The measurement is performed with a 7X or 10X magnifier fitted with a graticule. It is, however, necessary that the slot have well-defined edges.

This measurement gives only the relative location of the contact sets and not the absolute location. If it is important to know the latter, the inspector has to base his measurement on the reference system. The reader is therefore referred to Paragraph 6.2, particularly Paragraph 6.2.2.

14.7.11d Board Disposition

Failure to meet the registration requirements should be considered cause for rejection, and the fault is consequently rated as an A-defect. If the contact tabs are very wide—for example, 0.12 to 0.20 in. (3 to 5 mm)—the possibility of exempting the fault should be considered. This implies that the PCB designer goes through the tolerance chain, which might reduce the contact tab width and, in turn, reduce the gold consumption.

In the case of a small number of boards, a substantial saving should not be expected as the ratio between processing costs and the cost of the gold used for the entire connector is often around three to five, depending on the connector length. Thus, the saving from removing unused contact tabs is rather limited and in many cases not worth the trouble. In the case of a large number of boards such savings can, of course, be significant.

14.8 Areas Adjacent to the Edge Connector

At the same time he inspects the edge connectors, the inspector should check the

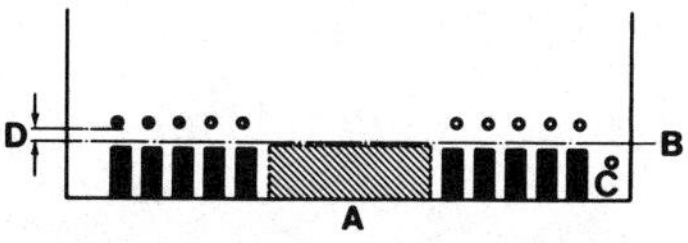

Fig. 14.18. Areas adjacent to edge connector.
A: Area which can be stripped of tin/lead together with the contact tabs.
B: Demarcation line for gold plated area.
C: A plated-through hole located beneath the demarcation line B can erroneously be stripped of tin/lead.
D: Advisable minimum distance between solder pads and the demarcation line B is 0.12 in. (3 mm).

conditions of the adjacent areas. If the board is designed as shown in Figure 14.18, there is a risk that the area A between the two edge connector sections also will be stripped of tin/lead. For example, the board designation or code number could be placed here, and as there is no metallic connection to the plating bar, the area will show bare copper, which may quickly corrode.

The inspector should also note whether there are plated-through holes C just around the demarcation line B of the gold plated area. See Figure 14.18. Such holes will be stripped of tin/lead if the PCB manufacturer does not protect them by suitable masking. The PCB designer is well-advised to keep designations, circuitry, and plated-through holes away from the connector area.

Frequently, it is necessary to place via holes just above the contact tabs in order to transfer some of the contact connections to the opposite side. Such via holes should be placed with a distance D (see Figure 14.18) of at least 0.12 in. (3 mm) between the solder pad and the demarcation line of the gold plated area. In this way, the PCB designer obtains full assurance of not having the tin/lead plating of such holes damaged when the tin/lead is stripped off the contact tabs.

14.9 Possibilities of Repair

Repair of damaged gold plated contact tabs is increasingly becoming more and more topical for the PCB manufacturers, at least for expensive boards from which the plating bar has been removed. One can adopt the same attitude as in Paragraph 17.5.10 concerning qualified repair of pattern defects.

14.9.1 Acceptability of Repaired Tabs

In order to keep the rate of rejection—and the prices—down, more and more PCB manufacturers are interested in repairing boards having various defects, including defective edge connectors. As far as can be seen, repairs will become more commonly accepted, but the customer should not accept repairs unless the manufacturer has documented the method and also proven his capability in this respect.

14.9.2 Repair Methods

For the inspector's information, some methods of repairing edge connector tabs are now described.

In the case of a particularly expensive board—for example, a large multilayer board with a single defective contact tab—it is sometimes acceptable to transplant a new contact tab. The defective contact tab, including a bit of the attached conductor, is machined out. A new contact tab with attached conductor is then machined out from another (defective) board and carefully glued in place with epoxy in the cutout. The electrical connections are re-established—for example, by welding a kovar strip across the gaps between the conductor ends. Acceptability of the procedure requires perfect workmanship, which means:

1. The new contact is flush with the board.
2. The distances to the adjacent contact tabs are correct.
3. The glue joint is filled and the glue kept away from the contact tabs.
4. The connections to the new contact tabs are established in a satisfactory manner.

The repair procedure is quite complicated and expensive, for which reasons it is used only in a few exceptional cases.

In the case of repairing a gold layer which is delaminated or damaged by scratching, it is possible to use the selective plating method called brush plating (swab plating). The method can also be of direct interest to the customer if a service job includes repair of a scratched or worn contact tab.

A detailed description is found in the IPC publication IPC-R-700B, "Suggestive Guide Lines for Modification and Repair of Printed Boards and Assemblies." In all simplicity the method is to establish, on a small scale, an electrolytic bath by means of a cotton wad saturated with electrolyte. The cotton wad is placed on a stylus and becomes the anode, whereas the contact tab is the cathode. A suitable current deposits a new gold layer as the cotton-wad anode is brushed back and forth across the contact tab.

14.9.3 Identification of Repairs

If the method is accepted by the customer, selectively plated contact tabs should be indicated by the PCB manufacturer so that they can be identified and inspected carefully by the inspector. The same criteria as for all other contact tabs apply to selectively plated contact tabs.

Chapter 15

Machining Inspection

15.1 Introduction

Until now we have dealt mainly with the internal characteristics of the board—the pattern, solder holes, plating, solderability, solder masks, etc. Now we shall examine the machining, which, in addition to the board contour, can also include cutouts and any holes too large to be drilled. This includes chamfering the board edge along an edge connector and milling a polarization slot.

Several points must be borne in mind when considering machining, such as the way machining operations are called up in the documentation, the way tolerances are stated for the finished board, and also the equipment used by the PCB manufacturer as well as various machining methods.

The inspector should possess sufficient knowledge of all these matters to be able to determine whether the delivered boards satisfy the quality requirements regarding the machining. Possibly, no specific requirements exist at all. In such cases the inspector should understand common and reasonable manufacturing practices, so that he can decide whether the board is acceptable.

Having to discuss quality problems without specifications to support one's stand is always undesirable; it is therefore strongly recommended that specifications be included to a reasonable extent.

15.2 General Quality Requirements

The machining must not cause scratching, crazing, or delamination of the board. A slight delamination along the outer edges can be accepted provided it does not extend more than 0.02 in. (0.5 mm) from the board edge. This implies that the distance between the board edges and the pattern (conductors, ground planes, etc.) exceeds 0.02 in. (0.5 mm).

The board edges should not be so uneven or frayed that splinters of glass weave can become embedded in the assemblers' fingers during board assembly.

Scratches can be another result of machining. Paragraph 4.6.2 addresses this problem.

15.3 Defining the Contour

The board contour can be defined in several ways, each influencing the choice of machining method. Before addressing the methods and obtainable results (Paragraph 15.4), this chapter describes the more common methods of defining the contour.

15.3.1 Contour Defined by Corner Marks on Master Film

Corner marks are the oldest and simplest way of defining the contour of the board, but they are also the least accurate. This

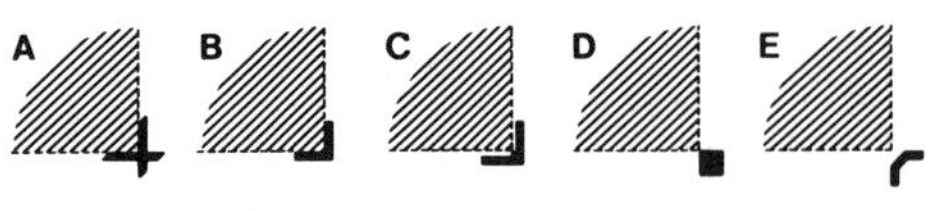

Fig. 15.1. Various types of corner marks.

is due to the fact that the contour is not defined with respect to internal tooling holes in the board. Machining based on the corner marks means visual alignment and all the inaccuracies implied by this method.

15.3.1a Residual Metal Strips

There are several types of corner marks, as shown in Figure 15.1. The most common types are shown in views A and B, but if the contouring is performed with any inaccuracy, a very thin strip of metal from the corner mark can be left along the board edge.

Because of its relatively low adhesion, the strip can fall off and cause short circuits if not removed. PCB manufacturers commonly remove such remnants of the corner marks. The inspector should always inspect the boards on a general sampling basis and watch for possible remnants. They should be removed during a 100 percent inspection, performed either by the PCB manufacturer or the customer's production department.

The types shown in Figure 15.1, views C, D, and E, leave no metal remnants along the board edges, and deserve to be used to a wider extent than they are now.

15.3.1b Superfluous Corner Marks

In addition, when the board contour is defined by the methods mentioned in Paragraphs 15.3.2 and 15.3.3, corner marks are frequently used even though they are superfluous. However, they do help one to visualize the contour, which can be important to the PCB designer as well as the PCB manufacturer.

The latter especially benefits when a current robber is to be laid around the board in order to achieve a more uniform plating thickness all over the board. The PCB manufacturer adds the current robber to the master film by means of lithotape placed slightly outside the corner marks.

15.3.1c Preferred Types of Corner Marks

Most PCB manufacturers prefer the corner mark shown in view E of Figure 15.1, and to a lesser extent that in view D, when contouring the board according to the methods of defining the contour described in Paragraphs 15.3.2 and 15.3.3. In this way, they can avoid soft chips of copper and tin/lead, which tend to clog some types of cutter bits.

15.3.1d Metal Edges

Another point is that metal edges easily become quite ragged or even frayed. This is of no importance in the case of corner marks, which are to be removed completely. However, it is important in the case of ground planes extending to the board edge. Such areas should always be withdrawn from the board edge at least 0.02 in. (0.5 mm).

This discussion clearly shows that it is poor practice, as well as superfluous, to tape or plot the total contour when preparing the artwork or the master film.

15.3.1e Internal Measuring Points

As the corner marks disappear during machining, the PCB designer can facilitate the inspection considerably by introducing internal measuring points on the board. It can be practical to place small squares in each corner at a standardized distance from the edge—for example, 0.04 in.

(1 mm), so that the inspector has fixed measuring points.

When contour checking is based on internal measuring points, the inspector should realize that small measuring errors can occur. Dimensional changes can cause the distance between the measuring points to change. Furthermore, the edges of the individual measuring points can be displaced as a result of the common edge displacement of the pattern.

The latter type of error can be eliminated by a double measurement, in which the distances from opposite sides of the square measuring points to the board edge are measured and the average value is calculated. This implies that the nominal distance is stated from the center of the square and not from its side.

15.3.1f Internal Cutouts

Internal cutouts can also be marked in several different ways as shown in Figure 15.2. Examples A and B have the disadvantages mentioned in conjunction with Figure 15.1. Examples C and D are therefore preferred. The PCB designer should notice that it is expedient to place aiming marks at the opposite edges of the board to visually aid the machining of the two sides of an interior corner. See Figure 15.2.

15.3.1g Dimensioning

The overall board dimensions—its length and width—must be stated on the master drawing, as tolerance indications without corresponding dimensions are meaningless. Incidentally, dimensioning the length

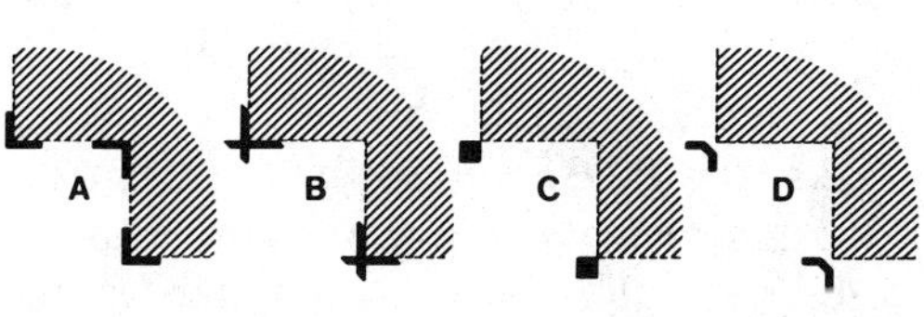

Fig. 15.2. Marking of cutouts by means of corner marks.

and width of the board (as mentioned in Paragraph 15.5.1) aids the PCB manufacturer in contouring the board.

15.3.2 Contour Defined by a Reference System

A more accurate definition of the board contour can be obtained with the use of coordinates in a reference system as shown in Figure 15.3. The zero point of the reference system can be chosen internally as well as externally with respect to the board contour. On account of measurements performed on the finished board, it is important that the reference system be clearly defined regarding the position of the zero point and the direction of the axes.

15.3.2a Tooling Hole Locations

The author advocates that the zero point be placed within the contour and, where possible, at the center of the solder pad of the major tooling hole T_1. The direction of the X-axis is to the center of the solder pad of the second tooling hole T_2; the X-axis should preferably be parallel with the board edge.

Naturally, the reference system is identical to the reference system described in Paragraph 6.2.2, and the same considerations concerning the location of the centers of the tooling holes T_1 and T_2 as well as the use of the tooling holes for measuring purposes are also valid here.

15.3.2b Advantages of the Reference System

The board contour is defined by the X-Y coordinates of the various details of the contour, such as corners and cutouts. See Figure 15.3. The method is considerably clearer than the traditional method of dimensioning a drawing, and it suits the operator far better when generating the NC

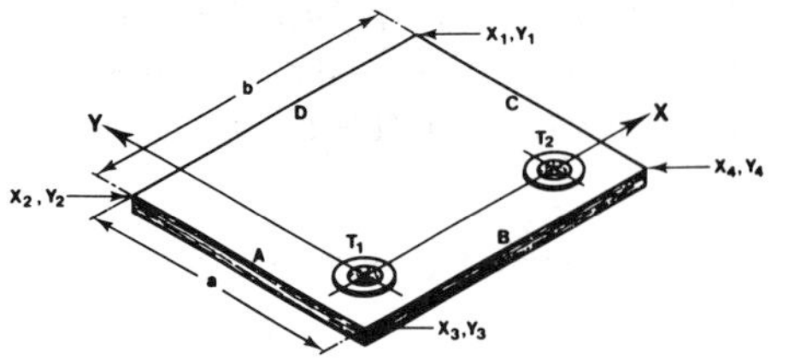

Fig. 15.3. Determining the board contour by means of coordinates in a reference system. T_1 = major tooling hole. T_2 = secondary tooling hole. X,Y = corner coordinates.

router program. It is, however, always practical to state the overall dimensions of the board as this makes it possible to check very quickly whether the board's length and width tolerances are met.

15.3.3 Contour Defined by Customer's Router Tape

The increasing use of CAD (Computer-Aided Design) has caused customers to deliver router tapes to the PCB manufacturer in order to save some of the tooling charges. As the plotter tape used for plotting the master film is derived from the same data base, the PCB manufacturer can achieve a very high degree of coincidence between the pattern and the contour.

Generating a router program, however, requires extremely close cooperation between the PCB manufacturer and the customer, as the latter must know the size of the router bit to be used and the routing technique, and whether the contour is routed in one pass with compensation for unavoidable cutter deflection or it is routed in two passes in the same track. On the first pass the board is slightly too large, and on the second pass—the cleanup pass—the desired contour is obtained and the edges smoothed and evened. Furthermore, the router tape must be compatible with the NC router machine.

All these conditions make it less attractive to deliver a router tape as documentation for the board contour. To this must be added that the customer assumes the responsibility of incorrect contouring if a fault is caused by a fault in the router tape. As a router program can be written in a fairly short time, the customer saves little by doing this himself.

15.4 General Methods of Contouring

The board can be contoured in several different ways, depending on a number of factors: economy, machines available, and the way in which the contour is defined. The following examples illustrate this.

When a board with a complex contour, possibly with internal cutouts, is ordered in large quantities, a blanking operation may be the most suitable method. Another possibility is NC routing on a multispindle machine, which in many cases is fully competitive with blanking. Both cases imply a well-defined contour, based on a reference system and not on corner marks alone.

For small quantities and a relatively regular contour without severe accuracy requirements, the PCB manufacturer may choose to shear the contour with a guillotine or to cut it with a circular saw, particularly when the contour is defined by corner marks only.

When the contour is a little more complicated—for example, with cutouts or nonparallel edges—the contour is usually based on a reference system and not on corner marks. If the PCB manufacturer does not have an NC router machine, he will probably choose to pin route the board, which is a routing process controlled manually with a template.

If the contour is defined by corner marks, the PCB manufacturer can introduce his own reference system, usually based on tooling holes outside the board contour, which will allow routing the contour.

The PCB manufacturer has many options for machining the contour. The greatest flexibility, which means independence of the PCB manufacturer's machinery complement, is attained by defining the contour on the basis of a reference system as well as corner marks.

Insofar as the PCB manufacturer can meet the accuracy demands of the contour and the smoothness of the edges, he should be allowed a free choice of contouring methods.

It is emphasized that the capability of a PCB manufacturer should not be assessed on the basis of his owning an expensive NC router machine. The crucial point is whether the PCB manufacturer has mastered his technique of contouring so well that he can meet the quality requirements and still offer competitive prices.

Next we shall look at the general methods of contouring and review the tolerances attainable. Having acquired the necessary background knowledge, the inspector should be able to assess whether the results correspond to the method of machining as well as the method of defining the contour.

15.4.1 Shearing

Shearing is performed by means of a guillotine. The contour accuracy depends on how accurately the operator can align the corner marks with the shear blade. Because of unavoidable parallax errors, the inspector should expect a tolerance of ±0.012 to ±0.020 in. (±0.3 to ±0.5 mm) per edge, depending on the skill and care of the operator. The sheared edge is usually fairly coarse.

15.4.2 Sawing

Sawing is always performed with a circular saw. Epoxy glass laminates are usually sawed with a diamond-steel bonded saw blade, whereas phonelic paper is usually cut with a carbide-toothed saw blade. The operator locates the board as precisely as possible to align the corner marks with the saw blade. The tolerance of visually aligned sawing is usually ±0.012 to ±0.020 in. (±0.3 to ±0.5 mm) per edge.

However, the operator may use a sawing jig, to which the board is attached with a tooling pin and hole system. By moving the sawing jig along a stop, which is adjusted with respect to the saw blade, the operator can obtain a relatively high degree of reproducibility and a tolerance of ±0.008 in. (±0.20 mm) per edge.

Sawing produces a relatively smooth board edge, as long as the saw blade has no play.

15.4.3 Blanking

Before the advent of modern NC router machines, the PCB manufacturer was nearly always forced to blank all complexly contoured boards manufactured in large quantities. Some of the disadvantages of using a blanking tool are the long lead time associated with the tooling, the high cost, and the fact that it is very difficult, perhaps impossible, to modify the contour. On the other hand, it is possible to obtain an accuracy of ±0.002 to ±0.004 in. (±0.05 to ±0.10 mm) per edge, if the tool is not too worn.

15.4.3a Board Expansion

Resilience of the base material causes the blanked board to grow a little, which should be taken into consideration when designing the blanking tool. Some types of phenolic paper laminates can be blanked only when warm, and the coefficient of thermal expansion should therefore be allowed for. Moreover, a special problem is that phenolic paper laminates do not always have exactly the same coefficient of expansion in the X and Y directions of the board.

15.4.3b Trend to NC Routing

Except for boards used by the consumer electronics industry for radios, TV, etc., the current trend is from blanking to NC routing. The exception is relatively small boards—for example, 1 × 2 in. (25 × 50 mm) or less—which are difficult to hold in place during routing; when quality requirements permit, blanking is often preferred.

15.4.3c Retention in Panel

Blanking tools can be built with return-to-the-blank, which means that after blanking the board is pressed back into the panel from where it was blanked and is now held in place by friction.

Blanking tools can also be made in such a way that the contour lines are just broken so that the board is pressed down a little into the panel and remains there, again held in place by friction. See Figure 15.4. The entire panel with the blanked, but retained, boards is delivered to the customer, who assembles and mass solders the panel as a unit, after which the individual boards are pressed out of the panel.

15.4.3d Edge Quality

Blanked boards usually have ragged edges, particularly when the blanking tool is worn so that it cannot perform a pure shearing, but rather a crushing along the fracture line. This implies that the 0.002 in. (0.05 mm) accuracy limit can be difficult to achieve on a long-term basis.

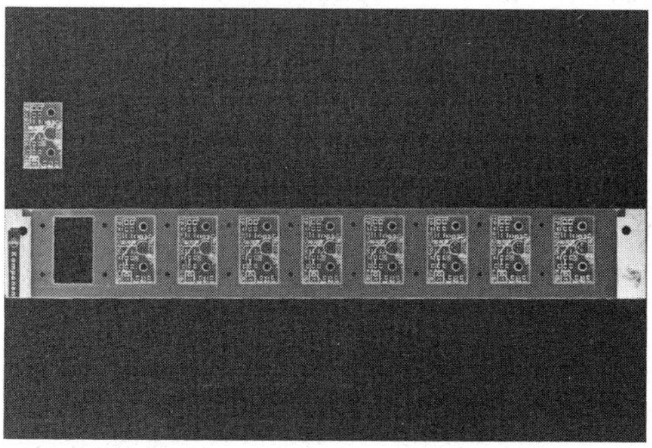

Fig. 15.4. Example of small boards being blanked and returned to the blank.

15.4.4 Routing

Routing, which gives the best results with regard to an even and smooth edge, can be done either manually or automatically. In the first case, the method is called "pin routing," and the operator attaches the board, or a stack of boards, to a template by means of tooling pins located on the template and internal tooling holes in the boards.

By tracking the template against a pilot pin located axially with the router bit, the operator can route the entire board circumference. The accuracy is ±0.004 to ±0.008 in. (±0.1 to ±0.2 mm) per edge, but it depends very much on the operator's skill and attentiveness.

Automatic NC router machines have gained increasing use, and it is frequently economical to use such expensive machines for routing boards even when the order quantity is relatively small and the contour simple. The cost of writing the program is fairly low, and under all circumstances much lower than the cost of a template. An accuracy of ±0.002 to ±0.004 in. (±0.05 to ±0.10 mm) per edge can be expected.

15.5 Attainable Tolerances of the Contoured Board

The tolerances stated in Paragraphs 15.4.1 through 15.4.4 are valid per edge; that is, the edge displacements to be expected as a consequence of the inaccuracies caused by the contouring method used. Now we address some typical examples and see what tolerance can be obtained for the

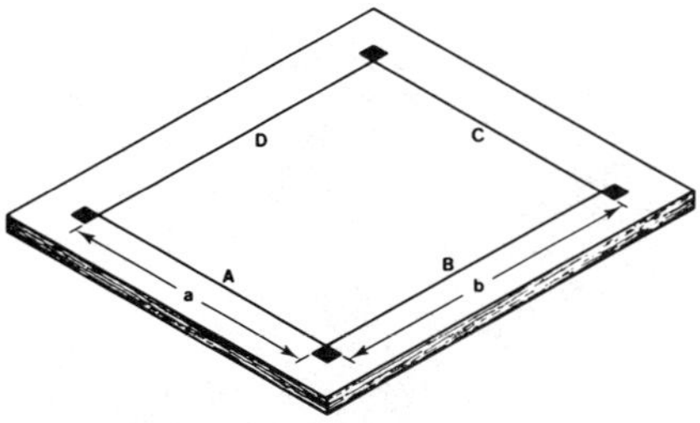

Fig. 15.5. Contouring of board based on corner marks.

overall dimensions of the board (length and width).

15.5.1 Tolerances with Contouring Based on Corner Marks

We first study the case where the contouring is based on corner marks alone. See Figure 15.5. The machining is accomplished by shearing or sawing with visual alignment. The long side B is machined first. When machining side A, the operator places the contoured side B against a stop perpendicular to the cutting direction of the machine.

The operator has more assurance of cutting a right angle at the corner AB—and at the remaining corners—than if side A, which may be quite short, is visually aligned relative to the corner marks. The sides C and D are machined similarly.

15.5.1a Undimensioned Boards

If the length and width of the board are undimensioned, the PCB manufacturer cannot be held responsible for exceeding any tolerances, as it is meaningless to speak of tolerances when there is no reference. Referring to normal manufacturing practices, the inspector may expect the following requirements to be met:

Displacement of edges A to D:
maximum ± 0.012 in. (± 0.3 mm)

Tolerance of length and width:
± 0.024 in. (± 0.6 mm)

15.5.1b Dimensioned Boards

When the length a and width b of the boards are given, the PCB manufacturer machines the sides A and B as previously described. The sides C and D are machined by placing the first machined edges A and B against a stop located parallel to the first cutting direction and at a distance b and a, respectively, from the cutting line. In this way the operator achieves the most accurate machining of the board. Consequently, the inspector can now anticipate that the contour fulfill the following requirements:

Displacement of edges A and B:
maximum ± 0.012 in. (± 0.3 mm)

Tolerance on width a and length b:
± 0.008 in. (± 0.2 mm)

Displacement of edges C and D:
maximum ± 0.020 in. (± 0.5 mm)

15.5.1c Boards Provided with Manufacturer's Tooling Holes

We shall now examine the case where the PCB manufacturer adds his own tooling holes at well-defined positions relative to the corner marks. The machining can now be done by shearing, sawing, or pin routing by means of a jig or template, the boards being attached via tooling pins on the jig or template. The inspector can now expect the contour to meet the following requirements:

Displacement of edges A and B:
maximum ± 0.008 in. (± 0.2 mm)

Tolerance on width a and length b:
± 0.016 in. (± 0.4 mm)

By comparison with the previous case, it is seen that the tolerance on the overall dimensions of the board has increased. On the other hand, all edges are located more accurately with respect to the pattern.

15.5.1d Contour Inspection

In all these cases, the inspector can check the edge contour by ascertaining the edge displacement with the master film as a reference. The film is laid over the board, and the edge displacement relative to the corner marks is measured with a 7X or 10X magnifier fitted with a graticule. If the length and width of the board are stated in the master drawing, they can be measured directly with a caliper.

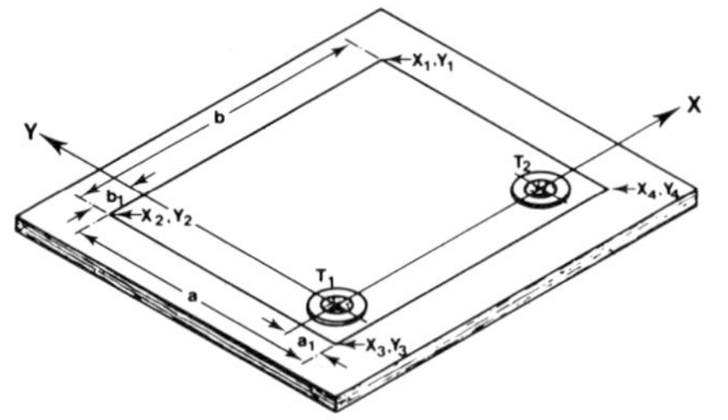

Fig. 15.6. Contouring of board based on a reference system.
T_1 = major tooling hole. T_2 = secondary tooling hole. X,Y = coordinates of the corners.

15.5.1e Board Disposition

The board should not be automatically rejected if the requirements are not met. In many cases, the exact location of the contour with respect to the pattern is of no particular importance. This is especially true when the board is to be mounted in a chassis with plenty of space around the board edges. In such cases, the inspector can waive even large excesses, and the fault is rated as a B-defect.

If the board is designed as a plug-in type of board—that is, it has an edge connector or a soldered-on connector and is guided into the rack by board guides—it is very important that the contour be placed correctly with respect to the connector system.

In the case of plug-in boards, it is assumed that the tolerance requirements are stated on the master drawing and that they are consistent with the obtainable results. Exceeding a tolerance is cause for rejection and the fault is consequently rated as an A-defect. A 100 percent screening will frequently be necessary when the inspector finds that the tolerances are exceeded.

15.5.2 Tolerances with Contouring Based on a Reference System

Here the board contour is defined unambiguously—that is, it is dimensioned with respect to the pattern which, for simplicity's sake, is considered to be represented by the reference system. The contour is defined by corner coordinates or by traditional dimensioning. See Figure 15.6. Contouring can be accomplished either by blanking or by routing. Both methods are addressed below.

Routing can be done in several ways as described in Paragraph 15.4.4, but a common weakness of this method is the relatively thin router bit, frequently only 3/32 in. (2.4 mm) or 1/8 in. (3.2 mm) in diameter. The thin bit deflects and reduces the accuracy of the machining. This problem can be solved by routing at a low feed rate to avoid overloading the router bit.

Another method is to make two passes, the first being a coarse contouring, and the second a fine contouring. When making the second pass, there is not much material to cut, so the loading of the router bit is quite low. Therefore, the contouring accuracy, as well as the edge smoothness, is correspondingly higher. We shall now address methods in actual use.

15.5.2a Manual Pin Routing

Nearly all PCB manufacturers have the necessary equipment for manual pin routing. This method requires attaching the board to the router template. The board should therefore be provided with internal tooling holes which can advantageously be located diagonally. Internal tooling holes are normally provided when the contour is defined by a reference system. The inspector can therefore expect the following requirements to be met:

Displacement of edges A to D:
maximum ±0.006 to ±0.008 in.
(±0.15 to ±0.20 mm)

Tolerance on width a and length b:
±0.012 to ±0.016 in.
(±0.30 to ±0.40 mm)

15.5.2b NC Routing

A large number of PCB manufacturers already have procured NC router machines, and more will follow as the quality and economic advantages of these machines are demonstrated.

In some cases, the machine may be a combination router and drilling machine, but dedicated routers are more common. The boards are usually stacked two or three high. Compared with pin routing of one board at a time, a higher loading—and larger deflection—of the router bit could be expected.

The feed rate programming predetermines the router bit loading, which is much greater than in the case of pin routing. As before, a coarse and fine contouring can be executed, but it is also possible for one pass to suffice because the deflection of the router bit can be compensated for in the program. The following requirements should be met:

Displacement of edges A to D:
maximum ±0.002 to ±0.004 in.
(±0.05 to ±0.10 mm)

Tolerance on width a and length b:
±0.004 to ±0.008 in.
(±0.10 to ±0.20 mm)

15.5.2c Blanking

Blanking requires making a blanking tool corresponding to the board. The PCB designer should realize that blanking of even a relatively small board requires a relatively large press. The shear strength of epoxy glass laminates is frequently around 20,000 psi (pounds per square inch).

If a Eurocard sized 4 × 6.3 in. (100 × 160 mm) and 0.063 in. (1.6 mm) thick is to be blanked, the area to be sheared is 20.6 × 0.063 in. = 1.30 sq. in. (520 × 1.6 = 832 mm ~ 1.30 sq. in.). The necessary pressure becomes 1.3 × 20,000 = 26,000 lb or approximately 12,000 kilograms. When the blanking tool is provided with a spring-loaded stripper, twice as much pressure is required, which means at least 24,000 kilograms. Obviously, the press must be quite large.

In the case of phenolic paper laminates the shear strength is only 12,000 psi so that only 60 percent of the pressure necessary for blanking an epoxy glass laminate is needed. If the board is much larger than a Eurocard, the PCB designer should consult the PCB manufacturer regarding his blanking capacity.

The board should be guided by tooling pins during the blanking. As mentioned in Paragraph 6.2.2d, the corresponding tooling holes should be drilled in the first drilling operation so that the contour is not unnecessarily loaded with inadequate positional accuracy of the tooling holes. The inspector may expect the following requirements to be met:

Displacement of edges A to D:
maximum ±0.002 to ±0.004 in.
(±0.05 to ±0.10 mm)

Tolerance on width a and length b:
±0.004 to ±0.008 in.
(±0.10 to ±0.20 mm)

15.5.2d Measuring the Contour

Measurement of the board contour is based on the reference system, the contour being measured with respect to the tooling holes. The measuring technique is described in more detail in Paragraphs 6.2.2 and 6.4. Incidentally, the caliper shown in Figures 6.6 and 6.7 can be modified as shown in Figure 6.8, which seems most practical when checking the contour.

15.5.2e Board Disposition

The problems of rejection or exemption in the case of exceeded tolerances are exactly the same as described in Paragraph 15.5.1. The high degree of reproducibility achieved by blanking or NC routing makes it unnecessary to check the contour of more than one board.

Manual pin routing is operator-dependent, so the inspector should inspect a larger sample—for example, according to Special Inspection Level S3. For this reason, the inspector should know which machining method the PCB manufacturer uses.

Incidentally, the inspector should realize that although a board is machined by blanking or NC routing, this is no indication of a practical need of the very narrow tolerances associated with these methods. For example, the PCB manufacturer himself may find NC routing more advantageous than sawing.

15.5.2f Tab-Attached Boards

As the router bit can be lifted and lowered by the router program, it is possible to contour small, difficult-to-handle boards in such a way that the contoured boards remain attached to the production panel by small tabs as shown in Figure 15.7. This corresponds very closely to boards which are blanked with return-to-the-blank (Paragraph 15.4.3).

Detaching the individual boards after assembly and mass soldering requires a special punching tool for removing the tabs. The method is also advantageous to the PCB manufacturer as it is not nec-

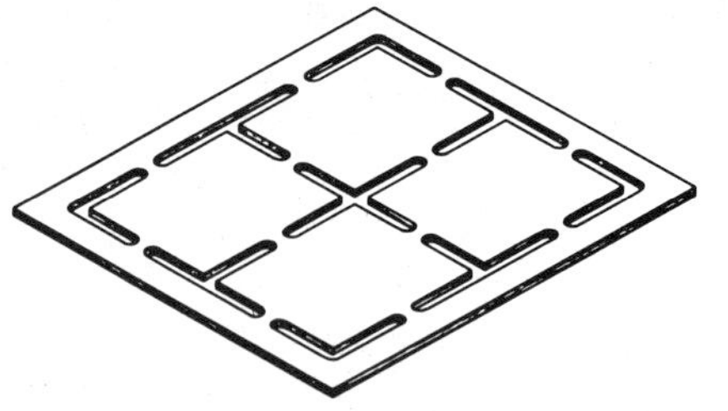

Fig. 15.7. Contouring of nondetached boards.

essary to pin the individual boards to the subplate of the machine. If there are many pins, it could be somewhat difficult to fit the production panel onto the subplate.

15.5.2g Ion Contamination

As a side step, attention is called to a very rare fault, which passes almost unnoticed by the majority of PCB manufacturers. In Paragraph 4.6.6c we described ion contamination of the board surface, which reduces the surface resistance dramatically.

When it is impossible to find the cause in the chemical processes, it can possibly be traced to the NC routing. Here the board is held down at the work area by means of a ring-shaped brush located concentrically around the router bit.

If the brush is too hard, it can remove a very small amount of tin/lead from those parts of the pattern it passes over and smears tin/lead over the board surface in an invisible layer of atomic thickness, causing the surface resistance to drop. The fact that the fault only appears in a band along the edge can be used as an indication of the nature of the fault.

15.6 Special Cutouts

An NC router is particularly well suited for making complex external as well as internal cutouts. Since the router bit can be lifted and lowered as needed, it can conveniently be brought to work inside

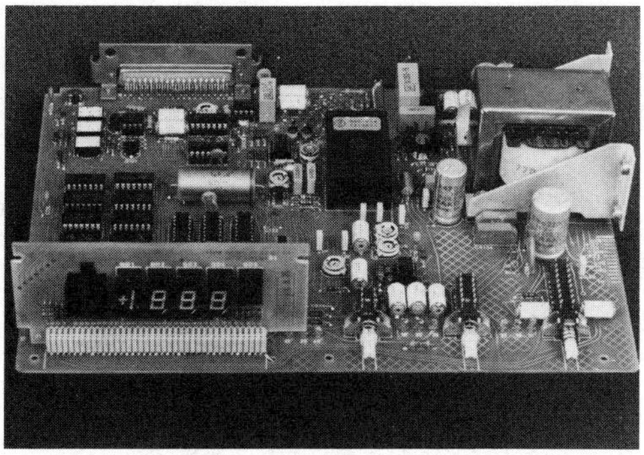

Fig. 15.8. Mounting a display board orthogonally with respect to a main circuit board by means of a tongue-and-slot technique.

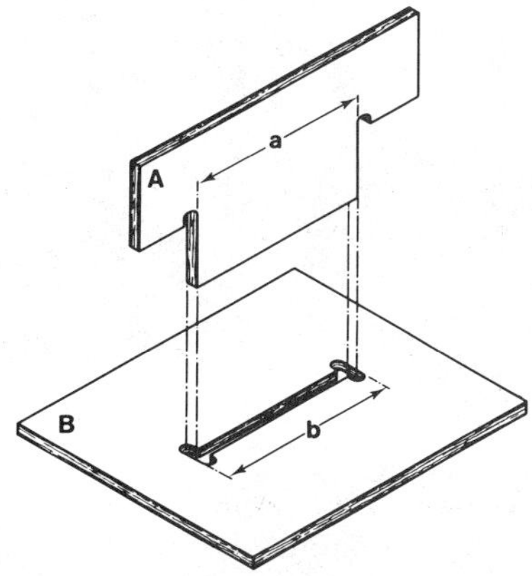

Fig. 15.9. Adapting the slot and tongue to NC routing.
A: Rounding removed from "corner". Width "a" of tongue is maintained at full insertion.
B: Slot terminated by orthogonal slots to give a defined length "b" of slot.

the board contour without damaging the board. Large circular holes which cannot be drilled are easily routed.

15.6.1 Internal Corners

Internal corners are rounded according to the radius of the router bit. This is usually acceptable, but in some cases it is necessary to have sharp corners. It is, however, not particularly practical to post-treat such corners manually—for example, by filing. The PCB designer should modify the corners so that they can be machined by routing. The following example illustrates this.

Two boards A and B are to be mounted at right angles to each other, guided by a tongue and slot as shown in Figure 15.8. The intent of this design is that the contact tabs of the two boards are soldered together after assembly. For this reason it is extremely important that the two boards are guided in a well-defined manner with respect to each other.

The tongue of the one board should have sharp corners, and the slot in the other board should have straight-edged ends. Unfortunately, this is impossible to accomplish with NC routing. Instead, a modified method has been chosen, as illustrated in Figure 15.9, which shows a tongue and slot in oversize. Obviously, this solution is easy to machine by NC routing, as it is only a question of a very small change in the program.

When defining the dimensions a, b, and c, machining tolerances of at least ± 0.004 in. (± 0.1 mm) should be considered, so that even under worst-case conditions, b is always a little longer than a—for example, 0.004 in. (0.1 mm).

15.6.2 Cutout Dimensioning

The PCB designer should always indicate the cutouts on the master drawing. A combination of traditional dimensioning and use of coordinates is frequently the most practical method. For example, the center point of an internal slot can be defined by coordinates, whereas the slot itself is dimensioned. See Figure 15.10.

15.6.3 Measuring the Cutout

Measurement of internal and external cutouts follows the same guidelines as described in Paragraph 15.5.2. In this case also, it is only necessary to check a single

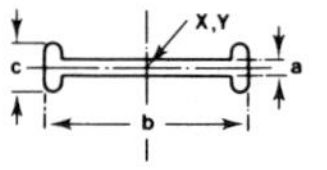

Fig. 15.10. Determining a slot by means of coordinates and ordinary dimensioning.

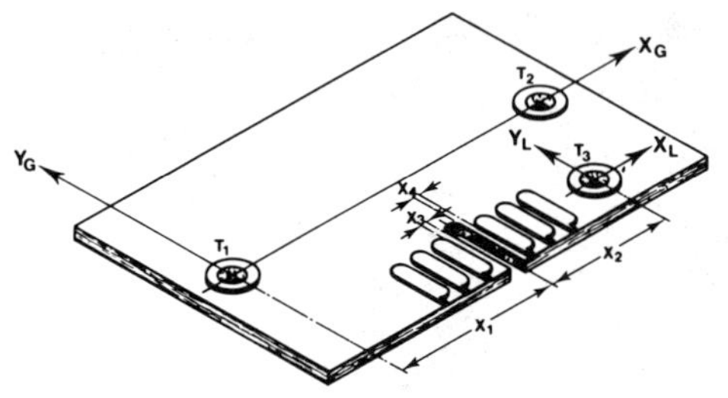

Fig. 15.11. Location of slot determined by:
I: A general reference system X_G, Y_G (x_1).
II: A local reference system X_L, Y_L (x_2).
III: A symmetrical location ($x_3 = x_4$).
T_1 = major tooling hole. T_2 = secondary tooling hole. T_3 = local tooling hole.

board since the method is highly reproducible.

15.6.4 Metal Remnants

The inspector should be aware of metal remnants left from internal corner markings. When routing the contour, corner marks as shown at A, Figure 15.2, imply a very small metal area in the corner itself as the corner is rounded, corresponding to the router bit.

15.7 Edge Connector Slots

Edge connectors are frequently provided with a slot which serves to guide the board with respect to the receptacle (locating slot) and/or preventing incorrect insertion of the board (polarization slot). The design conditions are addressed in Section 6.7 of Reference A12, Reading List, to which the reader is referred.

15.7.1 Slot Definition

It is important for the inspector to understand how the slot position is defined, as this can influence the conduct of the inspection. We shall deal with methods in general use. With reference to Figure 15.11, the slot location can be based on:

1. The general reference system X_G, Y_G (Figure 15.6) in which the coordinate X_1 indicates the position of the symmetry line of the slot. Alternatively, the symmetry line can be dimensioned with respect to the reference system X_G, Y_G.
2. A local reference system X_L, Y_L in which the coordinate X_2 indicates the position of the symmetry line of the slot. Alternatively, the symmetry line can be dimensioned with respect to the reference system X_L, Y_L.

   The local reference system X_L, Y_L is displaced in parallel to the general reference system X_G, Y_G and has its zero point in the center of the pad of the local tooling hole T_3. If the tooling hole is placed in line with the contacts, the inspector has a very convenient base from which to check the slot position.
3. A demand for a symmetrical position with respect to adjacent contact tabs—that is, X_3 equals X_4.

15.7.1a Slot Definition Relative to Board Edge

In no case is the slot definition based on the board edge. In principle, this would be incorrect as an edge connector slot inherently must be tied to the pattern—that is, the contact tabs of the edge connector.

15.7.1b Slot Definition Relative to Contour

A definition of the slot relative to the contour misleads the PCB manufacturer into using an inadequate contouring method

(Paragraph 15.7.2), and the customer will be in a weak position to blame the manufacturer when the slot is displaced as much as the board edge.

15.7.1c Slot Definition by a Reference System

It is strongly recommended that all slots be defined on the basis of a reference system which unambiguously represents the pattern. When designing the width, due consideration should be given to the tolerance of the width of the plug used in the receptacle. Also the machining tolerances attainable should be considered.

15.7.1d Shape of Slot End

Regarding the end of the slot, there are two possibilities. It can either be semicircular or straight, depending on whether the machining is performed by routing or sawing. When no specific requirement concerning the end of the slot is needed, both alternatives should be stated on the master drawing. Although the length of the slot is not generally critical, a tolerance should always be stated on the master drawing.

15.7.1e Distance from Slot to Contact Tabs

In order not to damage the adjacent contact tabs when machining the slot, the nominal distance from the edge of the slot to the nearest contact tabs should be at least 0.012 in. (0.3 mm). This allows enough margin to avoid affecting the contact tabs as a result of the inherent machining tolerances.

15.7.2 Machining of the Slot

In practice, there are two possibilities of machining: sawing and routing. Before the advent of NC routing machines, the slot was most commonly cut by sawing. It should, however, be noted that a PCB manufacturer who still saws the slot cannot be considered a worse manufacturer than one who uses an NC router machine.

15.7.2a Choice of Method

Disregarding the equipment at the disposal of the PCB manufacturer, the choice between sawing and routing depends primarily upon the shape and dimensions of the slot.

Usually the PCB manufacturer prefers to route the slot in the same operation as the contour in order to simplify the machining. Obviously, this requires that the slot be wider than the router bit used.

In most cases the slot is so narrow that it is impossible to use the same router bit as used for contouring the board. As the deflection of the router bit increases with decreasing bit diameter, most PCB manufacturers abstain from routing the contour with small bits.

There are two alternatives for machining the slot: either sawing on a special slotting machine or routing in a second routing operation with a smaller router bit. The choice depends on the shape of the slot; a completely regular slot will usually be sawed, whereas it is more convenient to route a slot having truncated (chamfered) corners at the entrance as shown in Figure 15.13.

Although a PCB manufacturer has an NC routing machine at his disposal, it is not always an advantage to use it for machining slots.

Slot sawing is usually done with a special saw, also called a slotting machine. It has a relatively small blade, actually a sort of side-milling cutter. In some cases, the sawing is done with the board held against a side stop and pushed into the cutter as shown in Figure 15.12.

This method transfers the edge inaccuracy to the slot location. Obviously, use

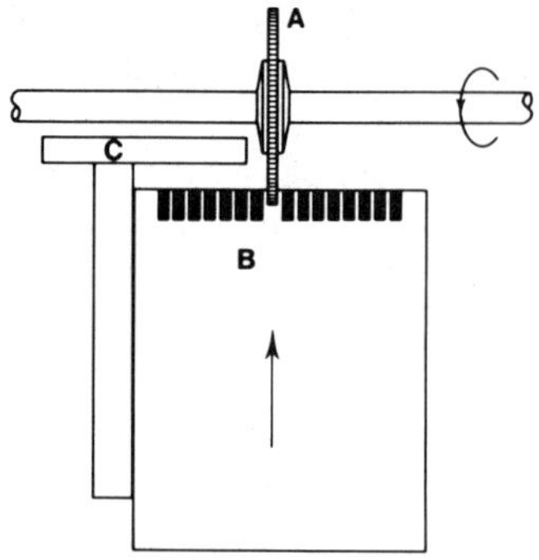

Fig. 15.12. Sawing the slot by means of a slotting machine.
A = side milling cutter. B = board to be provided with slot. C = stop determining sidewise position and depth of slot.

of a fixed side stop is undesirable even when the slot position accuracy requirements are relatively low.

By visually assessing the side stop adjustment, the operator can ensure that the board slot is located symmetrically with respect to the adjacent contact tabs. This corresponds to Method 3. See Paragraph 15.7.1.

In order to avoid individual alignment, the operator can also use a jig, to which the board is attached by means of guide pins mating with the tooling holes of the board. Now it is only necessary to adjust the side stop once in order to correctly locate the slots. The accuracy increases a great deal when the tooling holes are located close to the connector area, preferably with one tooling hole at each end of the connector.

The slot can also be routed, and as mentioned above, the routing will be part of the general contouring. In most cases the slot will be narrower than the router bit normally employed—3⁄32 in. (2.4 mm) or 1⁄8 in. (3.2 mm). The slot must therefore be routed in a second routing operation with a suitable router bit, often specially made for the job. Unfortunately, many manufacturers of connectors/receptacles do not adapt the plug width to commercially available standard sizes of router bits.

The program for routing the slot is written according to the aforementioned Method 1 or 2 and needs no further comments. See Paragraph 15.7.1.

15.7.2b Nominal Tolerances

When no tolerances of the location and dimensions are specified, the inspector should expect that the following requirements, based on good manufacturing practices, be met:

Tolerance on width of slot:
±0.004 in. (±0.10 mm)

Tolerance on length of slot:
±0.010 in. (±0.25 mm)

Displacement of slot centerline:
≤±0.004 in. (±0.10 mm)

The inspector must be able to measure the dimensional and positional errors of the slot. Usually a somewhat lower level of inspection—for example, Special Inspection Level S3—suffices.

15.7.2c Measuring the Slot Width

The width of the slot is ascertained with feeler gauges. It should not be measured by inserting round gauges into the slot perpendicular to the board, as some routed slots have grooved edges and the measurement may indicate that the slot is slightly larger than it really is. The inaccuracy is the nominal width less the thickness of the thickest feeler blade which can be inserted in the slot.

15.7.2d Measuring the Centerline Displacement

The centerline displacement of the slot can be ascertained in two ways. If the slot position is based on the demand for a symmetrical position with respect to the ad-

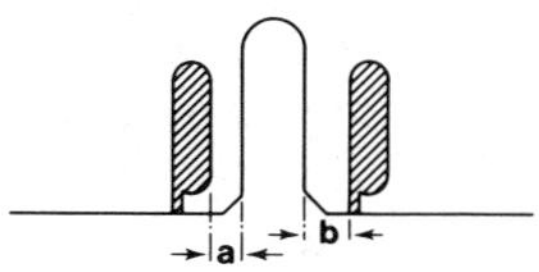

Fig. 15.13. Measuring the displacement ΔD of the slot.

jacent contact tabs (Method 3, Paragraph 15.7.1), the distances a and b from the edges of the slot to the adjacent contact tabs are measured with a 7X or 10X magnifier fitted with a graticule. See Figure 15.13.

The displacement ΔD is quite simply defined by the expression $\Delta D = \frac{1}{2}(a-b)$. If the slot position is defined by means of a reference system, the measurement is performed according to the methods described in Paragraph 6.4.8.

The special caliper with two conical points (Figure 6.7) is particularly suited, if a tooling hole occurs at one end of the edge connector, and the distance from the hole to the centerline of the slot is given. Due to the conical points, the caliper measures the exact distance mentioned above. The measurement is thus a practical utilization of the aforementioned Method 2, Paragraph 15.7.1, which comprises a local reference system with a tooling hole T_3 as shown in Figure 15.11.

15.7.2e Board Disposition

In principle, the board should be rejected if the tolerances are not met, because there is a risk of contact failure—for example, if the slot is wide enough to allow the board to be displaced by half the width of a contact tab. Waiving the fault requires the PCB designer to evaluate the tolerance chain in detail.

The best way is to base the assessment on the tooling holes of the board. Important factors are the pattern placement relative to the tooling holes, including an assessment of possible pattern displacements caused by possible dimensional changes, faults and instability of the master film, plating faults, and etching faults.

Important information for the evaluation is the actual width of the slot and the displacement of the slot centerline, whereas the slot length is seldom important, provided it is long enough without touching any circuitry.

An excellent method of evaluation is to make a special gauge to give a go/no-go indication of any connector dimensional failure. In course of time, a very high percentage of boards are undoubtedly rejected because of dimensional errors in the edge connector area—for example, an incorrectly located slot—even if the boards are in fact fully usable.

If the evaluation shows a risk of contact failure, the board is rejected and the fault rated as an A-defect.

15.8 Repair of Displaced Slot

Unfortunately, it is not uncommon for a connector slot to be milled a little too wide or displaced so much that the board is unusable and must be rejected.

As milling or sawing of the slot is one of the last operations performed by the PCB manufacturer, the fault means a great loss to the PCB manufacturer and a relatively long delay to the customer when the boards are to be remade.

It is possible to perform a proper repair so that remake can be avoided. The slot is closed on both sides with tape and the cavity filled with epoxy which cures overnight without any heat treatment. The next day the slot can be milled at the right position and the board delivered. The repair is fully reliable and relatively inconspicuous.

15.9 Chamfering of Connector Edge

In order to ease the insertion of the board into the fixed receptacle, the edge along

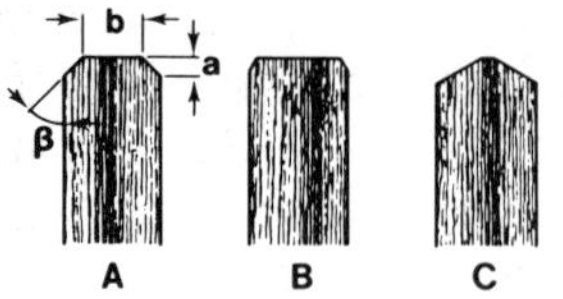

Fig. 15.14. Cross-sectional drawing of chamfered board.
A: Nominal case. B: Edges just "broken."
C: Heavy chamfering.

the connector should be chamfered. The chamfering has also the advantage of reducing the scraping of the contacts of the receptacle when the board is inserted.

The chamfering is often specified by the angle β and the dimension a as shown at A, Figure 15.14, which is drawn oversize for a 0.062 in. (1.6 mm) board.

A common specification is $\beta = 45 \pm 15°$ and $a = 0.012 \pm 0.004$ in. (0.3 ± 0.1 mm). The nominal case is shown at A, whereas B and C show the permissible limits of the chamfering. As a consequence of the wide tolerance, the edges can vary from 0.055 to 0.008 in. (1.4 to 0.2 mm).

With such large specification limits, it really serves no purpose to measure the chamfering. The inspector should merely check that the edge has been broken on both sides of the board. However, the chamfering must be performed in such a way that no glass fibers protrude from the machined surfaces.

Omission of chamfering is cause for rejection, and the fault is rated as an A-defect. The boards are returned to the PCB manufacturer, who normally can chamfer the connector edges very easily.

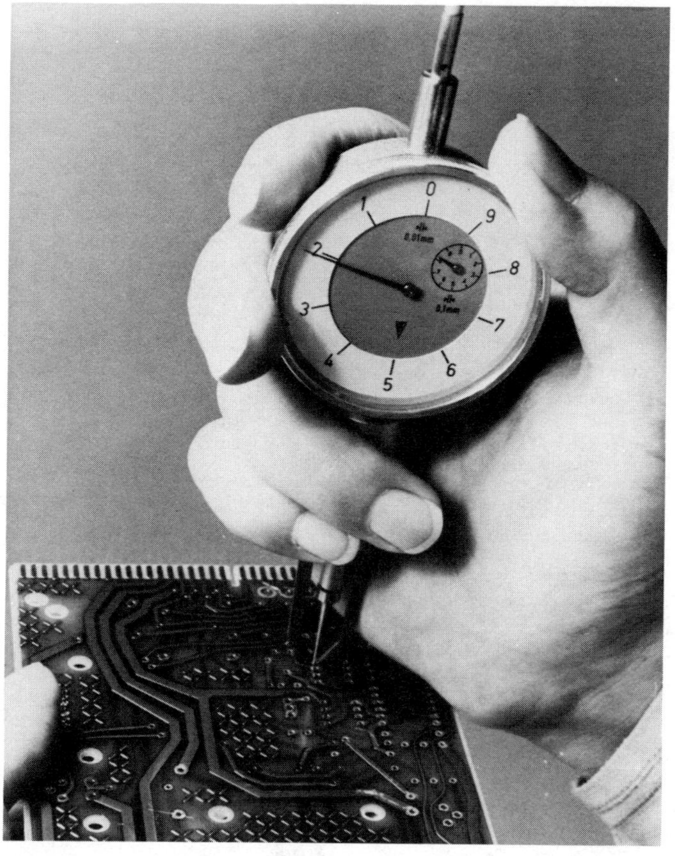

Part 3

Routine Inspection

Chapter 16

Measuring Equipment and Procedures

16.1 Introduction

Part 2 comprehensively dealt with a large number of defects. It described the defects, their causes, and their major influences on the finished board. The inspector should now understand practically all conceivable types of defects so that he can recognize and assess such defects.

The PCB designer should also have gained an understanding of the precautions that he should take in order to reduce the probability of manufacturing faults, and thereby facilitate the incoming inspection.

Now we shall translate the information presented in Part 2 into a more practical description of the routine inspection. Part 2 should be regarded as a reference, to which the inspector can refer for more detailed information on immediate problems. As we assume that the inspector is now familiar with the various types of defects, Part 3 is limited to brief mentions of the necessary inspections as well as practical inspection sequences.

Part 3 primarily addresses the incoming inspection of commercial boards of normal quality. This implies that there are no special requirements regarding the boards' possible use in military and space applications or other demanding conditions. In such cases, the demands are usually specified in every detail by the end user, and such matters are immaterial here.

The description of the routine inspection is organized according to procedures followed by the inspector. In this way, Part 3 is divided into various chapters. A great many of the assessments are performed simultaneously, particularly the visual ones, although they are described sequentially.

16.2 Inspection Record

It is most expedient to base the incoming inspection on an inspection record in the form of a checklist. This list contains most of the common types of defects, which are simply checked off in the respective boxes when found. The checklist supports the inspector in his work because it also provides a practical inspection sequence.

The completed inspection record can also be used as a practical means of feedback to the PCB manufacturer. Feedback of this kind is far more important to the PCB manufacturer than most customers realize, as the manufacturer receives accurate information about the customer's assessment of the boards. Therefore, it can also be important to report on minor defects which the PCB manufacturer otherwise would not learn of and therefore be unable to correct. Frequently a particular level of defect has to be exceeded before the inspector considers it worthwhile to complain, but a standardized inspection record makes reporting these defects much more convenient.

Figure 16.1 shows an example of an inspection record. The inspection record

ALL REFERENCE NUMBERS REFER TO PARAGRAPHS IN *QUALITY ASSESSMENT OF PRINTED CIRCUIT BOARDS* (ORDER NO. 10012) 1985, BISHOP GRAPHICS, INC.

COMPANY NAME AND ADDRESS

INSPECTION RECORD OF PCBs

PCB PART NO. ____ REV. ____
INSPECTED BY ____ DATE ____

PCB VENDOR ____ QTY. ____
DELIVERY NO. ____ ORDER NO. ____

TYPE OF BOARD: SINGLE-SIDED / DOUBLE-SIDED / MULTILAYER / NONPLATED-THROUGH / PLATED-THROUGH
DIMENSIONS OF BOARD: LENGTH ____ WIDTH ____

ABBREVIATION: O.O.T. = OUT OF TOLERANCE

II / S3 } INSPECTION LEVEL

	A-DEFECTS		B-DEFECTS	
AQL VALUES	%		%	
	SAMPLE	r	SAMPLE	r
NOMINAL				
REDUC. ↓				
EXTRA REDUC. ↓↓				

INDICATED BY ARROWS IN THE BOXES

USE OF THE BOXES: NUMBER OF A-DEFECTS, B-DEFECTS, ACTION TAKEN

- BLANK BOX = ACCEPTED
- R REJECTION
- W WAIVER
- M MENDING
- P PACKING
- A ADVANCE WAIVER
- ✓ DEFECTS NOT COUNTED

CONDITION OF SHIPMENT	
OUTER BOX INFERIOR	16.9.1
OUTER BOX DAMAGED	16.9.2
BOARD PACKING INCORRECT	16.9.3
BAG MATERIAL INCORRECT	16.9.4
BAG CLOSING INCORRECT	16.9.5
TISSUE PAPER INCORRECT	16.9.6
BATCH MIX-UP	16.9.7

INITIAL VISUAL INSPECTION	17
ISSUE VERIFICATION (INCORRECT ISSUE)	17.2
SOLDER SIDE	17.2
COMPONENT SIDE	17.2
SOLDER MASK	17.2
INSULATION MASK	17.2
COMPONENT NOTATION	17.2
INNER LAYER NO.	17.2
BOARD MARKINGS	17.3
COMPANY NAME MISSING	17.3.1
DATE MARKING MISSING	17.3.2
BOARDS OBSOLETE	17.3.2
INSPECTION STAMPS MISSING	17.3.3
APPROVAL MARKING MISSING	17.3.4
FLAMMABILITY INCORRECT	17.3.4
WATERMARKS INCORRECT	17.3.5
BASE MATERIAL	17.4
TYPE INCORRECT	17.4.1
MAKE INCORRECT	17.4.2
MAIN DIRECTION INCORRECT	17.4.3
MEASLING	17.4.4
CRAZING	17.4.5
BLISTERING	17.4.6
DELAMINATION	17.4.7
HALOING	17.4.8
WEAVE TEXTURE	17.4.9
WEAVE EXPOSURE	17.4.10
CURING INCOMPLETE	17.4.11
CRACKS	17.4.12
METALLIC INCLUSIONS	17.4.13
DISCOLORATION	17.4.14
DEPRESSIONS	17.4.15
OVERALL APPEARANCE OF BOARD	17.5
DENTS AND PITS	17.5.1
CRACKS IN COPPER FOIL	17.5.2
SCRATCHES	17.5.3
METAL PARTICLES	17.5.4
ETCHING FAULTS	17.5.5
SURFACE CONTAMINATION	17.5.6
HOLE DRILLING INCORRECT	17.5.7
PLATING TOO THICK	17.5.8
EXCESSIVE REPAIR WORK	17.5.10

DETAILED VISUAL INSPECTION	18
PATTERN	
SHORT CIRCUITS	18.2
INTERRUPTIONS	18.3
PERCENTAGE RULE NOT MET	18.4.1
DIMENSIONAL CHANGES O.O.T.	18.4.2
EDGE UNEVENNESS O.O.T.	18.4.3
INDENTATIONS O.O.T.	18.4.4
PROJECTIONS O.O.T.	18.4.4
VOIDS/PINHOLES	18.4.5
METAL PARTICLES	18.4.6
LOCATION OF HOLES	18
SOLDER HOLES O.O.T.	18.5
ANNULAR RINGS O.O.T.	18.5
MOUNTING HOLES O.O.T.	18.6
OTHER CONDITIONS	18
SOLDER HOLES CONTAMINATED	18.7
BURNING OF PATTERN	18.8.1
STEP PLATING	18.8.2
REFLOWING/SOLDER COATING	18.9
SURFACE UNEVEN/MATTE	18.9.1
SURFACE COARSE-GRAINED/DARK	18.9.1
SURFACE GRITTY/GREY	18.9.1
SURFACE WITH BLUE PLAGUE	18.9.2
ALLOY COMPOSITION O.O.T.	18.9.3
PATTERN EDGES UNCOVERED	18.9.4
SLIVERS/VISIBLE OVERHANG	18.9.5
REFLOWING INCOMPLETE	18.9.6
NONWETTING/DEWETTING	18.9.7
SOLDER HOLES BLOCKED	18.9.8/18.9.9b
CURTAINS ON CONDUCTORS	18.9.9a
WEBBING	18.9.9c/18.9.9d
SOLDER DROPS/BALLS	18.9.9g
ROLLER TINNING	18.10
SURFACE UNEVEN/MATTE	18.10.1
SURFACE COARSE-GRAINED/DARK	18.10.1
SURFACE GRITTY/GREY	18.10.1
SOLDER DROPS O.O.T.	18.10.2
COMPONENT HOLES CLOSED	18.10.3
BARE COPPER AT HOLE EDGE	18.10.4
PATTERN EDGES UNCOVERED	18.10.5
DROSS ON SURFACE	18.10.6
NONWETTING/DEWETTING	18.10.7
TIN PLATING	18.11
SLIVERS/VISIBLE OVERHANG	18.11.1
APPEARANCE UNACCEPTABLE	18.11.2
PROTECTIVE LACQUERING	18.12
LAYER UNEVEN/BLISTERED	18.12
ACCUMULATION OF LACQUER	18.12
HOLES BLOCKED	18.12
OTHER TYPES OF SURFACE PROTECTION	18.13
SOLDER AND INSULATION MASKS	18.14
MASK MISREGISTERED	18.14.1
OVERPRINTING OF PADS O.O.T.	18.14.2
COVERAGE INCOMPLETE	18.14.3
MASK APPERTURES BLURRED	18.14.4a
METAL PARTICLES IN MASK	18.14.4b
OTHER PARTICLES IN MASK	18.14.4c
AIR BUBBLES/UNEVENNESS	18.14.4d
SKIPPED AREAS IN MASK	18.14.4e
MASK ADHESION INSUFFICIENT	18.14.4f
CRACKS IN DRY FILM MASK	18.14.4g
CURING INCOMPLETE	18.14.4h
MASK THICKNESS TOO LOW	18.14.5
COMPONENT NOTATION	18.15
LEGIBILITY UNSATISFACTORY	18.15.1
ANNOTATION MISREGISTERED	18.15.2
PRINTING INK IN HOLES	18.15.2
ANNOTATION DISTORTED	18.15.3
COLOR INCORRECT	18.15.4
EDGE CONNECTORS	18.16
SURFACE UNEVEN	18.16.1
APPEARANCE UNSATISFACTORY	18.16.2
SCRATCHES	18.16.3
DEPRESSIONS	18.16.4
PINHOLES	18.16.5
NODULES/BLISTERS	18.16.6
BURNING	18.16.7
GOLD COVERAGE INSUFFICIENT	18.16.8
EDGES NOT COVERED	18.16.9
TRANSITION INCORRECTLY MADE	18.16.10
TRANSITION ZONE DISLOCATED	18.16.11
ADJOINING AREAS DAMAGED	18.16.12
EDGE UNEVENNESS O.O.T.	18.16.13
SHORT CIRCUITS	18.16.14
PROJECTIONS O.O.T.	18.16.15
INDENTATIONS O.O.T.	18.16.16
LIFTING OF TABS/BURRS	18.16.17
CRACKS IN GOLD LAYER	18.16.18
PRINTING INK ON TABS	18.16.19

ADHESION INSPECTION	19
GOLD OR GOLD/NICKEL	19.4
PLATING OF PATTERN	19.5
COPPER FOIL	19.6
SOLDER/INSULATION MASKS	19.7
COMPONENT NOTATION	19.8

MECHANICAL INSPECTION	20
BOARD THICKNESS O.O.T.	20.2
WARP AND TWIST O.O.T.	20.3
SOLDER HOLES DISPLACED	20.4
MOUNTING HOLES DISPLACED	20.5
TOOLING HOLES DISPLACED	20.6
NOM. HOLE DIA. O.O.T.	20.7
SOLDER HOLE DIA. O.O.T.	20.8
MOUNTING HOLE DIA. O.O.T.	20.9
TOOLING HOLE DIA. O.O.T.	20.10
BOARD DIMENSIONS O.O.T.	20.11
CONTOUR O.O.T.	20.12/20.13
CUTOUTS O.O.T.	20.14
EDGE CONNECTORS	20.15
SLOT WIDTH O.O.T.	20.15.1
SLOT LENGTH O.O.T.	20.15.2
SLOT LOCATION O.O.T.	20.15.3
CHAMFERING UNACCEPTABLE	20.15.4

MICROSCOPIC INSPECTION	21
VOIDS IN HOLE WALL	21.2
CRACKS IN HOLE WALL	21.3
INCOMPL. REFLOWING IN HOLES	21.4
HOLE WALL W. GROOVES	21.5.1
HOLE WALL W. NODULES	21.5.2
HOLE WALL CONTAMINATED	21.5.3
HOLE WALL W. GLASS FIBRES	21.5.4
HOLE WALL IN LANDLESS HOLES	21.5.5
OVERHANG BROKEN OFF	21.6

ELECTRICAL INSPECTION	22
SHORT CRCT./INTERR. (MAN. TEST)	22.2
SHORT CRCT./INTERR. (AUTO TEST)	22.3
INSULATION RESISTANCE	22.4
COPPER PLATING O.O.T.	22.5

SOLDERABILITY INSPECTION	23
POOR WETTING OF PATTERN	23.3.1
POOR WETTING OF HOLES	23.3.2
NONWETTING	23.3.3
DEWETTING	23.3.4
BLOWHOLES	23.3.5
WEBBING	23.3.6

ASSESSMENT AND DISPOSITION			24
ΣA	24.2	ΣB	24.2
A (R)	24.3	B (R)	24.3
A (W)	24.3	B (W)	24.3
A (M)	24.3	B (M)	24.3
BATCH ACCEPTED		BATCH REJECTED	

- BOARDS ACCEPTED WITH NO COMMENTS
- BOARDS ACCEPTED AFTER EXEMPTION
- BOARDS ACCEPTED AFTER REPAIR
- BOARDS REJECTED AND RETURNED

INSPECTION RECORD PAD ORDER NO: 20220, BISHOP GRAPHICS, INC.

Fig. 16.1. Inspection record (reduced size).

forms can be purchased in pads from the publisher. The individual sections of the record are explained in the following paragraphs.

The general principle is that the number of defects found during the inspection is indicated in the respective boxes for A- and B-defects. The number of A-defects as well as B-defects can be added to express the overall board quality.

In this and the following chapters, the descriptions of the various inspection procedures refer to selected parts of the inspection record shown in Figure 16.1.

16.3 Indispensable Measuring Equipment

The most common measurements of mechanical quantities require the following equipment, which is shown in Figure 16.2:

A. A standard caliper with a resolution of 0.005 in. (0.1 mm).
B. A large caliper with a span at least equal to the largest board dimension, and a resolution of 0.005 in. (0.1 mm). A typical span is 24 in. (500 mm for calipers with metric calibration).
C. A micrometer with a resolution of 0.0005 in. (0.01 mm) and a throat of approximately 2 in. (50 mm).
D. A 7X or 10X magnifier with graticule having a resolution of 0.005 in. (0.1 mm).
E. A set of cylindrical plug gauges as follows:

   0.02 to 0.06 in. in steps of 0.0025 or 0.001 in. (0.5 to 1.5 mm in steps of 0.05 or 0.02 mm)

   0.06 to 0.25 in. in steps of 0.005 in. (1.5 to 6.3 mm in steps of 0.10 mm)

   Punching needles can be purchased with fine tolerances and used as plug gauges.

F. A set of feeler gauges which can be combined in steps of 0.005 in. (0.1 mm).

The visual inspection is greatly facilitated and improved by using the following measuring equipment which in fact can also be characterized as necessary:

G. A 2X or 3X magnifier with built-in illumination to avoid undue fatigue of the inspector's eyes when inspecting fine details in the pattern (Figure 16.3).
H. A simple monocular microscope with magnifications up to 50X or

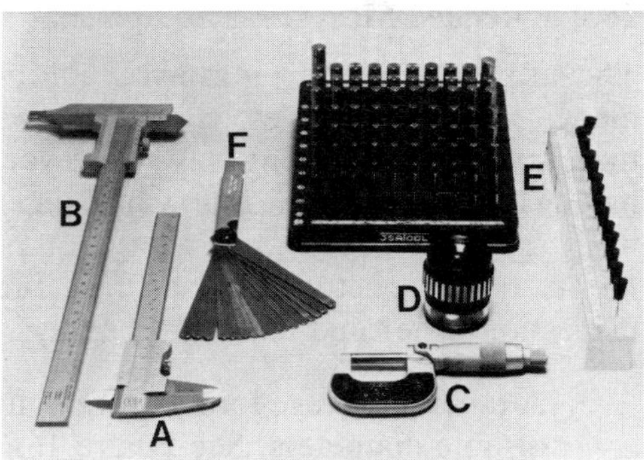

Fig. 16.2. Measuring equipment for mechanical quantities.

Fig. 16.3. 2X or 3X magnifier with built-in light source (to be used during the visual inspection).

Fig. 16.4. Simple monocular microscope with 50 to 100 times magnification (Courtesy Graticules Limited).

100X makes it possible to thoroughly inspect details of the pattern or master film (Figure 16.4). This microscope is somewhat superfluous if the stereo microscope mentioned in Paragraph 16.4, Item M, is available.

Fig. 16.5A. Magnifier with a special optical system for inspection of hole walls (Courtesy Zi-Tech Hole Inspector, Aikenwood Corp.).

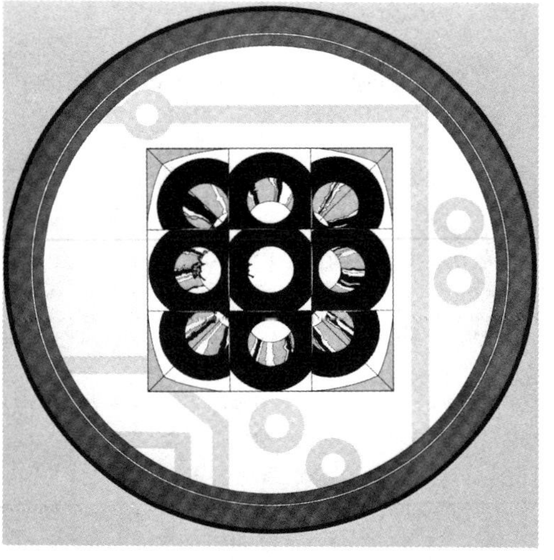

Fig. 16.5B. Example of hole wall seen through the magnifier shown in Fig. 16.5A.

I. A magnifier with a special optical system allowing a quick inspection of the entire hole wall (Figure 16.5).

16.4 Extended Measuring Equipment

The aforementioned measuring equipment—which is relatively cheap—allows the inspector to go a long way. However, having at his disposal the following, somewhat more advanced, measuring equipment, the inspector can perform the inspection faster and more thoroughly:

J. A taper pin gauge for measurement of hole diameters. See Figure 16.6 and the comments in Paragraph 6.7.2.

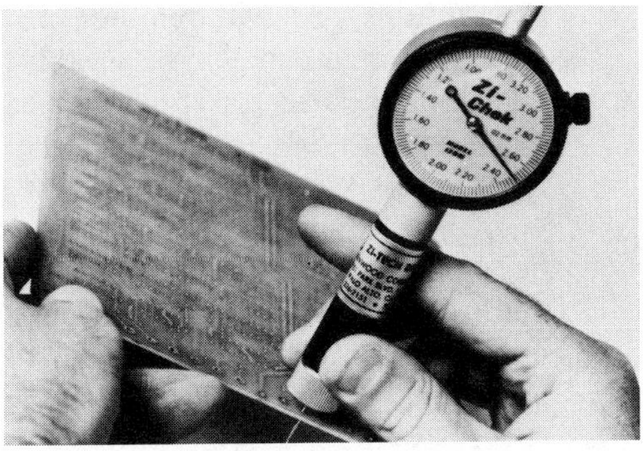

Fig. 16.6. Taper pin gauge for measuring hole diameters (Courtesy Zi-Check Gage, Aikenwood Corp.).

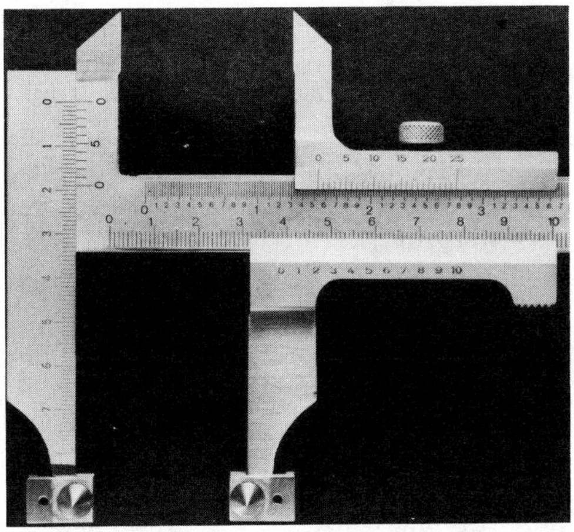

Fig. 16.7. Special caliper suitable for measuring center distances.

K. A special caliper as shown in Figure 16.7. The jaws of the caliper can be provided with cones so that the caliper is well-suited for measuring the center-to-center distance between two holes. Examples of using this caliper are given in Paragraphs 6.4 and 6.5, and also in Paragraph 15.7.2. The slide caliper, which has the type designation MULTIPRE DBP-PL, can be obtained from Friedrich J. Schall KG, D-7442 Neuffen/Wurtt., P.O. Box 80, Federal Republic of Germany.

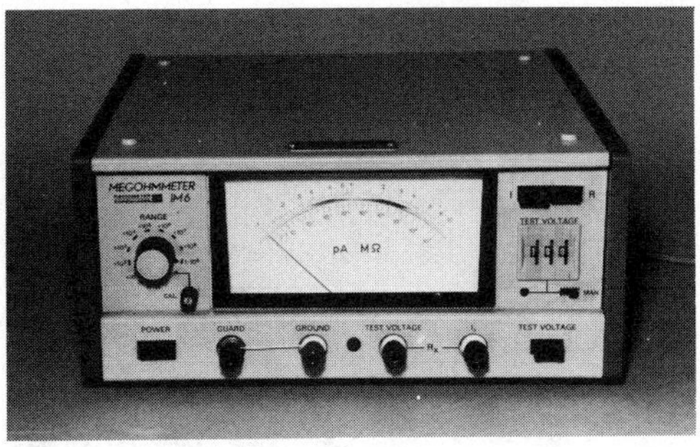

Fig. 16.8. Megohmmeter for measuring insulation resistance of printed circuit boards.

L. A megohmmeter for measuring insulation resistance (Figure 16.8). Application of the megohmmeter is described in Chapter 28.

M. A stereo microscope with a magnification up to 80 or 100 times, and provided with an illuminating device. See Figure 16.9. The stereo microscope gives the inspector good opportunities to perform relatively

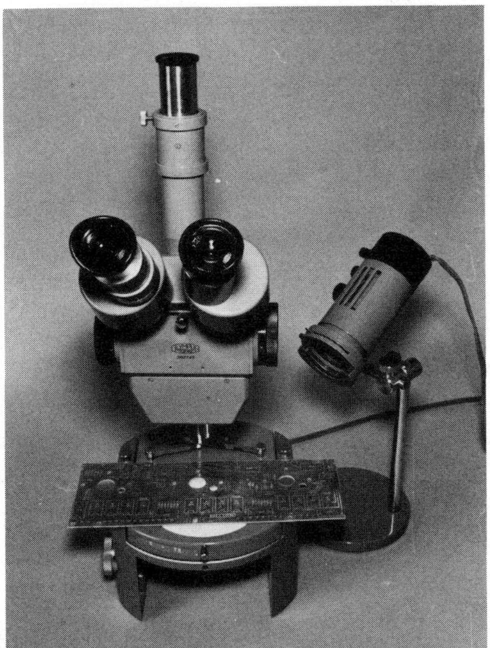

Fig. 16.9. Stereo microscope.

detailed investigations of the hole wall of plated-through holes, because stereo microscopes have a fairly wide range of definition in depth. The inspector can view the hole wall at a small angle and see a sharp image of practically the whole depth of the hole.

In the case of a microsection, the inspector can assess the plating thickness or the evenness of the drilled hole wall. If the microsection is prepared with a transparent epoxy, which implies no hard filler in the epoxy, the inspector can also view the inner surface of the hole wall and assess its evenness.

A more thorough investigation of microsections requires a metallurgical microscope as described under Paragraph 16.5, Item S. Chapter 25 describes preparation of microsections in both a fairly primitive way and a fully professional manner.

With the special fixture shown in Figure 16.10, the microscope can be given so large a throat that even large boards can be inspected. Figure 16.10 also shows the light table described in the next item.

N. A light table with tilted work plane consisting either of a matte glass or plastic work plate for inspecting master films—for example, for misregistration. It can also be used for inspecting boards for short circuits as most boards are somewhat transparent.

In addition, the light table can be modified for use in revealing voids in the walls of plated-through holes. The modification consists of replacing the work plate with a metal sheet provided with a slot as mentioned in Paragraph 7.3.3. The slot should be located on a level with the central fluorescent tube of the light table.

As tiny reflections can be easily confused with voids or pinholes in the hole wall, the inspector should take precautions to avoid reflections. The sides of the slot as well as the surface of the metal sheet should therefore be matte. The metal sheet can, for example, be covered with a matte self-adhesive plastic film, preferably black. Figure 16.10 shows a modified light table.

Fig. 16.10. Stereo microscope with a pillar stand used in conjunction with a light table.

16.5 Advanced Measuring Equipment

Sometimes it is desirable to perform more accurate measurements than the aforementioned measuring equipment makes possible. Therefore the incoming inspection department has to invest in the following more advanced and expensive measuring equipment:

O. A coordinate table (coordinatograph) with digital display and a measuring accuracy of 0.00005 in. (1 μm). Figure 16.11 shows an example; Paragraph 6.4 describes its use.

P. A digital height gauge with a measuring accuracy of 0.00005 in. (1 μm) can be used when measuring layer thicknesses. See Figure 16.12 and

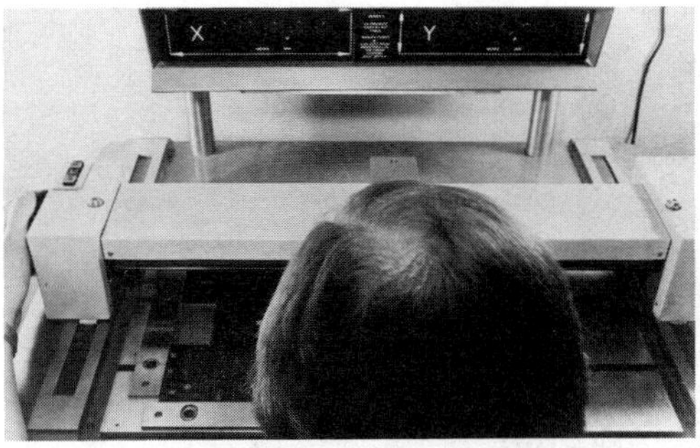

Fig. 16.11A. Digital measuring equipment.

Fig. 16.11B. Close-up of the optical unit of the digital measuring equipment shown in Fig. 16.11A.

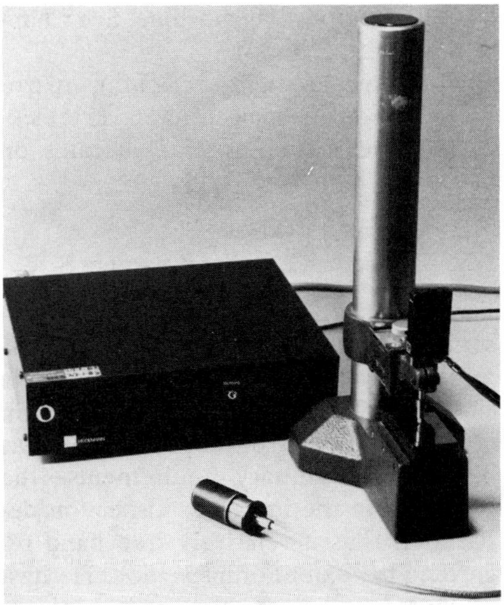

Fig. 16.12. Digital height gauge.

Fig. 16.13. Four-terminal measuring equipment for measuring copper plating thicknesses.

the description in Paragraph 8.7.2c.

Q. A four-terminal measuring instrument for measuring the thickness of the copper plating in plated-through holes. See Figure 16.13 and Paragraph 7.3.1d4.

R. A projector to be used when inspecting the pattern. See Figure 16.14 and Paragraph 5.1.7.

S. A metallurgical microscope as shown in Figure 16.15, having magnifications up to 1000X. The range of definition in depth is more limited than in a stereo microscope, particularly with high magnification. To overcome focusing problems, the metallurgical microscope is frequently designed as an inverted microscope—that is, the stage is placed at the top. The sample is placed with the plane and polished surface facing the stage which has a central viewing aperture. The focusing remains constant, regardless of how the stage is moved with respect to the objective. It is most expedient to have the microscope equipped with a camera so that the microsections can be photographed. The camera should have a built-in exposure meter and be of the instant type so that photographs are available in a few minutes.

Fig. 16.14. Projector for pattern inspection.

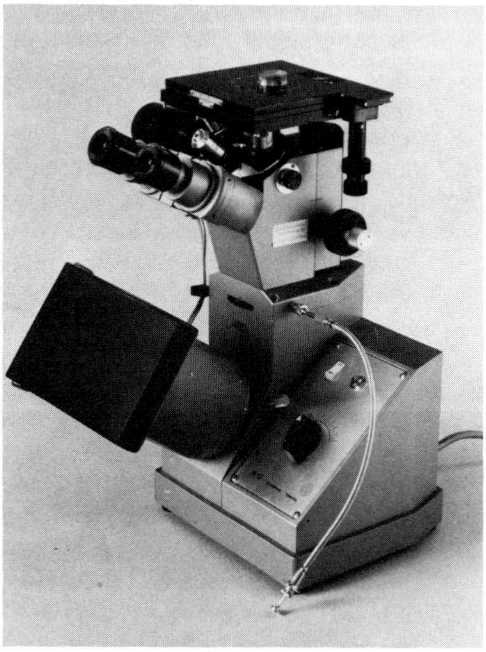

Fig. 16.15. Metallurgical microscope.

T. Equipment for preparing microsections, consisting of a diamond saw and a polishing machine. See Chapter 25.

U. A special light-sectioning microscope for measuring layer thicknesses. Chapter 26 elaborates on this.

16.6 Inspection Level and Sample Size

Paragraphs 1.2.6, 3.3.5, and 3.3.6 described sampling plans and how to determine the inspection level and the sample size. It is important to lay down that in commercial electronics production—as distinct from military requirements—the inspector in the incoming inspection department has a relatively free hand regarding the extent of inspection. He has, of course, to balance the costs of inspection against the losses which arise later in the production as a consequence of a few defective boards slipping undetected through the incoming inspection. This is addressed in Paragraphs 1.2.2 and 1.2.3.

The result of such considerations is generally that the inspector shall not perform a 100 percent inspection, but rather a sampling inspection of manageable size. It is therefore recommended that the inspector adopt the sampling plans described in MIL-STD-105D. The reason for this recommendation is not that it is particularly desirable to follow this standard to the extreme regarding the returning of a whole batch when the number of defective boards reaches the rejection figure. For the inspector there are two very important reasons for following the sampling plans:

1. The inspector has a well-supported sampling plan, according to which he can work on an accepted, professional basis.
2. The sampling plan provides the inspector with a reasonable protection against complaints of defective boards slipping through the incoming inspection.

As shown in Paragraph 1.2.5, a sampling plan always includes a certain risk that some defective boards will slip through the inspection. The board users must understand and accept this as a fact. The alternative is a 100 percent inspection which is so slow and expensive that it is rarely used. The only instances can be an especially critical board or a batch that has to be screened because of a more or less general defect found in the sample.

In the latter case, the extent of the work is highly reduced as only one parameter is to be inspected. Incidentally, a 100 percent inspection is no guarantee of preventing defective boards from slipping through the incoming inspection. The inspector will inevitably, during a prolonged and intensive inspection, tire and possibly overlook some defects.

We shall now determine the inspection levels to be used according to Paragraphs 2.2.1 and 2.2.2:

Special Inspection Level S3 for approved suppliers

General Inspection Level II for new or unstable suppliers

The quality is determined by the following AQL values, according to Paragraphs 1.2.5a and 1.2.6g:

AQL = 1 percent for A-defects

AQL = 2.5 percent for B-defects

The AQL values indicated are not chosen at random but serve to give the inspector a perception of a practical and usable quality level. The inspector can of course freely choose quite different values according to his own quality criteria.

The sample size is determined from the sampling plans in MIL-STD-105D or, more easily, from Tables 1.2 and 1.3 in Paragraph 1.2.6h.

It should be noted that if the delivery comprises boards from more than one production batch, each lot should be treated as an independent delivery as described in Paragraph 3.4.6b. If a delivery is supplemented with a few excess boards from the preceding delivery, the inspector should decide, based on the preceding inspection record, whether it is necessary to inspect those boards.

16.7 Data Recording

All substantial basic data of the delivery are recorded in the head section of the inspection record. See Figure 16.16. The following data are recorded:

1. The inspector's initials and the dates of the inspection and approval.

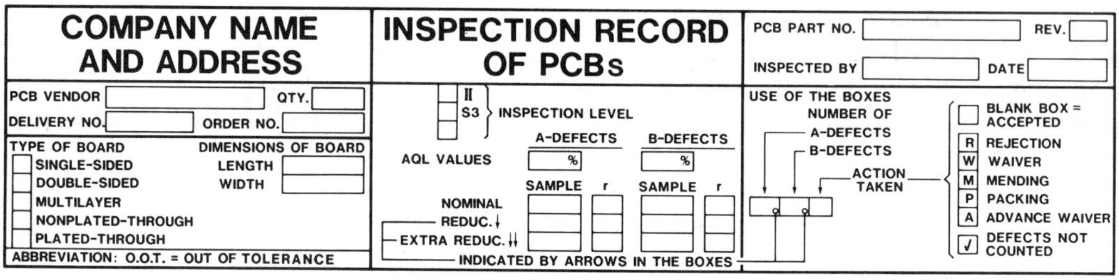

Fig. 16.16. Inspection record heading.

2. Identification of the board, expressed by its type or code number as well as its issue or change number.
3. Order data, expressed by the supplier name, order number and quantity received according to the packing slip or the invoice, or according to the inspector's own count.
4. It can be practical to state the most important parameters of the board. This can greatly influence a later assessment of the inspection results. For example, the misregistration of a solder mask also depends on the size of the board.
5. Sampling data, expressed by the inspection level used (either General Inspection Level II, or Special Inspection Level S3, or another inspection level). In addition, the AQL values selected—for example, 1 percent for A-defects and 2.5 percent for B-defects. The inspection level and the AQL values should appear in the purchase specification.

   The inspection level is based on knowledge of the supplier in question, either on an immediate perception of the supplier's capability of delivery and general quality level, or on the information contained in the supplier record (Paragraph 3.3.4). The latter procedure is, of course, more reliable.
6. Sample size, determined by the size of the batch, the inspection level, and the AQL values. The corresponding rejection figures should be stated.
7. Reduced sample size. During many of the inspections described later, the sample size is reduced from General Inspection Level II to Special Inspection Level S3 or even lower. In this way the inspection is facilitated. The reduced sample size is stated in the box "REDUC" and indicated by a downward arrow located immediately after the recorded number of defects.

   At a later assessment of the inspection results (see Chapter 24) it may be important to analyze the defects detected on the basis of a reduced sample size.

   In some cases the sample size is reduced further—for example, to one piece. This extra reduced sample size is stated in the box "EXTRA REDUC," and indicated by a downward double arrow immediately following the recorded number of defects.
8. The purchase specification will often include a clause stating that rejection of a delivery must be based on General Inspection Level II. As mentioned in Item 7, some partial inspections are based on a reduced or extra reduced sample size. This implies that the recorded number of defects can assume too low a value.

   Although the rejection figure has

not been reached, the inspector can, on the basis of a general assessment of the inspection, suspect that the delivery is so poor that it should be rejected.

As rejection requires that the inspection be based on General Inspection Level II, the inspector inspects the rest of the nominal sample for the most dangerous defects found on the basis of the reduced sample.

The corrected number of defects—that is, the total of the old and the new numbers—is stated in the box, and the arrow marking is removed. Obviously, the boxes should be filled in with pencil.

During inspection, it may become necessary to tighten the inspection—for example, by changing from Special Inspection Level S3 to General Inspection Level II; in this case a new form is used.

9. The area "Use of the Boxes" should not be used as this only explains how the form is to be used. Every type of defect has three boxes:

   In the first and second boxes the inspector states the numbers of A- and B-defects, respectively, found during inspection.

   In the third box the inspector states a letter to show the action taken: An "R" (Rejection) indicates that the board with the defect is rejected, whereas a "W" (Waiver) indicates that a waiver has been granted. An "M" (Mending) means that the defect has been remedied. A "P" (Packing) indicates a repacking of the boards (Paragraph 16.9) and an "A" (Advanced waiver) means that a waiver was granted in advance to the PCB manufacturer before delivery of the boards.

   A check mark indicates that the number of A- and B-defects, for one reason or another, has not been counted. Finally, an empty box means that no defect has been found. This facilitates the use of the form as it has only to be filled in when defects have been found.

16.8 Documentation

As mentioned in Paragraph 3.3, the inspector should have up-to-date documentation at his disposal. Among other things, this documentation includes:

1. A set of master films comprising:

   a. Solder side (positive and negative).
   b. Component side (positive and negative).
   c. Component notation (positive).
   d. Solder and insulation masks (positive).

   Note: In the case of multilayer boards, add master films of the inner layers.

2. A master drawing stating the individual specifications of the board, including dimensional information regarding the mechanical execution.
3. Engineering change orders.
4. Purchase specification.
5. Previous inspection records to indicate frequently recurring defects, if any. Such defects can indicate difficult production conditions or weak areas in the board design. Knowledge about such matters can aid the inspection considerably.

The inspector should verify that the master films mentioned in Item 1 show correct registration. This is ascertained in the following way: The films are overlaid two and two on a light table and registered with the registration marks. By taking a positive film of the solder side and a neg-

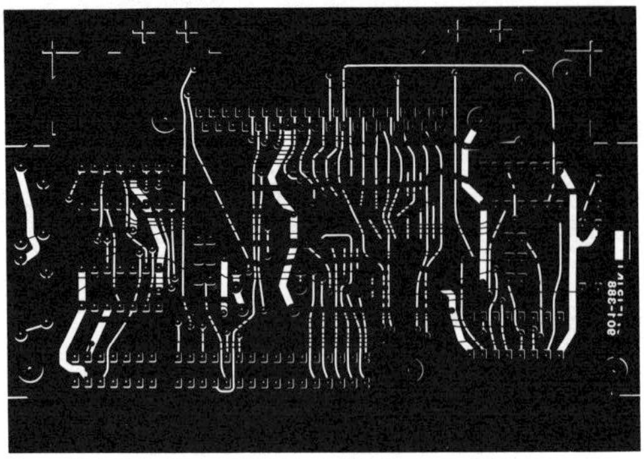

Fig. 16.17. Example of filmwork misregistration between the solder and component sides.

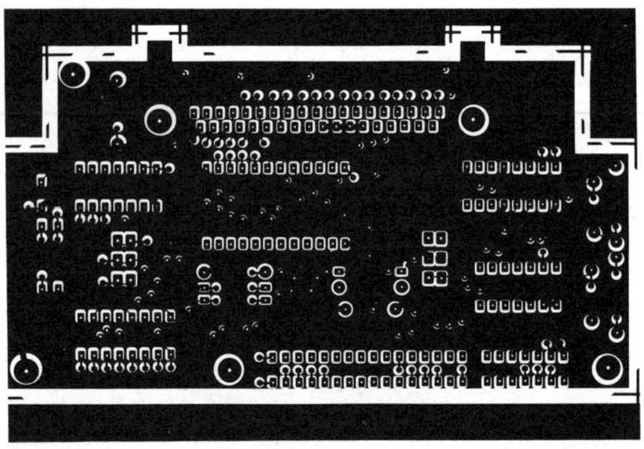

Fig. 16.18. Example of filmwork misregistration between the solder side and the solder mask.

tive film of the component side, no illuminated "half-moons" should be found around the solder pads, everything being black. Figure 16.17 shows an invented example of misregistration. A solder mask can be checked against the solder side in a similar manner. With a positive solder side film and a positive solder mask film, the inspector should see an illuminated ring with constant width around the pads. Figure 16.18 shows an invented sample of misregistration.

16.9 Condition of Shipment

In the case of an unsatisfactory condition of the shipment, the PCB manufacturer

CONDITION OF SHIPMENT				
			OUTER BOX INFERIOR	16.9.1
			OUTER BOX DAMAGED	16.9.2
			BOARD PACKING INCORRECT	16.9.3
			BAG MATERIAL INCORRECT	16.9.4
			BAG CLOSING INCORRECT	16.9.5
			TISSUE PAPER INCORRECT	16.9.6
			BATCH MIX-UP	16.9.7

Fig. 16.19. Conditions of Shipment. (Section of inspection record.)

should be given feedback (Paragraph 3.4.5). Damage to the outer packing or deviations from the required method of packing are entered in the inspection record, but are excluded when making the final quality assessment, which is the reason the A and B boxes are hatched.

Necessary repacking due to unsatisfactory packing of the individual boards (Items 3 to 6, following) is marked by the letter "P" (Packing) in the third box. Defects not causing any repacking can be marked with a check mark. The section of the inspection record shown in Figure 16.19 applies to the following conditions:

1. The consignment is packed in an inferior or unacceptable manner.
2. The outer packing is damaged when received. This is marked with a check mark whether or not the contents are damaged.
3. The boards are not packed individually.
4. The bags are not made of polyethylene.
5. The bags are not closed as required.
6. Tissue paper containing sulfur was used.
7. A mixture of different production batches is received. For the sake of a correct sampling inspection, different production batches (of the same type of board) must be segregated.

Chapter 17

Initial Visual Inspection

17.1 Introduction

The inspection should be organized with the simplest inspection procedures first. In this way many rejections occur early, before too much effort has been expended. Many procedures are visual, implying that the sample is assessed with the naked eye, possibly aided by a 2X to 3X magnifying lamp or a 7X to 10X magnifier. Inspection procedures requiring a microscope are addressed in Chapter 21.

17.2 Issue of the Pattern

It is important to verify that the manufacturer has used the correct issues of the master films, particularly for frequently revised boards. Because of the risk of mixing issues, the inspector takes a single board from the sample and verifies that the pattern, the solder and insulation masks, and the component notation comply with the order (Paragraph 3.4.4).

There are two procedures. The appropriate one depends on which method the designer used to identify the issue:

1. The issue of the master films is indicated on the master films themselves. The designation is therefore transferred to the board, where it appears in metal or as a screen print. The inspector simply checks that the issue markings on the board correspond to the issue designations on the master drawing.
2. The issue of the master films does not appear on the board. The inspector must therefore verify that the board corresponds to the master films specified. As Paragraph 3.4.7 describes in detail, the master films are overlaid on the board, after which possible discrepancies can be seen.

Incorrect issue of a board is cause for rejection and consequently rated as an A-defect. Regarding a possible waiver, the reader is referred to Paragraph 3.3.4. The inspector should assume a functional attitude—that is, giving high priority to the solder and component sides, and less to a solder mask. The inspection result is entered in the inspection record. See the segment shown in Figure 17.1.

The inspector should check whether the board sides have been turned (laterally reversed). A good indicator is reversed text as shown in Figure 17.2A. This type of defect, however, is very rare. More common is the application of a solder mask on the component side, usually as a result of an unclear specification on the master drawing.

Most users consider it impossible for a

INITIAL VISUAL INSPECTION	17
ISSUE VERIFICATION (INCORRECT ISSUE)	17.2
SOLDER SIDE	17.2
COMPONENT SIDE	17.2
SOLDER MASK	17.2
INSULATION MASK	17.2
COMPONENT NOTATION	17.2
INNER LAYER NO.	17.2
BOARD MARKINGS	17.3
COMPANY NAME MISSING	17.3.1
DATE MARKING MISSING	17.3.2
BOARDS OBSOLETE	17.3.2
INSPECTION STAMPS MISSING	17.3.3
APPROVAL MARKING MISSING	17.3.4
FLAMMABILITY INCORRECT	17.3.4
WATERMARKS INCORRECT	17.3.5

Fig. 17.1. Initial visual inspection. (Section of inspection record.)

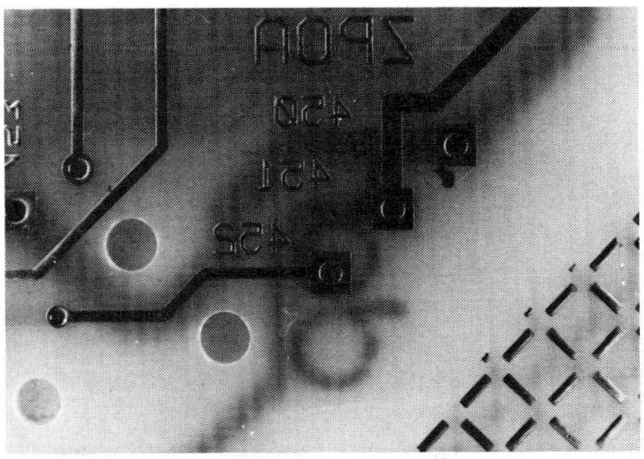

Fig. 17.2A. Example of laterally reversed pattern sides. An immediate indication is the reversed text. Note the small etching fault.

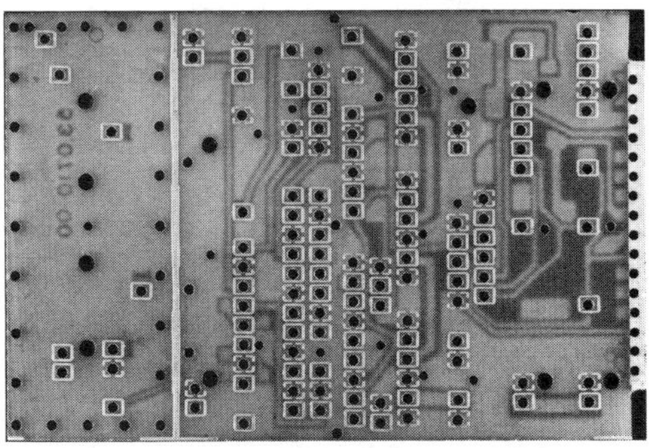

Fig. 17.2B. Board in "negative" execution.

PCB manufacturer to use a reversed set of master films—that is, positive films rather than negative films. Figure 17.2B shows an example of a board in a negative execution and with plated-through landless holes. Nevertheless, the board should be considered a curiosity.

17.3 Board Markings

General requirements regarding the marking of boards usually appear in the purchase specification, and those for a specific board on the master drawing. See Paragraph 3.4.9 and Figure 17.1, which shows the relevant part of the inspection record.

The inspector normally needs to inspect only a single board of the sample for markings. When defects or omissions are discovered, the whole sample should be inspected. For example, the requirements can comprise:

17.3.1 Identification

An identification of the PCB manufacturer by name, company logo, code, or other means (Figure 17.3).

17.3.2 Date Marking

A stamped or screen-printed date of manufacturer, frequently expressed by year and week number (Figure 17.3). The date marking tells the inspector whether the board is obsolete—for example, stored too long as an excess board from the preceding delivery. This can be important if the shelf life is limited—for example, to 3 months in the case of bare-copper boards with organic coating (lacquer).

Some companies have introduced a time limit for such boards, frequently 8 weeks. For plated-through and reflowed boards a considerably longer limit is acceptable—for example, 20 weeks.

Fig. 17.3. Identification of PCB manufacturer by means of company logo. Age of board appears from year and week markings.

17.3.3 Inspection Stamps

The PCB manufacturer's statement of having performed a final inspection—for example, with an inspection stamp as shown in Figure 17.4. It should be noted that a large number of PCB manufacturers no longer use inspection stamps. This may result from a change from a 100 percent inspection to a sampling inspection—for example, with an AQL value of 0.65 percent. Inspection stamps in such cases cause the customer to doubt the efficiency of the PCB manufacturer's final inspection department when defects are found.

17.3.4 Approved Markings

Special approval markings, such as a UL (Underwriters Laboratories Inc.) flammability rating. Such markings frequently are in a special code, making it extremely important for the inspector to know the code.

When the inspector does not know the code, he cannot ascertain whether the board meets the flammability requirements. See the marking example in Figure 17.5. A board not meeting the requirement is usually worthless. A higher flammability rating—for example, 94V-0 instead of 94V-1—is always acceptable.

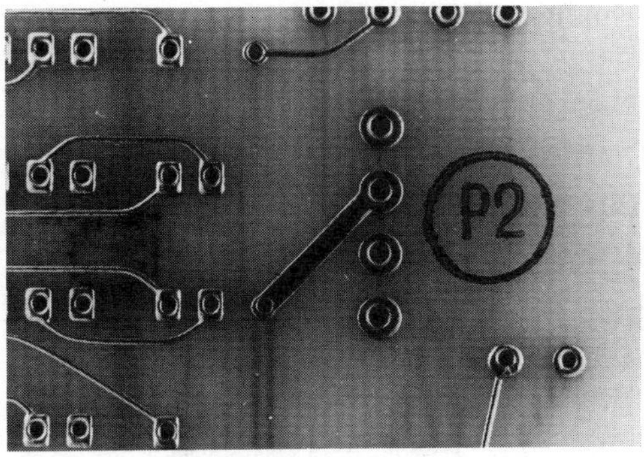

Fig. 17.4. Inspection stamps as proof of outgoing inspection.

Fig. 17.5. UL marking of flammability rating of board. The marking comprises an identification of the PCB manufacturer, type designation recorded by UL, and a closed but palpable code for the flammability rating of the board.

A potentially exemptable defect is that although the PCB manufacturer uses a base material of the correct flammability rating, he forgets to mark the boards. When the manufacturer confirms this in writing, the inspector can waive the fault.

17.3.5 Watermarks

When a flame-retardant/self-extinguishing base material is specified, the watermarks must be red (Paragraph 4.1). This rule applies to boards thicker than 0.020 in. (0.5 mm). The inspector should note that an inadequate flammability rating is usually an indisputable cause for rejection of the whole batch, making further inspection immaterial.

17.4 Base Material

The inspector next checks whether the base material is as specified. Figure 17.6 shows the relevant part of the inspection record. The requirements can, for example, comprise the following items:

17.4.1 Type of Base Material

The base material must be of a certain type—for example, epoxy glass, either G10 or FR4. The inspector can usually distinguish the various types from one another by the appearance.

The base material can also be identified by the shape and color of the watermarks. The inspector should therefore collect data sheets on the most common base materials and make a list of the appearance of the watermarks.

17.4.2 Make of Base Material

In addition to these requirements it can, in very rare cases, occur that a special make of the base material is specified. As the watermarks on the base material also identify the make, the inspector should have a list of approved watermarks.

17.4.3 Orientation of Pattern

The pattern must be oriented in a certain way relative to the main direction of the base material, which is identified by the watermarks. However, this requirement rarely occurs.

The inspector should now visually inspect the boards of the sample for the following defects, which are relatively rare. Since Paragraph 4.5 has described them in detail, together with rejection criteria, they are dealt with quite summarily here. The inspector should refer to Paragraph 4.5 and the referenced illustrations.

17.4.4 Measling

See Figure 17.7. If it is possible to waive the defect according to Items 1 to 4 of Paragraph 4.5.1, the defect is rated as a B-defect. Otherwise, it is an A-defect.

17.4.5 Crazing

See Figure 17.8. As stated in Paragraph 4.5.2, this defect is normally rated as an

BASE MATERIAL				17.4
			TYPE INCORRECT	17.4.1
			MAKE INCORRECT	17.4.2
			MAIN DIRECTION INCORRECT	17.4.3
			MEASLING	17.4.4
			CRAZING	17.4.5
			BLISTERING	17.4.6
			DELAMINATION	17.4.7
			HALOING	17.4.8
			WEAVE TEXTURE	17.4.9
			WEAVE EXPOSURE	17.4.10
			CURING INCOMPLETE	17.4.11
			CRACKS	17.4.12
			METALLIC INCLUSIONS	17.4.13
			DISCOLORATION	17.4.14
			DEPRESSIONS	17.4.15

Fig. 17.6. Base material. (Section of inspection record.)

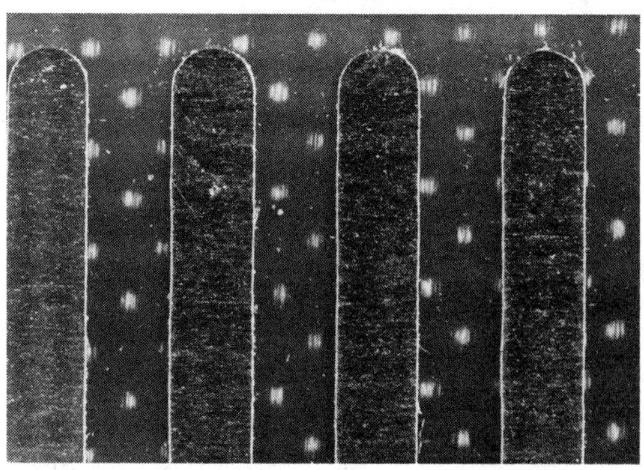

Fig. 17.7. Board with measling (discrete white spots below the surface).

A-defect. It can, however, be waived when the crazing is isolated from the circuitry. If so, it is rated as a B-defect.

17.4.6 Blistering

See Figure 17.9. The defect is dealt with in detail in Paragraph 4.5.3.

Boards for manual soldering: A limited occurrence with no lifted copper is rated as a B-defect. When the copper shows signs of lifting, the defect is rated as an A-defect.

Boards for mass soldering: Lifted copper is rated as an A-defect.

17.4.7 Delamination

See Figure 17.10. According to Paragraph 4.5.4, this defect is always rated as an A-defect.

17.4.8 Haloing

See Figure 17.11. As described in Paragraph 4.5.5, isolated haloing is rated as a B-defect. On the other hand, if the haloing touches the pattern, it is rated as an A-defect.

17.4.9 Weave Texture

See Figure 17.12. The defect (described in Paragraph 4.5.6) is not considered a

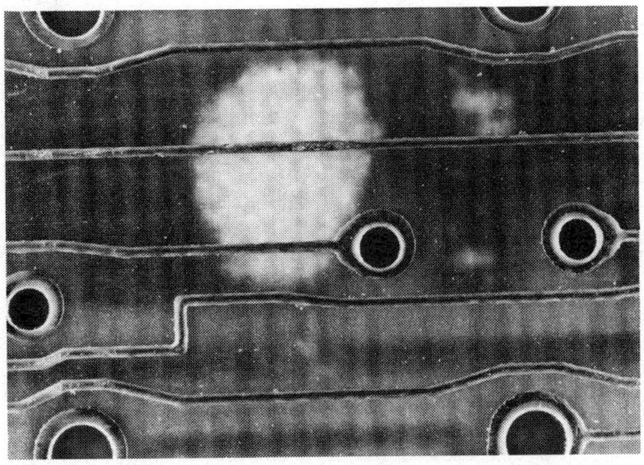

Fig. 17.9. Board with blistering (local separation between the glass weave layers, characterized by a small local swelling).

Fig. 17.10. Board with discoloration (widespread separation between the glass weave layers).

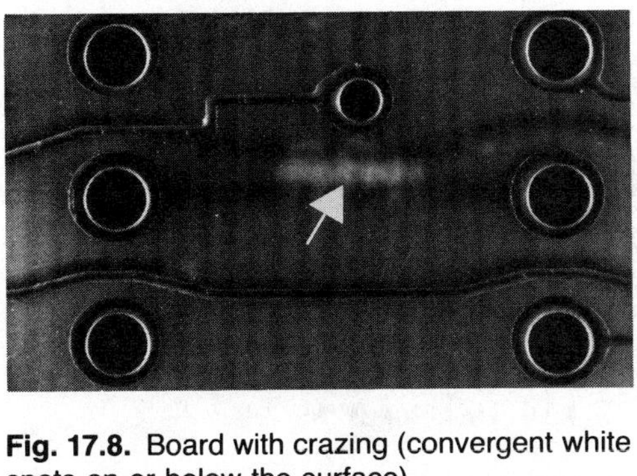

Fig. 17.8. Board with crazing (convergent white spots on or below the surface).

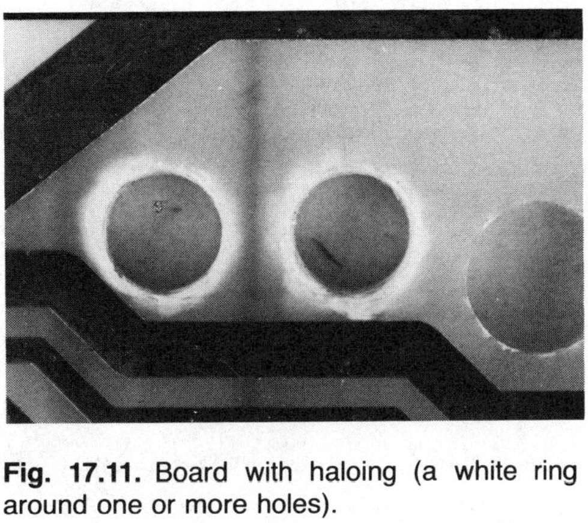

Fig. 17.11. Board with haloing (a white ring around one or more holes).

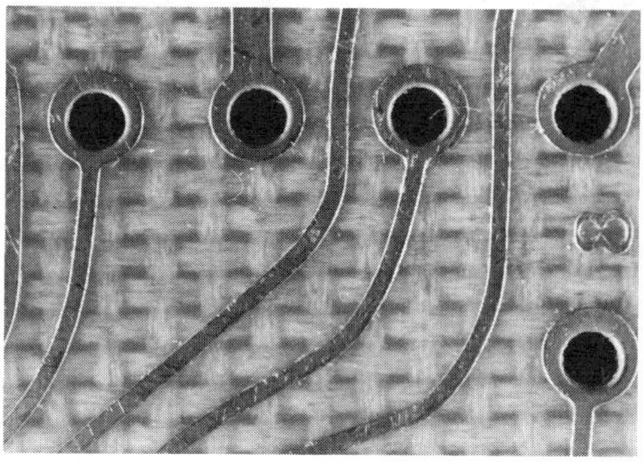

Fig. 17.12. Board with weave texture. (The glass weave is covered with epoxy, but since the layer is thin, the weave pattern is visible.)

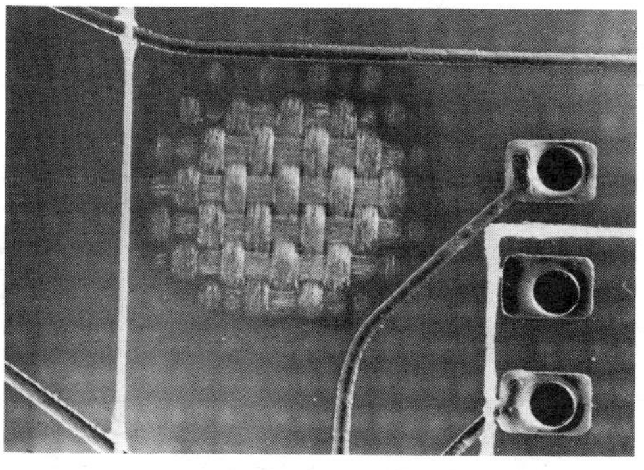

Fig. 17.13. Board with weave exposure. (The glass weave is not covered with epoxy.)

defect when the area in question is fully covered with epoxy. The occurrence can be marked in the inspection record with a check mark.

17.4.10 Weave Exposure

See Figure 17.13. Unlike weave texture, weave exposure is always, as described in Paragraph 4.5.7, cause for rejection and consequently rated as an A-defect.

17.4.11 Incomplete Curing

The defect (elaborated in Paragraph 4.5.8) is always cause for rejection and rated as an A-defect.

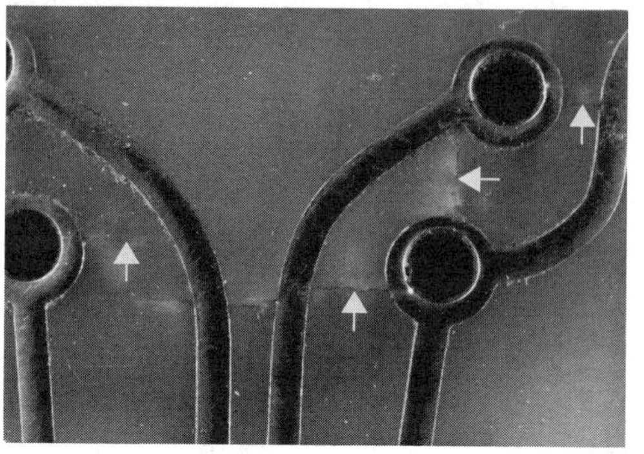

Fig. 17.14. Board with cracks.

17.4.12 Cracks

Various types of cracks are dealt with in Paragraph 4.5.9. Cracks extending through the board are usually rated as an A-defect. Cracks not extending through the board and meeting the requirements of Items 1 to 3 in Paragraph 4.5.9 are rated as a B-defect, whereas more serious cases are rated as an A-defect and cause for rejection. See Figure 17.14.

17.4.13 Metallic Inclusions

Small metallic inclusions occurring to a small extent are not entered in the inspection record. If they occur to a greater extent than specified in Paragraph 4.5.10, the board is rejected and the defect consequently rated as an A-defect.

The defect is extremely rare in ordinary single- and double-sided boards. In multilayer boards, on the other hand, it is not nearly so rare. The cause is usually an etching fault in one of the inner layers. See Figure 17.15. The assessment of such defects follows the guidelines given in Paragraphs 17.5.4 and 17.5.5. See also Paragraph 4.6.3.

17.4.14 Discoloration

A faint, locally occurring discoloration of the base material is rated as a B-defect,

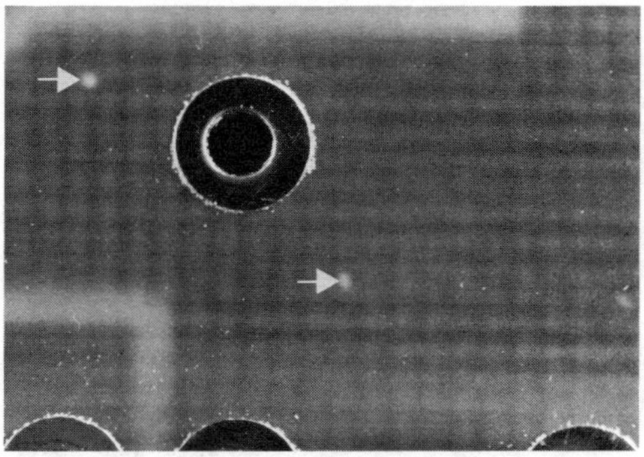

Fig. 17.15. Board with encapsulated metal particles. (The example shows an etching fault in one of the inner layers of a multilayer board.)

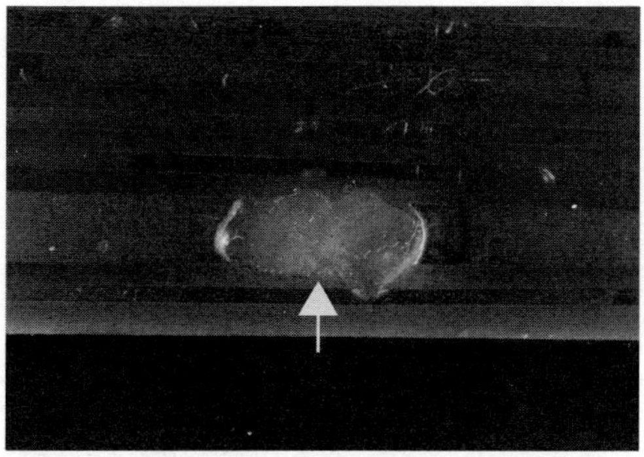

Fig. 17.16. Board with a hole in a conductor caused by a cavity in the base material

and the board is accepted. Widespread discoloration, starting as light brown, is cause for rejection, and the defect is consequently rated as an A-defect. See Paragraph 8.2.2.

A few customers specify such a high-temperature treatment of the boards that the base material assumes a light brown color. In such cases this is accepted with no remarks entered in the inspection record. The PCB manufacturer should have given the customer advance warning of the risk of discoloring the board.

17.4.15 Depressions

Depressions or cavities in the base material underlying the copper foil can cause holes in the etched pattern. Figure 17.16 shows where a conductor has covered the greater part of a depression on a bare-copper board. When the unwanted copper was removed during etching, the etchant ran into the cavity and etched the conductor from the unprotected underside.

In the example shown in Figure 17.16, this ruptured the conductor. The defect is so rare that it has not been listed in the inspection record. Should it occur, the fault is rated as an A-defect.

When the depression or cavity is sealed by an unbroken ground plane, there is no etching from the underside and the defect will most likely never be noticed.

17.5 Overall Appearance of Boards

The appearance of the finished pattern should also be inspected, following the part of the inspection record shown in Figure 17.17.

In the vast majority of cases a nonideal appearance is of only cosmetic importance since the board's functionality is not affected. The inspector should base his assessment primarily on the board functionality. This means that the board must meet all requirements regarding the circuitry, whereas cosmetic faults are of secondary importance, at least as long as they do not occur to an unreasonable extent.

			OVERALL APPEARANCE OF BOARD	17.5
			DENTS AND PITS	17.5.1
			CRACKS IN COPPER FOIL	17.5.2
			SCRATCHES	17.5.3
			METAL PARTICLES	17.5.4
			ETCHING FAULTS	17.5.5
			SURFACE CONTAMINATION	17.5.6
			HOLE DRILLING INCORRECT	17.5.7
			PLATING TOO THICK	17.5.8
			EXCESSIVE REPAIR WORK	17.5.10

Fig. 17.17. Overall appearance of board. (Section of inspection record.)

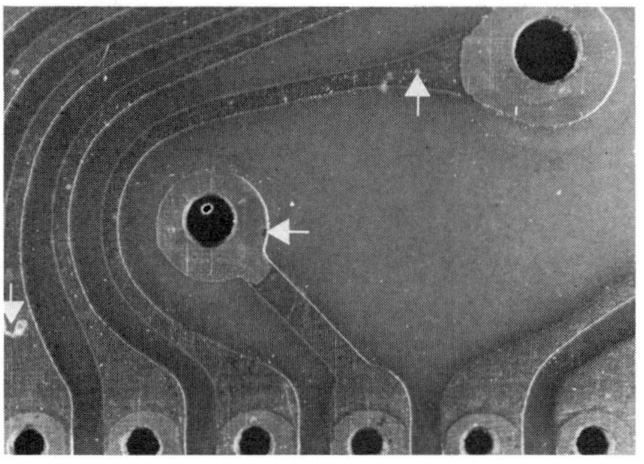

Fig. 17.18. Board with dents and pits in the pattern surface.

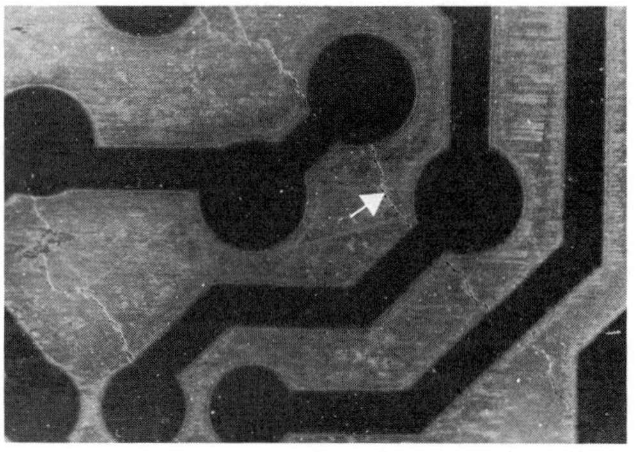

Fig. 17.19. Bare-copper board with small cracks in the copper foil. (The board was held in the production and did not reach the drilling stage.)

There are several conditions to be considered by the inspector:

17.5.1 Dents and Pits

A limited number of dents and pits in the pattern have only a modest cosmetic importance and are consequently not entered in the inspection record. The defect is not particularly common, and frequently occurs on only one side of the board. The cause can be small foreign bodies entering the lamination process and causing marks in the surface of the copper foil.

The fault is most noticeable on bare-copper boards as shown in Figure 17.18. A reflowed tin/lead layer usually conceals any dents and pits. A widespread occurrence of small dents or pits should be regarded as a B-defect. Since deep dents or pits in a conductor can mean the risk of rupture, the fault is rated as an A-defect.

17.5.2 Cracks in the Copper Foil

Small cracks in the copper foil are fortunately a very rare fault which, on the other hand, can greatly affect the reliability of the board. The defect will usually be found at the PCB manufacturer's process control, specifically at the inspection after imaging the pattern, which is performed as a 100 percent inspection. Figure 17.19 shows a board stopped at that inspection.

The fault is extremely rare. Should it, however, be found, it is rated as an A-defect and the board is rejected.

17.5.3 Scratches

Although it is quite hard to assess scratches in an objective manner, the inspector is always expected to perform a fair assessment. Paragraph 4.6.2 addresses the problem. See also the illustrations, referenced in the following subitems, which illustrate different degrees of scratching.

For boards with a tin/lead surface, and possibly provided with solder and insulation masks, the following conditions apply:

a. A limited and light scratching is not entered in the inspection record (Figure 17.20).

b. A widespread but light scratching with a perceptible cosmetic degradation of the board is rated as a B-defect. The scratches must not expose the underlying copper (Figure 17.21).

Fig. 17.20. Board with a limited and light scratching.

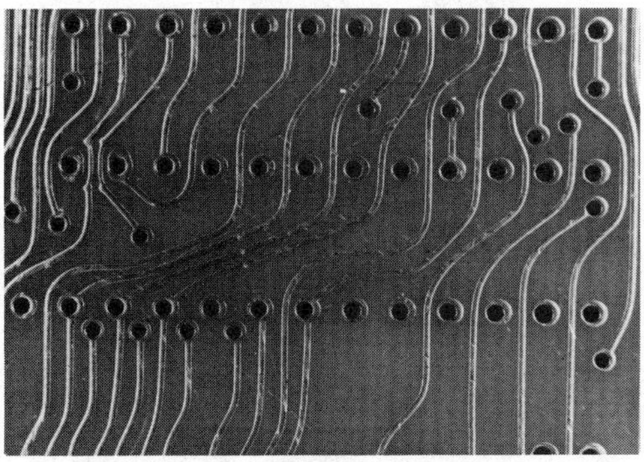

Fig. 17.21. Board with a widespread but light scratching, rated as a B-defect.

c. A deep scratch with visible copper is cause for rejection and consequently rated as an A-defect (Figure 17.22).

d. Scratches which result in tin/lead chips, either loose or adhering, imply a risk of short-circuiting the circuitry. Such scratches are an indisputable cause for rejection (Figure 17.23). The fault is rated as an A-defect.

17.5.4 Metal Particles

Unwanted metal particles should never be larger than about 0.04 by 0.04 in. (1 by 1 mm). See Figure 17.24. Larger metal par-

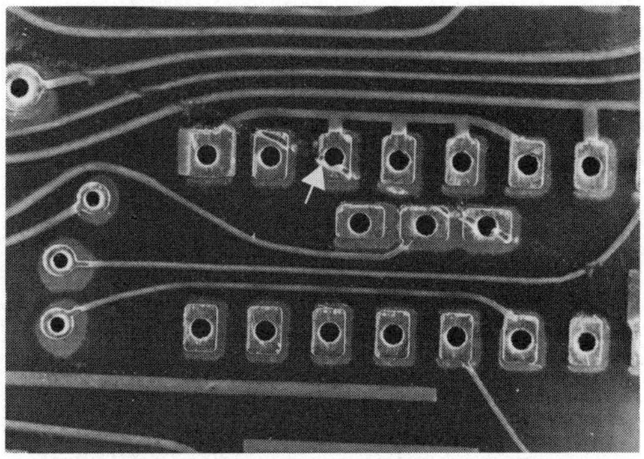

Fig. 17.22. Board with a deep scratch and visible copper at the bottom of the scratch is rated as an A-defect.

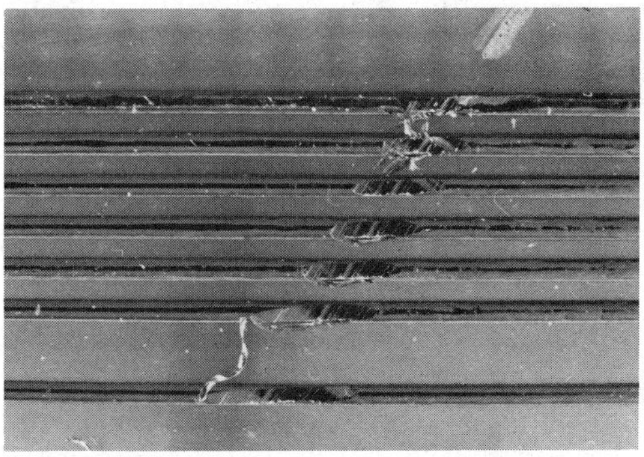

Fig. 17.23. Board with scratching resulting in adhering tin/lead chips which short-circuit some of the conductors. (The example justifies the use of solder and insulation masks.)

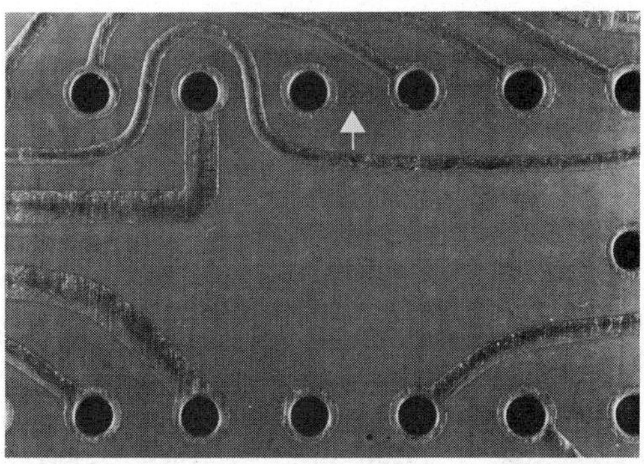

Fig. 17.24. Board with unwanted metal particles (0.04 × 0.04 in. = 1 × 1 mm).

ticles in active circuit areas are considered cause for rejection and rated as an A-defect.

It is acceptable for the PCB manufacturer to remove metal particles, if it is done neatly. The board surface must not be damaged, meaning there must be no weave exposure. If the weave is exposed, the inspector should refer to Paragraph 17.4.10.

Weave exposure can be exempted if such areas are small and occur in limited numbers isolated from the circuitry. The damage has been inflicted after the production processes and is therefore not so serious as described in Paragraph 4.5.7.

Paragraph 4.6.3d described how the inspector can assess unremoved metal particles according to the extent of their occurrence. In particular, the inspector should be aware of metal particles which reduce the actual insulation distance by more than 25 percent or, in areas carrying line voltage, reduce the effective insulation distance to less than 0.1 in. (2.5 mm). If such metal particles have not or cannot be removed, the defect is rated as an A-defect.

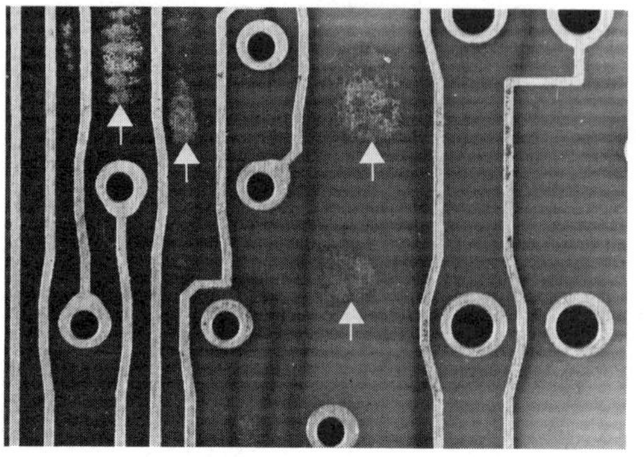

Fig. 17.25. Board with etching faults (0.125 × 0.125 in. = 3 × 3 mm).

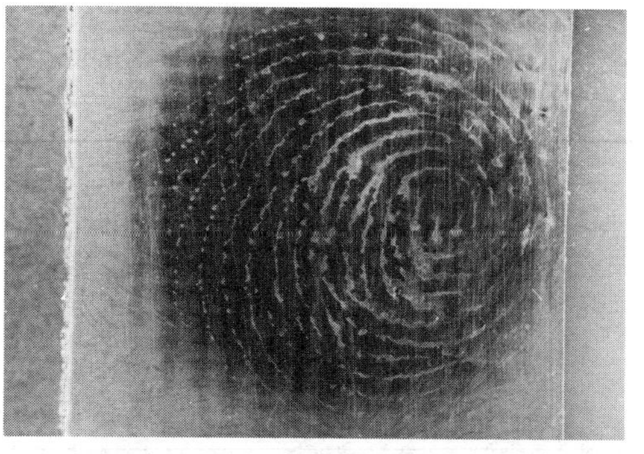

Fig. 17.26A. Board with fingerprints etched in the surface.

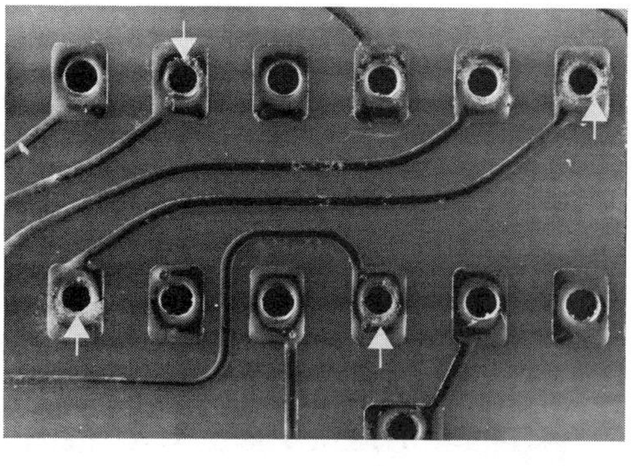

Fig. 17.26B. Board with acid stains. (The stained areas exhibit green corrosion products.)

17.5.5 Etching Faults

Etching faults should never be larger than 1/8 by 1/8 in. (3 by 3 mm) and they must not occur near active parts of the circuitry (Figure 17.25). Paragraph 18.2 addresses etching faults within the pattern (short circuits).

The PCB manufacturer may remove etching faults, but it must be done neatly without exposing the weave. See 17.5.4.

17.5.6 Surface Contamination

The surface must not be contaminated (Paragraph 4.6.5). The contamination can, for example, be grease, residues from plating resist, ordinary dirt, fingerprints, acid stains, or electrolyte stains. See Figure 17.26.

In rare cases, the contamination can come from the screen printing of the component notation; it is a somewhat trans-

Fig. 17.27. Board with a local epoxy contamination. (Transparent epoxy ink has overflowed the solder pads and downgraded the solderability.)

Fig. 17.29. Board with double hole drilling.

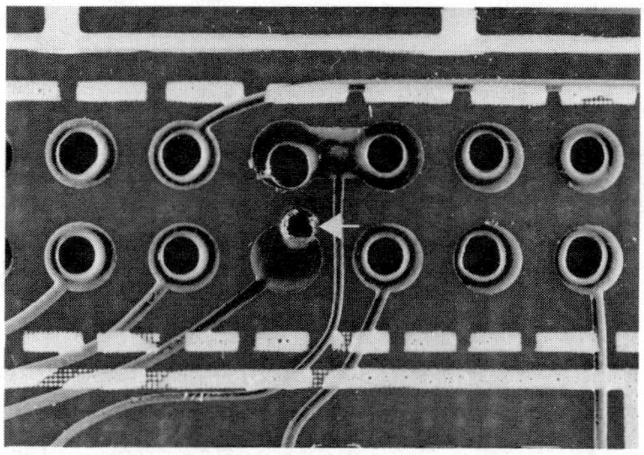

Fig. 17.30. Board with displaced hole drilling.

parent epoxy ink covering a small area of the pattern. Figure 17.27 shows such a contamination.

A limited contamination is rated as a B-defect if there is no sign of reduced surface resistance and/or reduced solderability. When in doubt, the inspector must test the surface resistance and solderability as described in Chapters 23 and 28. If the test results are affirmative, the fault is rated as an A-defect.

An unacceptably widespread contamination is always rated as an A-defect.

17.5.7 Drilling Errors

During the initial visual inspection, the inspector should verify that all solder pads have been drilled (Figure 17.28). Neither should there be a double drilling as shown in Figure 17.29 nor a displaced hole drilling as shown in Figure 17.30. Although such defects are quite rare, they can still occur.

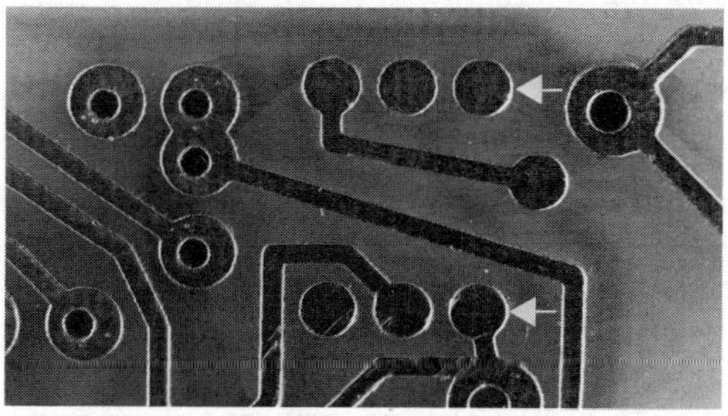

Fig. 17.28. Board with missing hole drillings.

A fourth and extremely rare type of fault is that the drill bit has not been lifted sufficiently above the main panel when moving the panel during the drilling process. This causes a groove in the board surface stretching from hole to hole as shown in Figure 17.31. Thin drill bits, however, will break before milling a groove so that the fault will presumably occur only in the case of fairly large holes—for example, with a diameter of 0.1 in. (2.5 mm) or more.

All these types of defects are cause for rejection and consequently rated as an A-defect.

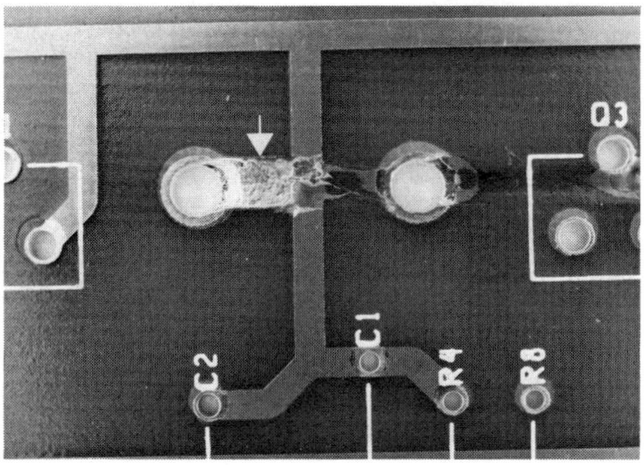

Fig. 17.31. Board with groove in the surface caused by insufficient lifting of the drill bit during movement.

17.5.8 Plating Thickness

Simultaneously with the inspection of the board's solder pads for missing or redundant holes, the inspector should check whether the plating thickness is excessive in certain areas of the board. This is most easily seen at the solder pads in open and unbalanced areas, particularly near the edge, where the solder pads can reach a considerable thickness (Figure 17.32). A large copper thickness should not be regarded as a real fault, but it indicates reduced control of the plating process. An unbalanced pattern contributes to this situation.

The inspector should note such overplated solder pads and check the hole diameter at the inspection described in Paragraph 20.8. The defect is entered in the inspection record with a check mark—that is, with no indication of the number of defects.

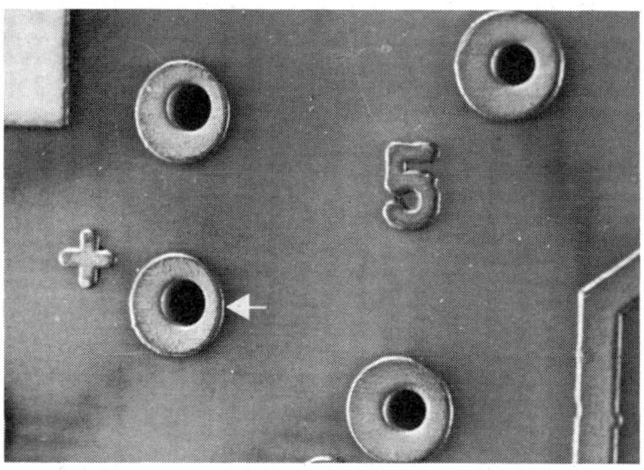

Fig. 17.32. Board with a very heavy plating in open areas.

Excessive plating can also be felt by touching the pattern, provided no solder and insulation masks have been applied. The inspector simply lets his fingers slide across the pattern. The pattern edges will then feel fairly sharp, also if the board is reflowed. The cause is frequently a plating resist which is too thin for the desired plating thickness. Therefore, we have a mushroom effect, in which the copper plating extends sidewise. This type of overplating should not be regarded as a fault, and it is therefore only entered in the inspection record with a check mark.

17.5.9 Unwanted Through-Plating

Coinciding with the inspection in Paragraph 17.5.8, the inspector should check whether nonplated mounting holes erroneously have been plated through. The fault can particularly occur at nonplated holes larger than 0.10 to 0.12 in. (2.5 to 3 mm) in diameter.

In a plated-through board, the cause of this fault is that the PCB manufacturer found it cheaper to drill the nonplated holes in the same drilling operation as the plated-through holes. He then tented the nonplated holes so that they were sealed by the dry film plating resist during the plating process. An incomplete sealing or a puncture of the dry film caused a more or less complete plating of the hole. The defect should be rated as a B-defect when the hole diameter is acceptable.

17.5.10 Repair Work

Repair work performed by the PCB manufacturer must be limited. The inspector should assess the general appearance of the board and distinguish between:

a. Very extensive repair work: A- or B-defect

b. Incomplete repair work: A-defect

c. Poor repair work: A-defect

It is difficult to give infallible guidelines for case a, as the assessment is subjective. The most reasonable solution is to adapt the assessment to the practice followed in the production department when executing engineering change orders on the boards. Examples are cutting conductors, drilling extra holes, establishing alternative conductor routings with wires, and so on.

Fig. 17.33A. Acceptable repair of interrupted conductor.

Fig. 17.33B. Limit of acceptable repair of interrupted conductor.

Fig. 17.33C. Unacceptable repair of interrupted conductor.

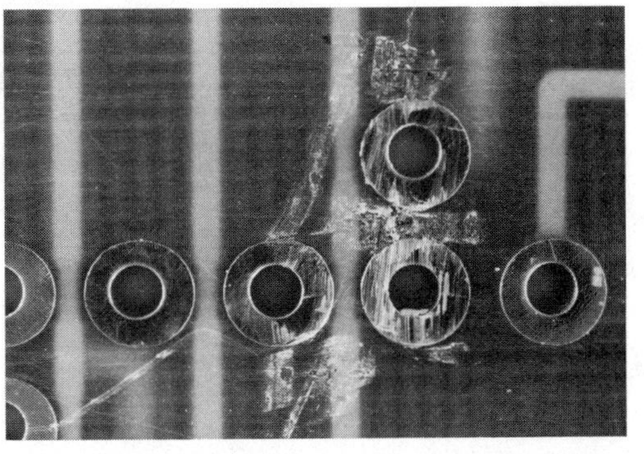

Fig. 17.34. Unacceptable removal of etching fault.

If urgently needed, the inspector can waive cases b and c and send the boards to the production department for repair.

Some customers accept interrupted conductors repaired by a kovar strip welded across the gap. This can be done with various degrees of skill as shown in Figures 17.33A to C. Figure 17.34 shows a somewhat clumsy removal of an etching fault.

Many PCB manufacturers are increasingly interested in having their customers accept repair work on defective boards, and more and more PCB manufacturers have invested in special equipment for PCB repairs.

As increasing pattern densities require

still narrower conductors and spacings, it is inevitable that defects occur more often in the form of open and short circuits in the pattern.

Since the cost of complex boards, particularly multilayer boards, is high, the rejection of boards because of these defects will cause the PCB manufacturer great losses. In the long run, such losses are passed on to the customers as a higher overhead. In order to avoid such a price increase, many PCB manufacturers have reached agreement with their larger customers that some defect types may be remedied by qualified repair work. In the future, the electronics industry will undoubtedly accept such repairs more than it does now.

Chapter 18

Detailed Visual Pattern Inspection

18.1 Introduction

After the initial visual inspection, the inspector carries out a pattern inspection on the full sample. The functional conditions described in this chapter should be assessed according to the part of the inspection record shown in Figure 18.1.

18.2 Short Circuits

Because short circuits between the pattern components (conductors and solder pads) caused by etching faults are among the most commonly occurring types of de-

DETAILED VISUAL INSPECTION	18
PATTERN	
SHORT CIRCUITS	18.2
INTERRUPTIONS	18.3
PERCENTAGE RULE NOT MET	18.4.1
DIMENSIONAL CHANGES O.O.T.	18.4.2
EDGE UNEVENNESS O.O.T.	18.4.3
INDENTATIONS O.O.T.	18.4.4
PROJECTIONS O.O.T.	18.4.4
VOIDS/PINHOLES	18.4.5
METAL PARTICLES	18.4.6
LOCATION OF HOLES	18
SOLDER HOLES O.O.T.	18.5
ANNULAR RINGS O.O.T.	18.5
MOUNTING HOLES O.O.T.	18.6
OTHER CONDITIONS	18
SOLDER HOLES CONTAMINATED	18.7
BURNING OF PATTERN	18.8.1
STEP PLATING	18.8.2

Fig. 18.1. Pattern, location of holes, and other conditions. (Section of inspection record.)

Fig. 18.2A. Short-circuited conductors caused by defective plating resist. (Tin/lead is seen on the defective areas.)

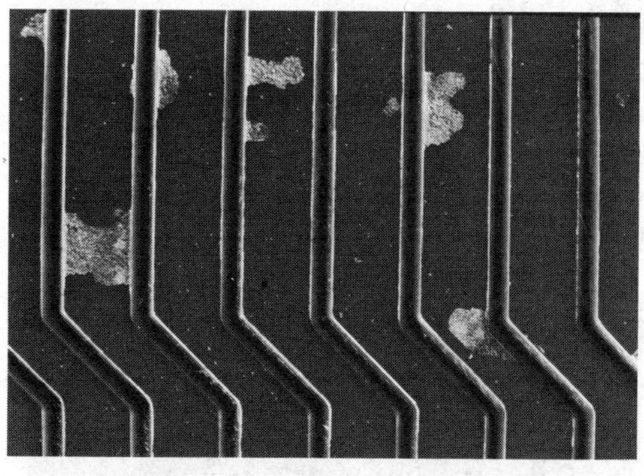

Fig. 18.2B. Short-circuited conductors and metal remnants caused by an incomplete removal of the plating resist. (Bare copper is seen on the defective areas.)

fects, the inspector should focus on such irregularities. They frequently occur between adjacent parallel conductors (Figure 18.2).

Short circuits are cause for rejection and rated as an A-defect. Paragraph 5.2.1c discusses the assessment of short circuits removed by the PCB manufacturer.

The inspector can consider exempting a defect which has been acceptably repaired. In that case, it may be rated as a B-defect.

Another type of short circuit is bridging caused by reflowing or solder-coating and hot-air leveling. Since such short circuits are easily removed with a soldering iron or preferably by solder extraction, they are rated as B-defects. Short circuits are usually very few and normally occur between adjacent conductors, particularly if the tin/lead plating is a little too thick.

18.3 Interruptions

Interruptions, which mainly occur at the conductors (Figure 18.3), are an obvious cause for rejection and rated as an A-defect. Paragraph 5.3 discusses the possibility of exempting the defect.

18.4 Pattern Defects

Defects at and along the edges of the pattern have been circumstantially addressed in Paragraphs 5.4 through 5.7, and the assessment criteria have been recapitulated in Tables 5.1 through 5.3.

Fig. 18.3. Interrupted conductor caused by a fault in the plating resist.

The inspector simplifies the routine inspection procedures as much as possible, following the aforementioned criteria only in doubtful cases. Instead, the inspection is based on the following simplified requirements:

18.4.1 Rule of Percentage

The main requirement is adherence to the rule of percentage, which means that all dimensions of conductors, solder pads, and insulation spacings must not be reduced to less than the stipulated percentage of the nominal dimensions as measured on the master film.

This book uses a 75 percent rule, but most military specifications and some company specifications use an 80 percent rule.

The requirement is valid for any combination of edge displacement (Paragraph 18.4.2), edge definition (Paragraph 18.4.3), indentations and projections (Paragraph 18.4.4), and voids and pinholes (Paragraph 18.4.5). When assessing the insulation distance between two conductors, the inspector should also consider possible metal particles between the conductors (Paragraph 18.4.6).

Violating the rule of percentage is cause for rejection and consequently rated as an A-defect. There are exceptions, however, such as the example in (Paragraph 18.4.7).

18.4.2 Dimensional Changes

In the case of bare-copper, reflowed, and hot-air leveled boards, the dimensions stated in Paragraph 18.4.1 must not be reduced or increased by more than 0.004 in. (0.10 mm) as a result of edge displacement.

Note that we are discussing the total dimension and not the edge displacement as stated in Paragraph 5.4.2, Requirement 2. Failure to meet the total dimension re-

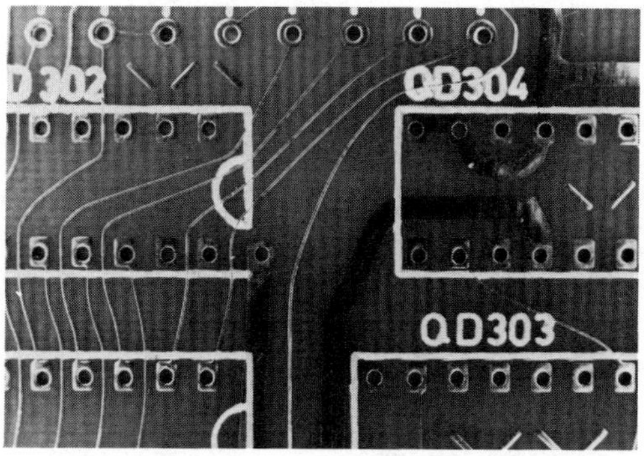

Fig. 18.4. Reduced pattern width caused by excessive etching. The edge displacement is about 0.003 in. (0.075 mm) which implies that conductors having a nominal width of 0.012 in. (0.3 mm) achieve an unacceptable width of 0.006 in. (0.15 mm) only.

quirement is rated as a B-defect provided the rule of percentage is met.

Outgrowth greater than 0.002 to 0.004 in. (0.05 to 0.10 mm) is not especially common. The most common fault is reduced width caused by excessively heavy or prolonged etching. Figure 18.4 shows how conductors with a nominal width of 0.012 in. (0.30 mm) have been reduced to approximately 0.006 in. (0.15 mm). As the rule of percentage has been violated, the fault is rated as an A-defect.

18.4.3 Edge Unevenness

Unevenness along the pattern edges, also called waviness, must be less than 0.004 in. (0.10 mm) peak-to-valley and can be accepted with no remarks entered in the inspection record.

Unevenness up to 0.008 in. (0.20 mm) is rated as a B-defect, whereas unevenness exceeding 0.008 in. (0.20 mm) is cause for rejection and consequently rated as an A-defect. See Figure 18.5.

The inspector is referred to Requirement 2 in Paragraph 5.5.2 regarding exemption of larger unevenness in a few exceptional cases.

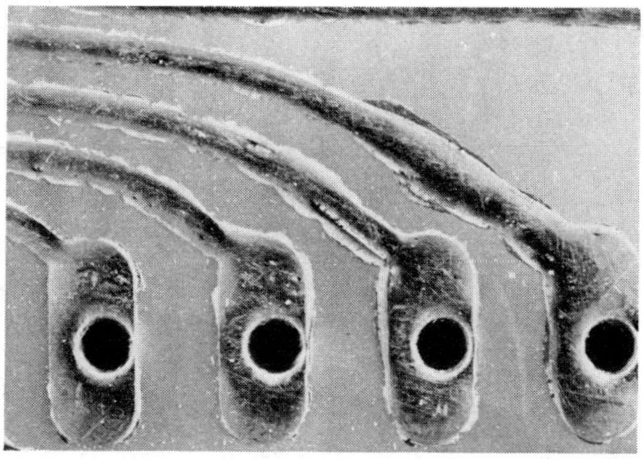

Fig. 18.5. Uneven pattern. (Note the displaced hole drilling.)

18.4.4 Indentations

Localized unwanted indentations and projections along the pattern edges can be accepted without remarks in the inspection record, provided the following requirements are met:

a. The indentation depth or the projection height shall be less than 0.04 in. (1 mm), and the length shall be less than 0.2 in. (5 mm). For solder pads, the length is also limited to 25 percent of the periphery.
b. The average number of indentations and projections along the pattern edges shall not exceed one defect per 2 in. (50 mm) of edge length.

Violating these requirements is cause for rejection and consequently rated as an A-defect. See Figures 18.6, 18.7, and 18.8. The inspector usually sees only a few cases of indentations and projections in a delivery; these faults are usually due to an inferior image transfer.

18.4.5 Voids and Pinholes

Visible, open voids and pinholes in the pattern can be accepted without remarks in the inspection record, provided the following requirements are met:

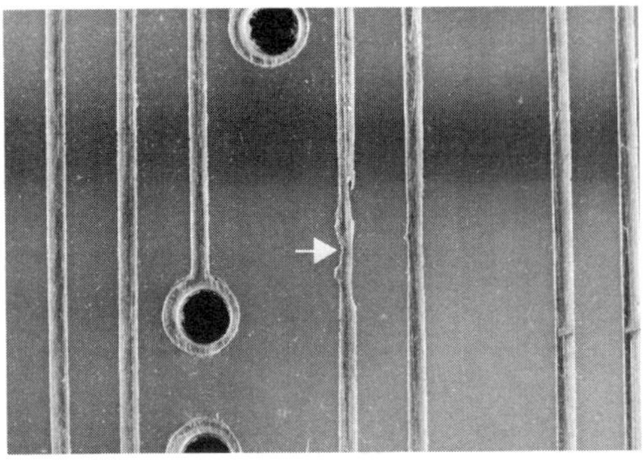

Fig. 18.6A. Indentation of conductor.

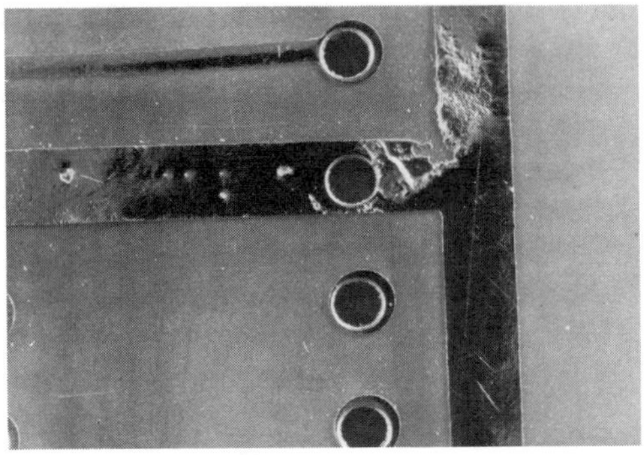

Fig. 18.7A. Indentation at branch point of conductor.

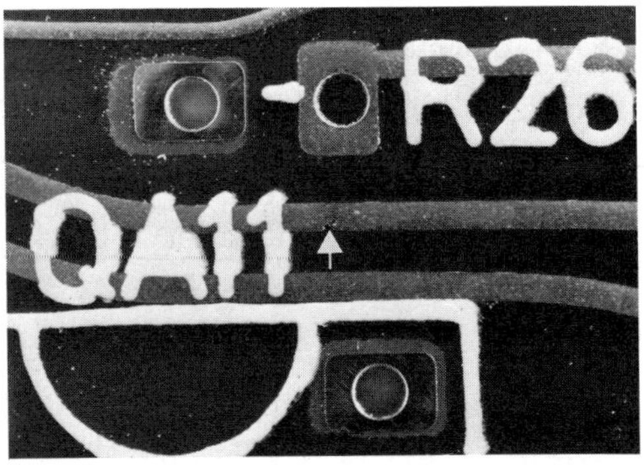

Fig. 18.6B. Indentation of conductor. The fault is easily overlooked.

Fig. 18.7B. Indentation conductor, close to rupture. (The fault is caused by incomplete development of the plating resist.)

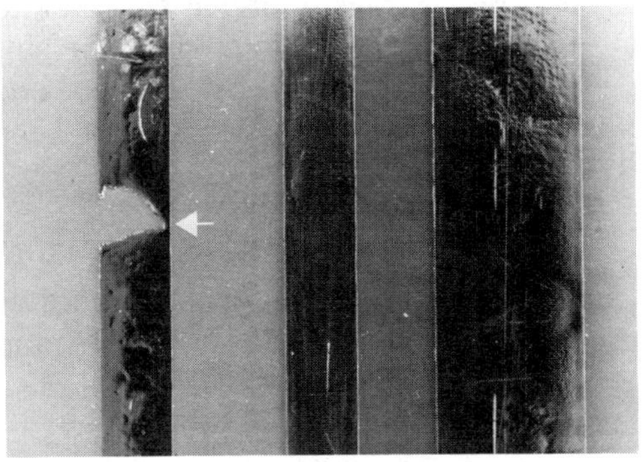

Fig. 18.6C. Indentation of conductor, close to rupture.

In a solder pad, the size of a void or pinhole must be less than 0.006 in. (0.15 mm). No more than one void or pinhole per solder pad is permitted, and none in the transition zone between the solder pad and the conductor. In plated-through holes, no void or pinhole is permitted at the rim of the solder hole. See Figure 18.9.

In a conductor, the diameter of the void or pinhole must be less than 20 percent of the conductor width, and the average number of voids or pinholes in a conductor shall not exceed one void or pinhole per 2 in. (50 mm) of conductor length. See Figure 18.10. Violation of these requirements is cause for rejection and consequently rated as an A-defect.

On rare occasions the inspector can see circumferential holes in the solder pads

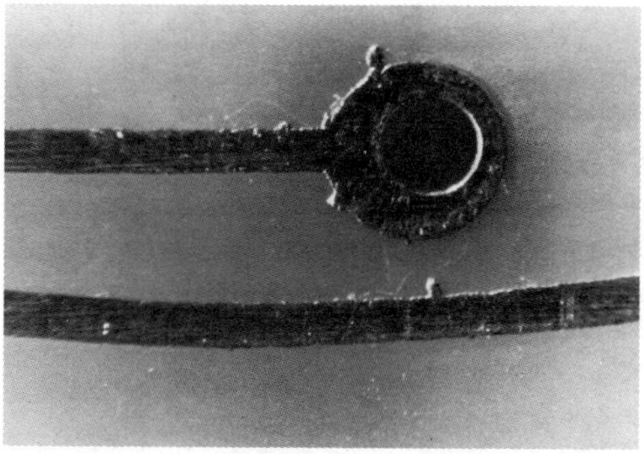

Fig. 18.8A. Projections along conductor and solder pad. (The requirement regarding minimum spacing is not fulfilled.)

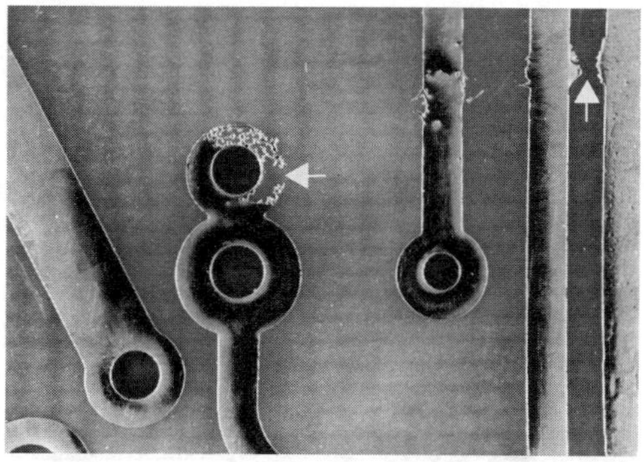

Fig. 18.9B. Solder pad nearly etched away. (Note the projections between the conductor and the ground plane.)

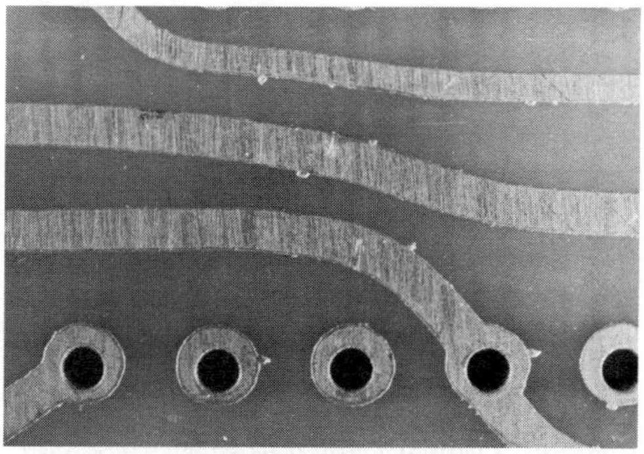

Fig. 18.8B. Projections along conductor and solder pads. (The projections are not unacceptably large.)

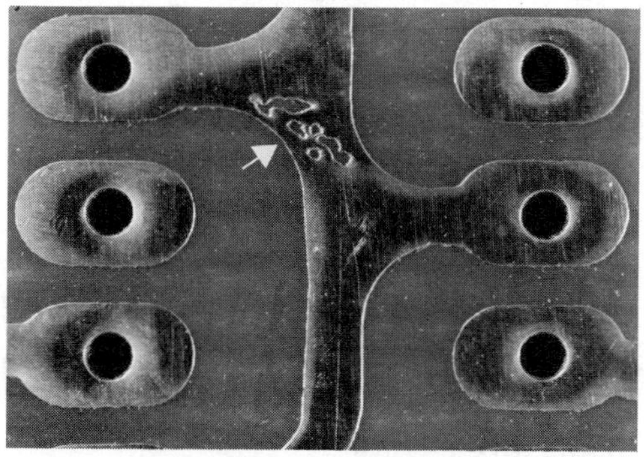

Fig. 18.10A. Conductor with voids.

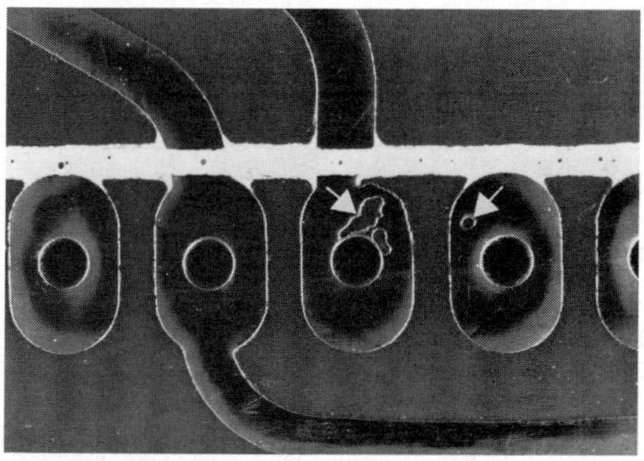

Fig. 18.9A. Solder pads with voids.

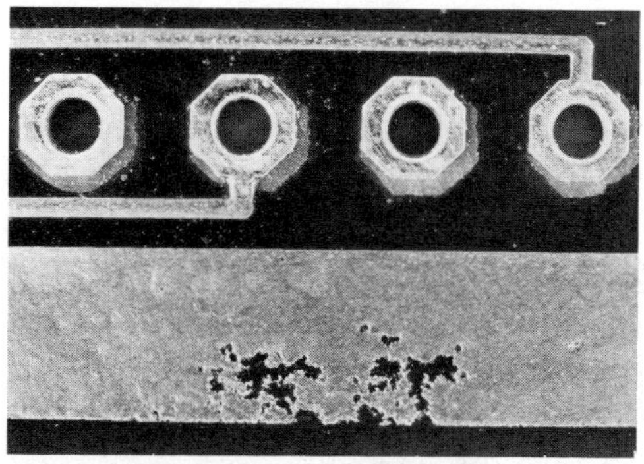

Fig. 18.10B. Conductor with voids.

of plated-through holes, usually at the transition from solder pad to barrel. See Figure 18.11. The cause is usually heavy brushing or scrubbing of the electroless copper, which is removed from the rim of the solder hole, making copper plating impossible at such locations. This defect

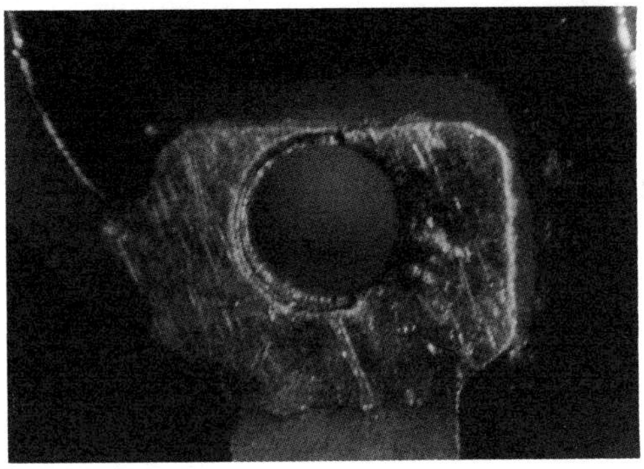

Fig. 18.11. Circumferential hole in solder pad at the rim of the solder hole.

is an indisputable cause for rejection and consequently rated as an A-defect.

18.4.6 Metal Particles

Metal particles in functional circuit areas—for example, between two conductors—shall not reduce the effective insulation distance more than permitted by the rule of percentage. Figure 18.12 shows examples of metal particles between conductors.

18.4.7 Inspector's Attitude

Since assessment of pattern defects is always somewhat subjective, the inspector is advised to take a functional attitude. Figure 18.13 clarifies this recommendation. A commonly accepted practice is necking down a wide conductor where it passes between two adjacent solder pads. The width can be reduced, for example, by 50 percent over a length of 1/8 in. (about 3 mm). Just outside the intended indentation a nick reduces the width by 40 percent over a length of 0.010 in. (0.25 mm). When the nick is reasonably rounded with no risks of notch effects—meaning that the nick cannot continue in a crack later on—it seems absurd to reject the board on grounds of violating the 75 percent rule.

Fig. 18.12. Metal particles between conductors.

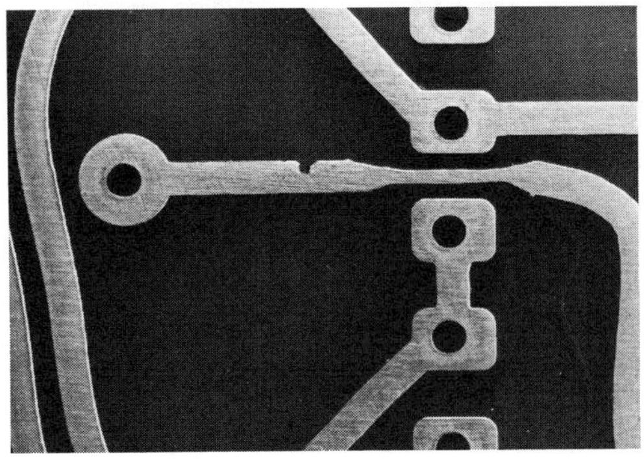

Fig. 18.13. A controversial case: A small nick very close to where the conductor has been necked down. Is the nick acceptable?

Unfortunately, such nicks have caused rejections although intended nearby indentations were considerably larger.

The inspector should get more satisfaction in performing an intelligent assessment and waiving such a defect when it is of no functional importance. The alternative is to reject the board for an insignificant violation of the quality requirements.

This requires, however, a change in attitude towards the inspector's responsibilities. In most electronics companies, the inspector's role is traditionally limited to ascertaining whether or not the boards conform to the purchase specification. All waivers must be submitted to a supervisor or to the Material Review Board. As this is frequently an inconvenient, burdensome process, the result is usually unnecessary rejections.

Incidentally, the inspector can aid the change in attitude by acquiring a broader background knowledge. An alternative is to record—and occasionally submit—examples of formally correct, but objectively unnecessary rejections.

18.5 Location of Solder Holes

In most cases, the drilling accuracy of the solder holes—their location within the solder pads—can easily be assessed from the appearance of the annular rings as discussed in Paragraph 5.9. As the placement is assessed relative to the pattern, the inspector should be aware of the assembly method used in the production department.

In boards for manual assembly, the absolute hole location is not particularly critical. In contrast, the absolute hole location is extremely important for automatic component insertion. Paragraphs 6.3 and 6.4 elaborate on this topic. It is therefore necessary to perform a supplementary inspection of the board by measuring the absolute hole positions (Paragraph 20.4.1).

The visual assessment is based on the assumption that the annular rings are quite narrow. The inspection is easier in the case of plated-through boards because it is very common to have a nominal width of 0.010 to 0.012 in. (0.25 to 0.30 mm) of the annular ring of IC solder pads. If the annular ring is broken, the drilling is inferior because the eccentricity is of the same order of magnitude as the nominal width of the annular ring.

First, the inspector checks the purchase specification. Most specifications require that the annular rings be unbroken, but some allow a certain number of broken annular rings—for example, 1 percent of the solder pads.

Naturally, the documentation material must have a good registration (Paragraph 16.8). According to many purchase specifications, the PCB manufacturer must object if the documentation cannot meet the requirements of the purchase specification.

Although the full sample size is used when inspecting the boards, the inspector should not check the location of every hole, but seek a general impression of whether the holes have been drilled with sufficient accuracy.

This is best done by scanning the entire board. It seems reasonable to study the outer areas a little more carefully, but the inspection of even a fairly large board—for example, 8 by 12 in. (200 by 300 mm)—should require no more than a minute or so.

The rule of no more than 1 percent of broken annular rings should primarily be regarded as a safety valve for the PCB designer and draftsman as well as the PCB manufacturer (Paragraph 5.9). It is normally unnecessary for the inspector to count the number of broken annular rings, as the number is usually either very small or very large.

The inspector should realize that special conditions can occur in boards for automatic component insertion. The solder holes are frequently oversized in order to allow safe insertion of components. For several reasons, this implies very narrow annular rings, and under certain circumstances the inspector will find that the board has been designed with intentionally broken annular rings. In such cases the inspector must disregard the annular ring criterion and accept the delivery.

The inspector can use the following criteria (cf. Paragraphs 5.9.3 and 5.9.4):

Plated-through boards:

Assuming a nominal annular-ring width of 0.010 to 0.012 in. (0.25 to 0.30 mm), the width must not be less than 0.002 in. (0.05 mm) on the finished board.

Broken annular rings are accepted at a maximum of 1 percent of the solder pads. When the thickness of the base material is less than 0.032 in. (0.8 mm), broken annular rings are accepted at 2 percent of the solder pads.

These percentages apply for board dimensions up to 10 in. (250 mm). For every 4 in. (100 mm) dimensional increase beyond 10 in. (250 mm), the acceptable number of broken annular rings doubles, the maximum acceptable number being 10 percent of the number of solder pads.

No more than 25 percent or 90 degrees of the periphery of the solder hole may fall outside the solder pad. Breakouts must not occur closer than 0.002 in. (0.05 mm) to a conductor attached to the solder pad.

If these requirements are met, no remarks are entered in the inspection record. Broken annular rings up to twice these requirements, with a maximum of 10 percent, are rated as a B-defect. A larger number, or more than 25 percent (90 degrees) of the periphery of the solder holes falling outside the solder pads, are cause for rejection and consequently rated as an A-defect. See Figure 18.14.

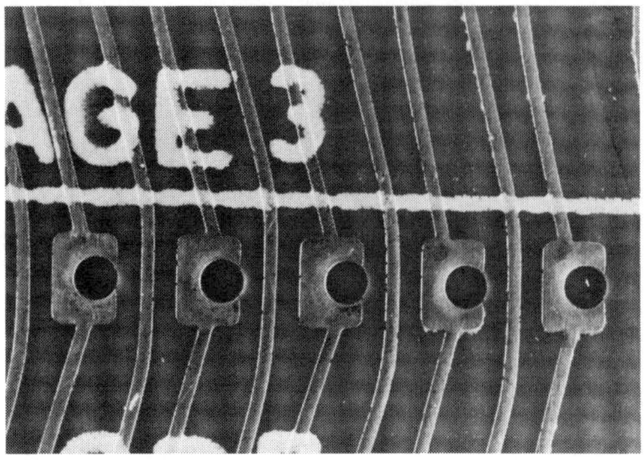

Fig. 18.14A. Broken annular rings of plated-through holes. More than 90° of the circumference of the solder hole is outside the solder pad.

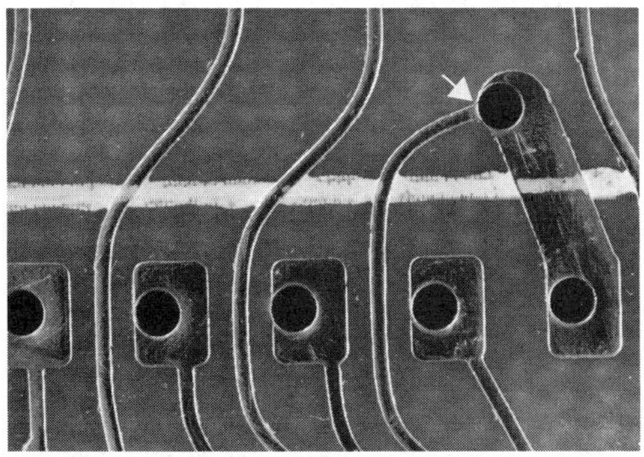

Fig. 18.14B. Broken annular rings of plated-through holes. Less than 90° of the circumference of the solder hole is outside the solder pad, but a conductor is attached to the solder pad just where the annular ring is broken.

Nonplated-through boards:

Assuming a nominal annular ring width of 0.020 in. (0.50 mm), the width must not be less than 0.006 in. (0.15 mm) on the finished board. Broken annular rings are cause for rejection and rated as an A-defect. See Figure 18.15.

Note: In entertainment electronics and similar equipment it is not always economical to use plated-through boards. However, more and more integrated circuits are being used in entertainment electronics, and furthermore, the boards are quite dense. The PCB designer is therefore compelled to follow the design rules for plated-through boards when designing boards which are not plated through. In such cases, obvious from the dimensions of the solder pads and the holes, the inspector is compelled to inspect the annular rings according to the criteria for plated-through boards.

If the annular rings are wider than stated above, the visual assessment is more difficult. The inspector can, however,

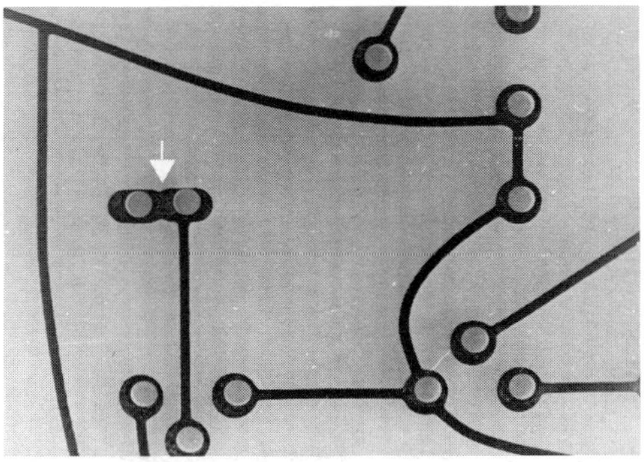

Fig. 18.15A. Unmistakable breakout of annular rings of nonplated-through holes.

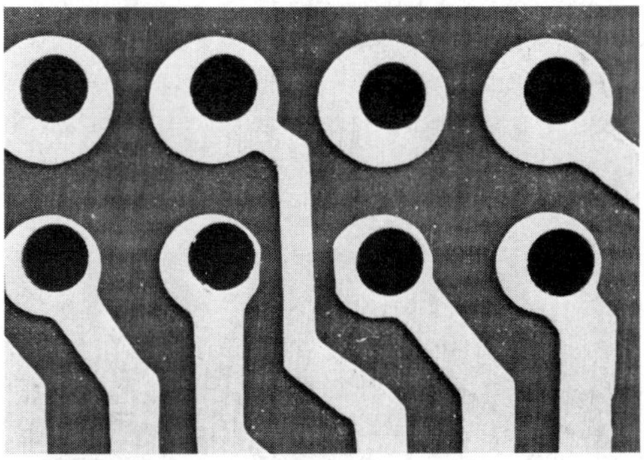

Fig. 18.15B. Minimum width of the annular ring is 0.0024 in. (0.06 mm) and, as a matter of fact, too small.

measure the eccentricity with the magnifier; he chooses a few obvious examples and determines the displacement as described in Paragraph 6.3.5. Displacements up to the previously-stated amount are accepted—that is, 0.008 to 0.010 in. (0.20 to 0.25 mm) for plated-through boards, and 0.014 in. (0.35 mm) for nonplated-through boards.

Paragraph 6.6 describes repair methods and exemption possibilities.

18.6 Location of Mounting Holes

The mounting hole location is assessed visually when measurement is impossible. This implies that the mounting holes were not dimensioned relative to a reference system (Paragraph 6.5.3).

In cases where the mounting holes are defined by means of solder pads on the board, the inspection follows the guidelines given for component holes. If the solder pads have disappeared during the drilling, the inspector uses the master film as a reference, laying it over the board and registering it with respect to the solder pads of the pattern. He can then visually assess the hole location, possibly with the aid of the magnifier, as previously described.

In other cases the mounting holes are determined from a drill tape, but if the inspector does not know the coordinates, he must use the master film method previously described.

The inspector can generally accept displacements up to ± 0.012 in. (± 0.3 mm) on boards having dimensions up to 16 in. (400 mm). When this tolerance is exceeded, the fault is rated as an A-defect, assuming that the purchase specification states this tolerance.

It is frequently possible to exempt a violation since the mounting holes can be moved a little by suitable machining (filing or redrilling to a larger diameter). The inspector should be aware that in some cases the mounting holes guide the board in the jig of an automatic component insertion machine. Especially in the latter case, the same tolerance is required, in principle, for the location of the mounting holes as for the solder holes.

18.7 Contaminated Holes

At the same time he inspects the solder hole locations, the inspector should verify that the solder holes are not contaminated by dust, metal chips, or brown or black crusts. See Figure 18.16. Limited contamination is rated as a B-defect and the delivery is accepted, provided it can be cleaned—for example, by compressed air.

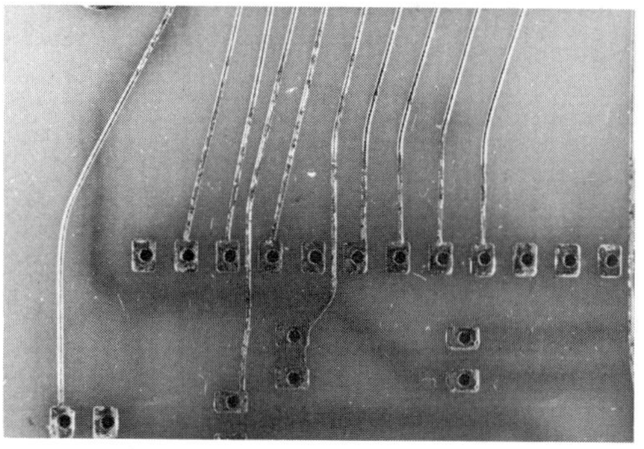

Fig. 18.16. Black crusts in solder holes and on pattern.

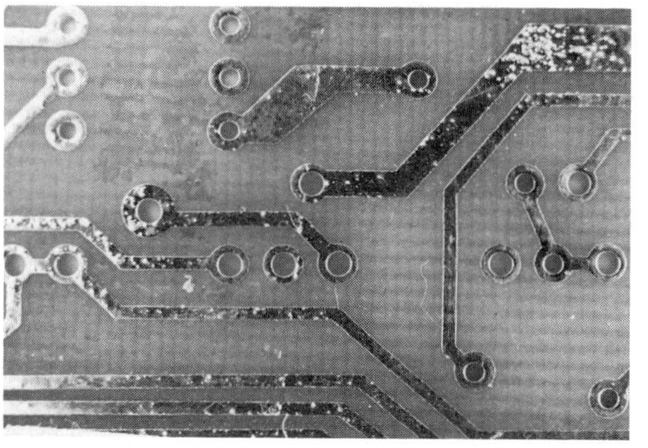

Fig. 18.17. Burning of pattern.

The delivery can also be returned to the PCB manufacturer for a thorough cleaning. Widespread contamination which reduces the solderability is rated as an A-defect and the boards are rejected.

Screen-printing ink in the holes is an indisputable cause for rejection, and the fault is therefore rated as an A-defect (Paragraph 7.4.3).

18.8 Plating Faults

Because of adverse plating conditions, the plating faults described in the following paragraphs can occur. These faults are assessed visually.

18.8.1 Burning

As described in Paragraph 7.3.8, burning shows up as a chain of small nodules along the edges of the pattern as well as on the pattern. See Figure 18.17. The distribution can vary somewhat across the board because of variations in the pattern density.

Unfortunately, reflowing does not always clearly indicate wetting problems as it can be difficult to see possible dewetting on the uneven surface of the nodules. Solderability problems can easily occur on burned boards; in the case of widespread burning of the solder pads, the board is rejected and the fault rated as an A-defect. Moderate burning, away from the hole edges, is rated as a B-defect and the board is accepted. See Figures 18.18 and 18.19.

Fig. 18.18. Moderate burning of pattern edges but not of the solder hole rims.

Fig. 18.19. Serious burning of pattern. Note in particular the burning of the solder hole rims.

18.8.2 Step-Plating

Paragraph 7.3.9 discussed this in detail. The decisive factor is whether the copper meets the thickness requirement on the solder pads as well as in the plated-through holes. An inspection requires microsectioning and subsequent measurement of the copper thickness in the thin areas. See Paragraph 25.5.3a for acceptance criteria. Examples of microsections of plated-through holes are shown in Figures 25.55-31 through 25.55-33 in Chapter 25. Fortunately, step-plating is an extremely rare fault with the plating baths currently in use.

18.9 Reflowing and Solder-Coating with Hot-Air Leveling

In Paragraphs 18.9 through 18.13 we address the visual inspection of the board's surface treatment. The inspector should follow that part of the inspection record shown in Figure 18.20. This inspection is performed concurrently with the other visual inspections and is based on the same sample size. The boards are inspected for the following conditions:

18.9.1 Surface Requirements

General requirements of the tin/lead layer are that the surface be bright and even with a fine-grained and homogeneous structure without dark areas. Paragraph 8.3.4 describes different types of surface defects which can cause reduced solderability. If the inspector suspects the boards of suffering from such defects, he can arrange for a solderability test as described in Chapter 23.

When a surface defect reduces the solderability, the defect is rated as an A-defect. If the solderability is not noticeably reduced, the boards are accepted and the fault rated as a B-defect.

Gritty and grey areas are accepted when

REFLOWING/SOLDER COATING				18.9
			SURFACE UNEVEN/MATTE	18.9.1
			SURFACE COARSE-GRAINED/DARK	18.9.1
			SURFACE GRITTY/GREY	18.9.1
			SURFACE WITH BLUE PLAGUE	18.9.2
			ALLOY COMPOSITION O.O.T.	18.9.3
			PATTERN EDGES UNCOVERED	18.9.4
			SLIVERS/VISIBLE OVERHANG	18.9.5
			REFLOWING INCOMPLETE	18.9.6
			NONWETTING/DEWETTING	18.9.7
			SOLDER HOLES BLOCKED	18.9.8/18.9.9b
			CURTAINS ON CONDUCTORS	18.9.9a
			WEBBING	18.9.9c/18.9.9d
			SOLDER DROPS/BALLS	18.9.9g
ROLLER TINNING				18.10
			SURFACE UNEVEN/MATTE	18.10.1
			SURFACE COARSE-GRAINED/DARK	18.10.1
			SURFACE GRITTY/GREY	18.10.1
			SOLDER DROPS O.O.T.	18.10.2
			COMPONENT HOLES CLOSED	18.10.3
			BARE COPPER AT HOLE EDGE	18.10.4
			PATTERN EDGES UNCOVERED	18.10.5
			DROSS ON SURFACE	18.10.6
			NONWETTING/DEWETTING	18.10.7
TIN PLATING				18.11
			SLIVERS/VISIBLE OVERHANG	18.11.1
			APPEARANCE UNACCEPTABLE	18.11.2
PROTECTIVE LACQUERING				18.12
			LAYER UNEVEN/BLISTERED	18.12
			ACCUMULATION OF LACQUER	18.12
			HOLES BLOCKED	18.12
OTHER TYPES OF SURFACE PROTECTION				18.13

Fig. 18.20. Reflowing/solder coating, roller tinning, tin plating, protective lacquering, and other types of surface protections. (Section of inspection record.)

they occur to a limited extent on conductors and ground/voltage planes, but not on solder pads. The largest acceptable extent is:

- For a gritty surface: 5 percent of the tin/lead surface
- For a grey surface: 10 percent of the tin/lead surface

If these limits are observed, the defect is rated as a B-defect; otherwise, the defect is rated as an A-defect.

18.9.2 Blue Plague

A very unusual fault is the so-called blue plague. As the name suggests, the tin/lead surface has a blue discoloration caused by refraction in a very thin layer of tin oxide. The layer thickness is said to be about 1000 angstroms (0.000004 in. or 0.1 μm).

The blue plague arises under certain conditions such as high humidity and temperature in conjunction with a clean, chemically active and tin-rich surface. Nevertheless, since it does not reduce the solderability of the board or cause corrosion, no remarks are entered in the inspection record, but the occurrence should be check-marked.

Although the black and white reproduction of the illustrations prevents showing an example of the blue plague, the inspector will have no doubts when he sees such a case.

18.9.3 Tin/Lead Alloy Composition

To a certain degree, the alloy composition can be assessed during visual inspection (Paragraph 8.3.3). In the case of solder-coated and hot-air leveled boards, the alloy composition is always 60/40.

If the board meets the requirements in Paragraph 18.9.1 regarding a bright and even surface with a fine-grained structure, the alloy composition will be around 60/40 as desired. A distinctive sign is that the freeze lines are very thin and nearly invisible.

If feather-shaped crystalline structures with a random orientation occur, the tin content is probably above 75 percent, and the PCB user can expect soldering problems during mass soldering.

If the surface is coarse-grained with wide and grey freeze lines and light-grey areas, the tin content is too low, which can also cause soldering problems during mass soldering.

The final decision regarding rejection implies a practical solderability test in the soldering machine as described in Chapter 23.

18.9.4 Coverage of Pattern Edges

Whether the inspector should inspect the board for tin/lead coverage of the pattern edges depends on the purchase specification.

When coverage is required, the inspection is performed most easily by observing the surface at a very low angle. A reddish tinge of copper indicates an insufficient edge coverage. In principle, failure to meet a possible requirement is considered an A-defect and the board is rejected. When considering a possible waiver, the inspector should refer to the general tendency towards not requiring full coverage (see Paragraph 8.5.1).

18.9.5 Slivers/Overhang

While inspecting the pattern edges for tin/lead coverage, the inspector should look carefully for visible traces of broken-off overhang (slivers). This also applies when solder and insulation masks are applied to the board (Paragraph 8.4).

A more thorough examination with a microscope is only performed when the visual inspection shows a divergent alloy composition (Paragraph 8.3.3) or various surface faults (Paragraph 8.3.4). In such cases, complete reflowing is difficult, and the risk of overhang is great.

Overhang which has not been melted down is cause for rejection and consequently rated as an A-defect. When solder and insulation masks have been applied, the overhang is locked in place. As shown in Figure 18.21, the overhang can easily be halfway broken loose before the mask is applied. The slivers (broken-off overhang) can come dangerously close

Fig. 18.21. Broken-off conductor overhang encapsulated in solder mask. The fault causes an effective short circuit.

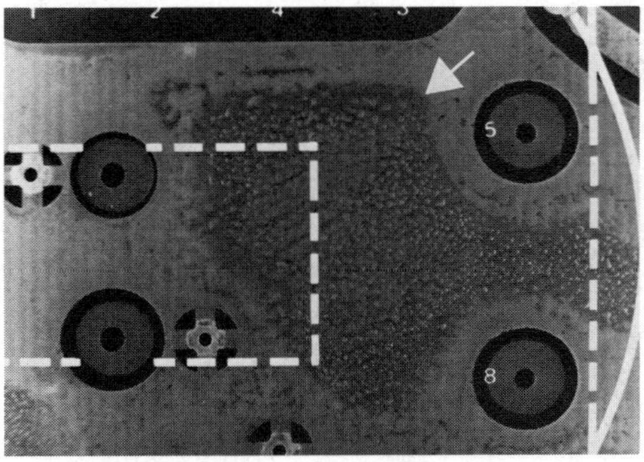

Fig. 18.22. Incomplete reflowing (melting) of large metal areas.

to other parts of the pattern and cause short circuits.

One obvious method of inspection is to study the board while it is lit from below—for example, by placing it on a light table. If short circuits (slivers) are present, the inspector will see them as dark lines in the translucent base material (epoxy glass).

Unfortunately, this method is useless if a ground plane on the other side of the board blocks all light transmission. The photo shown in Figure 18.21 was taken with bottom illumination and clearly shows two short circuits. With top illumination, the short circuits cannot be seen with the naked eye, and just barely with a microscope.

A more reliable method is to test bare boards with an automatic test equipment. Upon ascertaining that there are no short circuits caused by incomplete melting of the overhang, the inspector can waive the fault. If this is the case, the fault is rated as a B-defect.

It is assumed that the test voltage is at least three times the working voltage of the circuitry and never less than 40 volts. It is furthermore assumed that a 100 percent inspection is performed.

18.9.6 Degree of Reflowing

The inspector must also check whether the board has been reflowed over the entire surface. Incomplete reflowing is usually a warning that the PCB user will encounter mass soldering problems. It is cause for rejection and rated as an A-defect.

The inspector can issue a waiver for incomplete reflowing (melting) of larger metal areas such as voltage/ground planes. Solder pads which are thermally insulated from such metal areas must, however, be fully reflowed. In such cases, the fault is rated as a B-defect. See Paragraph 8.7.1 and Figure 18.22.

18.9.7 Wetting Problems

Another criterion of good solderability is the tin/lead wetting of the surface. See the discussion of the contact angle in Paragraph 8.7.2a.

Whether the board is tin/lead plated and reflowed, or solder-coated and hot-air leveled, nowetting is unacceptable. See the example in Figure 18.23A. Nonwetting is an indisputable cause for rejection and rated as an A-defect.

When dewetting occurs on solder pads and in solder holes, the fault is considered

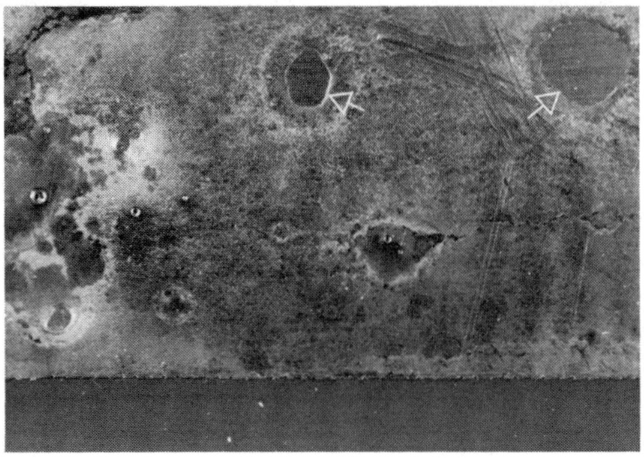

Fig. 18.23A. Example of nonwetting found on large metal area.

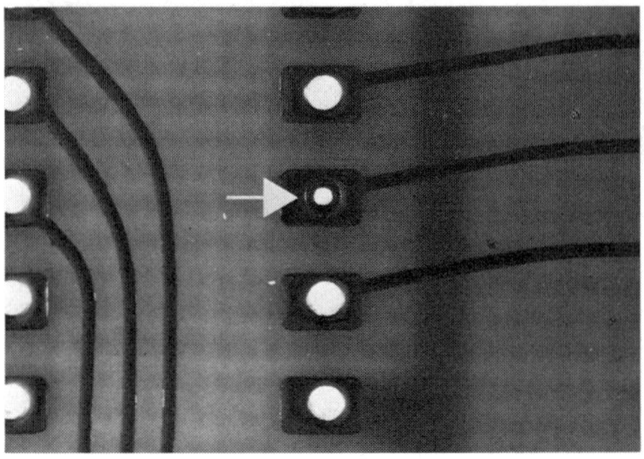

Fig. 18.24. Blocked hole caused by tin/lead which has run down into the hole during reflowing.

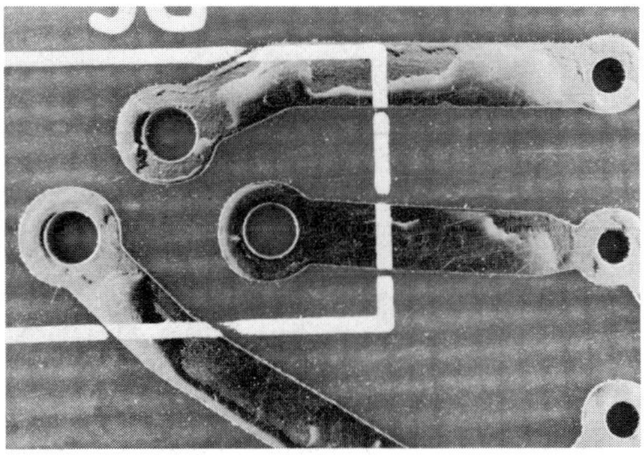

Fig. 18.23B. Example of dewetting found on conductor.

as an absolute cause for rejection and rated as an A-defect.

Dewetting occurring on 25 percent of the area of a conductor or a larger metal area can be accepted provided no solder joints are to be made at such places. See the example in Figure 18.23B. The defect is rated as a B-defect. Dewetting of more than 25 percent of the area is rated as an A-defect.

18.9.8 Blocked Solder Holes

Molten tin/lead running down into the solder holes from the surface of the solder pads, possibly also from the attached conductors, causes the holes to become fully or partially blocked. The root of the fault is usually excessive tin/lead plating. See Figure 18.24.

When the fault occurs sporadically, it is rated as a B-defect. Tin/lead blockage is easily removed by solder extraction methods. The whole batch should be checked for this fault unless the production department agrees to clean blocked holes during assembly. Widespread blockage is rated as an A-defect and the delivery is returned to the PCB manufacturer for repair.

18.9.9 Hot-Air Leveling Defects

The fact that solder coating with subsequent hot-air leveling is a relatively new process is a good reason to discuss some faults characteristic of this method.

18.9.9a Curtains on the Conductors

A not-too-unusual problem is the formation of curtains on the conductors (Figure 18.25A). The appearance very much resembles dewetting, but the board probably can be soldered satisfactorily.

As a guide when assessing the surface, the inspector should note that in the case of curtains, the contact angle of the tin/lead is smaller than in the case of dewetting. These curtains, mostly found on

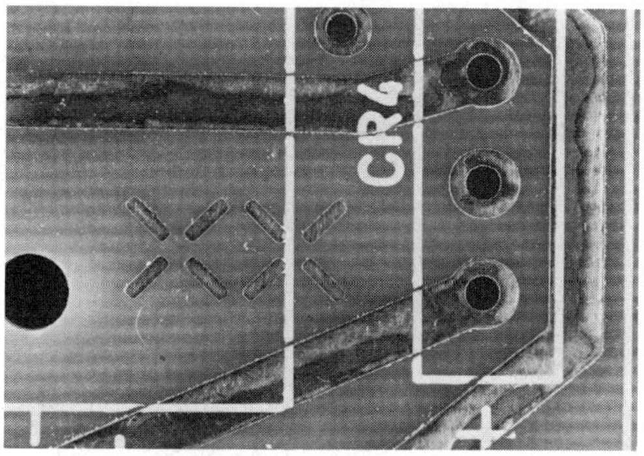

Fig. 18.25A. "Curtains" found on the conductors of solder-coated and hot-air leveled board.

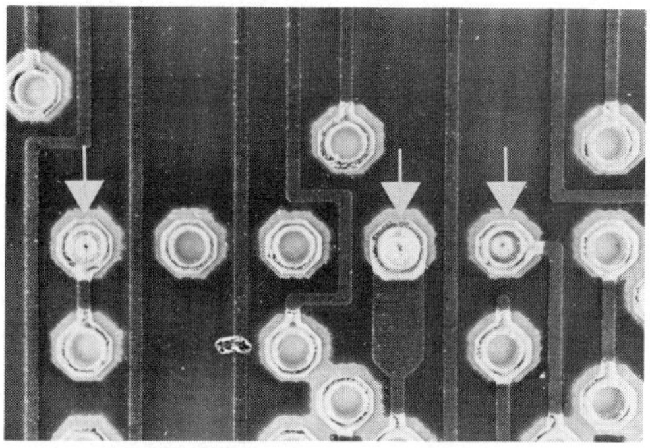

Fig. 18.25B. Blocked holes in solder-coated and hot-air leveled board.

wide conductors, are caused by the air jet from the air knives. No remarks are entered in the inspection record.

18.9.9b Blocked Solder Holes

Particularly during hot-air leveling, blocked holes can occur in a band across the board. See Figure 18.25B. This is usually due to partial clogging of the air-knives. Paragraph 18.9.8 describes assessment of this fault.

18.9.9c Webbing Caused by Surface Treatment

When the laminate surface (butter coat) is slightly roughened by excessive chemical action or mechanical scrubbing, web-

Fig. 18.25C. Fine-threaded webbing found on solder-coated and hot-air leveled boards.

bing can occur. This form of webbing is frequently so fine-threaded or occurs as such small drops that it can hardly be seen with the naked eye. See Figure 18.25C. The inspector should therefore use a 7X magnifier. Webbing is an indisputable cause of rejection and consequently rated as an A-defect.

18.9.9d Webbing Caused by Incomplete Curing

Webbing can also be due to incomplete curing of the solder mask so that the tin/lead adheres to the surface in spite of the action of the air-knives. See the glaring example shown in Figure 18.25D.

It should be noted that the tendency to

Fig. 18.25D. Widespread webbing found on solder-coated and hot-air leveled boards.

Fig. 18.25E. Nonwetting and dewetting found on solder-coated and hot-air leveled boards.

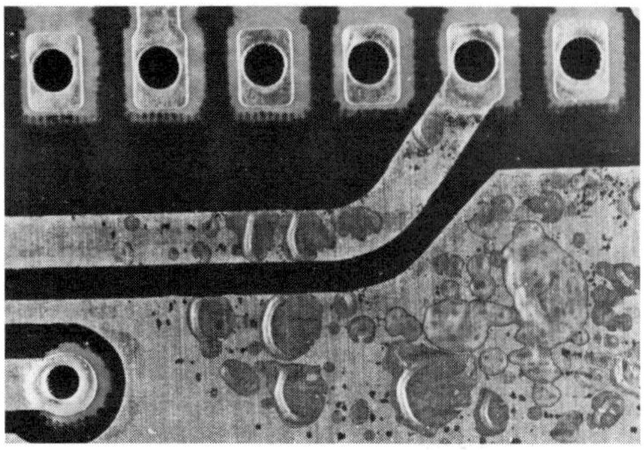

Fig. 18.25G. Blistering of solder mask as a result of impurities on copper surface of solder-coated and hot-air leveled boards.

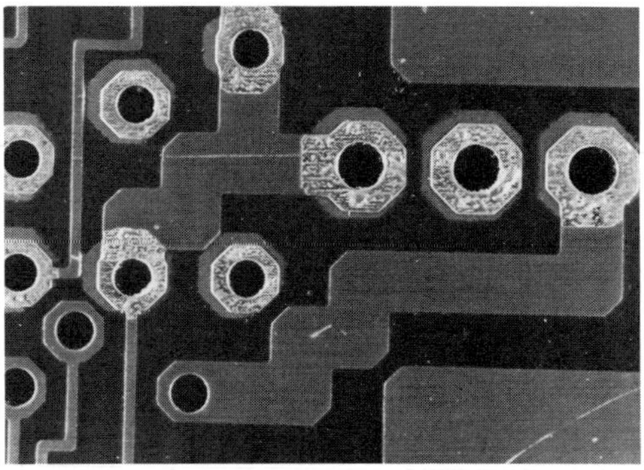

Fig. 18.25F. Dewetting found on solder pads as a result of washing off an erroneously screen-printed solder mask.

webbing appears already during manufacturing the board, whereas webbing in the case of an ordinary tin/lead plated and reflowed board does not appear until mass soldering. Webbing is an indisputable cause for rejection and rated as an A-defect.

18.9.9e Wetting Problems

Nonwetting and dewetting are caused by a contaminated copper surface during solder coating. Figure 18.25E shows examples of both faults.

Paragraph 11.5.5 mentioned that an erroneously screen-printed solder mask should never be washed off in an attempt to salvage the board.

A solder-coated board is frequently only copper plated (Paragraph 9.5), and because the copper surface is uneven, it is practically impossible to completely remove the epoxy ink, even with the solvent used for two-pack epoxy inks.

Tiny remnants in the pores and general unevenness cause dewetting when the board is soldered. See the example in Figure 18.25F. Paragraph 18.9.7 gives assessment criteria for nonwetting and dewetting.

18.9.9f Contamination of Copper Surface

Impurities (such as hand lotion) on the copper surface and encapsulated under the solder mask can cause the mask to blister or wrinkle during the solder-coating process. See Figure 18.25G.

In the case of ordinary tin/lead plated boards, it is impossible to distinguish a similar fault from the usual wrinkling of the solder mask, but if the blistering occurs when solder-coating the boards, the fault lies with the PCB manufacturer.

If the PCB user accepts a wrinkled solder mask on top of tin/lead plated conductors, it seems a little irrational to rate blistering or wrinkling of a solder mask on solder-coated boards worse than a B-

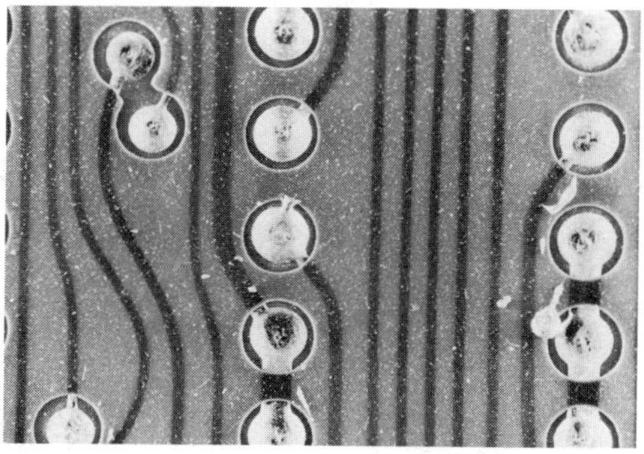

Fig. 18.25H. Formation of drops on surface of solder-coated and hot-air leveled boards.

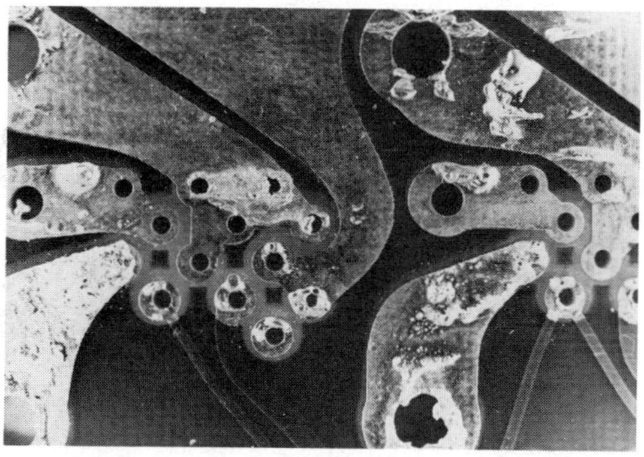

Fig. 18.26. Roller tinned boards with a very uneven surface.

defect, assuming that the fault occurs locally and to a modest extent.

When the fault is widespread, it is rated as an A-defect and the board is rejected. The inspection record reference is 18.14.4d (Air Bubbles in the Mask).

18.9.9g Solder Drops/Solder Balls

Solder drops or balls on the surface as shown in Figure 18.25H are caused by evaporated flux mixed with solder. See Paragraph 9.4.2d. Since there is a high probability of short circuits, the fault is regarded as cause for rejection and consequently rated as an A-defect.

18.10 Roller Tinning

When the board is roller tinned, the inspector should apply the following criteria:

18.10.1 Surface Requirements

The appearance of the surface is inspected according to the same guidelines as described in Paragraph 18.9.1. Figure 18.26 shows an example of an uneven surface.

18.10.2 Solder Drops

If an edge connector is specified as roller tinned, no solder drops are allowed on the contact area which is inserted in the receptacle.

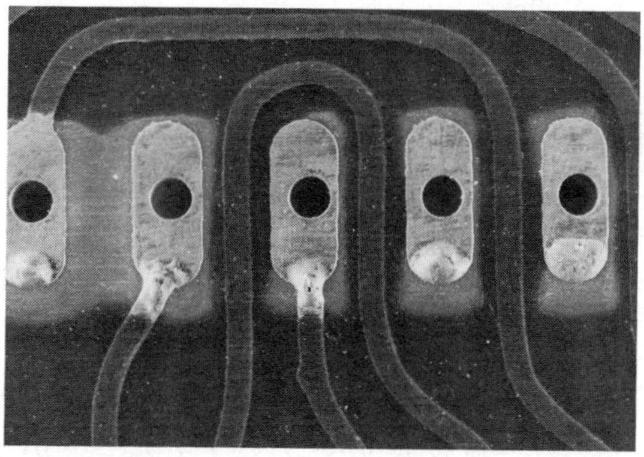

Fig. 18.27. Roller tinned board with formation of drops on the solder pads.

Drops at such places are rated as an A-defect. The boards are returned for a renewed roller tinning, preferably in the opposite direction. Figure 18.27 shows an example of drop formation.

18.10.3 Closed Component Holes

Closed component holes caused by roller tinning are accepted as a B-defect, provided there are no more than two closed holes per 16 sq. in. (100 cm^2) of board area.

A larger incidence of closed holes is rated as an A-defect. The boards are returned or repaired in the production de-

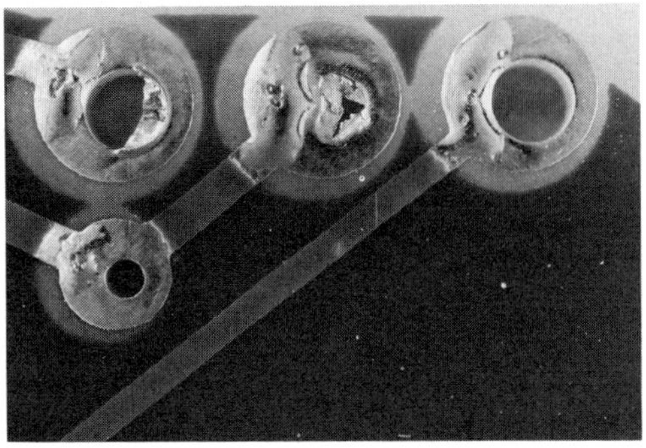

Fig. 18.28. Roller tinned board with closed component holes.

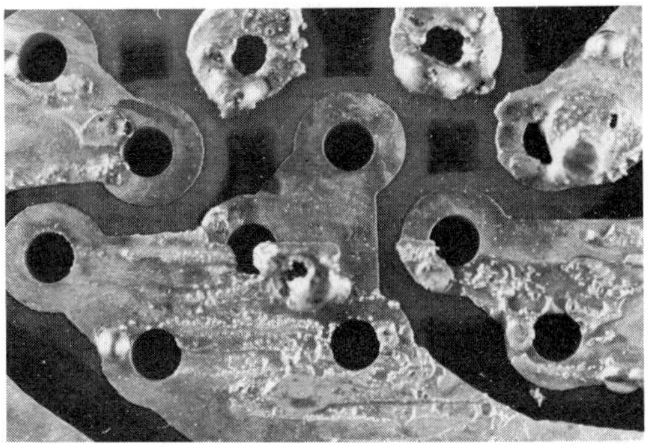

Fig. 18.29. Roller tinned board exhibiting nonwetting.

partment by piercing the closed holes with an awl. See Figure 18.28.

18.10.4 Execution of Hole Drilling

The inspector should also check whether the hole drilling has been performed as required in the purchase specification—that is, drilling before or after the roller tinning. He does this by checking whether there is tin/lead on the internal copper edge of the solder pad. The inspector should use his 7X or 10X magnifier.

Noncompliance with this requirement is, in principle, cause for rejection and therefore an A-defect. However, this applies only where the hole drilling is performed after roller tinning the board, so that bare and oxidized copper is present at the hole edge.

If experience shows that the type of solder flux used prevents soldering problems, the inspector can waive the fault. Alternatively, he can perform a solderability test. See Paragraph 9.6.4.

18.10.5 Coverage of Pattern Edges

Coverage of the pattern edges is treated in the same way as described in Paragraph 18.10.4. See also Paragraph 9.6.5.

18.10.6 Dross on Surface

No dross, caused by roller tinning, should be found on the surface. Such occurrence is cause for rejection and consequently rated as an A-defect. The boards are returned for cleaning.

18.10.7 Wetting Problems

A visual inspection can give some clues to the solderability of the board. See Paragraph 18.9.7. Figure 18.29 shows an example of nonwetting.

18.11 Tin Plated Boards

As mentioned in Paragraph 9.8.1, tin plating is little used today, so the inspection of this type of surface will not be described in detail. If such an occasion arises, the inspector should be aware of the following criteria:

18.11.1 Slivers/Overhang

As described in Paragraph 18.9.5, the inspector should verify that no overhang, fully or partially detached as slivers, is present. By passing a thin needle along the pattern edges, he can easily see whether the overhang in question is large enough to break off.

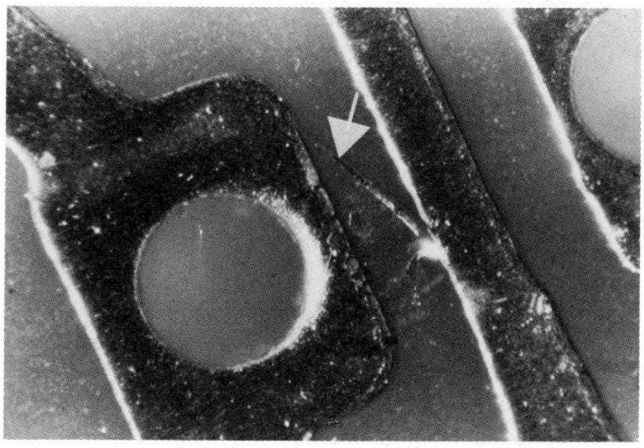

Fig. 18.30. Tin plated board with partly detached overhang.

The production method implies some overhang, but the processes must be controlled to prevent dangerous overhang; in other words, the overhang must be difficult or impossible to break off. If not, the board is rejected and the fault rated as an A-defect. Figure 18.30 shows an example of a tin plated board with partially detached overhang along a conductor.

18.11.2 Surface Appearance

The surface appearance gives no reliable indication of the solderability. As mentioned several times in Part 2, it is easy to plate tin (just as tin/lead) on top of nonsolderable areas. This inevitably causes nonwetting during mass soldering.

Unless the inspector is familiar with the PCB manufacturer's record regarding satisfactory solderability, he should draw a board from the sample and perform a solderability test as described in Chapter 23.

18.12 Protective Lacquering

The inspector should assess whether the applied layer of protective lacquer is tolerably even without flecks, blisters, thick accumulations of lacquer, or blocked holes. The examination is easiest to perform with the board tilted a little, so that the light is reflected from the surface.

It is difficult to give unambiguous criteria for rejection other than to state that absence of lacquer coverage or visibly thick accumulation of lacquer can cause assembly problems. Such faults are rated as an A-defect.

A visual inspection cannot reveal whether the board solderability is satisfactory. The protective lacquer can be a type which polymerizes strongly in time. This implies that it is difficult for the flux to dissolve the lacquer within the short time available during mass soldering, preventing the flux from cleaning the copper surface sufficiently. The result is inferior solderability.

Neither can the inspector tell by visual inspection whether any oxidation and contamination of the copper surface has been removed by suitable chemical action immediately before applying the protective lacquer.

When the inspector has little practical experience with the solderability to be expected from the PCB manufacturer in question, the only reliable solution is to test-solder a board. Chapter 23 describes the methods and criteria of such an assessment.

It should be noted that some methods of pretreatment and certain types of protective lacquers only preserve the solderability of the boards for perhaps a month, while the normal storage time may be 3 months or more. A solderability test can therefore be misleading if the boards are stored for a month or more after the solderability test.

18.13 Other Types of Surface Protection

For the types of surface protection mentioned below, visual inspection cannot reliably reveal the board's solderability. In the case of boards with abnormal surface appearance—for example, with a faded, matte, or gritty surface—the inspector can solder-test a single board. The types of surface protection in question are:

Gold plating of the entire pattern

Tin/nickel plating, possibly with a thin gold plating on top

Immersion tin

These types of surface protection will not be discussed further.

18.14 Solder and Insulation Masks

The visual inspection also includes the solder and insulation masks, which can be either screen printed or photo printed (photopolymer masks). The inspector should realize that screen-printed masks generally should be inspected more thoroughly than photopolymer masks because of the greater risk of faults.

In Paragraphs 18.14.1 through 18.14.3 below, we address three main types of faults; the first usually causes the latter two. Therefore, the inspector should only enter the number of A- or B-defects caused by the primary fault. In order to give the PCB manufacturer the best possible information, resulting secondary faults can be indicated by check marks.

When inspecting the boards, it is most practical for the inspector to follow the part of the inspection record shown in Figure 18.31, which also includes the component notation discussed in Paragraph 18.15.

SOLDER AND INSULATION MASKS				18.14
			MASK MISREGISTERED	18.14.1
			OVERPRINTING OF PADS O.O.T.	18.14.2
			COVERAGE INCOMPLETE	18.14.3
			MASK APPERTURES BLURRED	18.14.4a
			METAL PARTICLES IN MASK	18.14.4b
			OTHER PARTICLES IN MASK	18.14.4c
			AIR BUBBLES/UNEVENNESS	18.14.4d
			SKIPPED AREAS IN MASK	18.14.4e
			MASK ADHESION INSUFFICIENT	18.14.4f
			CRACKS IN DRY FILM MASK	18.14.4g
			CURING INCOMPLETE	18.14.4h
			MASK THICKNESS TOO LOW	18.14.5
COMPONENT NOTATION				18.15
			LEGIBILITY UNSATISFACTORY	18.15.1
			ANNOTATION MISREGISTERED	18.15.2
			PRINTING INK IN HOLES	18.15.2
			ANNOTATION DISTORTED	18.15.3
			COLOR INCORRECT	18.15.4

Fig. 18.31. Solder and insulation masks and component notation. (Section of inspection record.)

18.14.1 Misregistration of Mask

The most common fault is that the mask is displaced with respect to the pattern. The displacement can be measured as described in Paragraph 11.4.2 and shown in Figure 11.7. The fault can show up in the following three variants:

a. A misregistration, which means that in principle the mask is displaced by the same amount over the entire board with respect to the pattern (Paragraph 11.3.1)
b. A displacement which varies in extent over the board with respect to the pattern (Paragraph 11.3.2)
c. A combination of a and b

In the case of screen-printed masks, the inspector views the board and notes where the largest displacements between the solder pads and the mask apertures occur. This will frequently be in the outer zones of the board. The largest displacement is then measured.

When the displacement is less than 0.010 in. (0.25 mm), no remarks are entered in the inspection record. Displacements up to 0.016 in. (0.40 mm) are rated as a B-defect, whereas larger displacements are cause for rejection and consequently rated as an A-defect. See the examples in Figure 18.32.

A waiver will, in all probability, cause a certain amount of repair work after mass soldering. The inspector is therefore rec-

Fig. 18.32A. Screen-printed solder mask exhibiting satisfactory register. Conductors between IC solder pads are covered by the solder mask.

Fig. 18.32B. Screen-printed solder mask with a small and fully acceptable misregistration.

Fig. 18.32C. Screen-printed solder mask with a large and unacceptable misregistration. The solder pads exhibit overlapping, and one side of the conductors is uncovered.

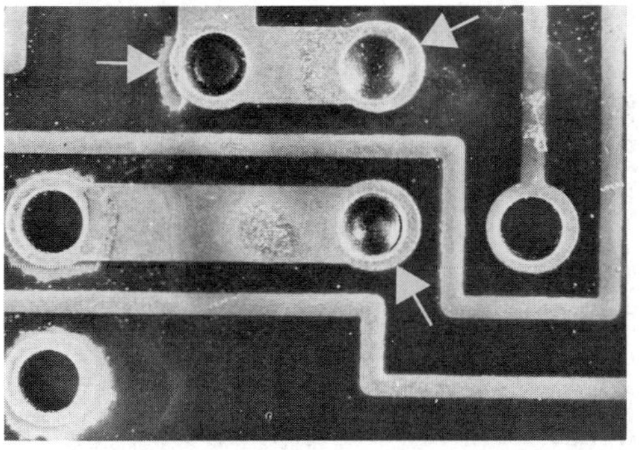

Fig. 18.33. Board with three solder holes filled with screen printing ink.

ommended to consult the assembly department before granting a waiver. It is a natural precondition that no screen printing ink has run down into the solder holes. See Figure 18.33.

It is easy to inspect a photopolymer mask because the fault variants a, b, and c occur very seldom. The inspector has only to ensure that the mask apertures are located concentrically with respect to the solder pads. See Figure 18.34.

A displacement up to 0.004 in. (0.10 mm) is accepted with no remarks entered in the inspection record. Since the photopolymer mask, in all probability, has been chosen in order to achieve a good

Fig. 18.34A. Dry film solder mask exhibiting good registration. Note the coverage of the conductors between the solder pads.

Fig. 18.34B. Liquid-film solder mask exhibiting good registration. Note the coverage of the conductors between the solder pads.

Fig. 18.35. Overprinting of solder pads with solder mask.

protection against solder bridging, displacements larger than 0.004 in. (0.10 mm) are rated as an A-defect.

The inspector should realize that the permanent dimensional change of the board can amount to ±0.05 percent as a worst-case value. In the case of very large boards, a dimensional change in this order of magnitude will highly affect the registration of the masks. A possible exemption should be followed up by negotiations with the PCB manufacturer regarding the use of a stabilized base material. Responsibility for the misregistration lies with the PCB manufacturer when he has expressed no reservations concerning permanent dimensional changes.

18.14.2 Overprinting of Pads

Another effect of the aforementioned main fault can be an unintended coverage of the solder pads, usually combined with misregistration and/or displacements. The inspector must therefore be familiar with the requirements of the purchase specification. He should also realize that intentional coverage of via holes can occur.

Unintentional overprinting (coverage) of solder pads occurs particularly in conjunction with screen-printed masks. Some purchase specifications state that overprinting is a direct cause for rejection and rated as an A-defect, whereas in other ones, overprinting is allowed. A limitation can, for example, be that overprinting is accepted only to the middle of the annular rings, and only valid for 25 percent of the solder pads. If so, no remarks should be entered in the inspection record. See the example in Figure 18.35.

When the purchase specification makes no reservation regarding overprinting, the inspector faces a serious problem. He can insist on full solderability of the board, including the requirement that mass soldering completely fill the solder holes.

In any case, these requirements can be interpreted in such a way that the solder mask is not allowed to reach the hole edge. This would imply a great risk of printing ink flowing down into the holes (Paragraph 11.5).

With photopolymer masks, the risk of unintentional coverage (overlapping) of the solder pads is quite small, insofar as the nominal clearance is at least 0.004 in. (0.1 mm). See Paragraph 12.6.

The inspector can quickly inspect the outer zones of the board for overlapping. As photopolymer masks cannot flow down into the solder holes, the inspector should take a functional attitude toward overlap-

Fig. 18.36. Apertures in dry-film solder masks "shrink" a little.

Fig. 18.37. Unacceptable misregistration of screen-printed solder mask. The solder pads are overprinted more than halfway towards the solder hole, and one side of the conductors are fully uncovered where they pass by the solder pads.

ping—for example, by accepting overlapping to the middle of the annular ring of the solder pads.

This acceptance is particularly necessary when the designed nominal clearance is zero, because the mask apertures shrink slightly. See Figure 18.36 and Paragraph 12.3.1, which deals with special methods of enlarging the size of the mask apertures in the solder mask film.

As a matter of fact, it will be extremely rare that an inspector is compelled to rate overlapping of a dry film solder mask higher than a B-defect.

18.14.3 Coverage of Pattern

Another effect of the previously mentioned main fault could be incomplete coverage of conductors running very close to solder pads, or conductors running between pairs of IC pads. In practice, this fault occurs only when the mask is screen-printed (Paragraph 11.4.1), where a displacement or misregistration of 0.01 in. (0.25 mm) is nearly unavoidable. See Figure 18.37.

When the conductor is 0.012 in. (0.30 mm) or more away from the edge of the mask aperture (as measured on the master film), it should be covered. In high-density boards it is extremely difficult to obtain this distance, and the PCB manufacturer cannot be blamed in the case of incomplete coverage.

Unfortunately, it is quite common for a PCB manufacturer to be criticized for missing coverage of a conductor side as mentioned above. The inspector, and not least the PCB designer, must learn that screen printing of such solder masks is actually about the limit of what is possible, and screen printing qualities will always fluctuate somewhat.

18.14.4 Appearance of Mask

The inspector also has to make sure that the following quality requirements are met. Paragraph 11.8.1 gives a detailed description.

18.14.4a Mask Definition

All edges and mask apertures must be sharply defined without any noticeable blur. The mask apertures must have the intended shape and size. See Figure 18.38. The assessment, however, is of a quite subjective nature, and the decisive criterion is whether a risk of solder bridging exists. See Paragraph 18.14.3 above.

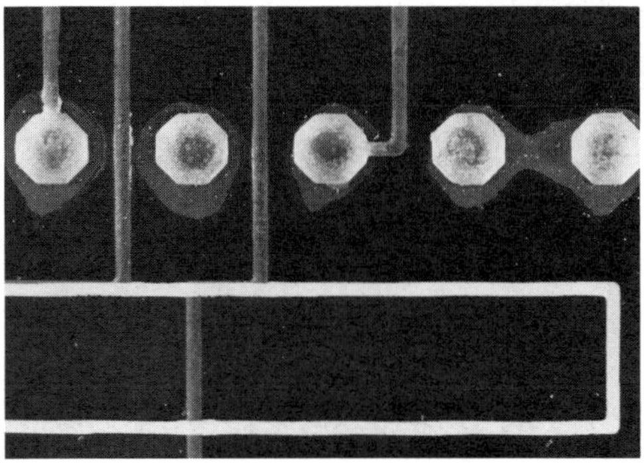

Fig. 18.38. Distorted solder mask apertures. The risk of solder bridging is, however, quite low.

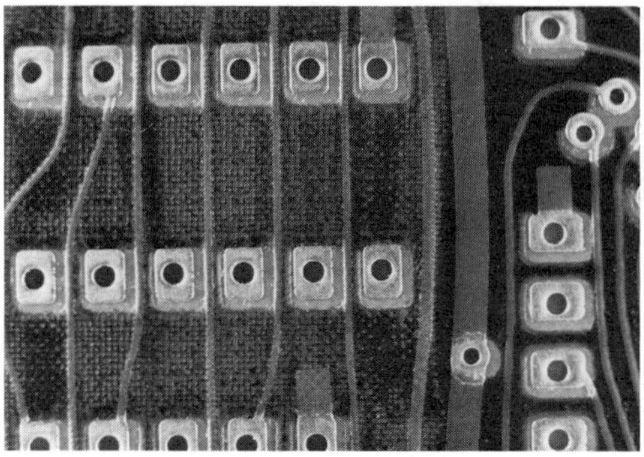

Fig. 18.40A. Insufficient application of screen-printing ink so that the screen weave is marked in the mask surface.

18.14.4b Metal Particles in the Mask

Metal particles included in the mask should only occur to a very limited extent—for example, one particle per 16 sq. in. (100 cm^2). Exceeding this value is rated as an A-defect.

18.14.4c Foreign Particles in the Mask

Nonmetallic foreign particles included in the mask (such as a textile thread) are accepted with no remarks entered in the inspection record, provided the occurrence is limited to a single particle. See Figure 18.39.

Fig. 18.40B. Close-up of the solder mask shown in Fig. 18.40A.

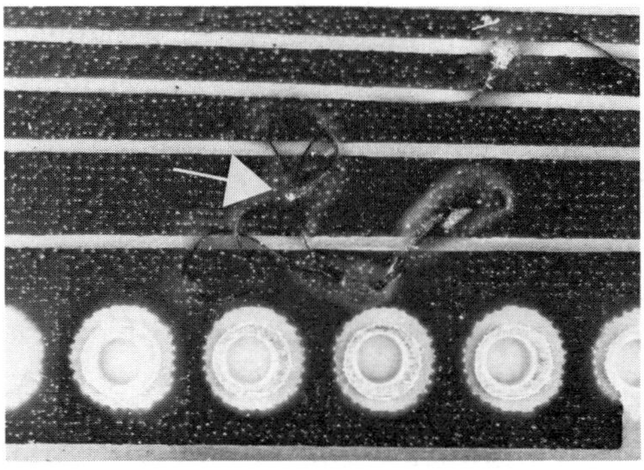

Fig. 18.39. Inclusion of textile thread in screen printed solder mask.

18.14.4d Air Bubbles in the Mask

The surface of the mask must be even and without large air bubbles. Pinholes are normally no significant problem and need no entry in the inspection record.

In rare cases the surface of the mask is not closed because the printing ink was applied too sparsely. This shows up in a clear visual impression of the screen weave, possibly in a band across the board. See Figure 18.40. This fault is an obvious cause for rejection and rated as an A-defect. See Paragraph 11.8.1c for assessment criteria.

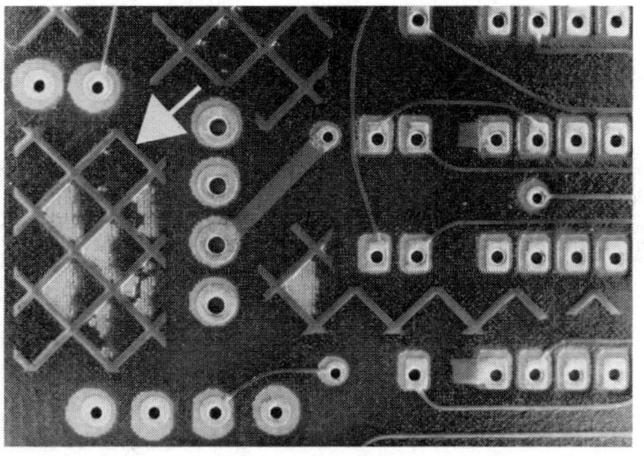

Fig. 18.41. Unintentional holes in screen printed solder mask can be accepted in nonfunctional areas.

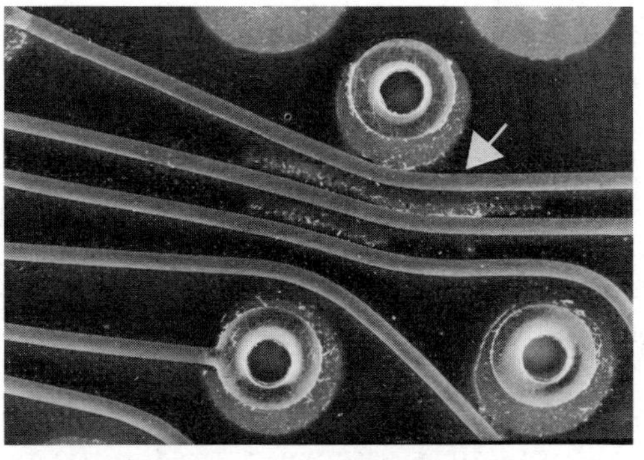

Fig. 18.42. Lacking coverage with solder mask between conductors running in parallel.

18.14.4e Skipped Areas in the Mask

Unintentional openings or holes (skipped areas) in the mask can be accepted provided they occur in nonfunctional areas. No remarks are entered in the inspection record. See Figure 18.41.

In functional areas unintentional openings and holes can be accepted if they occur to a moderate degree—for example, 2 faults per 16 sq. in. (100 cm^2)—with little risk of solder bridging during mass soldering. The fault is rated as a B-defect. Figure 18.42 shows that it can be difficult to get the printing ink right down between parallel conductors to the base material.

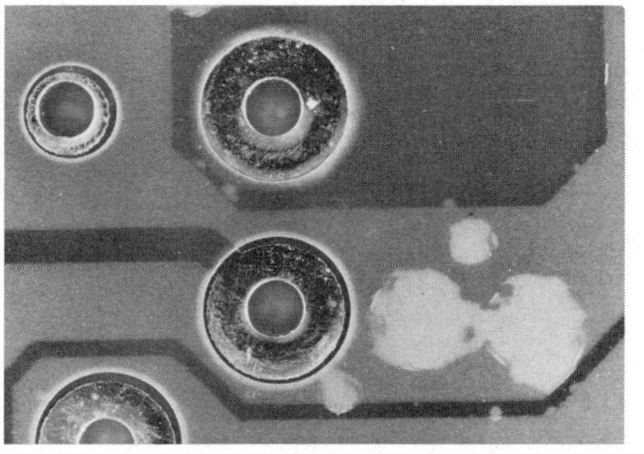

Fig. 18.43. Reduced adhesion of dry-film solder mask is indicated by lighter areas in the mask.

Unintentional openings and holes in greater numbers are rated as an A-defect, but the inspector can always issue a waiver. It is, however, possible to return the boards for reprinting of the solder and insulation masks.

18.14.4f Adhesion of the Mask

Light areas in the mask can indicate reduced adhesion. The light color is usually caused by air pockets beneath the mask. See Figure 18.43. The board is therefore subjected to a tape test as described in Paragraph 19.3.

18.14.4g Cracks in Photopolymer Masks

Photopolymer masks are inspected for tendency to crack on top of the conductors, particularly along the conductor edges. The reason is probably that the dry film is too thin compared to the height of the pattern. Cracking is cause for rejection and the fault is rated as an A-defect.

18.14.4h Curing of the Mask

The mask must be fully cured and in no way soft or sticky. Although the inspection is visual, the inspector can easily perform a pencil test on the curing as

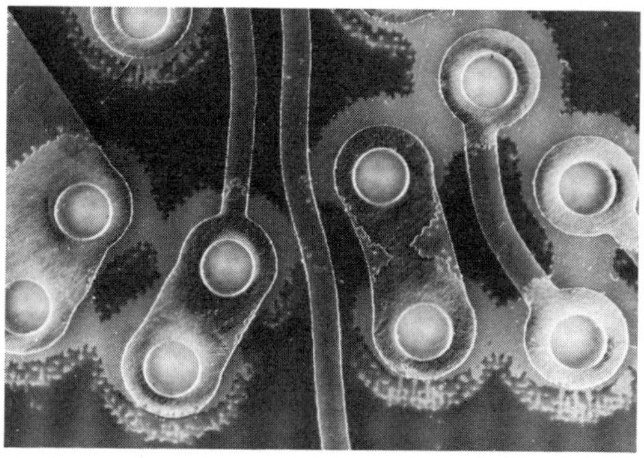

Fig. 18.44. Thin screen printing of a solder mask is clearly seen at the areas around the solder pads.

Fig. 18.45. Thick screen printing of a solder mask is indicated by a steep and well-defined edge of the mask apertures.

described in Paragraph 11.8.3. Incomplete curing is cause for rejection, and the fault is rated as an A-defect.

18.14.5 Thickness of the Mask

Naturally, it is impossible to measure the thickness of a screen-printed mask during the visual inspection, but the appearance of the mask immediately around the solder pads can give a certain indication.

Insufficient thickness is usually characterized by a fairly undefined edge of the apertures—that is, the edge is uneven and very flat—frequently with shadows behind the edge in the form of more or less uncovered areas. The printing conveys an obvious impression of being thin. See Figure 18.44. In contrast, a steep and distinctly defined edge of the mask aperture gives a clear impression of a thick and ample screen printing. See Figure 18.45.

Suspicion of an overly thin application can be confirmed or disproven only by a direct measurement of the mask thickness as mentioned in Paragraph 11.6.3.

Approval of the thickness depends on the requirement of the PCB specification. The thickness, however, is not always specified numerically, but based on the vague wording that a solder mask applied to a tin/lead surface must be able to withstand mass soldering. This implies a solder test.

The inspector as well as the PCB user must realize that wrinkling of the solder mask is unavoidable. The degree of wrinkling depends on the width of the underlying conductors. See the considerations in Paragraph 11.6 and see Figure 18.46, which shows various degrees of mask wrinkling.

A practical assessment method is to perform a visual inspection of the mask application around the solder pads, as in the above description, and compare the results to past experience with mask adherence and wrinkling, possibly supplemented by a solder test.

Inadequate mask thickness is cause for

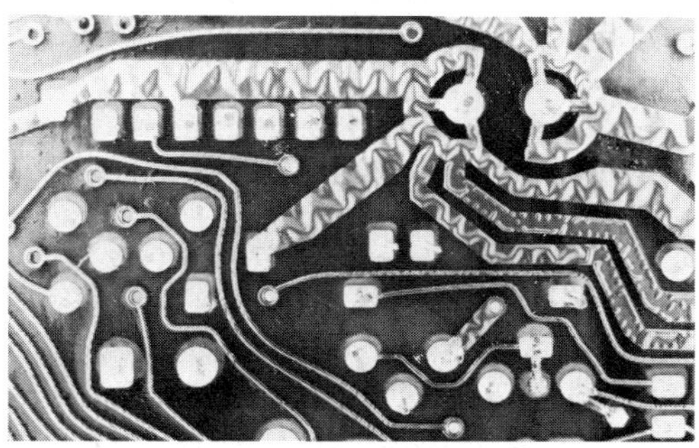

Fig. 18.46. Wrinkling of solder mask after mass soldering of tin/lead plated board.

rejection and consequently rated as an A-defect. However, the problem is not critical as the boards can be returned for reprinting.

18.14.6 Defective Tenting of Via Holes

When solder-coated and hot-air leveled boards have via holes which are sealed (tented) with dry film in the form of solder and insulation masks, the inspector should inspect these holes very carefully. The reason is that the sealing of a few holes can be either missing or defective as a consequence of applying the masks under heat and vacuum.

Sealing which is broken on one side only will make it difficult to have the unprotected copper in such via holes covered with solder. The holes will not be soldered very well and they will not be soldered at all when the sealing is missing on the component side. Finally, the sealing of some of the holes can easily be punctured during the later handling when the board is assembled and tested.

Because of the considerable risk of corrosion of the copper plating of such unprotected via holes, the board is considered rejectable and consequently the fault is rated as an A-defect.

Limited occurrence of this fault can be exempted if the production department agrees to perform a repair. The best method seems to be closing the open holes with epoxy ink. Closure by manual post-soldering can strain the copper plating so much that barrel cracking occurs, particularly when the soldering time is extended because the copper has oxidized. Soldering also implies that the sealing on the opposite side will be broken. Even if the inspector decides to exempt the defect, it should be rated as an A-defect.

For these reasons, the PCB designer is advised to avoid tenting via holes with dry-film solder and insulation masks on solder-coated and hot-air leveled boards.

18.14.7 Defective Tenting of Plated-Through Holes

During imaging of the solder and insulation masks, unwanted light reflections can occur at some of the plated-through holes. The dry film tenting the holes at this stage can be polymerized locally so that small flecks of dry film are found in some of the plated-through holes of the finished boards.

Moderate occurrence—for example, in 5 percent of the holes—is rated as a B-defect, whereas a higher percentage is rated as an A-defect and the board is rejected.

18.15 Component Notation

The requirements for component notations are usually less exacting than for solder and component masks. Primarily, the inspector should note whether it is possible to reliably identify the components. The functional requirements are:

18.15.1 Legibility

The component designations must be legible, which means a high definition and no blurring (Figure 18.47). It will, however, be necessary to accept a certain amount of blurring when the notation is printed on top of conductors.

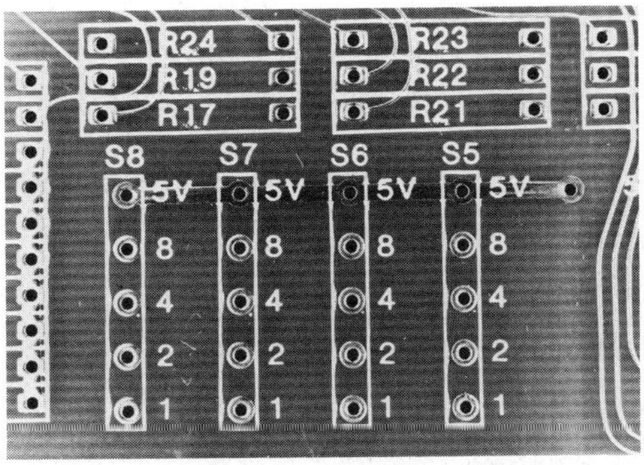

Fig. 18.47. Distinct component notation.

Fig. 18.48. Indistinct component notation.

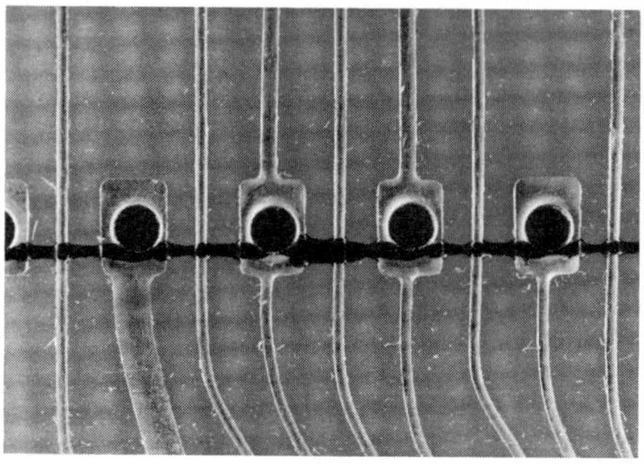

Fig. 18.49. Displacement of component notation.

The inspector tests the legibility by checking whether he can distinguish the individual figures of the following pairs from each other:

3 and 8, 5 and 6, 7 and 9

Reduced legibility is rated as a B-defect. See the example in Figure 18.48. It is here assumed that the character height is at least 0.06 in. (1.5 mm) and the line thickness at least 0.012 in. (0.3 mm).

18.15.2 Misregistration

Even a fairly large misregistration, amounting to 0.016 or 0.020 in. (0.4 or 0.5 mm), usually causes no identification problems regarding which holes to use during assembly. The misregistration is therefore accepted with no remarks entered in the inspection record.

Misregistration, however, must not cause the printing ink to flow down into the solder holes. See Figure 18.49, which shows a displacement of an IC outline right to the edge of the solder holes.

Extreme cases of printing ink in plated-through holes are rated as an A-defect. Ink in nonplated-through holes is rated as a B-defect, provided that it does not interfere with insertion of component leads.

18.15.3 Distortion

Contour lines and leader lines must not be distorted—for example, by being unintentionally curved. The inspector should assume a functional attitude, which means that he enters no remarks in the inspection record. In extreme cases, the fault is rated as a B-defect.

18.15.4 Incorrect Color

If the component notation has been printed in the wrong color, the fault is rated as a B-defect. The fault is not especially rare, particularly if a nontraditional color (other than black or white) is specified.

18.16 Edge Connectors

The initial inspection of edge connectors is a visual inspection of the surface, which yields much information about the quality. The inspector should follow that part of the inspection record shown in Figure 18.50. See also Chapter 14.

Since an edge connector represents the board's electrical connection to its surroundings, the full sample is inspected according to General Inspection Level II. In order to best study the details, the in-

			EDGE CONNECTORS	18.16
			SURFACE UNEVEN	18.16.1
			APPEARANCE UNSATISFACTORY	18.16.2
			SCRATCHES	18.16.3
			DEPRESSIONS	18.16.4
			PINHOLES	18.16.5
			NODULES/BLISTERS	18.16.6
			BURNING	18.16.7
			GOLD COVERAGE INSUFFICIENT	18.16.8
			EDGES NOT COVERED	18.16.9
			TRANSITION INCORRECTLY MADE	18.16.10
			TRANSITION ZONE DISLOCATED	18.16.11
			ADJOINING AREAS DAMAGED	18.16.12
			EDGE UNEVENNESS O.O.T.	18.16.13
			SHORT CIRCUITS	18.16.14
			PROJECTIONS O.O.T.	18.16.15
			INDENTATIONS O.O.T.	18.16.16
			LIFTING OF TABS/BURRS	18.16.17
			CRACKS IN GOLD LAYER	18.16.18
			PRINTING INK ON TABS	18.16.19

Fig. 18.50. Edge Connectors. (Section of inspection record.)

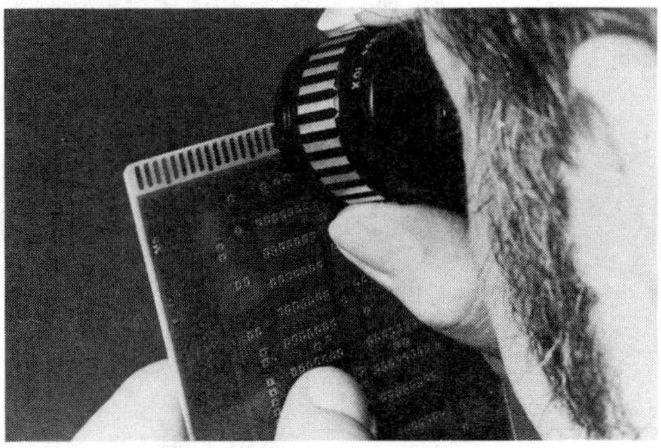

Fig. 18.51. Inspection of edge connector by means of a magnifier. The board is held in a slanting manner against the light so that the best possible light reflection is achieved.

spector should use a 7X or 10X magnifier (Figure 18.51). He should tilt the board so that the light is reflected by the edge connector. The inspection includes the following guidelines, which should be supplemented by a tape test as described in Paragraph 19.3.

18.16.1 Surface

The surface must be smooth and even with no visible traces of grinding in the sub-

Fig. 18.52A. Visible grinding traces in the contact surface.

Fig. 18.52B. Very uneven contact surface.

structure. This inspection is performed without the magnifier. Visible traces of grinding as shown in Figure 18.52 are cause for rejection, and the fault is rated as an A-defect.

18.16.2 Discoloration

The surface must be reasonably bright (Paragraph 14.2.2). Discoloration of the contact tabs (yellowish brown to reddish brown) must not occur, either partially or over the entire tab. Discolored contact tabs are cause for rejection and rated as an A-defect. See Figure 18.53.

18.16.3 Scratches

In principle, the surface must be free of scratches, although a very light and su-

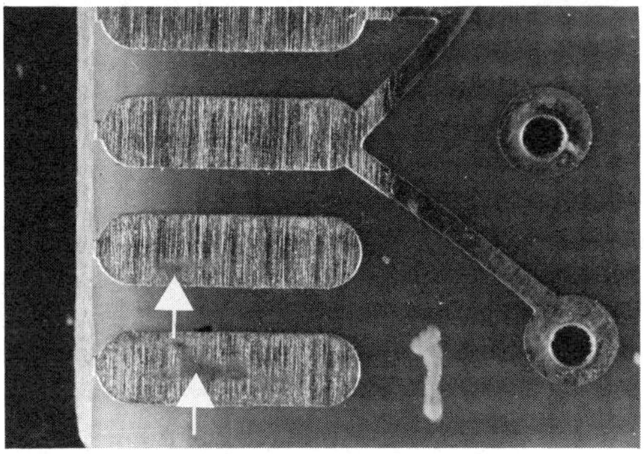

Fig. 18.53. Discoloration of contact surface.

Fig. 18.55. Scratch in contact surface with uncovered copper layer.

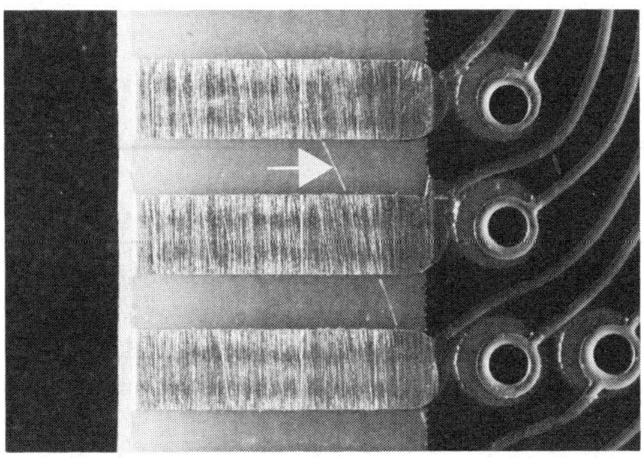

Fig. 18.54. A very light and superficial scratch in the contact surface. The gold layer is intact and protects the underlying copper. This was checked by means of a gas exposure test with hydrogen sulphide, cf. Para. 27.2.5.

Fig. 18.56. Sharp-edged depressions in contact surface.

perficial scratching can be accepted with no remarks entered in the inspection record. See Figure 18.54. Heavy scratching is always cause for rejection and rated as an A-defect, whether the scratch is embossed in the tab with the gold layer preserved or the underlying metal is exposed. See Figure 18.55.

18.16.4 Depressions

Sharp-edged depressions as shown in Figure 18.56 are an obvious cause for rejection and consequently rated as an A-defect.

Round, shallow depressions with a maximum diameter of 0.006 in. (0.15 mm) are accepted with no remarks entered in the inspection record, provided they occur to a limited degree—for example, an average of one depression per contact tab. Figure 18.57 shows an example.

If the depressions are deep or occur in large numbers (Figure 18.58), the fault is rated as an A-defect, and the board is rejected. The fault is usually caused by pores in the nickel layer as a result of adhering hydrogen bubbles. Paragraph 18.16.5 discusses this in detail.

18.16.5 Pinholes

The contact tabs are inspected for pinholes. Visible pinholes are not allowed within the effective contact area. One pin-

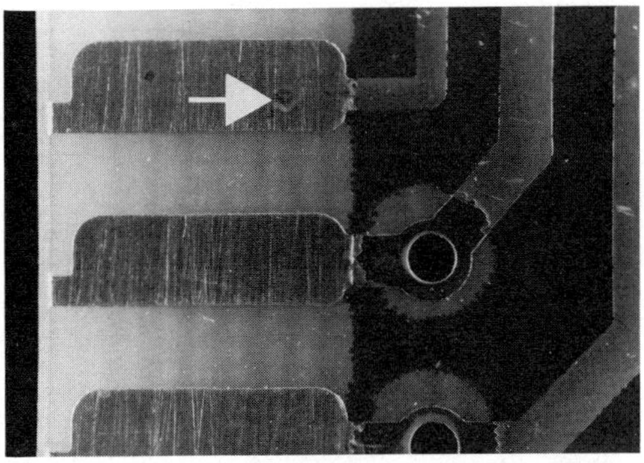

Fig. 18.57. Small depressions in contact surface.

Fig. 18.58. Widespread occurrence of depressions in contact surface.

hole per contact tab can be accepted, provided it is less than 0.004 in. (0.1 mm) in diameter and occurs outside the effective contact area or on unused contact tabs. Exceeding these requirements is cause for rejection, and the fault is rated as an A-defect. Figure 18.59 shows an extreme example of pinholes in the nickel layer.

Such pinholes are usually caused by the formation of hydrogen bubbles during the nickel-plating process. If the bubbles adhere to the surface, the nickel plating will be hampered at such places.

18.16.6 Nodules and Blisters

Nodules and/or blisters are not allowed on the contact tabs. See Figures 18.60 and

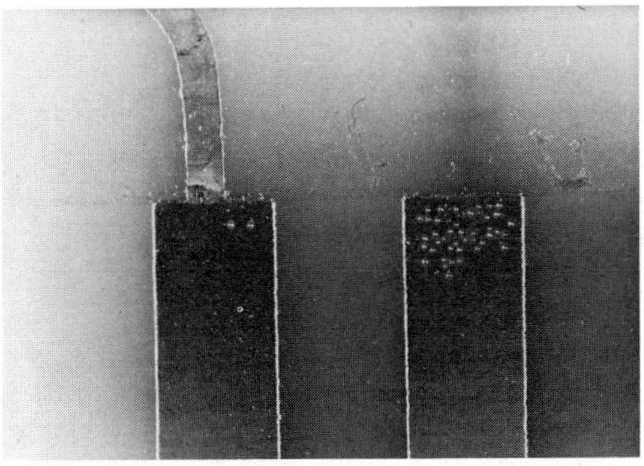

Fig. 18.59. Pinholes in nickel layer. (The board was held in production.)

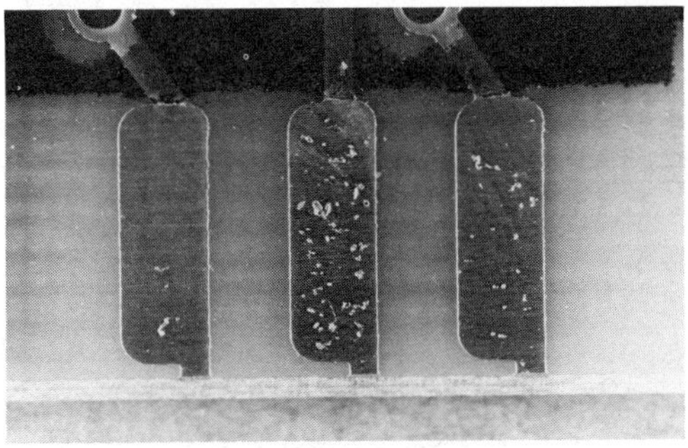

Fig. 18.60. Small nodules on contact surface.

Fig. 18.61. Blisters on contact surface.

18.61. They are cause for rejection and the fault is rated as an A-defect.

18.16.7 Burning

Burning along the contact tab edges (Figure 18.62) is also an obvious cause for

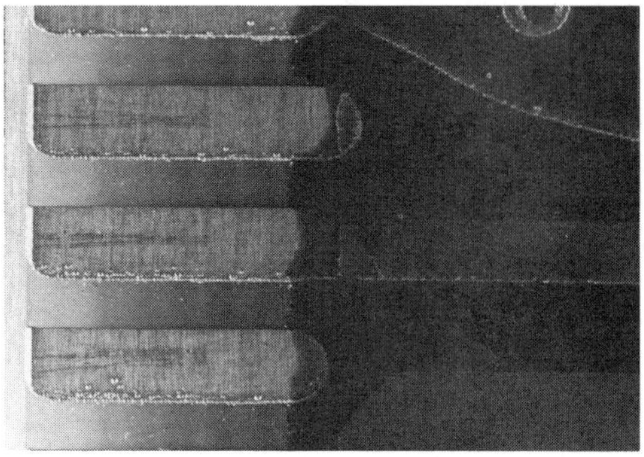

Fig. 18.62. Burning along contact tab edges.

Fig. 18.63. Incomplete coverage of contact surface with gold.

rejection, and the fault is rated as an A-defect. Since the burning originates from the copper plating, it also occurs along the edges of the conductors attached to the contact tabs. This means that when the pattern is burned, the inspector should always very carefully inspect the contact tabs for burning.

18.16.8 Gold Coverage

All contact tabs must be completely covered with gold. Insufficient coverage (Figure 18.63) is cause for rejection, even when occurring in spots. The fault is rated as an A-defect.

18.16.9 Edge Coverage

Except for the machined edges of the contact tabs, all edges of the contact tabs must be covered with gold (and nickel). Insufficient edge coverage is normally cause for rejection and consequently rated as an A-defect. It is assumed that the PCB specification requires full edge coverage.

Paragraph 14.5 describes an alternate method of gold plating, where the edges remain in bare copper. The edges can also be gold-plated without an underlying barrier layer of nickel.

When no requirements regarding coverage of the edges are stated in the PCB specification, the inspector must accept such executions, and no remarks are entered in the inspection record. The PCB specification should be tightened to clearly define any requirements of edge finish.

18.16.10 Transition Zone

The transition zone between the contact tab and an attached conductor must also be inspected. Paragraph 14.6 comprehensively describes the various types of execution. PCB specifications usually forbid exposed copper in the transition zone.

The inspector should verify that the transition zone is in agreement with the required execution, if specified. A violation of the requirement is rated as a B-defect insofar as the actual execution is satisfactory. If the end user has set certain execution requirements, it will be necessary to get an exemption from him. If an exemption is not possible, all boards are rejected and the fault is rated as an A-defect.

When no special requirements are given, the inspector checks for exposed copper in the transition zone. Exposed copper is cause for rejection and the fault is always rated an an A-defect. See Figure 18.64.

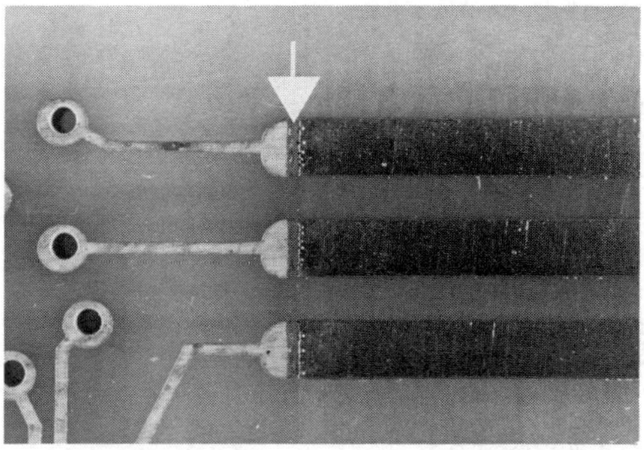

Fig. 18.64. Exposed copper in transition zone between contact tab and attached conductor.

18.16.11 Location of Transition Zone

Furthermore, the inspector has to verify that the transition zone is located at the right spot. It usually lies at the upper edge of the contact tabs, but as the location is not marked by the PCB designer, it can move back and forth a little.

In extreme cases, a screen-printed sealing band can be displaced so much that copper at the transition zone is exposed. It can also happen that some of the solder holes or via holes have been stripped of tin/lead and show bare copper. Such faults are an obvious cause for rejection and consequently rated as an A-defect. See Figure 18.65.

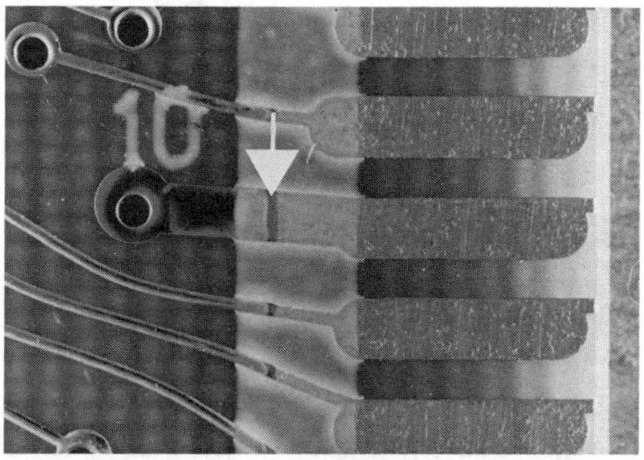

Fig. 18.65A. Correct location of screen-printed sealing band. The gap between the gold plating and the tin/lead plating is covered by the sealing band.

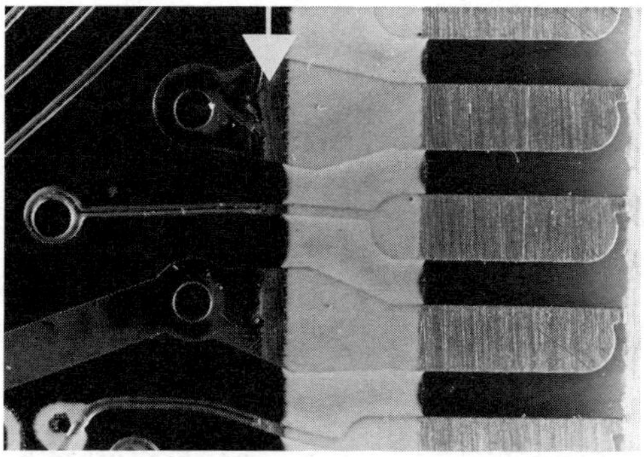

Fig. 18.65B. Displaced location of screen-printed sealing band. Exposed copper occurs outside the sealing band.

18.16.12 Adjoining Areas

Areas adjoining the edge connector should be checked for damage revealed as exposed copper. During the taping preceding the tin/lead stripping, contact numbers can easily be overlooked and consequently stripped of tin/lead; as they usually have no metallic connection to the plating bar, the numbers will appear in copper on the finished board. See Figure 18.66.

Whether the fault should be rated as an A-defect or a B-defect depends on the board application (environmental requirements) and the location of the num-

Fig. 18.66. Contact numbers at the end of an edge connector appear in bare copper because of unfortunate stripping of the tin/lead layer.

Fig. 18.67. Unevenness along contact tab edges.

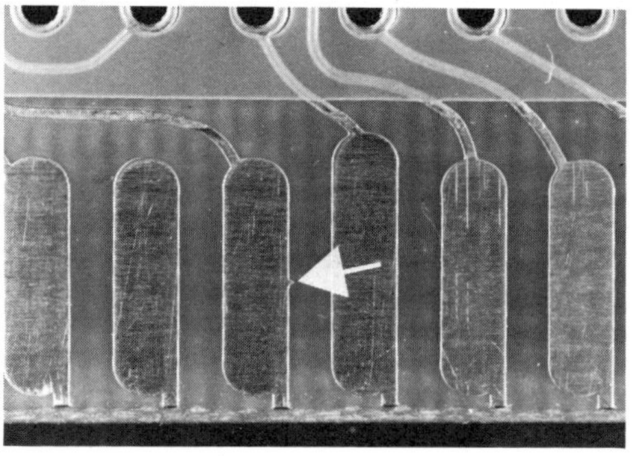

Fig. 18.69. Projection along edge of contact tab.

bers relative to the contact tabs. A radical method of repair is to remove the numbers with a knife, and have them screen printed.

18.16.13 Edge Unevenness

Unevenness along the contact tab edges (Figure 18.67) must not exceed 0.004 in. (0.1 mm) peak-to-valley. If so, it is accepted with no remarks entered in the inspection record. Unevenness larger than 0.004 in. (0.1 mm) is always cause for rejection and rated as an A-defect.

18.16.14 Short Circuits

Short circuits between the individual contact tabs (Figure 18.68) are an indisputable cause for rejection and rated as an A-defect. If the requirement of full edge coverage is maintained, short circuits cannot be removed as this will leave exposed tab edges.

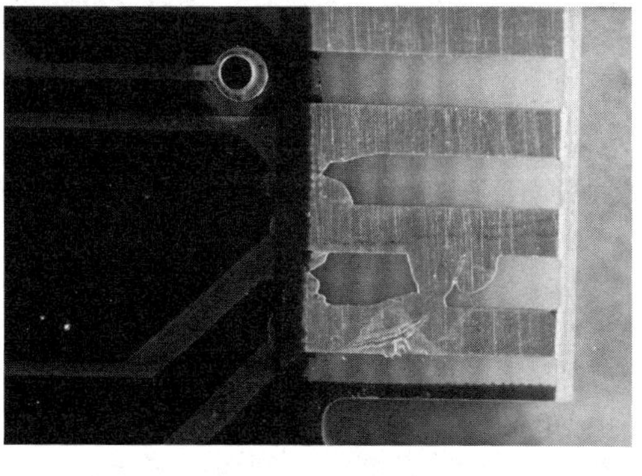

Fig. 18.68. Short circuits between the contact tabs.

18.16.15 Projections

One projection of maximum 0.008 in. (0.2 mm) per contact tab can be accepted, provided that 75 percent of the nominal spacing (insulation distance) remains, and the baseline of the projection is less than 0.02 in. (5 mm). Exceeding one of these conditions is cause for rejection and the fault is rated as an A-defect. See Figure 18.69.

Edge connectors with a pitch of 0.05 in. require a careful calculation of the actual connector system before criteria for approval and rejection can be established.

18.16.16 Indentations

One indentation per contact tab (Figure 18.70) can be accepted, provided the indentation does not affect the effective contact area. The indentation should never be deeper than 10 percent of the nominal width of the contact tab, and the baseline of the indentation must be less than 0.2 in. (5 mm). Exceeding one of these conditions is cause for rejection and the fault is rated as an A-defect.

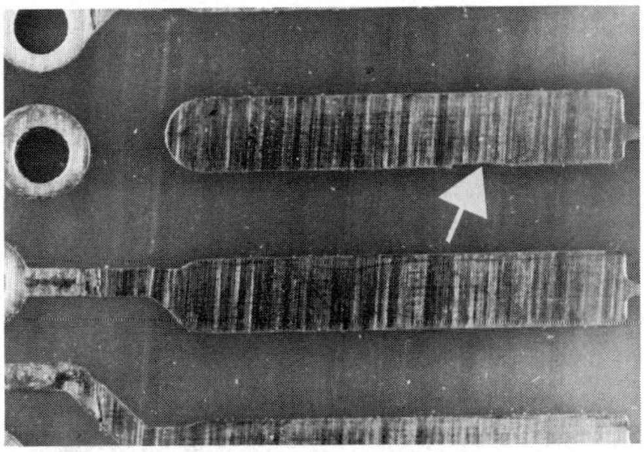

Fig. 18.70. Indentation along edge of contact tab. Note the visible grinding traces in the contact surface.

Fig. 18.72. Burrs at milled slot used for the assembly of a daughter board.

18.16.17 Burrs and Lifting of Tabs

The inspector must also check that no lifting of the contact tabs occurs, or that no burrs are present where the edge connector has been chamfered. Such faults are an obvious cause for rejection, and the fault is rated as an A-defect. See the example shown in Figure 18.71.

Burring is particularly dangerous when the assembly method described in Paragraph 15.6.1 and shown in Figure 15.8 is followed. Although the contact tabs, located at both sides of the milled slot and used for perpendicular assembly of the daughter board, are tin/lead plated, the inspector must inspect the milling carefully for burrs. See Figure 18.72.

Fig. 18.71. Burrs caused by chamfering the edge connector.

The PCB manufacturer might wish to shorten the nominal length of the contact tabs so that they end a little before the slot—for example, 0.012 to 0.020 in. (0.3 to 0.5 mm). This can, however, impede the soldering as it can be difficult to bridge the gap, particularly if the milling is slightly displaced.

18.16.18 Cracks in Gold Layer

Cracking of the gold/nickel layer is an indisputable cause for rejection, and the fault is rated as an A-defect. In particular, the inspector should realize that in case the board contour has been blanked or sheared, the mechanical action can introduce a tendency to cracking at the end of the contact tabs. In such cases, a more thorough investigation should be carried out with a microscope.

18.16.19 Printing Ink on Tabs

A rather special form of contact contamination can be the adherence of very small drops of printing ink (Figure 18.73). These drops come from the application of the board's solder and insulation masks.

Because of possible electrical contact problems, such drops found within the effective contact area are cause for rejection and consequently rated as an A-defect.

Fig. 18.73. Small drops of solder mask ink sticking to the contact surfaces.

The fault is very rare and only mentioned here for the sake of completeness. The PCB manufacturer can possibly salvage the boards by dissolving the epoxy drops, but this implies a limited number of drops.

18.16.20 Dislocated Plating Bar

Because of a design error regarding the location of the plating bar, the contact tabs can be mutually short-circuited by a remaining strip of the plating bar (Figure 18.74).

The plating bar should of course have been located outside the board contour. In the actual case, no chamfering was specified, which otherwise would have removed the remaining strip of plating bar.

Fundamentally, the fault is an A-defect, but as the design is very unprofessional and presumably very rare, the fault is not included in the listing given in the inspection record.

18.16.21 Pinholes in the Transition Zone

Paragraph 18.16.5 discussed the inspection of contact tabs for pinholes. As shown in Figure 18.75, small depressions in the gold plated layer can occur just below the transition zone. These depressions usually originate from a defect in the underlying nickel layer. The hydrogen bubbles developed during nickel plating have possibly been caught by the tape used for masking the adjacent area. As the depressions do not continue fully down through the nickel layer, and they are furthermore located outside the effective contact area, no remarks are entered in the inspection record. Incidentally, the occurrence is quite common in most makes of boards.

18.16.22 Repair of Plating

Repairing damaged plating on an edge connector is expected to become more common in years to come. In this respect, the same view can be taken as mentioned in Paragraph 17.5.10 regarding a qualified

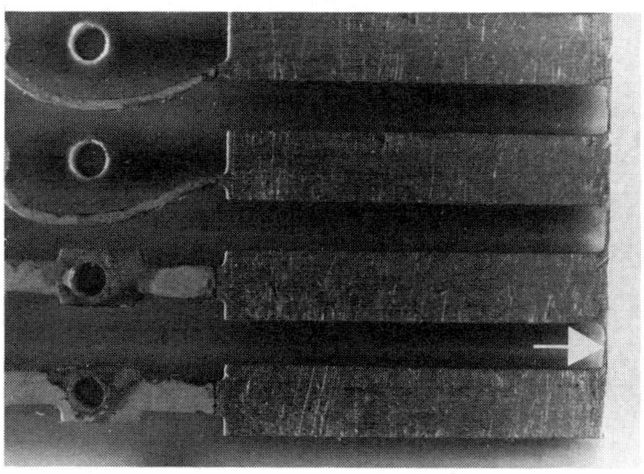

Fig. 18.74. Incomplete removal of plating bar results in short-circuited connector tabs.

Fig. 18.75. Small depressions in the gold plating just below the transition zone.

repair of pattern defects. See also Paragraph 14.9 concerning repair work.

Whether it is allowed at all to perform a repair of the plating on even a few contact tabs depends on the customer's purchase policy as embodied in his purchase specification.

Under no circumstances should repairs be performed without the customer's knowledge and consent. A repaired contact tab should always be marked and the board separated from other boards in the delivery, so that the inspector can easily locate and inspect it.

Chapter 19

Adhesion Inspection

19.1 Introduction

A tape test immediately follows the visual inspection. This test reveals the adhesion of the plating to the underlying copper and/or the presence of overhang along the pattern edges. The inspector can also evaluate the adhesion between the solder and insulation masks to the underlying surface.

Being a fast test, the tape test is most expediently done before the more comprehensive inspection of the board machining. Reduced adhesion is usually cause for rejection.

19.2 Prerequisites of a Tape Test

19.2.1 Tape Specifications

As indicated by the name, this test is performed by means of a tape. Some specifications (for example, MIL-P-55110C) prescribe the use of a type II, class A tape according to Federal Specification L-T-90, which translates into a self-adhesive, transparent cellulose acetate tape with a matte surface. L-T-90 specifies an adhesion corresponding to 12 oz. per in. of width (0.013 kg/mm).

IPC's Test Manual Methods, IPC-TM-650, specifies the same type. On the other hand, UL796 (from Underwriters Laboratories, Inc.) prescribes an adhesion of 35 $\pm$5 oz. per in. of width (0.039 $\pm$0.006 kg/mm). British Standard BS9760 specifies a self-adhesive cellulose tape with an adhesion between 7.2 and 7.8 N per 25 mm of tape width. This corresponds to approximately 27 oz. per in. of width (0.030 kg/mm). IEC 326 specifies a transparent self-adhesive tape with an adhesion of at least 2.2 N per 10 mm of tape width, which corresponds to approximately 20 oz. per in. of width (0.022 kg/mm).

Obviously, there is a certain variation of adhesion among the tapes prescribed. In addition, the pressure with which the tape is applied to the board is crucial to the result. The effective adhesion of the tape depends on whether it is applied by pressing it against the board with the fingers or with a rubber roller, which makes the test somewhat uncertain.

In practice, these variations are of no great importance. Experience shows, in fact, that either the plated layer adheres so well that it cannot be peeled off no matter how good the tape is, or the plating is so inferior that it can be peeled off by even a tape with relatively low adhesion.

It may prove a little difficult to find a tape having the adhesion specified; tape 691 from the 3M company is recommended. The adhesion of this tape, 30 oz. per in. of width (0.034 kg/mm), is as close as possible to the value specified in UL796.

19.2.2 Significance of Test

If the plating cannot be peeled off by the tape, its adhesion is most likely satisfactory. The assessment is unaffected if the tape adhesion is higher than specified in the other standards mentioned. In principle, the PCB specification should state the required adhesion, preferably by specifying the type of tape to be used.

Several PCB specifications state an adhesion between the copper cladding and the base material of at least 1.4 kN/m = 0.14 kg/mm, corresponding to 125 oz. per in. of width. However, the inspector should not try to measure the adhesion of the pattern to the base material with a tape having even more adhesion than specified in Paragraph 19.2.1. Adjacent areas not covered by the tape will restrain the area which is to be tested, and the rigidity of the copper will resist peeling. Therefore, a tape test reveals nothing about the adhesion of the pattern to the base material.

Only very narrow conductors with a heavy undercut will show a peeling-off effect. In most cases, a lower limit to the conductor width is specified—for example, 0.031 in. (0.8 mm)—when testing the adhesion so that a tape test cannot be used. The adhesion of the copper foil to the base material can be tested only by a regular peel test. See Paragraph 5.8.

19.3 Execution of a Tape Test

The tape should be ½ in. (12.7 mm) wide and about 2 in. (50 mm) long, plus enough to hold on to during the test. The board is carefully degreased—for example, with chlorothene—and the tape placed across the area to be tested as shown in Figure 19.1.

The tape is pressed onto the board over its length of 2 in. (50 mm) to eliminate trapped air. An ordinary eraser, rubbed over the tape, is an excellent aid. The tape is now pulled off the board with a quick pull, perpendicular to the board, as shown in Figure 19.2. The tape is then examined for adhering particles, a clean tape as shown in Figure 19.3 being the desired result.

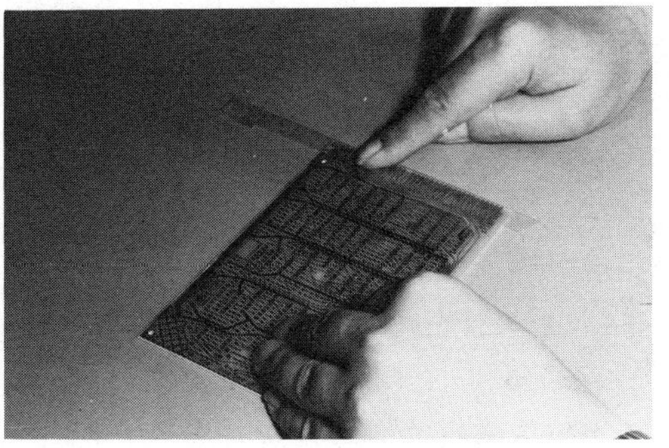

Fig. 19.1. Execution of a tape test. The tape is applied and pressed on to the board.

Fig. 19.2. Execution of a tape test. The tape is peeled off the board with a quick pull.

Fig. 19.3. Execution of a tape test. The tape is examined for adhering particles.

ADHESION INSPECTION	19
GOLD OR GOLD/NICKEL	19.4
PLATING OF PATTERN	19.5
COPPER FOIL	19.6
SOLDER/INSULATION MASKS	19.7
COMPONENT NOTATION	19.8

Fig. 19.4. Adhesion inspection. (Section of inspection record.)

The entire sample should always be subject to the tape test when the boards are delivered by a new supplier, with whom the inspector has no experience regarding adhesion.

When the boards are delivered by a known and recognized supplier over a longer period of time and the inspector has found the adhesion satisfactory, the extent of the tape test can be reduced to a single board. For a large batch, the tape test should include a few boards.

Tape tests are performed as described in the rest of this chapter. Obviously, a fresh tape is used for each test. The inspector should follow the part of the inspection record shown in Figure 19.4.

19.4 Adhesion of Nickel and Gold Platings

On boards having edge connectors, the tape test is performed across as many connector tabs as possible. The test should include the transition zone between the contact tabs and their attached conductors as well as the machined ends of the contact tabs. If the tape cannot completely cover the contact tabs, the test must be performed in two parts.

When gold or nickel and gold from the contact tabs adheres to the tape, the board is indisputably rejectable, and the defect is therefore rated as an A-defect. In such cases an obvious precaution is a 100 percent inspection (screening of the batch by an extended tape test). All boards showing delamination of the gold or nickel and gold layers are rejected and returned. See

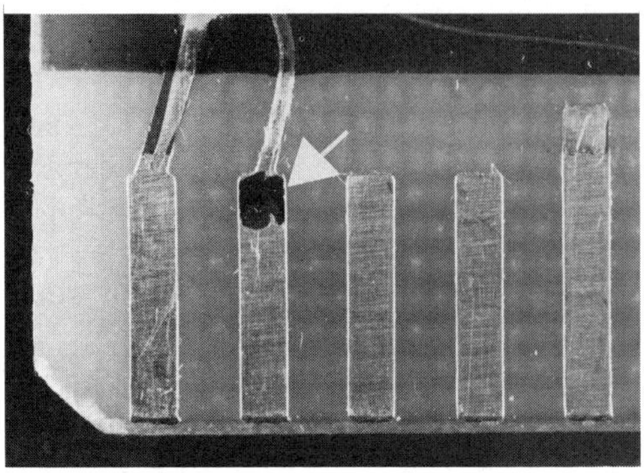

Fig. 19.5. Edge connector exhibiting delamination of the gold plating.

the description of sources of error in Paragraph 14.4. A tendency towards delamination usually shows up at the transition zone (Figure 19.5).

19.5 Adhesion of Copper and Tin/Lead Platings

Although the most serious adhesion problems occur at edge connectors, the adhesion of the plating to the pattern can, in rare cases, be a problem.

A similar tape test is therefore carried out on the pattern, assuming no solder or insulation mask has been applied, or that the mask has sufficiently large apertures to allow the tape test to be performed on the solder pads. The inspector should distinguish between two fundamentally different faults:

1. A direct delamination caused by reduced adhesion of the copper plating and/or the tin/lead layer, the latter being applied by normal tin/lead plating or by solder coating with subsequent hot-air leveling.

   The defect is revealed by torn-off metal areas as shown in Figure 19.6.
2. Torn-off overhang (slivers) occurs at places where the tin/lead alloy

Fig. 19.6. Torn-off copper plating.

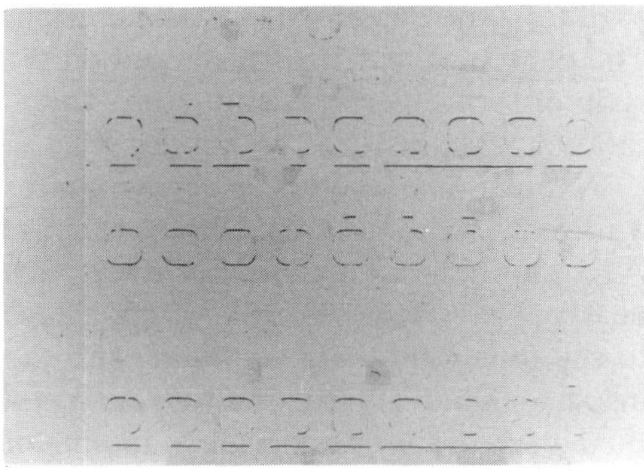

(**Fig. 19.7.** Tearing-off tin/lead overhang (slivers).

composition, for one reason or another, falls outside the specified range. The increased melting temperature of the tin/lead imposes the risk of incomplete melting of the tin/lead overhang. Incorrect setting of the reflow machine—for example, insufficient heating—can also cause the same effect. See the discussion in Paragraphs 5.10, 8.3.3, and 8.7.1a. From such places the tape test will tear off the tin/lead overhang in the form of slivers as shown in Figure 19.7.

Torn-off material on the tape is an obvious cause for rejection, and the fault is consequently rated as an A-defect.

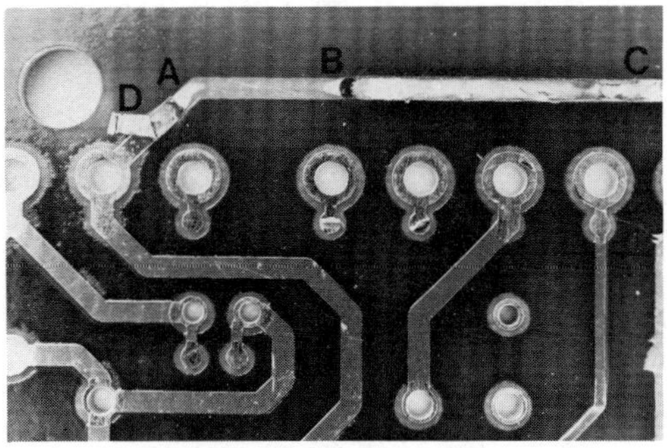

Fig. 19.8. Conductor peeled off over the stretch A-B. The conductor exhibits severe undercutting caused by heavy etching. The copper foil is tapered at A and B. Over the stretch B-C and at D and E the copper plating is peeled off the copper foil.

19.6 Adhesion of Copper Foil with Undercut

As already described in Paragraph 19.2.2, a heavy etching implies that narrow conductors adhere only weakly to the base material. Such conductors can in some cases become so loose that they are, in fact, only held in place by the overlying solder or insulation mask.

Incidentally, the printing ink can creep under the conductors and lift them slightly along the edges, which can be seen quite clearly. The tape test may peel such conductors off the base material (Figure 19.8). The fault is an indisputable cause for rejection and always rated as an A-defect.

19.7 Adhesion of Solder and Insulation Masks

A tape test is performed on solder or insulation masks only in those few cases where the mask appearance suggests reduced adhesion. The inspector is referred to Paragraph 11.8.2, which describes in detail the test procedure and adhesion measurement of both screen-printed and photopolymer masks.

If the flux used to reflow the board was not completely removed before the masks

Fig. 19.9. Peeling off a solder mask where it is screen-printed on top of conductors. Tape with adhering solder mask is seen beside the board.

were applied, it can reduce the adhesion at tin/lead covered areas. It is therefore possible to peel off the solder mask where it overlies the conductors (Figure 19.9).

This defect is not necessarily a direct fault of the PCB manufacturer because the flux manufacturers may slightly change the flux composition without informing the PCB manufacturer.

The fault is not always cause for rejection. Mass-soldering the board can possibly result in better adhesion because of the restructuring of the tin/lead layer, including changes in the surface characteristics. If so, the fault is rated as a B-defect, otherwise as an A-defect.

19.8 Adhesion of Component Notations

As mentioned in Paragraph 13.3.2, certain circumstances can make it difficult to obtain satisfactory adhesion of the component notation to a photopolymer mask. Since this is a processing fault, most probably recurring throughout the batch, the sample size can be reduced to a single board.

The table in Paragraph 11.8.2 can be used for a rough estimate of the adhesion. The degree of tearing-off is based on the ratio of torn-off figures and letters to the total number involved in the tape test.

A rejection of the entire batch requires a formal inspection of the total sample. A component notation, however, will nearly always be of secondary importance to the functioning of the board, and the inspector should consider a waiver. When the adhesion survives normal handling of the board, a waiver should be issued and the fault rated as a B-defect. A simple test can consist of scratching the printing ink with a fingernail.

Chapter 20

Mechanical Inspection

20.1 Introduction

The inspector follows the visual inspections and the tape test (Chapters 17 through 19) with the mechanical inspection of the board. He should follow that part of the inspection record shown in Figure 20.1.

			MECHANICAL INSPECTION	20
			BOARD THICKNESS O.O.T.	20.2
			WARP AND TWIST O.O.T.	20.3
			SOLDER HOLES DISPLACED	20.4
			MOUNTING HOLES DISPLACED	20.5
			TOOLING HOLES DISPLACED	20.6
			NOM. HOLE DIA. O.O.T.	20.7
			SOLDER HOLE DIA. O.O.T.	20.8
			MOUNTING HOLE DIA. O.O.T.	20.9
			TOOLING HOLE DIA. O.O.T.	20.10
			BOARD DIMENSIONS O.O.T.	20.11
			CONTOUR O.O.T.	20.12/20.13
			CUTOUTS O.O.T.	20.14

Fig. 20.1. Mechanical inspection. (Section of inspection record.)

20.2 Board Thickness

The base material must usually meet a certain thickness tolerance as stated in the PCB specification. The inspector's attention is called to the edge connector area, to which a closer tolerance frequently applies (Paragraph 4.3.2d).

The thickness tolerance appears in Table 4.1. The inspector checks the thickness with a micrometer as shown in Figure 20.2. The copper foil is included in the thickness in the case of ordinary base materials. The inspector is referred to Paragraph 4.3, which discusses the problem of whether an out-of-tolerance thickness is to be rated as an A-defect or as a B-defect. In summary, exceeding the thickness tolerance is assessed according to the board application as stated below:

Boards without edge connectors:
  B-defect
Boards with edge connectors:
  A-defect
Boards for microwave applications:
  A-defect

Fig. 20.2 Measuring the board thickness by means of a micrometer.

20.3 Warp and Twist

The base material must comply with the warp and twist requirements. The measurement is performed as described in

Paragraph 4.4 and illustrated in Figure 20.3. Typical requirements are listed in Tables 4.3 and 4.4.

When checking boards with the maximum dimension less than 12 in. (300 mm), the inspector can usually expect warp and twist (W/T) to meet the following requirements, where t indicates the board thickness:

in.	in./in.
$t \leq 0.031$:	$W/T \leq 0.015$
$0.031 < t \leq 0.062$:	$W/T \leq 0.010$
$0.062 < t \leq 0.125$:	$W/T \leq 0.007$

mm	mm/mm
$t \leq 0.8$:	$W/T \leq 0.015$
$0.8 < t \leq 1.6$:	$W/T \leq 0.010$
$1.6 < t \leq 3.2$:	$W/T \leq 0.007$

Exceeding these values by up to 50 percent is rated as a B-defect, and over 50 percent as an A-defect. It is recommended that the inspector pay careful attention to the remarks in Paragraph 4.4 regarding the influence of the board design on the warp and/or twist. Furthermore, he should adopt a functional attitude when evaluating the importance of the defects.

Fig. 20.3. Measuring the warp of the board.

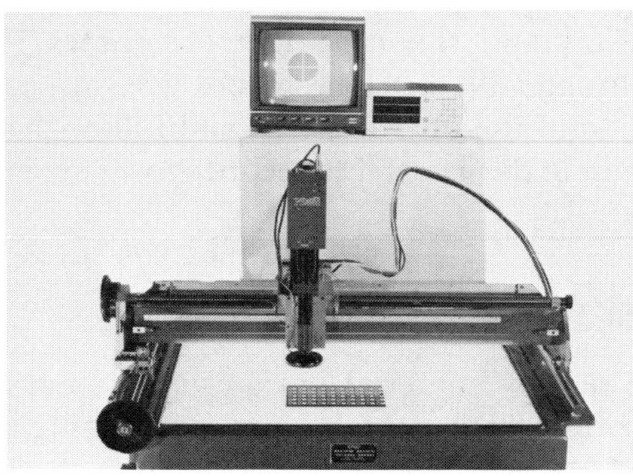

Fig. 20.4. Measuring the location of the solder holes by means of a digital coordinate table.

20.4 Location of Solder Holes

20.4.1 Inspection versus Insertion Methods

As mentioned in Paragraph 6.4, automatic component insertion makes the accuracy of the solder hole locations critical. Because of the risk of a permanent dimensional change of the board, it is not enough to assess the hole location with respect to the solder pads.

Since the entire pattern, including the solder holes, may have changed (Paragraph 6.4), holes drilled in the centers of the solder pads do not prove correct absolute positioning. The inspector should therefore check the hole positions by measurement. The most suitable equipment for this task is a digital coordinate table as shown in Figure 20.4. An alternative is to use a caliper with cones attached to the jaws as shown in Figure 20.5. Paragraph 6.4.8 details the procedure.

It is usually unnecessary to measure the solder hole locations on boards for manual assembly. As described in Paragraph 18.5, the location can be assessed well enough by the appearance of the annular rings.

The solder hole positions must be thoroughly inspected when the boards will be populated by automatic component insertion. The inspection level is therefore General Inspection Level II.

Fig. 20.5. Measuring the location of the solder holes by means of a caliper equipped with cones on the jaws.

Exceeding the tolerances is rated as an A-defect, and an AQL value of 1 percent is chosen. In certain cases it may be necessary to increase the demand regarding the AQL value, but this will depend on experience.

The inspector, however, should know of the new generation of automatic component insertion machines with a special tracking facility. Based on an optical measurement of the location of the solder holes relative to the insertion points, the board is automatically repositioned so that the solder holes are located exactly at the insertion points.

When the production department has such an automatic component insertion machine at its disposal, the requirements regarding the location of the solder holes are reduced and correspond on the whole to the values for boards for manual component insertion. Also, the solder holes no longer need to be oversized in order to compensate for mislocation.

20.4.2 Extent of Inspection

Obviously, it is both impracticable and superfluous to measure the locations of all the solder holes of the sample. The inspector instead selects a few solder holes at each corner and in the center region of the board, and measures their position. The measured values are then compared with the nominal values which appear on the coordinate list of the solder holes.

If the PCB manufacturer has performed raster digitizing (Paragraph 6.3.6), the nominal hole locations will be at the intersections of the grid lines, and the locations can therefore be calculated as multiples of the 0.05 or 0.10 in. pitch.

A similar calculation can be made in the case of CAD boards, because the computer normally performs the layout in increments of 0.025 or 0.050 in. The calculation should be made by the PCB designer who, in collaboration with the inspector, selects certain checkpoints and indicates these on the master drawing together with the tolerance.

20.4.3 Identifying Nominal Hole Locations

When a master drawing does not include any information regarding tolerances, the inspector can anticipate that a positional tolerance of ± 0.003 in. (± 0.08 mm) will be enough to ensure insertion. By increasing the oversize of the solder holes (Paragraph 10.3.6 of Reference A12, Reading List), the requirement for positional tolerance may be somewhat reduced.

The inspector should always obtain information from the PCB designer regarding the necessary positional tolerances so that rejections have valid causes.

The inspector should always include possible tooling holes in the inspection, because such holes are often used when attaching the board to the fixture plate of the automatic component insertion machine. The positional tolerance of the tooling holes is therefore the same as that of the solder holes. See Paragraph 20.6.

20.4.4 Board Disposition

Exceeding the positional tolerances should be considered as an absolute and indisputable cause for rejection, and the fault is rated as an A-defect. A rejection implies, of course, that the positional tolerances have been specified and have been accepted by the PCB manufacturer.

20.5 Location of Mounting Holes

When the locations of the mounting holes have been determined by the dimensioning on the master drawing, the measurement is most conveniently made with respect to the reference system. In principle, the measurement is carried out in the same way as described in Paragraph 20.4.1.

When the board is later mounted in a chassis, it is frequently possible to allow a displacement of ±0.012 in. (±0.3 mm). See Paragraph 18.6. Measuring with the special caliper shown in Figure 20.5 normally suffices. Out-of-tolerance locations are treated as described in Paragraph 18.6.

In the case of boards for automatic component insertion, the inspector should, however, realize that in many instances the mounting holes are used to attach the board to the fixture plate of the automatic component insertion machine. Such mounting holes require a tighter positional tolerance, frequently ±0.002 to ±0.003 in. (±0.05 mm to ±0.08 mm). Exceeding this tolerance is an absolute cause for rejection, and the fault is rated as an A-defect. A rejection implies, however, that the positional tolerances have been specified and accepted by the PCB manufacturer.

20.6 Location of Tooling Holes

The tooling holes must be correctly located if they are to be used as the basis of a measurement or for attaching the board to the fixture plate of an automatic component insertion machine. A convenient basis for assessing the correct location of a tooling hole is the appearance of the annular ring. This means that the width of the annular ring should be even all around.

20.6.1 Tooling Hole Deviations

See the remarks in Paragraphs 6.2.1 and 6.2.2 about digitizing with reference to the center or circumference of the solder pads. Some PCB manufacturers do not always use the specified procedure when determining and drilling the tooling holes. The author has seen several examples of such deviations, most of them caused by the fact that the requirements were overlooked or misunderstood.

The major tooling hole, which is the zero point (datum) of the reference system, must necessarily be determined by very careful digitizing. The two secondary tooling holes are used when adjusting the master film on the digitizing table so that the X- and Y-axes coincide with the directions of travel of the digitizing table.

Deviations can occur while the locations of the two secondary tooling holes along the X- and Y-axes are being determined. Perhaps a request for a determination based on the coordinates is not followed, and the tooling holes in question are determined by digitizing with a risk of mislocation, especially in the case of manually prepared artwork.

It can also happen that the PCB manufacturer drills the tooling holes in a second drilling operation instead of the first. A possible reason is not understanding that it is possible to use plated-through tooling holes. This deviation can cause a serious displacement.

20.6.2 Permanent Dimensional Change

Finally, the inspector must consider the permanent dimensional change of the board. He must therefore check that the tooling holes have the correct mutual distances. This is done as described in Paragraph 20.4.

20.6.3 Board Disposition

Tooling holes must not be displaced by more than ±0.003 in. (±0.08 mm). If the boards are for automatic component insertion, exceeding this value is regarded as cause for rejection and the faults are rated as an A-defect. In the case of boards for manual component insertion, the inspector should differentiate a little: Displacements up to ±0.008 in. (±0.2 mm) are a B-defect because bare-board testing in automatic test equipment, unlike automatic component insertion, will not be hampered. Larger displacements are regarded as an A-defect, and the board is rejected.

20.7 Nominal Hole Size

The inspector should verify that the solder holes as well as the mounting holes are of the correct nominal sizes, as discussed in Paragraph 6.7. The measurement can be made accurately enough with a tapered plunger gauge as shown in Figure 20.6. Cylindrical plug gauges can of course always be used, although the measurement may be more troublesome to perform.

20.7.1 Solder Holes

The inspector draws a board from the sample and inspects the holes. The extent of the checking depends on the number of different hole sizes, and the difference between sizes. The inspector can make a better decision if he studies the master drawing (drill drawing) to assess the possibility of errors when digitizing the board. The clearer and better-arranged the board (drill drawing) is, with regularly occurring hole groups, few hole sizes and conspicuous diameter variations, the less the extent of inspection required.

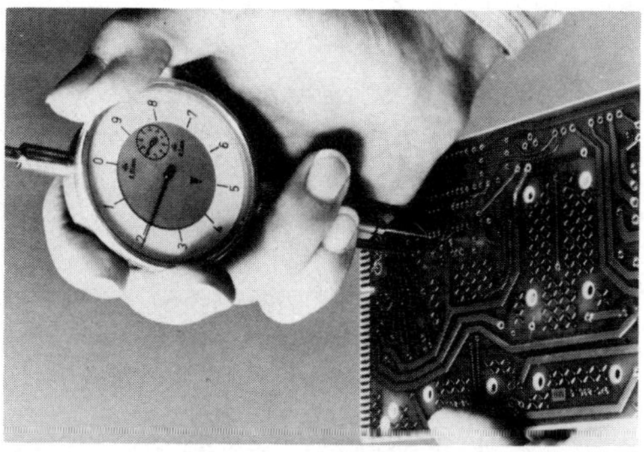

Fig. 20.6. Measuring the nominal hole diameter by means of a tapered plunger gauge.

The following rules mainly apply to first orders or to orders with new issue levels. For repeat orders from the same PCB manufacturer, the drill tape remains unchanged, so the inspection can be limited to just one hole within the various hole sizes. Whether it is a repeat order can be seen from the preceding inspection record.

1. IC holes—frequently 0.031 in. (0.8 mm)—are checked by measuring the diameter of one hole per IC hole group in a selection of 10 IC hole groups. For the inspector, it is quicker to view the remaining IC hole groups to check whether the annular rings look like those of the holes that have been measured.
2. Other holes which occur in groups are checked by measuring the diameter of one hole per group. Based on the previous remarks concerning the clarity of the drill drawing, the inspector should ascertain how many of the hole groups should be checked.
3. Isolated holes imply a substantially higher possibility of faults, and in principle all such holes should therefore be checked. The holes can be solder holes for discrete components, frequently 0.031 or 0.043 in. (0.8 or 1.1 mm), and a few larger holes—for example, 0.063 in. (1.6 mm).

The inspector can often assess the hole sizes from the width of the annular rings by viewing the board and noting abnormally wide or narrow annular rings. When such rings are found, the hole diameters obviously should be checked.

In principle, an incorrect nominal hole diameter is regarded as an A-defect. When a solder hole is oversized by 0.004 to 0.008 in. (0.1 to 0.2 mm), the inspector can usually waive the fault. It is, on the other hand, more difficult when the solder hole is too small, particularly in the case of a plated-through hole.

In many of these cases, it is possible to insert the component lead, but a waiver requires a thorough examination of the conditions. It will be necessary to ascertain how much the hole diameter in question varies from board to board to make certain that the waiver does not impede the assembly.

These considerations apply to manually assembled boards, whereas boards for automatic component insertion require a correct hole size.

In some cases the hole may be redrilled as described in Paragraph 6.6, Items 1, 2, 3, 4, and 6. A waiver implies that the fault occurs to only a moderate extent so that redrilling is practicable. The fault usually occurs systematically, somewhat reducing the workload of redrilling.

20.7.2 Mounting Holes

Mounting holes are dealt with in much the same way as solder holes. In general, it should be possible to exempt holes which have been drilled up to 0.012 in. (0.3 mm) over the nominal diameter. Undersized holes can usually be redrilled to the correct size without any problems. In both cases the fault is rated as a B-defect.

If the mounting holes are used instead of tooling holes for attaching the board to the fixture plate of an automatic component insertion machine, the inspector cannot simply exempt an incorrect hole size. Redrilling frequently causes an unacceptable displacement.

A better solution would be to adapt the guide pins to the holes, whether they are too small or too large. This requires the fault to occur in the entire batch of boards. In principle, the fault should be rated as an A-defect, even when exempted.

20.8 Diameter Tolerance of Solder Holes

Having made sure that the solder hole sizes are correct, the inspector verifies that the diameter tolerance is met.

20.8.1 Extent of Inspection

Experience shows that solder holes, in general, will be very much alike. The inspector can therefore deviate from the normal sampling plan and limit the sample size to two to five boards, depending on the batch size (Paragraph 6.8.5a).

The inspection focuses on the smallest hole size, frequently 0.031 in. (0.8 mm) in diameter, since that is where the major tolerance problems are found. Via holes are often somewhat smaller—for example, 0.024 in. (0.6 mm)—but as long as they are used as feed-through connections only, the diameter tolerance is less important.
clude via holes. As a consequence, the PCB specification may be reduced on this point because the PCB manufacturer can face unnecessarily severe problems in meeting the diameter tolerance for no functional reason.

The number of solder holes to be inspected should now be determined:

1. With nonplated-through boards and plated-through boards with a balanced pattern, the inspector uses

Special Inspection Level S3 (Paragraph 6.8.5a).

The sample size, as well as acceptance and rejection numbers, corresponding to an AQL value of 1 percent are taken from Table 6.1, Paragraph 6.8.5a. This applies to small solder holes up to 0.04 in. (1.0 mm).

The solder holes to be examined must be distributed over the entire sample with five to ten solder holes per board. The inspector may select isolated solder holes, which have notoriously more difficulty in meeting the diameter tolerance.

Having found that the small solder holes meet the tolerance requirements, the inspector proceeds with the larger solder holes. As the diameter tolerance is usually slightly lower and the plating conditions better than in the case of small solder holes, the larger solder holes will probably also meet the requirements. The inspector can therefore reduce the extent of inspection to a few solder holes within each hole size—for example, five or ten of each.

2. When the boards are plated-through, but the pattern has not been balanced, the risk of hole diameter variations increases. In principle, the inspection is carried out as previously described, but the inspector uses Special Inspection Level S4. The acceptance and rejection numbers corresponding to an AQL value of 1 percent appear in Table 6.1, Paragraph 6.8.5a.

20.8.2 Inspection Procedure

The inspection is carried out with a set of cylindrical plug gauges as specified in Paragraph 16.3, Item E. The diameter of the solder holes is defined as the size of the largest plug gauge which can pass through the solder hole (Figure 20.7). The diameter must fall within the tolerance range stated in Table 20.1.

Fig. 20.7. Determination of the diameter tolerance of the solder holes by means of cylindrical plug gauges.

To avoid damage to the hole wall, the inspector should always take care when inserting the gauge. For example, nodules can be torn off and cause cracking at a later stage (Paragraph 7.3.7).

Since the sample size is very limited, the inspector can compensate a little by including some of the most vulnerable solder holes, which means holes singly located with respect to the pattern.

20.8.3 Board Disposition

Exceeding the specified diameter tolerance is, in principle, an A-defect. When the solder holes are a little too large—for example, 0.004 to 0.008 in. (0.10 to 0.20 mm) over the upper tolerance limit—the inspector can issue a waiver without risk. When the solder hole is a little below the minimum—for example, 0.004 in. (0.10 mm) too small—it is also frequently possible to issue a waiver.

The inspector should contact the PCB designer since some locations may have smaller margins between the component lead and the solder hole. The reason could be a desire to use as few hole sizes as

Table 20.1. Tolerances on hole diameters as a function of hole sizes and types. (d indicates the finished hole diameter.)

Types of holes	Hole diameters/tolerances			
	$d \leq 0.04$ in. $d \leq 1$ mm	$0.04 < d \leq 0.25$ in. $1 < d \leq 6.3$ mm	$0.25 < d \leq 0.4$ in. $6.3 < d \leq 10$ mm	$d > 0.4$ in. $d > 10$ mm
Plated-through, nonreflowed	± 0.004 in. ± 0.10 mm	± 0.005 in. ± 0.13 mm	± 0.008 in. ± 0.20 mm	
Plated-through, reflowed	+ 0.006/− 0.004 in. + 0.15/− 0.10 mm	± 0.006 in. ± 0.15 mm	± 0.008 in. ± 0.20 mm	
Nonplated-through, drilled	± 0.002 in. ± 0.05 mm	± 0.004 in. ± 0.10 mm	± 0.008 in. ± 0.20 mm	± 0.100 in. ± 0.25 mm
Nonplated-through, pierced	± 0.004 in. ± 0.10 mm			

possible to hold down costs. See also the remarks in Paragraph 6.8.5b.

20.9 Diameter Tolerance of Mounting Holes

Checking the diameter tolerance of the mounting holes is performed according to the same guidelines as described in Paragraph 20.8. The sample size is just as moderate as that given for the larger solder holes, which means 10 mounting holes within each hole size.

Since mounting holes can be nonplated-through as well as plated-through, the inspector must distinguish between the different types, as stated in Table 20.1.

Like all other holes in the board, mounting holes must meet the specified tolerances. The inspector can, however, issue a waiver according to the rules set out in Paragraph 20.7. Failure to meet the diameter tolerances is rated as a B-defect.

20.10 Diameter Tolerance of Tooling Holes

In general, tooling holes occur to a very limited extent—for example, two or three holes per board. Therefore, the inspection can, without additional cost, include all tooling holes in the sample. The inspector should check the diameter tolerance, particularly when the tooling holes are used for attaching the board in the jig of an automatic test equipment, or to the holding fixture of an automatic component insertion machine.

The same tolerances as stated for the other holes in the board apply, and the inspector is referred to Table 20.1. Violation of the diameter tolerances is rated as an A-defect, and no waiver should be issued.

20.11 Overall Board Dimensions

Whether the contour of the board is based on corner marks or on a reference system, the overall dimensions are usually indicated on the master drawing together with tolerances. In some cases the tolerances are stated in conjunction with the dimensioning, and in other cases the tolerances are included in a table on the drawing.

The inspector measures the length and width of the board with a caliper and verifies that the stated tolerances are met (Figure 20.8). See also the attainable ma-

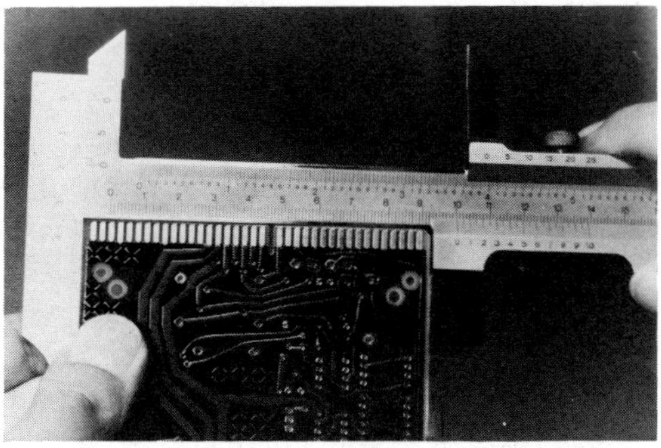

Fig. 20.8. Measuring the overall dimensions of the board by means of a caliper.

chining tolerances specified in Paragraph 15.5.

When the tolerances are exceeded, the inspector should consider or investigate whether a rejection is necessary. To a high degree, this depends on the way the board is to be used.

1. When the board is mounted in a chassis, it is generally possible to accept minor deviations from the tolerances, insofar as the board does not have to fit tightly in a frame.

   In this case, it is possible to exempt minor violations of the tolerances—for example, up to 0.020 in. (0.5 mm). This implies that the pattern itself is not affected by an inward edge displacement of the contour. The fault is rated as a B-defect and will probably necessitate a 100 percent inspection of the batch.
2. In the case of a plug-in board, the inspector can generally assume that the contour affects the board guidance, and a waiver should not be issued.

   The fault is rated as an A-defect, and the board is rejected. When the board is too large, the PCB manufacturer can often salvage it by rerouting. A 100 percent inspection of the batch is usually necessary.

   A possibility of exempting the fault exists when the board is mounted in a frame which takes over the guidance function. In such cases the fault is rated as a B-defect.

The inspector should realize that measuring the overall dimensions of the board does not ensure that the contour is located correctly. Particularly in the case of plug-in boards, it is necessary to check the contour location with respect to the pattern, especially the connector system, whether it be an edge connector system or a two-part connector system.

20.12 Contour Location Determined by Corner Marks

When the contour is determined by corner marks, the corner marks disappear during machining. The inspector can therefore detect possible edge displacements only by comparing the board with the master film, unless the PCB designer has placed measuring points internally on the board as mentioned in Paragraph 15.3.1e.

The inspector lays a positive master film over the board and lines it up with the pattern. Using a magnifier with graticule, he measures the edge displacements as the deviations between the internal side of the corner marks and the board edges. See Figure 20.9.

When no tolerances regarding edge displacements have been stated, the inspector should expect the following tolerances to be met. See the discussion in Paragraph 15.5.1 regarding attainable tolerances.

1. Contouring based on corner marks alone:

   Displacement of all edges:

   ± 0.012 in. (± 0.3 mm)
2. Contouring based on corner marks,

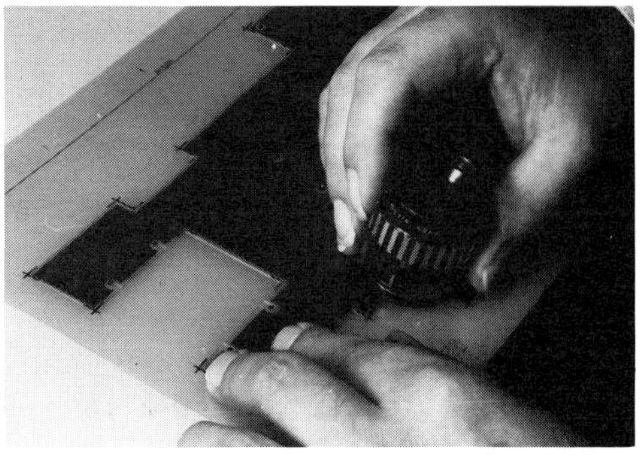

Fig. 20.9. Measuring the displacement of the contour with respect to the corner marks of the master film. The measurement is carried out by means of a magnifier with graticule.

but combined with machining of two sides against a stop:
Displacement of first two machined edges:

±0.012 in. (±0.3 mm)

Displacement of opposite edges:

±0.020 in. (±0.5 mm)

3. Contouring based on PCB manufacturer's own tooling holes:
Displacement of all edges:

±0.008 in. (±0.2 mm)

Method 2 is most often used. Provided no edge location tolerances have been specified, the inspector must accept edge displacements of ±0.012 in. (±0.3 mm) and ±0.020 in. (±0.5 mm), respectively.

The possibilities of exemption are the same as in Paragraph 20.11, Items 1 and 2. The inspector should realize that when the contour of plug-in boards is based on corner marks, any determination is so vague that it should be avoided. The PCB designer should always use a reference system as this unambiguously determines the machining tolerances.

20.13 Contour Location Determined by a Reference System

20.13.1 Measuring the Locations

With a reference system, the contour is locked to the pattern which, in this context, is represented by the tooling holes. All measurements of the contour location are therefore based on the tooling holes, and most precisely carried out by means of a coordinate table as illustrated in Figure 6.4 or 20.4.

If a coordinate table is not available, the measurement is performed by means of a caliper, preferably the special type (Paragraph 16.4, Item K), as shown in Figure 20.10.

In agreement with Paragraphs 6.2.2b and 6.2.2c, the tooling hole is in principle defined by the imaginary center of the solder pad of the tooling hole, and not by the center of the drilled hole. When measuring the location by means of a coordinate table, the rim of the solder pad is used. The special caliper implies that the cone is centered in the drilled hole, but this can mean a slight error.

When the tooling holes are drilled in the first drilling operation (which should be specified in the master drawing) and the hole size is reasonably large—for example, 0.12 in. (3 mm)—the tooling holes

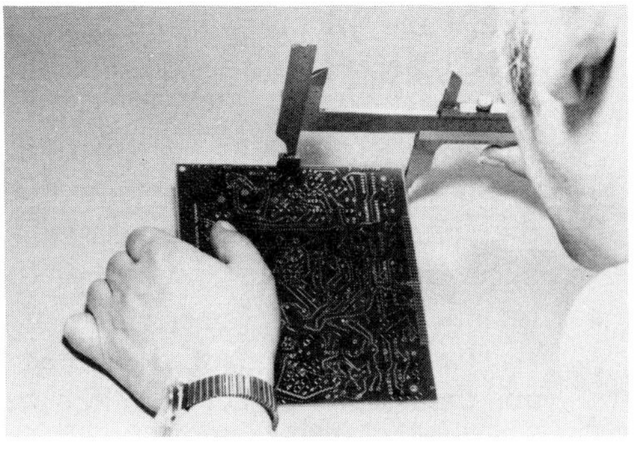

Fig. 20.10. Measuring the location of the contour with respect to the reference system by means of a special caliper.

can be expected to lie correctly within ±0.002 in. (±0.05 mm).

The inspector now measures the distance from the main tooling hole to the board edges which should, of course, be dimensioned on the master drawing or determined by coordinates.

20.13.2 Permanent Dimensional Change

When the board has been subject to a permanent dimensional change during manufacturing, the edges are displaced with respect to the pattern. The worst-case value of a permanent dimensional change is ±0.05 percent.

Measuring a possible dimensional change almost certainly requires a coordinate table. The inspector ascertains the coordinates of the tooling holes by a very careful measurement and calculates the mutual distances. These are compared to the dimensioning on the master drawing; the difference corresponds to the dimensional change.

As mentioned in Paragraphs 15.5.2b and 15.5.2c, it is possible to obtain contouring tolerances (edge displacements) of ±0.002 to ±0.004 in. (±0.05 to ±0.10 mm) by blanking or NC routing.

In the light of this discussion, the inspector should leave a certain margin due to a possible permanent dimensional change. A reasonable tolerance of the edge displacement relative to the pattern is ±0.008 to ±0.012 in. (±0.2 to ±0.3 mm), depending on the size of the board. This implies NC routing (most commonly used today) or blanking. With manual routing (pin routing), the tolerance tends towards ±0.012 in. (±0.3 mm).

When a narrower tolerance is necessary, it must be indicated on the master drawing so that the PCB manufacturer can take any necessary measures (Paragraph 6.4).

There seems to be a tacit agreement within the PCB industry to ignore permanent dimensional changes in the capability specifications. However, increasing attention is being paid towards this phenomenon, presumably enforced by the growing use of automatic component insertion machines.

20.13.3 Board Disposition

The possibilities of exemption are the same as mentioned in Paragraph 20.11, Items 1 and 2. The inspector should not necessarily perceive the use of a reference system as tantamount to requiring very narrow tolerances regarding the edges. The same is also true for using an NC routing machine.

When the contour exceeds the tolerances, the inspector should always assess the practical need for narrow tolerances before rejecting the batch.

20.14 External and Internal Cutouts

So far we have tacitly perceived the contour as regular, meaning a rectangular or square shape. However, many boards have external and/or internal cutouts and the location and size of such cutouts must of course also be inspected. This is performed in accordance with the same guidelines described in Paragraphs 20.12 and 20.13.

20.14.1 Internal Corners

Internal corners of such cutouts can present difficulties if the PCB designer has not taken the machining possibilities into account (Paragraph 15.6.1).

A common fault is to indicate sharp internal corners. When routing the contour, the corner will be rounded corresponding to the diameter of the mill bit, frequently ⅛ in. (3.2 mm) or 3/32 in. (2.4 mm). It is usually possible to waive such deviations, but a waiver should be preceded by an investigation of the actual requirement for sharp corners.

Deviations should, in principle, not be rated as a defect, since sharp corners are inconsistent with normal machining practice. When there is a real requirement for sharp corners, the batch is returned for a manual post-machining.

The inspector should in such cases urge the PCB designer to indicate an acceptable radius of curvature—for example, by stating $0 < R \leqslant 0.063$ in. (1.6 mm), or just $R_{max} \leqslant 0.063$ in. (1.6 mm). This makes it possible for the PCB manufacturer to use the standard size of router bit, with a diameter of either 1/8 in. (3.2 mm) or 3/32 in. (2.4 mm).

Roundings can be checked with cylindrical plug gauges, but it is easier to use a 7X or 10X magnifier. The normal reticle with a linear scale should be replaced by one showing circles in various sizes.

20.14.2 Proximity of Cutouts to Conductors

A not quite unusual fault is that an internal cutout is too close to a conductor. In some cases the cause is that the PCB manufacturer has made the cutout too deep, but more frequently the cause seems to be that the PCB designer or the PCB draftsman has underestimated the size of the cutout and placed the conductor too close to the edge (Figure 20.11).

The inspector is now faced with the problem of whether to exempt the fault. The decision should not depend on which party is responsible for the fault. An exemption implies that the conductor has not been damaged, and also that the insulation distance from the conductor to the edge has not become inadmissibly small or that the insulation can be improved. Finally, the conductor must show no sign of lifting from the base material. A solder mask is advantageous in such cases as it assists in keeping the conductor in place.

When these requirements are met, the inspector can exempt the fault, which is therefore rated as a B-defect. If not, the fault is rated as an A-defect, and the board is rejected.

The PCB manufacturer should, under all circumstances, have evaluated the first contoured board. If the cause was a design fault, the PCB manufacturer should have stopped further production and informed the customer.

Figure 20.12 shows a relatively clumsy cutout which can be exempted because it meets the functional requirements.

20.15 Edge Connectors

The machining tolerances for the edge connector area are usually somewhat closer than for the remaining board con-

Fig. 20.11. Conductor unacceptably close to a cutout.

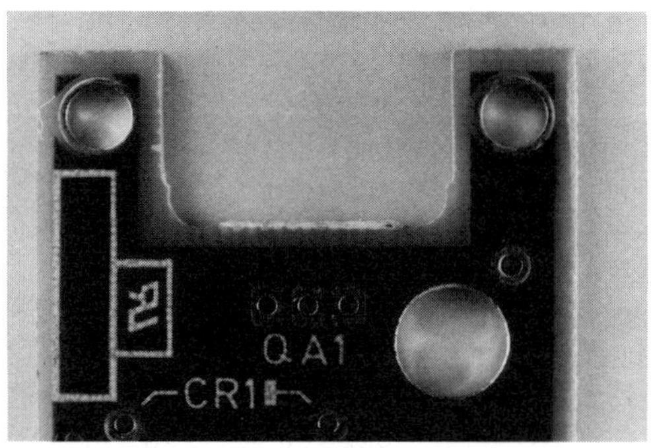

Fig. 20.12. Inaccurately executed cutout.

			EDGE CONNECTORS	20.15
			SLOT WIDTH O.O.T.	20.15.1
			SLOT LENGTH O.O.T.	20.15.2
			SLOT LOCATION O.O.T.	20.15.3
			CHAMFERING UNACCEPTABLE	20.15.4

Fig. 20.13. Edge connector. (Section of inspection record.)

tour to ensure a more precise mating with the receptacle. The inspector should use the same sampling plan as used in the preceding mechanical inspection. The inspection is performed in agreement with the part of the inspection record shown in Figure 20.13.

Paragraphs 15.7.1 and 15.7.2 have thoroughly addressed the slot of the edge connector. We shall next describe how to inspect the length, width, and location of the slot. Finally, we shall discuss the chamfering inspection.

20.15.1 Slot Width

The slot width is measured by feeler gauges. The inspector finds the thickest combination of feeler blades which just fits into the slot with no pressure applied (Figure 20.14). The fault is calculated as the difference between the thickness of the feeler blades and the nominal slot width.

Fig. 20.14. Measuring the width of the connector slot by means of a feeler gauge.

Frequently, a rather narrow tolerance is specified for the slot width, so exceeding this tolerance should, in principle, be rated as an A-defect. For very good reasons, the inspector cannot exempt a slot which is too narrow. When the boards are urgently needed, the assembly line can remedy the fault by touching up the slot edges with a small file. This particularly applies when the fault is not found in the incoming inspection and the boards have been assembled. In general, bare boards should be returned to have the slot remachined.

It is more problematic to exempt the fault when the slot is too wide. It is usually necessary for the PCB designer to recalculate the tolerance chain of the entire connector system. It is frequently possible to waive tolerances exceeding the specified values by as much as 0.002 to 0.004 in. (0.05 to 0.10 mm) in the case of a finer connector pitch p (0.1 in $\leqslant$ p $\leqslant$ 0.125 in.) and somewhat more in the case of a coarser connector pitch (p $\geqslant$ 0.150 in.).

20.15.2 Slot Length

The tolerance range of the length (or depth) of the slot is generally a little wider than that of the width, and the length is therefore measured accurately enough with a caliper.

It is possible to exempt even relatively large faults, which are rated as a B-defect. The criteria are that the slot is neither so long that it cuts into the pattern (Figure 20.15), nor so short that the board cannot be seated in the receptacle (an easily-checked criterion). When these criteria are not met, the fault is rated as an A-defect.

20.15.3 Slot Location

Just as important as the slot width is the slot location, which should always be measured with respect to the contact tabs.

Fig. 20.15. Interruption of conductor caused by too deep a slot.

Measurements with respect to the board edges are misleading because of the machining tolerances of the contour.

20.15.3a Measuring Methods

The measurement is made according to the method with which the slot location has been defined (Paragraph 15.7.1), and the inspector uses the same type of measuring equipment as when measuring the location of the contour. It is especially practical, when using a coordinate table, to determine the location of the slot in continuation of the other measurements made with the table.

1. When the slot is defined with respect to a general or local reference system, it is easiest to use a coordinate table. See Paragraph 6.4.8.

   By aiming at the slot, which means to set the bull's-eye in a symmetrical position with respect to the slot edges, the position of the centerline of the slot can be displayed as a coordinate.

   The difference between the measured value and the nominal value indicates the slot displacement.

   It is also possible to perform the measurement with a caliper. When the primary tooling hole is placed on line with the edge connector (Figure 15.11), it is particularly easy to perform the measurement, especially when the special caliper with two cones is used.
2. When the slot is specified to be located symmetrically with respect to the adjacent contact tabs, the distances a and b from the slot edges to the tab edges are measured with a 7X or 10X magnifier with reticle. The displacement ΔD is then calculated as

$$\Delta D = \tfrac{1}{2} (a - b)$$

20.15.3b Board Disposition

Exceeding the tolerance of the slot location is rated as an A-defect, and the board is rejected. It will probably require a 100 percent inspection of the batch.

A moderate deviation from the tolerances can be exempted, but this implies a thorough assessment of the tolerance chain of the connector system, including a consideration of a possible specification for an effective contact area (Paragraph 14.7.2 and Figure 14.14).

With narrow contact tabs, even a slight displacement of the slot causes the contact points to be moved more or less outside the effective contact areas. As mentioned in Paragraph 14.7.10, a moderate occurrence of pinholes and pores outside the effective contact area are accepted, so the slot must be correctly positioned. See Figures 20.16 and 20.17.

20.15.4 Chamfering of Edge Connector

The master drawing will usually indicate a chamfering of the edge of the connector. As mentioned in Paragraph 15.9, commonly specified tolerances of chamfering imply such large variations of the profile (Figure 15.14) that the inspection should

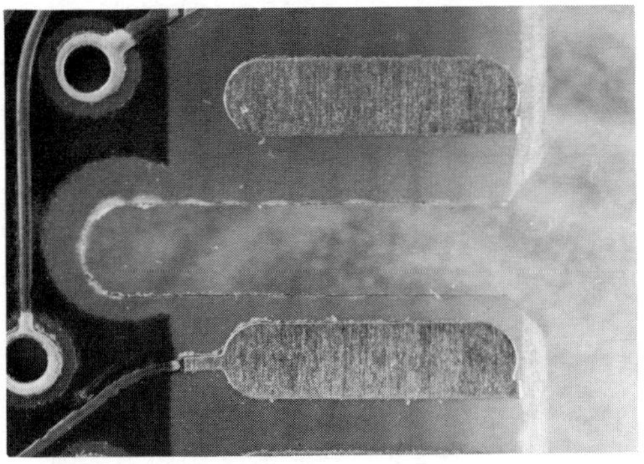

Fig. 20.16. Unsymmetrically located slot.

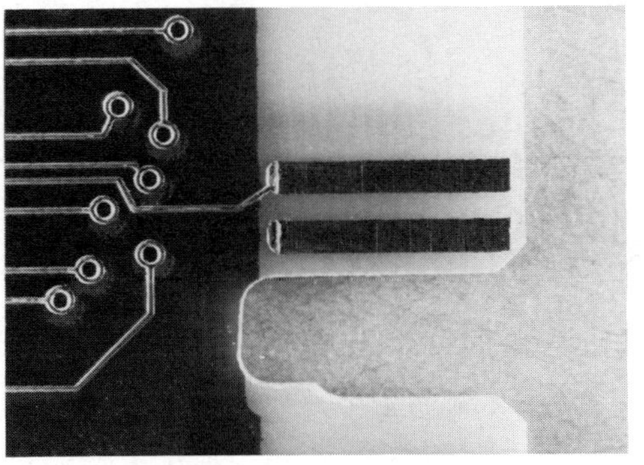

Fig. 20.17. Slot milled inaccurately.

be limited to visually verifying that the edge has been broken on both sides of the board. It should furthermore be verified that the chamfering does not go too far into the contact areas, normally not more than 0.016 in. (0.4 mm) from the edge.

Chapter 21

Microscopic Inspection

21.1 Introduction

Most of the pattern details are inspected with the naked eye, possibly aided by a 2X or 3X illuminated magnifier or a 7X to 10X magnifier with reticle. Although a stereo microscope can also be used, the magnification should not be so large that it reduces the viewing field or causes the inspector to lose the overall image and become overly involved in small details of no real importance.

The microscope is mainly used to indirectly inspect a board by means of microsections, as described in Chapter 25. It can, however, also be a tool for a direct and nondestructive investigation of certain pattern details, which is the subject of this chapter. See the part of the inspection record shown in Figure 21.1.

With the inspection described in Paragraphs 21.2 to 21.5, the inspector uses the measuring setup described in Paragraph 7.3.3c. The attainable light intensity is sufficient for the inspector to observe defects in the hole wall, but too low for acceptable photographic reproduction of the details.

Another problem is the limited depth of field, which makes it impossible to achieve a photographic reproduction with an acceptable definition over the entire hole depth. Therefore, the author has chosen to photograph the holes using direct illumination but no microscope.

MICROSCOPIC INSPECTION	21
VOIDS IN HOLE WALL	21.2
CRACKS IN HOLE WALL	21.3
INCOMPL. REFLOWING IN HOLES	21.4
HOLE WALL W. GROOVES	21.5.1
HOLE WALL W. NODULES	21.5.2
HOLE WALL CONTAMINATED	21.5.3
HOLE WALL W. GLASS FIBRES	21.5.4
HOLE WALL IN LANDLESS HOLES	21.5.5
OVERHANG BROKEN OFF	21.6

Fig. 21.1. Microscopic inspection. (Section of inspection record.)

21.2 Voids in the Hole Wall

A requirement for good quality is that boards with plated-through holes have good solder holes. A first indication of this is that no voids, or only a very limited number, are present in the hole wall.

As mentioned in the introduction to Chapter 7, the hole quality is usually quite uniform throughout the entire batch. This means that the plating is either good in all solder holes or it is inferior, whereas a mixture of good and inferior hole platings is extremely rare.

This uniformity justifies a relatively low inspection level, and we shall inspect at the same level as indicated for the inspection of hole diameters in Paragraph 6.8.5a. This implies that only a small number of solder holes are inspected, and that the holes inspected are distributed all over the sample. The inspection is based on the measuring setup described in Para-

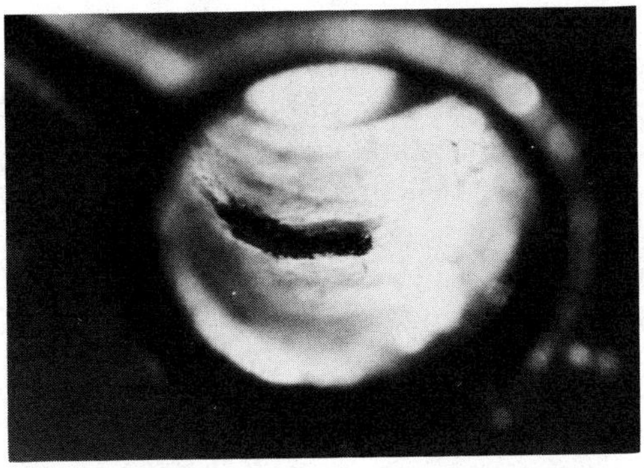

Fig. 21.2. Voids in the hole wall of a plated-through hole.

graph 7.3.3c. This inspection is quite revealing, and since the setup is so simple that it can be implemented in a few hours, this method is highly recommended.

It is most convenient to perform the examination by moving the board over the work table of the setup, so that the solder holes can be studied in turn under the microscope. The magnification should not exceed 10X or 20X.

Whether the board is reflowed or not, voids appear quite clearly as luminous spots in the normally dark solder holes (Figure 21.2). It is crucial that no light be transmitted from the under side into the hole being examined. The hole should be slightly displaced with respect to the slot in the work table (Paragraph 7.3.3c).

When solder holes with voids in the hole wall are found in the limited sample, the inspector should extend the sample to attain a better basis for assessment. The holes may be chosen in a single board of the sample. According to Paragraph 7.3.3d, the assessment criteria are as follows:

1. No more than three voids per solder hole in a maximum of 2 percent of the solder holes.
2. The total area of voids per solder hole shall not exceed 5 percent of the area of the hole wall.
3. There may be no voids at the pad-to-barrel transition.

The inspector will quickly know whether the board can be accepted or must be rejected. Exceeding just one of these requirements is cause for rejection, and the fault is consequently rated as an A-defect.

The inspector continues his investigation by examining additional boards; he will thereby quickly be able to confirm the validity of the experience that the quality of the holes is either satisfactory or generally inferior. When the full sample exhibits the same inferior result, the whole batch should be rejected.

21.3 Cracks in the Hole Wall

While inspecting the hole walls for voids, the inspector also checks for cracks, particularly circumferential cracks. A crack is indicated by a band of light in the dark hole wall.

Cracks usually occur in the transition zone between the plated-through hole and the solder pad; this is commonly called "corner cracking." See Figure 21.3.

Cracking is an indisputable cause for rejection, and the fault is rated as an A-defect. The inspector should not waive

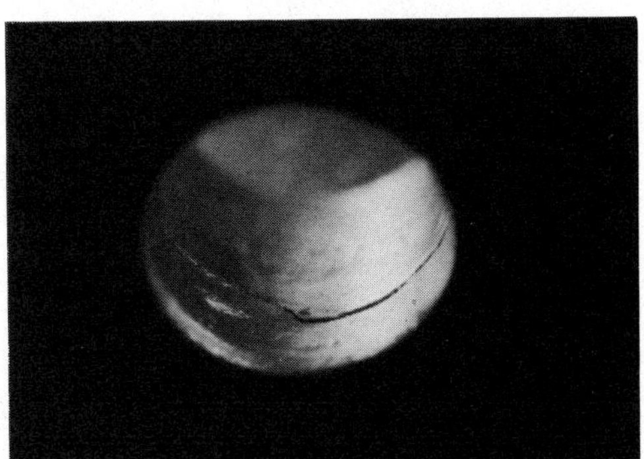

Fig. 21.3A. Circumferential crack in the hole wall of a plated-through hole, caused by thermal expansion of the base material.

Fig. 21.3B. Circumferential hole in the hole wall of a plated-through hole. This defect should not be mistaken for the crack shown in Fig. 21.3A since it is caused by an etching fault.

even short cracks, as they tend to grow in time.

This method cannot reveal microcracks.

21.4 Incomplete Reflowing of Solder Holes

While inspecting for voids in the solder holes, the inspector should also check whether the hole walls have been reflowed to their full depth. As mentioned in Paragraph 8.7.1a, an incorrect tin/lead alloy composition can cause an incompletely reflowed area halfway down in the solder holes (Figure 21.4). Such holes are often called "black holes."

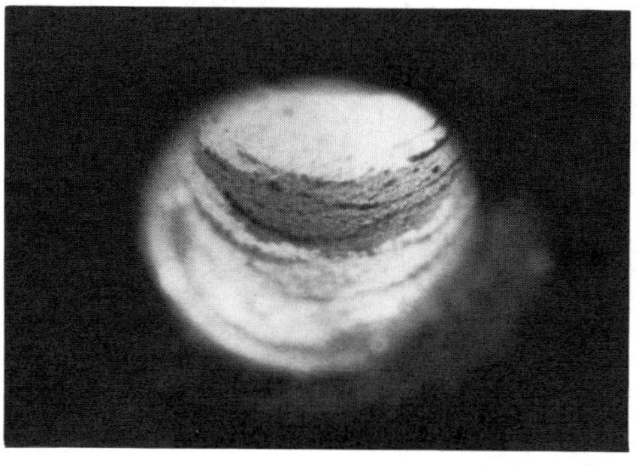

Fig. 21.4. Incompletely reflowed ring-shaped area halfway down in a plated-through hole, usually called a "black hole."

The incidence of black holes is extremely rare. As they reflect incorrect process conditions and can cause soldering problems, the inspector should reject such boards and rate the fault as an A-defect. Any possible exemption will depend on a solder test with subsequent microsectioning.

21.5 Irregular Hole Wall

By illuminating the board surface, the inspector can examine the hole wall for various irregularities, using the measuring setup previously described.

21.5.1 Hole Wall with Drilling Grooves

A hole wall which, even after reflowing, exhibits rounded but marked traces of drilling grooves, indicates poor hole drilling (Figure 21.5). Because such cases imply a risk of cracking of the copper layer (Paragraph 7.3.5), microsections should be prepared for a closer examination of the irregularity. See Chapter 25.

21.5.2 Hole Wall with Nodules

The inspector cannot always expect to find smooth and shiny plated-through holes. Because of process faults, small and harmless irregularities can occur in the form of small nodules on the hole walls.

Fig. 21.5. Hole wall of a plated-through hole with distinct traces of drilling grooves.

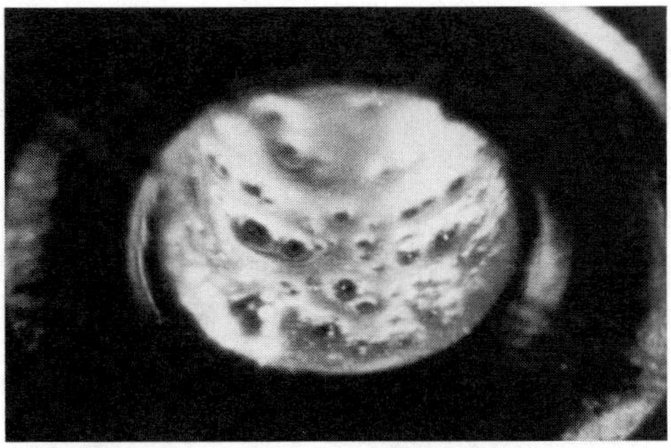

Fig. 21.6. Hole wall of plated-through hole with nodules.

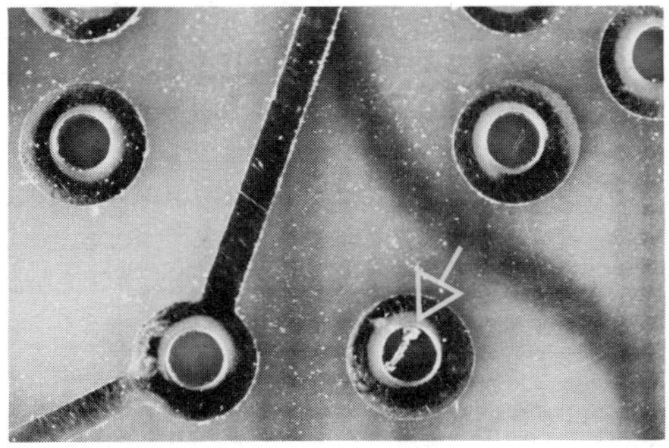

Fig. 21.7. Hole wall of reflowed plated-through hole with protruding glass fiber which has been plated together with the hole wall.

Only in the case of large nodules, projecting more than 0.008 to 0.012 in. (0.2 to 0.3 mm), should the board be rejected and the defect rated as an A-defect. The reason is that component insertion is hampered, or the nodules break off during component insertion. Smaller nodules are rated as a B-defect and the boards are accepted. See Figure 21.6.

21.5.3 Hole Wall with Contamination

Although the solder holes are examined for contamination during the visual inspection, the inspector should notice any contamination in holes examined according to Paragraphs 21.5.1 and 21.5.2.

Dust, drill chips, and other foreign particles in the solder holes are rated as a B-defect since they can be easily removed with compressed air.

Flux residues from reflowing, in the form of small brown or black crusts, are also rated as a B-defect, provided they do not occur at an excessive rate (Paragraph 7.4.2).

Screen-printing ink in solder holes is cause for rejection and rated as an A-defect. See the remarks in Paragraph 7.4.3 concerning responsibility. The inspector should note that the via holes of many boards have been intentionally overprinted, which implies that screen-printing ink will be found in such holes.

21.5.4 Hole Wall with Protruding Glass Fibers

Protruding glass fibers are caused by inferior hole drilling and are plated with the rest of the hole wall. See Figure 21.7 and Paragraph 7.2.7. The inspector should be aware of the fundamental difference between nodules and protruding glass fibers.

When larger nodules occur, the inspector frequently finds many solder holes with nodules, and each hole usually has several nodules. Protruding glass fibers generally occur in a very limited proportion of solder holes, if they occur at all, and there will probably be only a single glass fiber protruding into such holes.

Although the glass fiber is metallized, it can break off at its root when the component lead is inserted, and possibly leave a void in the hole wall. This is, however, accepted according to the criteria of assessing voids in the hole wall.

Reflowing can cause an accumulation of tin/lead at the root of the glass fiber so that the hole becomes partially blocked by tin/lead. The inspector should therefore check whether such a glass fiber will interfere with assembly.

He can, for example, insert a suitable cylindrical gauge in a few of the solder holes and note whether the glass fiber breaks reasonably easily, and also note whether it leaves a reasonably small void in the hole wall (Paragraph 7.3.3c). If so, and the number of solder holes with protruding glass fibers is less than 1 percent of the number of holes, the fault is rated as a B-defect and the board is accepted. If not, the fault is rated as an A-defect and the board is rejected.

A thread in a solder hole need not always be a metallized glass fiber. It can also be a brush fiber originating from the brushing or scrubbing of the copper foil immediately before the electroless copper was deposited. Inadequate cleaning of the production panel after the brushing or scrubbing can leave brush fibers in the solder holes so that they are plated in the same way as protruding glass fibers.

21.5.5 Hole Wall in Landless Plated-Through Holes

When, as an exception, landless plated-through holes occur in a board, the inspector should inspect the hole wall of a limited number of such holes—for example, corresponding to Special Inspection Level S3 (Table 6.1 of Paragraph 6.8.5a).

It seems reasonable to require the plating to cover at least 70 percent of the hole wall depth. The hole wall must meet the same requirements valid as ordinary plated-through holes, however, with the limitation that the requirements of Paragraphs 21.2 to 21.5.4 only apply to the fully plated part of the hole wall—that is, to a depth of 50 percent of the hole, starting from the solder pad. See Figure 21.8.

21.6 Broken-Off Overhang

Paragraph 18.9.5 describes a visual inspection of the board for attached as well as fully or partially detached slivers. Paragraph 19.5, Item 2, and Paragraph 19.3 describe a tape test.

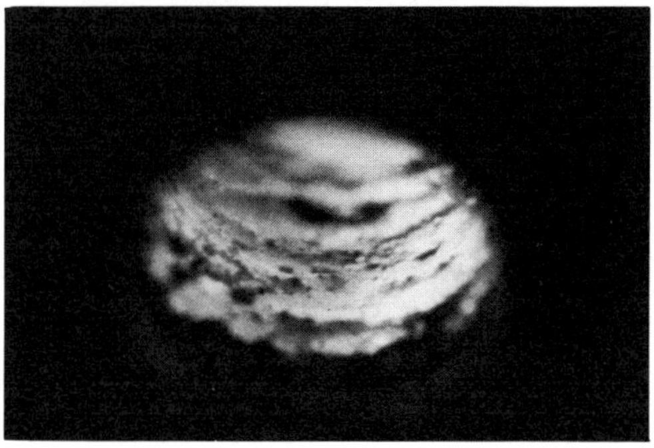

Fig. 21.8. Hole wall of reflowed plated-through landless hole. The plating stops inside the hole.

Although it is extremely rare to find fully or partly broken-off slivers causing short circuits, the author knows of unfortunate instances of completely assembled boards which had to be scrapped because of such faults. The cause was primarily that the boards had solder and insulation masks, which rendered the visual inspection with magnifier insufficient and the tape test useless.

It is therefore a wise precaution for the inspector to take a little extra time to inspect the surface of a sample board with the microscope, especially when the board is provided with solder and insulation masks.

Since the detached overhang appears as very fine slivers—for example, with a cross-section of 0.001 by 0.001 in. (25 by 25 μm)—a fairly intense surface illumination and a magnification of 10X to 20X are necessary.

Illuminating the under side of the board is especially necessary when the board is provided with solder and insulation masks. In this way a detached overhang appears very clearly as a dark shadow (Figure 21.9).

Cracking along the conductor edges as shown in Figure 21.10 is another infallible fault indication. The cracks can be difficult to see under the masks, so a strong surface illumination and a reasonable

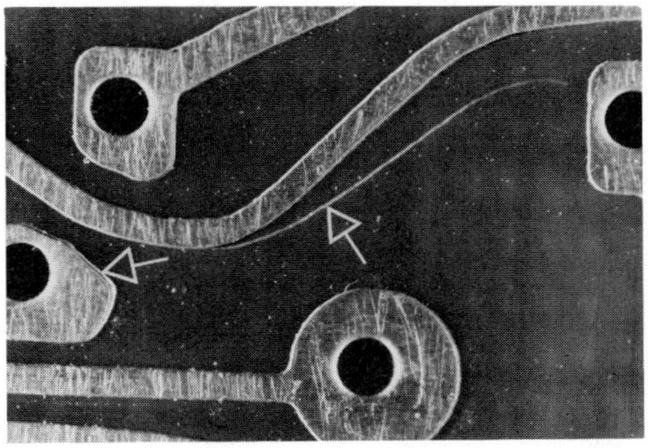

Fig. 21.9. Board with fully or partly broken-off overhang along the conductor edges.

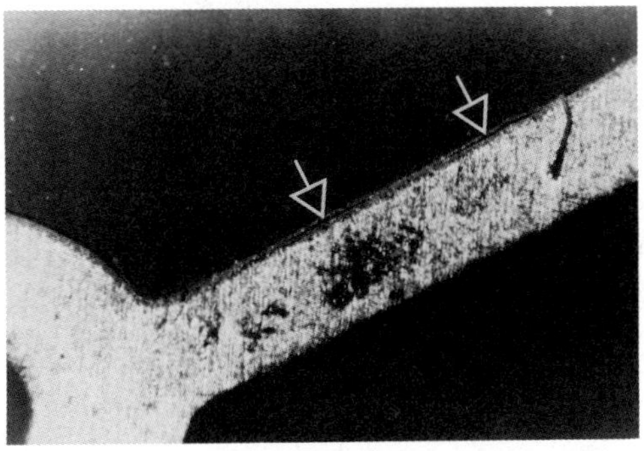

Fig. 21.10. Board with cracks along the conductor edges.

magnification of not less than 10X to 20X is necessary.

Broken-off overhang (slivers) or cracking along the pattern edges are an indisputable cause for rejection, and the fault is always rated as an A-defect. When such faults are found, the total batch should be inspected.

Chapter 22

Electrical Inspection

22.1 Introduction

The electrical inspection mainly ensures that the board meets the electrical requirements of the electronic circuitry. This chapter addresses only nondestructive methods, which the incoming inspection department can use on a routine basis. The inspector should not slavishly follow that part of the inspection record shown in Figure 22.1, as not all tests listed are equally desirable or feasible in routine work.

ELECTRICAL INSPECTION				22
			SHORT CRCT./INTERR. (MAN. TEST)	22.2
			SHORT CRCT./INTERR. (AUTO TEST)	22.3
			INSULATION RESISTANCE	22.4
			COPPER PLATING O.O.T.	22.5

Fig. 22.1. Electrical inspection. (Section of inspection record.)

22.2 Manual Test for Short Circuits and Interruptions

An earlier discussion established that short circuits and interruptions are not especially unusual. The inspector can, in theory, perform an electrical measurement of the sample boards in order to see whether short circuits or interruptions were overlooked during the visual inspection.

Complex boards primarily need such an inspection, but it requires so many measurements as to be impractical. Short circuits are detected by a resistance measurement from every conductor or solder pad to adjacent conductors and solder pads. Interruptions are identified by continuity measurements over the full length of each conductor.

This precaution could easily mean several hundred measurements per board, and as the frequency of faults is very low, a manual inspection is seldom worthwhile. Therefore, it is used only when special conditions favor such an inspection—for example, inspection of multilayer boards with internal voltage and ground planes. Here it is more practicable to check whether there are any unintended short circuits between the solder pads and the internal planes.

Short circuits and interruptions found by a manual test are rated as an A-defect, as described in Paragraphs 18.2 and 18.3.

22.3 Automatic Test for Short Circuits and Interruptions

This book does not attempt to address the principles of automatic testing of bareboards for short circuits and interruptions in the pattern. Should the occasion arise, such a test would be considered a 100 percent screening. It would also require large expenditures for the necessary equipment and for the development and maintenance of the test programs and jigs.

Since some PCB users do not wish to invest in this type of automatic test equipment, and others do not wish to load existing equipment with bare-board testing, it has become relatively common for PCB manufacturers to offer automatic testing of the boards before delivery.

This makes it possible to ensure that no short circuits or interruptions occur in the pattern. The customer usually pays a lump sum for the test program and the jig, plus a per-board fee. Especially for complex boards (including multilayer boards), this can be an economically attractive practice.

It should be emphasized that it is difficult for the test to reveal "almost" short circuits and "almost" interruptions, although testing with increased voltage and current can improve the confidence level. However, the incoming inspection department is still required to assess the general quality of the boards.

It has recently become more common to perform an automatic testing of the boards after assembly and soldering, preferably of the inserted components. See Figure 22.2. It also is usually possible to test the pattern for complete interruptions or short circuits. Particularly the latter is of great importance because there is always some risk that short circuits in the form of solder bridges have formed during soldering.

Fig. 22.2. Automatic testing of bare multilayer board for short circuits and interruptions. The multilayer board is placed in the test jig shown at the right, and the test equipment in the background makes a printout of the result.

Short circuits and interruptions which can be referred directly to the pattern are assessed as A-defects, as described in Paragraphs 18.2 and 18.3, even though they are detected long after the receiving inspection has been performed and the boards accepted.

22.4 Test of Insulation Resistance

It is seldom necessary to routinely test the insulation resistance of the board. When a measurement is required, it is best done with a megohmmeter; the test voltage is often 100 volts dc, and in some cases 500 volts dc.

A common requirement is that the insulation resistance between any two parts of the pattern—for example, between two parallel conductors—must be greater than 300 megohms, and in some cases 500 megohms. Chapter 28 details the use of a megohmmeter for measuring resistance.

In order to illustrate the order of magnitude, the insulation resistance has been measured between the points A and B of an epoxy glass fiber board as shown in Figure 22.3, and the result has been stated in the caption.

It is rare to find a minimum value of the insulation resistance stated in a PCB specification. As shown in Chapter 28, the insulation resistance depends on the length and spacing of the conductors, as well as the environmental conditions, primarily the relative humidity and temperature. It is therefore difficult to specify a general value, but normally the insulation resistance should not be less than 10^4 to 10^5 megohms for pattern designs occurring in practice.

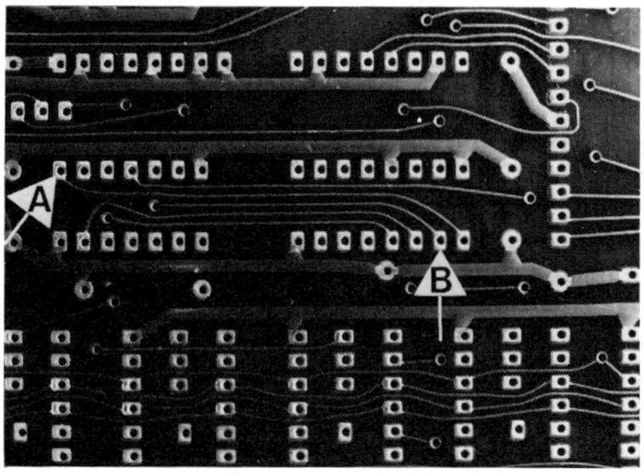

Fig. 22.3. Example of the magnitude of the insulation resistance between two parallel conductors. The measuring points are A and B.

Conductor length:	1.2 in. (30 mm).
Average spacing:	0.028 in. (0.7 mm).
Insulation resistance:	5 x 10^7 MΩ.
Test voltage:	100 V dc.

Conversion of the test result to other conductor lengths and spacings can be performed according to the guidance given in Paragraph 28.5.

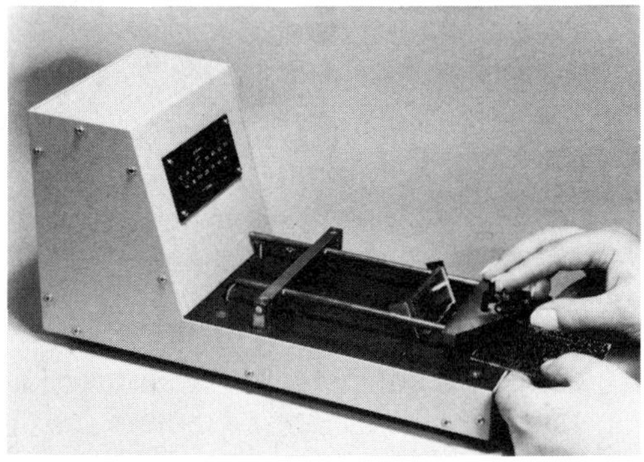

Fig. 22.4. Measuring the copper plating thickness of plated-through holes.

As previously mentioned, measuring insulation resistance is not part of the daily routine inspection, except for black screen-printed masks, which should always be checked. The PCB manufacturer can, for example, have changed to another supplier of printing ink, resulting in a drastic drop in the insulation resistance to an unstable value of a few megohms. It is not necessary to check more than one board of the sample.

22.5 Test of Copper Plating Thickness

The most important test which the inspector should perform routinely is, in the author's experience, ascertaining the thickness of the copper plating in plated-through holes. The test can be done with the test setup mentioned in Paragraph 7.3.1d3. See Figure 22.4.

Because of problems regarding contacting and current distribution in the examined holes, the test should not be considered a definitive measurement, but rather a nondestructive test, which can give a quick indication of abnormal or insufficient plating.

This test should be made on a sample of all batches. The extent should be as indicated in Paragraph 6.8.5a for balanced plated-through boards, possibly reduced to only 25 solder holes.

Unfortunately, most commercially available measuring instruments for checking the resistance of plated-through holes are extremely expensive. The reason presumably is that they are usually available as universal instruments adaptable to a variety of measuring parameters, primarily hole diameter and board thickness. Furthermore, it is possible to display the thickness of the copper plating either in mils or microns.

A more moderately priced equipment (Paragraph 7.3.1d4), which is designed for just one hole diameter (0.031 in. = 0.8 mm) and one board thickness (0.062 in. = 1.6 mm), falls in a far lower price range, within the reach of most electronics companies.

As the test has some uncertainty (Paragraph 7.3.1d3), we shall accept plating thicknesses down to 0.0008 in. (20 μm). No remarks are therefore entered in the inspection record. The actual part of the inspection record is shown in Figure 22.1. Thicknesses below 0.0008 in. (20 μm) are rated as an A-defect, and the board is returned to the PCB manufacturer for a closer investigation. As mentioned in Paragraph 7.3.1d3, a waiver requires microsectioning so that the hole quality can be assessed. See Chapter 25.

Chapter 23

Supplementary Solderability Inspection

23.1 Introduction

During the visual inspection, described in Chapter 18, the inspector assesses the board's solderability. This inspection mainly uses the appearance of the reflowed tin/lead surface to indicate the solderability. See Paragraphs 18.9.3 and 18.9.7.

Although this assessment is based on fairly clear criteria, the inspector will occasionally feel some uncertainty about the evaluation. It is recommended, under these circumstances, that he arrange a solderability test on one of the sample boards. Although such a test is not considered part of the routine inspection, it is included here because it lies on the borderline between the routine inspection and the special inspections mentioned in Part 4.

23.2 Solderability Test

A major requirement of printed circuit boards is that they can be soldered satisfactorily. It is obviously preferable to test the solderability on the production equipment, under the same conditions as normal production. This gives an immediate indication of the probable result of production-soldering the boards.

A disadvantage is, of course, that the inspector cannot express the solderability by a number, but in case of a complaint he can return the soldered board as proof of the reduced solderability. Naturally, the soldering machine must be correctly adjusted and operated (Paragraph 10.2.5).

The sample board is soldered in exactly the same way as all production boards except that no components are inserted. Immediately before the trial soldering, the sample board should, in the same way as the production boards, be baked in an oven for 1 hour at 230°F (110°C). The trial soldering is carried out as shown in Figure 23.1.

If the trial soldering is unsatisfactory, the inspector should inform the PCB manufacturer of the following soldering parameters:

Fig. 23.1. Trial soldering a board in a soldering machine. The vehicle with the sample board is passing the solder bath and is just starting to lift.

SOLDERABILITY INSPECTION				23
			POOR WETTING OF PATTERN	23.3.1
			POOR WETTING OF HOLES	23.3.2
			NONWETTING	23.3.3
			DEWETTING	23.3.4
			BLOWHOLES	23.3.5
			WEBBING	23.3.6

Fig. 23.2. Solderability inspection. (Section of inspection record.)

1. Preheating time and temperature
2. Fluxing (type of flux and degree of activation: R, RMA, or RA)
3. Soldering time and temperature
4. Type of soldering machine

In addition, the inspector should return the soldered board with a description of the soldering faults causing the rejection.

23.3 Solderability Assessment

The board's solderability is evaluated by visual inspection. In order to avoid overly subjective assessments, the inspector must examine the board for those faults characteristic of inferior solderability. If such faults do not occur, the solderability is satisfactory. The inspector should follow that part of the inspection record shown in Figure 23.2 when assessing the trial soldering.

23.3.1 Poor Wetting of Pattern

The inspector evaluates solderability by first looking at the soldered surface of the pattern. Reduced solderability shows up as a somewhat uneven surface, possibly with a number of solder drops (or lumps) on the conductors and solder pads. The unevenness and the drops exhibit large contact angles (sharp transitions). The solder does not show a true tendency to wetting the surface, in that it does not properly flow over the surface. This is called poor wetting. See Figure 23.3 and Paragraph 8.7.2.

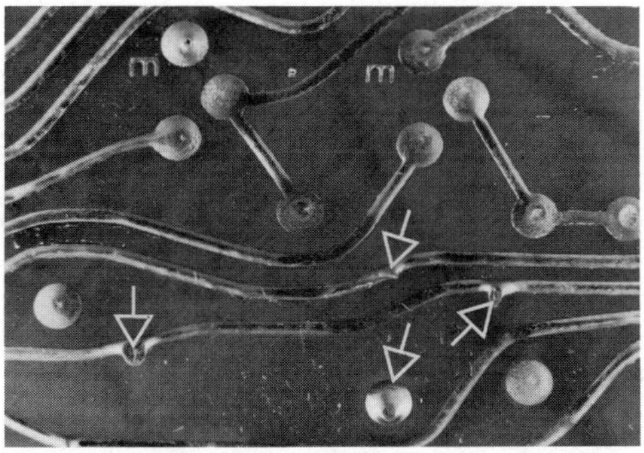

Fig. 23.3. Board with reduced solderability. Quite a number of solder lumps occur on the conductors and solder pads.

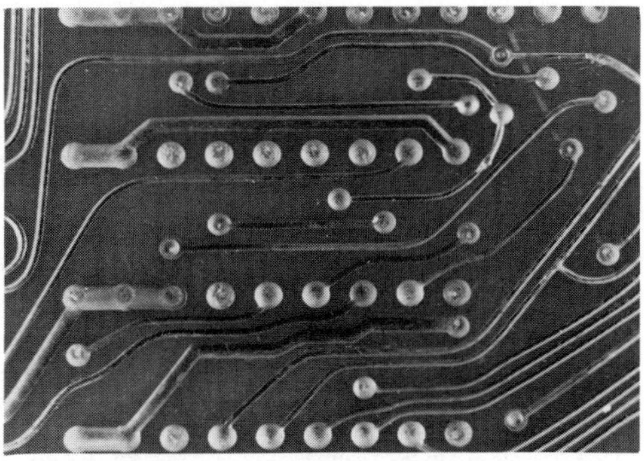

Fig. 23.4. Board with good solderability.

Satisfactory wetting is characterized by a flat appearance of the soldered surface—that is, the contact angle is small as shown in Figure 23.4.

Poor wetting is rated as an A-defect. Since the entire batch probably suffers from this fault, all boards are returned.

23.3.2 Poor Wetting of Plated-Through Holes

The wetting propensity can also be assessed from the distance the solder flows into and up through the plated-through holes. It is possible to set up requirements of various degrees of severity regarding the filling of holes.

Military requirements are the most severe, since wetting of the hole wall and

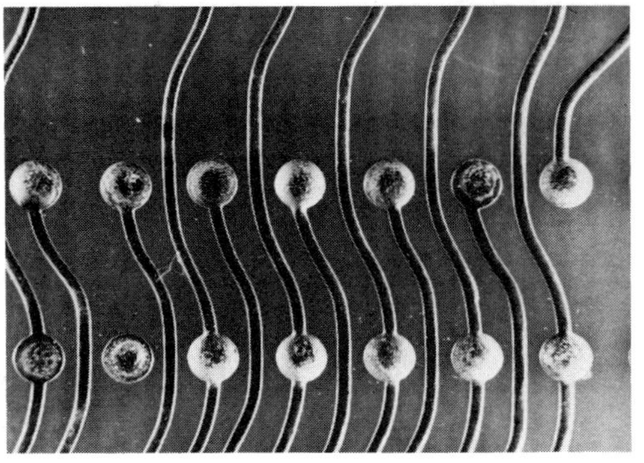

Fig. 23.5. Board exhibiting good filling of the solder holes after mass soldering. Component side of board is shown.

the solder pad of the component side is required—that is, a total filling of all solder holes. This protects the circuitry against possible cracking in the barrel of the plated-through hole (barrel cracking), as well as in the transition zone between the solder pad and the barrel (corner cracking).

For the inspection suggested here, we shall reduce this requirement somewhat and only demand a total filling of 95 percent of the solder holes. Total filling means wetting of the hole wall (barrel) right up to the rim, but not necessarily wetting of the solder pad. When this is met, no remarks are entered in the inspection record. See Figure 23.5.

Reduced wetting occurs when only 90 to 95 percent of the solder holes are totally filled. This is rated as a B-defect. Poor wetting occurs when less than 90 percent of the holes are totally filled. The fault is rated as an A-defect and the batch is returned.

It should be noted that several conditions other than a tendency to inferior wetting can reduce the filling. The empty holes reduce the capillary effect somewhat. Moreover, a poor pattern design with wide conductors or large metal areas can cause excessive heat transfer from the solder pad on the component side, permitting the temperature to drop too fast at the upper part of the barrel.

These requirements can of course be changed to meet the needs of the company. The improved copper baths currently in use deposit much more ductile copper than the older types of baths. In the author's experience, the improved baths and the aforementioned requirements are enough to ensure the reliability of commercial electronic equipment used under benign operating conditions.

23.3.3 Nonwetting

Nonwetting appears both as large and small areas with visible copper. The copper surface very clearly repels the solder, resulting in a large contact angle between the nonwetted area and the solder. See Figure 23.6 and Paragraph 8.7.2b. Nonwetting is rated as an A-defect and the batch is rejected.

When assessing the soldered surface of large boards with an organic protective coating, the inspector can easily confuse nonwetting with apparent nonwetting, which occurs particularly with drag soldering machines.

Gas pockets cause apparent nonwetting by preventing the solder from making contact with the pattern at the pocket locations. Gas pockets can form because of

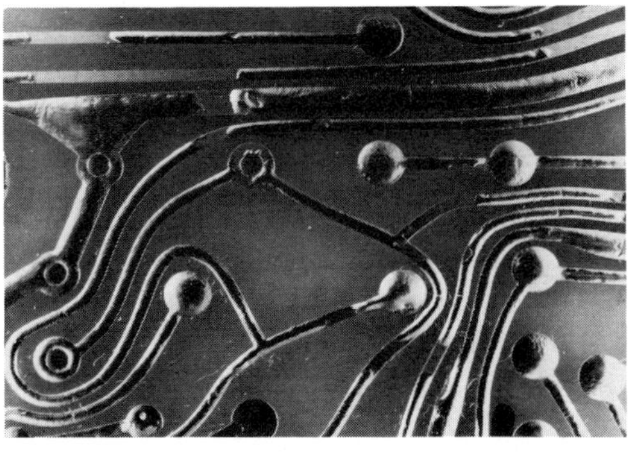

Fig. 23.6. Board exhibiting nonwetting after mass soldering.

Fig. 23.7. Gas pockets occurring during mass soldering.

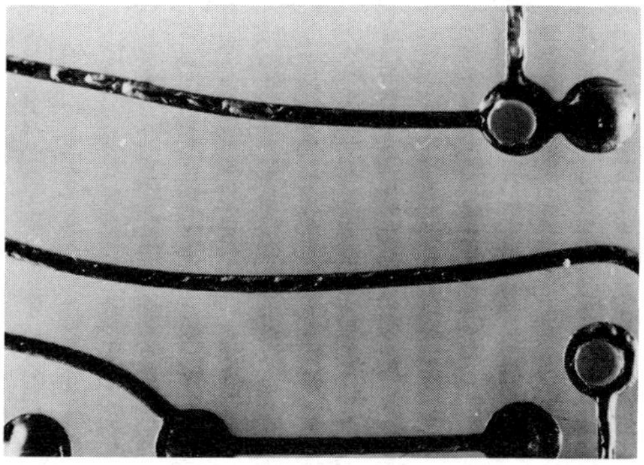

Fig. 23.8. Board exhibiting dewetting after mass soldering.

excessive fluxing combined with preheating for too short a time or at too low a temperature. The tendency increases in board areas lacking "ventilation"—that is, without solder holes or mounting holes through which the gasses can escape.

On bare-copper boards having an organic protective coating, but no solder mask, gas pockets appear clearly in the form of apparent nonwetting. The red unsoldered copper indicates, as shown in Figure 23.7, the extent of the gas pockets. Gas pockets can also appear beneath tin/lead-plated or solder-coated boards. Since it is usually difficult to distinguish between unsoldered areas and the commonly soldered surface, one cannot talk of nonwetting in terms of the previous definition. Based on the appearance of the flux, the inspector can possibly see the extent of the gas pocket. Apparent nonwetting is easily verified by resoldering the board. If the area of suspected nonwetting then disappears—that is, it is wetted by the solder—this is a case of apparent nonwetting.

23.3.4 Dewetting

Even when the solder can wet the surface, it may pull back to form round or protracted accumulations with fairly large contact angles. Since the solder has left a very thin layer, no copper is visible in areas with dewetting. See Figure 23.8 and Paragraph 8.7.2b.

Dewetting of the solder pads is cause for rejection and the fault is rated as an A-defect. On the other hand, dewetting of up to 25 percent of the area of a conductor or a ground/voltage plane is accepted, provided no soldering has to be done in the dewetted areas. This exception is not valid for soldering terminals—for example, those intended for flat packs. Dewetting exceeding 25 percent is rated as an A-defect, and the batch is rejected.

23.3.5 Blowholes

By themselves, blowholes as shown in Figure 23.9 indicate nothing of the board's

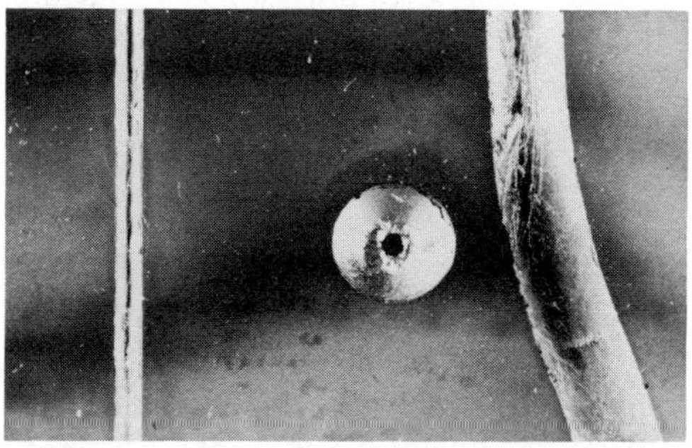

Fig. 23.9. Solder fillets with blowholes after mass soldering.

solderability, but they form part of the general assessment of the trial soldering. As mentioned in Paragraph 10.7.4a, blowholes can be caused by certain PCB manufacturing processes, as well as inferior mass soldering. If no blowholes occur in the production boards soldered immediately before and after the sample board, the fault is probably caused by the board itself. We can therefore use the following criteria when assessing the results of the trial soldering.

Blowholes with craters less than 0.01 in. (0.25 mm) in diameter are accepted with no remarks entered in the inspection record, provided blowholes are not present in more than 2 percent of the solder fillets.

Blowholes in up to 5 percent of the solder fillets are rated as a B-defect, and the batch is accepted. When the blowholes occur in more than 5 percent of the solder fillets or exceed 0.01 in. (0.25 mm) in diameter, the fault is rated as an A-defect and the batch is rejected and returned.

23.3.6 Webbing

Another important phenomenon to be assessed is webbing. Here the solder adheres to the base material in the form of thin threads, which can cause short circuits. See Figure 23.10.

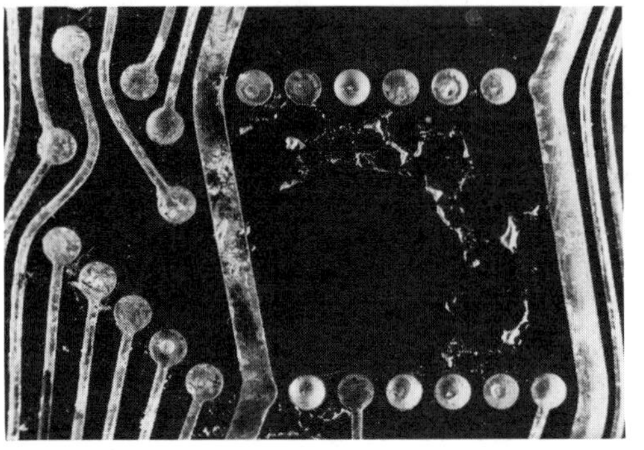

Fig. 23.10. Board exhibiting webbing after mass soldering.

Webbing can be caused by the PCB manufacturer's processes, or it can be due to mass soldering. Paragraph 10.7.1 discusses this fault and describes procedures for fixing the responsibility for the fault.

When the fault is due to a soldering process, the process can always be changed so that the fault is avoided. On the other hand, when it is caused by incomplete curing of the base material or by a screen-printed solder mask, the PCB manufacturer is responsible.

Unfortunately, webbing on a dry-film solder mask is not a rare fault. Although the inspector cannot prove that it is caused by lack of post-curing of the dry film (Paragraph 12.2.1e), there are good reasons for holding the PCB manufacturer responsible.

Moderate webbing, corresponding to a few adhering threads which can be removed without difficulty, is rated as a B-defect. Widespread webbing, which is difficult to remove, is rated as an A-defect and the batch is rejected and returned.

The inspector should be aware of the difference between webbing and dross on a board. Dross consists of impurities in the solder pot of the soldering machine, usually tin and lead oxides. Before performing the trial soldering, the inspector should ensure that the solder pot is free of dross. In a wave soldering machine the solder should be cleaned regularly of dross, and in a drag soldering machine with stagnant solder and a large surface, each board vehicle should skim the dross from the solder surface. Figure 23.11 shows an example of dross deposited on a board.

23.4 Process-Dependent Soldering Defects

When evaluating the board's solderability, the inspector should realize that it can be difficult to fix the responsibility for some types of soldering faults. Faults such as

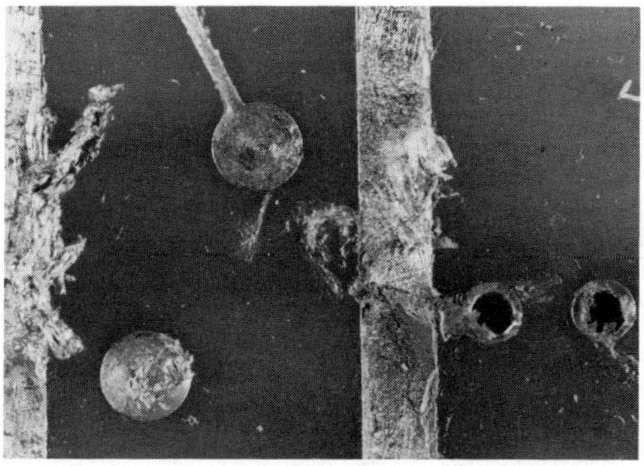

Fig. 23.11. Board with deposited dross after mass soldering, caused by insufficient cleaning of the surface of the solder bath.

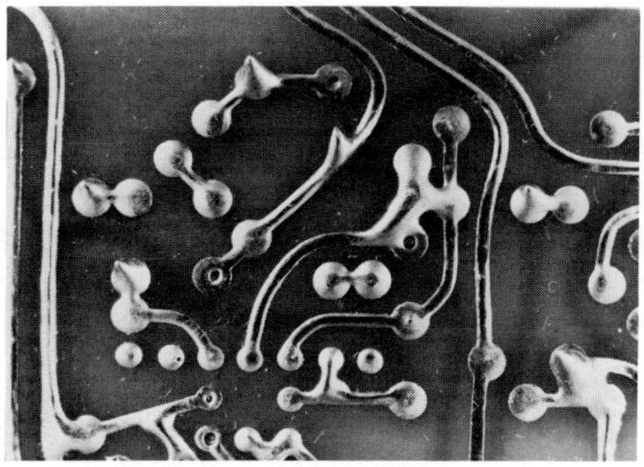

Fig. 23.13. Board exhibiting icicles after mass soldering.

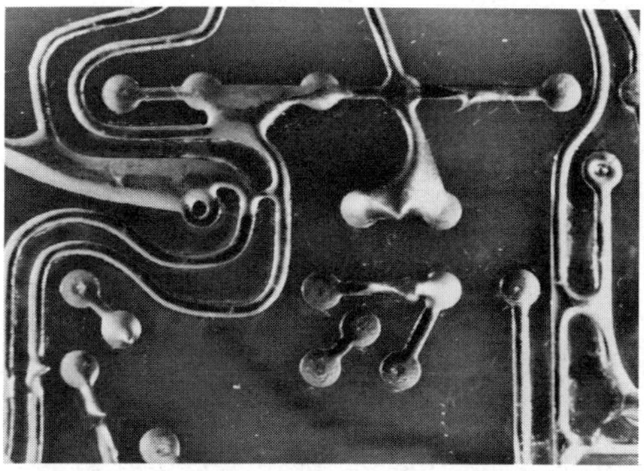

Fig. 23.12. Board exhibiting solder bridging after mass soldering.

solder bridging and icicles can be due to reduced board solderability, but they can just as well be due to inferior mass soldering or to a design error.

If such faults appear to only a very moderate extent, the inspector should not include them in his complaint. The previously mentioned fault indications will doubtless also occur, and the inspector should not complicate the matter with disputable faults.

23.4.1 Solder Bridging

Solder bridging is direct short-circuiting between adjacent conductors and/or solder pads by the solder as shown in Figure 23.12. As mentioned in Paragraph 10.7.2, it is usually caused by incorrect adjustment of the soldering machine.

Another cause can be poor design resulting in a dense pattern, frequently in the form of conductors which are closely spaced, and worse, perpendicular to the direction of board travel through the soldering machine. Such faults do not warrant remarks in the inspection record.

23.4.2 Icicles

Icicles, as shown in Figure 23.13, can be a common soldering fault. This cause is usually incorrect adjustment of the soldering machine. See Paragraph 10.7.3. If this is the case, the PCB manufacturer obviously cannot be blamed for the fault, and no remarks are entered in the inspection record.

The inspector should note, however, that when the soldering machine operates correctly and other boards are soldered satisfactorily, icicles can indicate poor solderability. If this is the case, the poor solderability will usually be accompanied by poor wetting of the pattern (Paragraph 23.3.1).

Chapter 24

Evaluation of Inspection Results

24.1 Introduction

The purpose of inspecting a PCB in the incoming inspection department is to keep defective printed circuit boards out of component stock. This requires the PCB batch to meet numerous specifications, some of which can be assessed objectively, and some only subjectively.

24.1.1 Uncertainty of Requirements

The quality requirements cannot always be expressed in clear and measurable quantities with associated tolerances. Since many quality requirements must be verbally described, the inspector needs a wide background knowledge in order to fairly assess them.

24.1.2 Rationale of Sampling Inspections

This book has repeatedly emphasized that a 100 percent inspection is normally uneconomical. The inspection is instead performed on a sample representing the whole batch. For this reason the sample must be inspected carefully.

24.1.3 Importance of Systematic Inspection

By following the inspection record and the comprehensive instructions given in Chapters 16 to 23, the inspector can systematically examine the board for more than 150 different phenomena, representing nearly all types of defects which can occur on printed circuit boards. With the inspection record, the inspector can more realistically evaluate the overall batch quality.

24.1.4 Primacy of Functional Considerations

All descriptions of the inspection procedures emphasize functional requirements, giving only secondary importance to the cosmetic requirements.

The inspector should remember that the most important thing for the PCB user—the assembly line as well as the test department—is to receive boards which can implement the electronic circuitry. Small cosmetic faults are normally of little importance, since the finished equipment is presumably sold on the basis of its specifications.

24.1.5 Disposition

As described in this chapter, the fate of the batch—its acceptance or rejection—can be determined in several ways. The inspector should follow that part of the inspection record shown in Figure 24.1.

ASSESSMENT AND DISPOSITION					24
ΣA		24.2	ΣB		24.2
ΔA (R)		24.3	ΔB (R)		24.3
ΔA (W)		24.3	ΔB (W)		24.3
ΔA (M)		24.3	ΔB (M)		24.3

Fig. 24.1. Assessment and Disposition. (Section of inspection record.)

Again it is emphasized that even if the incoming inspection is a sampling inspection, all defective boards found during the inspection can be returned to the supplier for replacement.

24.2 Disposition when Using a Sampling Plan

Based on the entries in the inspection record, the A- and B-defects found during the inspection are counted and entered in the upper blanks, ΣA and ΣB of the record (Figure 24.1).

24.2.1 Using Rejection Figures

The batch is accepted when the A- and B-defects are fewer than the respective rejection figures. When only one of the A- or B-defect numbers reaches or exceeds the rejection figures, the whole batch can be rejected and returned. It is then the PCB manufacturer's duty to thoroughly screen the batch and replace the boards sorted out, or possibly repair them. Sometimes the customer chooses, because of time pressures, to perform the screening and possibly also the repair.

It should be noted that the quality evaluation must be based on the number of defective boards and not on the total number of defects. A board should be recorded as defective only once, even if it has two or more A-defects—for example, an interrupted conductor and a missing drill hole. Similar considerations apply to B-defects as well as to a mixture of A- and B-defects.

24.2.2 Color-Coding of Boards

The individual boards of the sample are color-coded the first time an A- or B-defect occurs, preferably with different colors for the two categories. This makes it easy to count the number of boards which, within the two categories of defects, enter into the assessment.

24.2.3 Sampling Agreements

The procedure outlined implies that the purchasing department has an agreement with the PCB manufacturer regarding the criteria for evaluating the batch, including the sampling plan as determined by the inspection level and AQL values. To prevent disputes over the rejection of a batch, the purchasing department should make sure that the PCB manufacturer really understands the agreement.

24.2.4 Self-Elected Sampling Plans

The incoming inspection department can always approve a batch on the basis of a less severe, self-elected sampling plan. This increases the risk of a defective batch slipping through the incoming inspection, but in the long run it can be an economically sound measure to use a lower level of inspection, with a smaller sample size (Paragraph 1.2.3).

When it seems necessary to reject a batch, the final decision must be based on the agreed-on sampling plan. The smaller sampling sizes as described for certain partial inspections can cause problems when too few A- and B-defects are recorded.

When the number of defects is already exceeded, or is expected to exceed, the rejection number, there is no need to increase the sample size. When the number of defects approaches the rejection figure without actually reaching it, the sample size should be increased to the nominal size and the total number of defects should be counted. The extended sample frequently needs to be inspected only for the types of faults found in the reduced sample, which reduces the workload.

24.2.5 Application of the Sampling Plan

This system can be quite rigid, and in extreme cases it can require that the batch be rejected so that it is not placed in use. It is rarely used as described here in nonmilitary electronics production.

24.3 Disposition after Omitting Certain Faults

One of the main component procurement problems of an electronics company is having printed circuit boards of acceptable quality arrive when promised and in the right quantity. Otherwise, the customer cannot meet his production schedule. For this reason, most companies producing nonmilitary electronic products are comparatively willing to compromise regarding the severity of the quality assessment if reliability is not affected.

24.3.1 Screening

It is not normal practice to return a batch if the number of defects exceeds the rejection number stated in the agreed-on sampling plan. A screening, possibly followed by a repair of the defective boards, is more likely.

24.3.2 Use of Manufacturer's Rejects

It is not unknown for an electronics company which has received an incomplete batch to ask the PCB manufacturer to send all retained boards in order to see whether they can be used to bring the batch up to its full delivery size in spite of defects.

24.3.3 Use of Fault Record

It is difficult to prescribe universal rules for using the fault record to assess the quality, which the inspector must do in order to decide whether the batch can be accepted or must be rejected.

An obvious procedure is to weigh the faults using the ΔA and ΔB blanks in the inspection record (Figure 24.1). The inspector fills in the blanks with the number of A- and B-defects within each of the following categories:

(R) Rejection
(W) Waiver
(M) Mending (repairs)

This approach gives a clearer picture of the batch quality so that the inspector can make a better decision. First of all, he can disregard all B-defects, since they are irrelevant to the functioning of the board according to the definition given in Paragraph 1.2.6f.

He can also disregard all repairable A-defects because, after correction, they are no longer critical nor do they interfere with the functioning of the board.

Finally, it is quite conceivable that the inspector has waived certain borderline cases of A-defects, which means, for example, that the violation of a tolerance has been so moderate that a waiver is justified. The inspector can therefore disregard A-defects for which a waiver has been issued.

The remaining defects are therefore the critical and important A-defects of category R (rejection), and the number of these faults is the foundation of the final decision. The following cases occur:

1. The number of category R defects does not reach the rejection number, and the batch is accepted without further debate. If the defects

within the categories R and M show a wide distribution of types, the faults presumably appear randomly and screening the batch is not cost-effective.

When the recorded defects center about a single type of fault, or a very few types of faults, the batch may contain some fairly consistent faults, and it is expedient to screen the batch for such faults. The effort will presumably be rather limited since the inspector has only to inspect the boards for the indicated fault(s).

2. The number of defects of category R reaches or exceeds the rejection number, and requires an analysis of the distribution of fault types.

   If the recorded types of faults within the categories R and M show a wide distribution, the quality of the batch is doubtful. A screening in order to sort out rejectable boards and boards which can be repaired would require inspecting the entire batch, not only for the recorded types of faults, but according to the total inspection procedure. This can mean so much work that most inspectors choose to return the whole batch.

   When the recorded defects concentrate on a single type of fault, or a very small number of fault types, the same procedure as outlined in Item 1 is followed.

24.3.4 Disposition

As mentioned in Paragraph 24.2, the disposition of the delivery should be based on the number of defective boards and not on the total number of defects. The individual boards of the sample are therefore color-coded only the first time a critical A-defect of category R occurs, and only such color-coded boards affect the assessment. The total number of recorded defects is, however, also of interest since the inspector gets a more detailed picture of the PCB manufacturer's reliability.

24.4 Disposition without a Sampling Plan

Many PCB users have no agreement with the PCB manufacturer regarding how closely a batch must meet the requirements of the PCB specification. In other words, no quality level has been established, and a sampling plan does not exist. The reason could possibly be that the customer is afraid of a price increase when trying to reach an agreement regarding the quality level.

24.4.1 Lack of a Sampling Plan

As long as the user is satisfied with the quality of the batches, which means they meet the PCB specification, lack of an agreement causes no problem. It is, on the other hand, a common experience that even the best PCB manufacturer occasionally has bad luck during production, or he may also have an abnormally low quality level for a longer period. And then, just to make things worse, the customer finds another problem in addition to the quality problem.

24.4.2 Sampling Inspection

Although there is no agreement regarding the use of a sampling plan when inspecting a batch, it seems most expedient to perform the incoming inspection according to the guidelines given in Chapters 16 to 23. This means a real sampling inspection, which can give the inspector a more detailed picture of the batch quality.

24.4.3 Customer-Manufacturer Negotiations

It will be difficult for the customer to require the fulfillment of a certain quality level. He is therefore obliged to negotiate with the PCB manufacturer on the basis of a vague idea of equity with respect to the price.

The PCB manufacturer presumably is fairly willing to accommodate the customer, since he does not want to lose a customer or mar his own reputation. There is, of course, a limit, beyond which the PCB manufacturer will not go in order to avoid establishing a precedent.

If the batch is defective, the PCB manufacturer will probably be willing to take it back for screening, repair, or replacement. It is implied here that the sample indicates that it is impossible to meet the quality level stated in Paragraph 16.6, which any competent PCB manufacturer in all probability can be expected to meet.

24.4.4 Lack of a PCB Specification

The problem is still worse if there is no real PCB specification. If that is the case, it will be advantageous for the inspector to concentrate on objectively measurable faults in order to avoid disagreements about degrees of subjectively assessed faults. The inspector should therefore exclude all unimportant B-defects which are of a strictly cosmetic nature.

The remaining faults are primarily critical A-defects, which in principle should not occur in printed circuit boards, whether or not a PCB specification exists. A few examples of such indisputable faults are:

1. Short circuits/interruptions in the pattern
2. Solder holes with incorrect nominal diameter
3. Edge connectors with flaking gold plating
4. Exceeding the indicated dimensional tolerances

When the number of indisputable and unacceptable A-defects in the sample indicates that the quality level discussed in Paragraph 16.6 will not be met, the PCB manufacturer should be willing to take the batch back for screening, repair, and/or replacement.

The inspector and the purchasing department clearly are in a weak position relative to the PCB manufacturer. The only real solution is to set up a reasonable PCB specification and reach an agreement regarding the quality level.

24.5 Reporting to the PCB Manufacturer

In the preceding paragraphs we have seen how the inspection result is evaluated and how the decision whether to accept or reject the batch is made. The easiest way of communicating the decision to the PCB manufacturer is to use the inspection record described in the preceding chapters. As shown in Figure 24.2, part of the inspection record contains blanks for the relevant possible decisions as well as blanks for indicating details about the decision reached.

24.5.1 Disposition Decision

The inspector can distinguish between two primary decisions as shown in the record, the decision being marked by a cross in the appropriate blank:

BATCH ACCEPTED		BATCH REJECTED	
BOARDS ACCEPTED WITH NO COMMENTS			
BOARDS ACCEPTED AFTER EXEMPTION			
BOARDS ACCEPTED AFTER REPAIR			
BOARDS REJECTED AND RETURNED			

Fig. 24.2. Batch Accepted/Batch Rejected. (Section of inspection record.)

1. The batch is accepted.
2. The batch is rejected.

These two opposite decisions reflect the formal and rigid determination when using the sampling plan as described in Paragraph 24.2. As mentioned in Paragraphs 24.3 and 24.4, the fate of the batch will usually be decided in another and less rigorous manner.

The decision usually takes a relatively differentiated form. The inspection record therefore offers the possibilities of reporting the following four cases which, of course, can be combined. The sum of the quantities indicated in Items 3 to 6 must naturally correspond to the total batch size.

3. Quantity of boards accepted without special comments. In practice, this number appears as the number of boards not included in Items 4, 5, and 6.
4. Quantity of boards accepted after exempting unimportant faults.
5. Quantity of boards accepted after repair.
6. Quantity of rejected boards returned to the PCB manufacturer.

Also in the formal cases, Items 1 and 2, a further specification may be needed, primarily concerning the number of defective boards being rejected and returned.

24.5.2 Feedback to the PCB Manufacturer

The PCB manufacturer benefits from feedback showing the customer's perception of his board quality. It is therefore recommended that the inspector send a copy of the inspection record to the PCB manufacturer, even when the batch is accepted.

Even in accepted batches, it is not surprising for the inspector to find some boards with faults of one type or another, but this does not invalidate the general evaluation of the batch. It can, however, be important for the PCB manufacturer to receive feedback concerning faults which have apparently slipped through his own finished-goods inspection.

A standardized inspection record facilitates the PCB manufacturer's interpretation of the information given, just as the inspector's reporting is considerably eased. The author hopes that the inspection record described in this book will be used as a convenient and accurate means of communication between the customer and the PCB manufacturer, to the benefit of both parties.

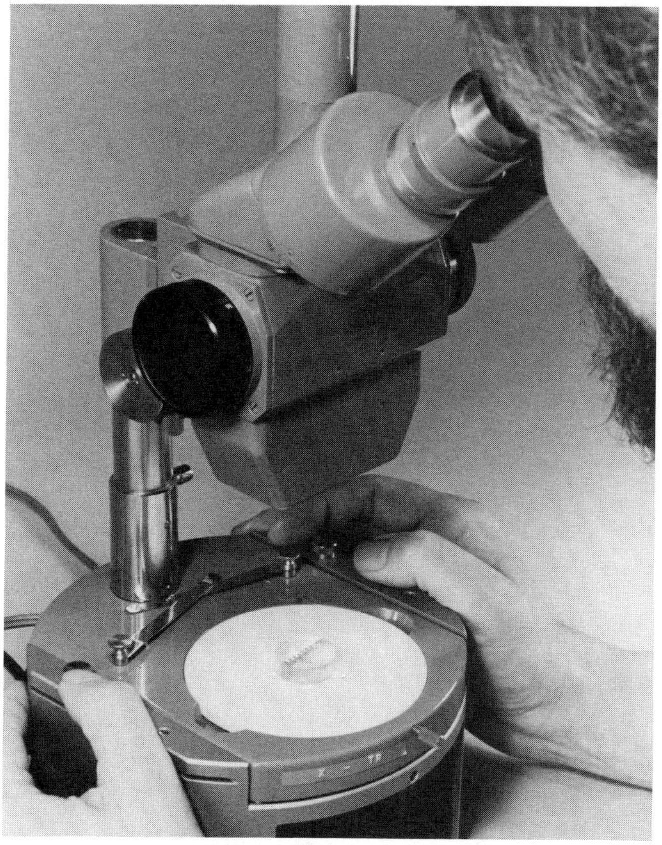

Part 4

Special Inspection Methods

Chapter 25

Preparation and Assessment of Microsections

25.1 Introduction

Parts 2 and 3 repeatedly emphasized microsectioning as the key to a detailed quality assessment of printed circuit boards, especially of plated-through holes and edge connectors. Figure 25.1 is a photograph of the microsection of a plated-through hole as seen under a microscope. The layer structure of the hole plating is clearly visible.

The microsection is simply a cut through a specimen. To allow the operator to handle the specimen and to protect its edges during preparation, the specimen is encapsulated in epoxy. The surface of the cut is then ground and polished, and finally etched in order to remove metal smear (caused by polishing) from the surface.

The inspector can study all details of the plating structure, including the interface between the various layers, with a microscope and thereby assess the quality.

Since microsectioning is frequently considered a very difficult process requiring expensive equipment, many inspectors refrain from preparing their own microsections. A way out is to leave the task to a company or laboratory having the necessary expertise, but this is not the same as having an in-house facility.

This chapter presents two procedures by which the inspector can prepare microsections. Although the first procedure is quite simple and requires only a relatively moderate investment in equipment, the results are adequate for a satisfactory assessment. They are not, however, always satisfactory for photography and reproduction. The other, more advanced, procedure requires considerably more equipment. In return, it offers microsections with a higher visual quality, so that the inspector can take clearer photographs.

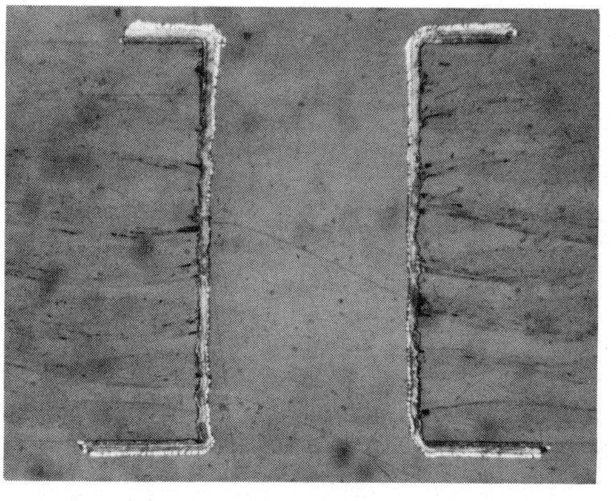

Fig. 25.1. Typical example of microsection of plated-through hole.

This chapter should dispel some of the mystery regarding microsectioning so that inspectors themselves will be encouraged to prepare microsections.

There can, however, be variations in the sample grinding and polishing sequences, which are usually selected by personal preferences and/or by adaptation to the personal techniques of an individual metallographer. There is always some latitude in the selection of materials and sequences, which provide equally adequate metallographically prepared samples.

25.2 The Simple Procedure

We shall now describe how to prepare microsections of plated-through holes using a minimum of special equipment. The procedure, which is outlined schematically in Figure 25.2, comprises the following steps:

- Defining the specimen
- Bulk-cutting the specimen
- Mounting (encapsulating) the specimen
- Turning the molding in a lathe
- Grinding the cut
- Polishing the cut
- Etching the cut

The following paragraphs describe the individual steps in detail.

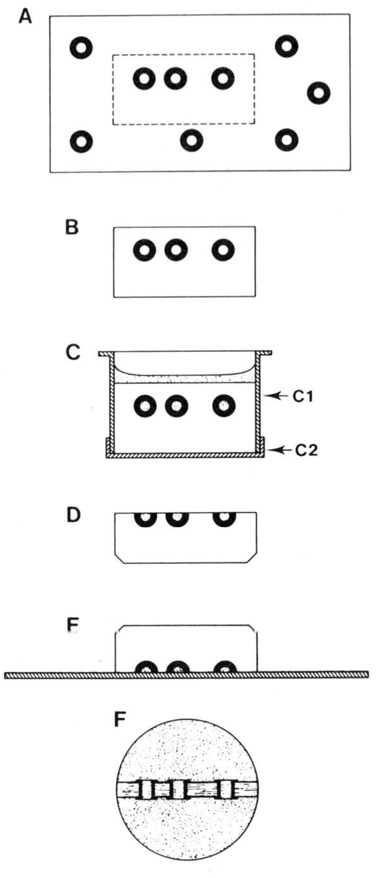

Fig. 25.2. Schematic drawing showing the procedure of microsectioning.
A: Defining the specimen by scribing its contour. (Paragraph 25.2.1)
B: Bulk cutting the specimen. (Paragraph 25.2.2)
C: Encapsulating (molding) the specimen in epoxy (Paragraph 25.2.3). C1: Upper part of the mold. C2: Detachable bottom.
D: Turning the molding in a lathe. (Paragraph 25.2.4)
E: Grinding and polishing the cut (the turned surface). (Paragraphs 25.2.5 and 25.2.6).
F: Etching the cut. (Paragraph 25.2.7).

25.2.1 Defining the Specimen

The first task is, of course, to define the size of the specimen. In order to keep the specimen upright in the mold without using a holding device, its length should equal the inner diameter of the mold.

Commercially available molds have diameters of 1 in. (25.4 mm) or 1.25 in. (31.8 mm). The width of the specimen can, for example, be 0.625 in. (16 mm).

The cut nearest the holes of interest should be kept 0.08 to 0.12 in. (2 to 3 mm) from the outer edge of the solder pads. Because of later machining, it is important that both lengthwise cuts of the specimen are parallel to the line through the centers of the holes to be microsectioned. Therefore, these holes should be located in line.

The specimen contour is marked on the board—for example, with a scriber. See Figure 25.3.

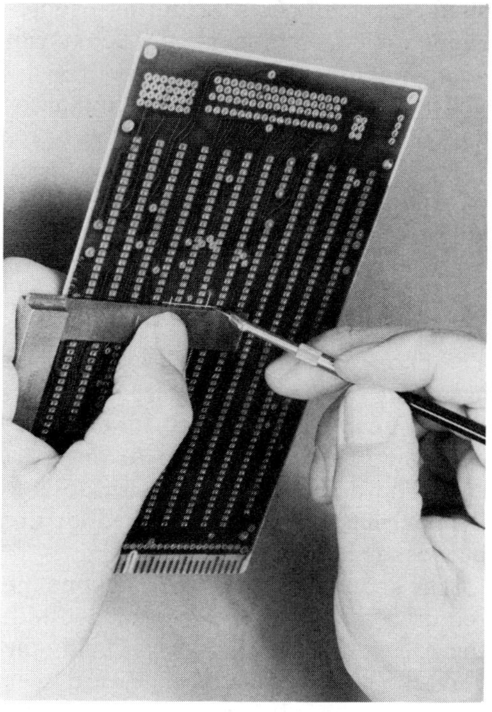

Fig. 25.3. Marking the contour of the specimen.

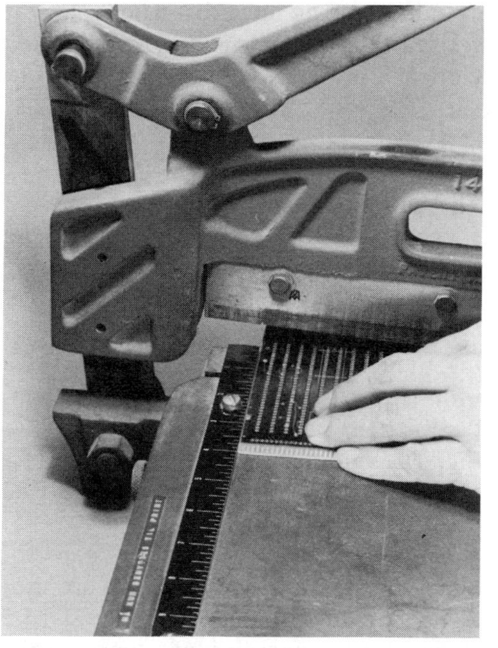

Fig. 25.4. Shearing the specimen by means of a hand-operated guillotine in a heavy-duty execution.

25.2.2 Bulk-Cutting the Specimen

The specimen is now cut out of the board, using available tools. Using a hacksaw is the simplest way, but if the specimen is located near the center of the board, cutting may be somewhat difficult. Most in-house machine shops have a circular saw or bandsaw available. However, a small specimen—for example, 0.625 by 1 in. (16 by 25.4 mm) requires great care to avoid operator injury. A hand-operated guillotine as shown in Figure 25.4 may prove more expedient for cutting small specimens.

The length of the specimen should be a little larger than the inner diameter of the mold—for example, 0.004 to 0.008 in. (0.1 to 0.2 mm) larger. This makes it possible to adjust the length with a file or coarse abrasive paper, so that the specimen can be placed in a stable position in the mold. It is, however, also possible to fix the specimen by means of a piece of double-sided adhesive tape placed on the bottom of the mold. In this case, the length of the specimen can be reduced.

Because the encapsulated specimen (the molding) will be turned on a lathe, the specimen must not distort the mold into an oval shape.

After filing the edges, the inspector thoroughly cleans the specimen of chips and other debris. He then degreases it in acetone, which has a bipolar characteristic—that is, it removes water-soluble as well as organic contamination.

25.2.3 Mounting the Specimen

The specimen is now to be encapsulated. To identify the specimen, a small typed label is placed on the bottom of the mold, typed side down. Then the specimen is placed in the mold with the holes to be examined facing upwards. The inspector verifies that the specimen fits stably and vertically in the mold.

25.2.3a Molding Materials

The molding is best performed by using a two-component epoxy, preferably the slow-curing type, which requires a curing time of approximately 8 hours at 77°F (25°C). The inspector can shorten the curing time by varying the mixing ratio a little. In some cases, however, the temperature rise may crack the finished molding. The mold itself is usually made of a thermoplastic material and may also be damaged.

Since epoxy has a fairly low degree of adhesion to polyethylene, molds are usually made of this material. Figure 25.5 shows some examples of commercially available molds. They consist of a ring with a detachable, close-fitting bottom.

25.2.3b Material Quantities

The first step is to calculate the amount of epoxy necessary for filling the mold a little above the upper edge of the specimen—approximately 0.35 fl. oz. (10 ml) for the small mold and approximately 0.56 fl. oz. (16 ml) for the large one (Paragraph 25.2.1). The mold should not be filled completely because during a subsequent vacuum treatment, the epoxy will foam violently and possibly run over.

The vacuum treatment is not absolutely necessary, although it serves to remove air from the specimen and the epoxy to reduce the risk of air bubbles in the holes, which will degrade the result.

25.2.3c Mixing and Pouring

The epoxy components are apportioned according to the manufacturer's instructions. The mixing ratio may, for example, be eight parts epoxy resin to one part epoxy hardener by volume. Disposable hypodermic syringes can be used for this purpose (Figure 25.6).

The syringe is filled slightly beyond the required volume; air is then expelled with the excess so that the syringe holds the correct amount. The syringes are then emptied into a plastic cup and the ingredients mixed for two minutes with a wooden spatula. The syringes can be reused many times without cleaning if they are always used with the same epoxy component.

The mixture, which has a pot life of approximately 30 minutes, is now poured into the mold (Figure 25.7). When the epoxy reaches the holes which are to be examined, the pouring is continued on one

Fig. 25.5. Examples of commercially available molds.

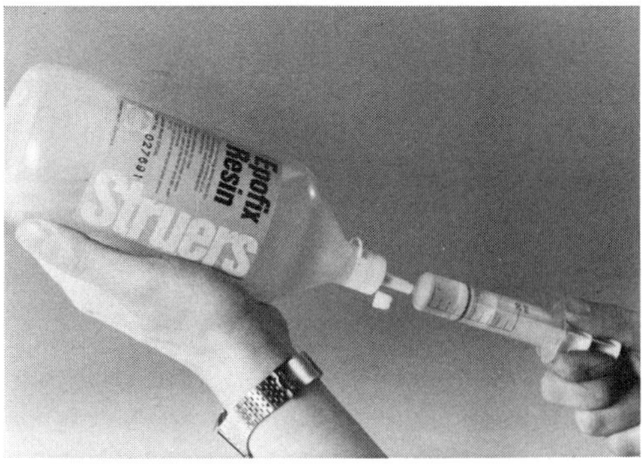

Fig. 25.6. Gauging the epoxy resin by means of a disposable hypodermic syringe.

Fig. 25.7. Pouring the epoxy into the mold.

side of the specimen only, which permits the epoxy to flow through the holes and remove most of the air bubbles.

25.2.3d Vacuum Treatment

As mentioned earlier, a small amount of air can escape from the specimen, and the epoxy collects a small amount of air from the stirring. However, the air is released so slowly that small air bubbles may become entrapped in the cured molding.

A vacuum treatment for a few minutes to release the air is strongly recommended. The mold can be placed in a desiccator, which is, in effect, a large glass bell jar with a ground-in lid. Figure 25.8 shows a desiccator connected to a mechanical vacuum pump. A much cheaper solution is to use a water ejector pump as shown in Figure 25.9.

Fig. 25.8. Example of a desiccator connected to a mechanical vacuum pump.

Fig. 25.9 Example of a water ejector pump which can be connected to a water tap.

Air is now drawn from the desiccator and the vacuum is held for a few minutes, after which the vacuum is released. The air released from the specimen and the epoxy may foam onto the mold surface, and in some cases overflow. It is therefore advisable to place the mold on a piece of cardboard to avoid spilling epoxy on the bottom of the desiccator, which will be difficult to clean.

A prolonged vacuum treatment serves no purpose and may increase the risk that too much hardener will evaporate. Should this occur, the molding will not harden completely, so that grinding and polishing becomes difficult or even impossible.

If any foam present does not collapse when the desiccator is opened, it can be removed with a small spatula. The mold

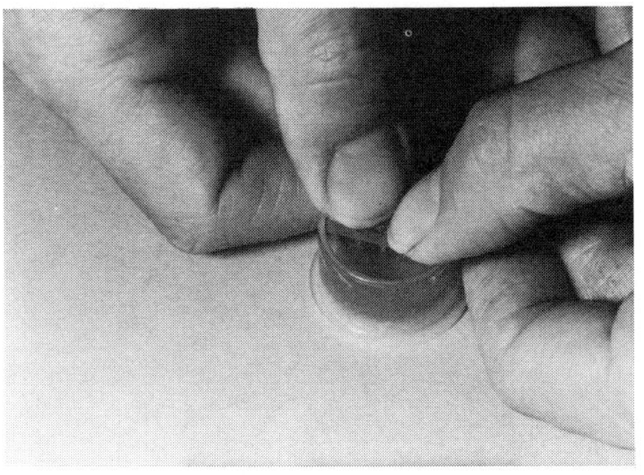

Fig. 25.10 Removing the molding from the mold. The detachable bottom is removed and the molding is pressed out of the mold.

is then left undisturbed during the curing of the epoxy.

25.2.3e Specimen Removal

When the curing is complete, usually in about 8 hours, the encapsulated specimen is removed from the mold (Figure 25.10). Specimen removal is somewhat easier if the mold is sprayed with a release agent before filling, but this is not absolutely necessary.

25.2.3f Cold-Mounting Materials

Cold-mounting materials are also available. These materials have quick-setting characteristics and low exothermic properties (low heat generation during setting). They are particularly useful when working with solder alloys that have heat-sensitive structures. However, the most popular mounting material seems to be the aforementioned epoxy material.

25.2.4 Turning the Molding in a Lathe

We now must cut through the holes which are to be examined. It is neither particularly easy nor desirable to saw through the molding and the holes with a hacksaw, since the cut will usually be uneven and will probably be dislocated.

25.2.4a Benefits of Turning

It is far easier and preferable to turn the molding in a lathe. With this method, the cut becomes relatively even, and post-treatment in the form of a very coarse grinding is unnecessary. Furthermore, in most electronics companies, a suitably qualified inspector has access to a lathe in the machine shop.

Initially, this method may seem somewhat crude. The author has, however, had extremely good results with it. The only requirement is that the turning be done fairly carefully—that is, with a small chip when close to the holes.

25.2.4b Effects of Hard Fillers

Turning the top of the molding requires that the molding be transparent so that the operator can clamp it correctly and keep his bearings during the turning. In some cases a hard filler—for example, aluminum oxide—is added to the epoxy to harden the molding (Paragraph 25.3.3a). This causes the sharp edges of the specimen to be retained without rounding, which may otherwise occur during polishing.

However, adding a hard filler makes turning impossible, so when following the simple procedure we must accept a slight rounding of the edges. In return, we can see the progress of the turning. Incidentally, transparent epoxy allows the inspector to observe the cylindrical part of the hole wall in the finished microsection, and not only the cross-sectioned hole wall.

25.2.4c Clamping the Molding

The molding is clamped in the lathe chuck and the turning started. See Figures

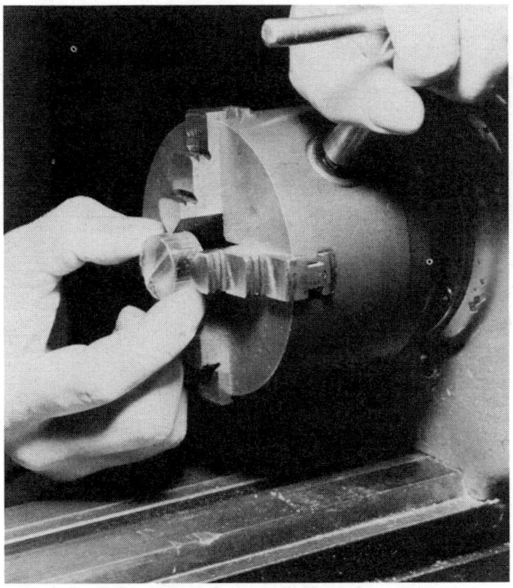

Fig. 25.11A. Clamping the molding in the chuck of a lathe.

Fig. 25.11B. Turning the molding in a lathe.

25.11A and 25.11B. Since the molding may not always be completely cylindrical, clamping the specimen in the chuck so that it is correctly oriented can be a little difficult. The following criteria must be met in order to attain a perfect microsection when finished:

1. The centerline through the holes to be examined must be perpendicular to the longitudinal axis of the lathe, independently of the chuck position.
2. The specimen must be parallel to the longitudinal axis of the lathe.

When these requirements are met, the finished cut is located correctly with respect to the center plane through the holes. It is therefore very important that the bulk cutting and the encapsulation, as well as the clamping, be performed very carefully.

25.2.4d Turning Procedure

Turning is carried out with a turning tool ground for plastic materials. The size of the chips should be relatively small. To avoid damage to the holes, the size of the turnings should be reduced further when the turned surface approaches the holes.

When the turned surface just touches the rim of the solder pads (A, Figure 25.12), the turning is stopped and the clamping criteria are checked. If the diameter of the solder pads is known and the hole drilling is reasonably concentric, the distance to continue the turning in order to reach the right cut can easily be calculated. In practice, the turning should stop a little before this point—for example, 0.008 to 0.012 in. (0.2 to 0.3 mm)—to allow for subsequent grinding.

The operator continues the turning with very fine turnings, observing the holes continuously. The holes will appear as bands in the rotating molding. With some practice, the operator can make a good assessment of how far to continue the

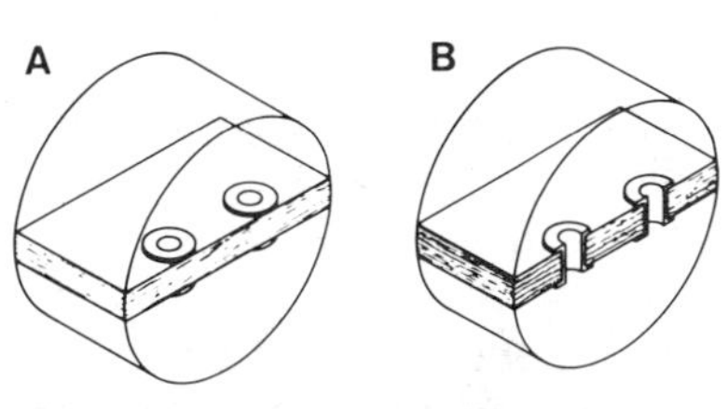

Fig. 25.12. Criteria of turning the mold. A: The turned surface is just touching the solder pads of the hole. B: The turning is stopped a little before reaching the hole centers.

turning. Turning a predefined distance from the pad rim seems unnecessary and can actually yield a slightly inferior result.

When the turned surface is just a little outside the center plane through the holes (B, Figure 25.12), the turning is stopped and the molding removed from the chuck. Since the chips contain glass particles, the operator should clean the lathe carefully to avoid unnecessary wear.

25.2.4e Coarse-Grinding as an Alternative

An alternative to turning is coarse-grinding, which quickly removes material. The concave and uneven top side of the molding makes it difficult to keep the ground surface perpendicular to the specimen while grinding. The greater the amount of material removed, the greater the risk of a skewed surface or a surface with several facets.

When the inspector lacks access to a lathe, he can only perform a grinding. It simplifies matters to mold the specimen so that the holes to be examined face downwards. Since the bottom of the mold is relatively flat, grinding will be easier. By bulk-cutting a little closer to the holes, the operator needs to grind away less material.

25.2.5 Grinding the Cut

The turned surface of the molding must now be ground. This requires that:

1. The cut is brought into correct position, coinciding with the plane through the hole centerlines. This is a prerequisite to correctly measuring the plating thickness.
2. Turning grooves, scratches, etc., are gradually removed by grinding, so that the subsequent polishing will produce a bright and smooth surface.

Fig. 25.13. A simple setup for manual grinding of the mold. The abrasive paper is clamped to a U-shaped support bracket by means of two small plates. Grinding is performed in a development tray filled with just enough water to cover the abrasive paper. Grinding can also be performed in a sink under running water.

25.2.5a Abrasives

The grinding is performed with abrasive papers, preferably under running water. A simple setup as shown in Figure 25.13 can be used. The following grit sizes are suitable: 240, 320, 400, and 600. These grit sizes correspond to the American Standard and are approximately 45, 30, 24, and 14 μm, respectively.

In the European Standard FEPA, the series of abrasive paper is 220, 320, 500, and 1000. The corresponding grit sizes are approximately 65, 46, 30, and 18 μm, respectively. In practice, the difference between these two standards is hardly perceptible, and we shall use the American Standard in the following description.

25.2.5b Grinding Procedure

Within each stage, the molding must always be ground in the same direction, which means that the molding should not be rotated. In the subsequent grinding stages, the molding is turned 90 degrees between each stage. This makes it easier to see when the scratches from the previous grinding stage have been removed.

Thorough cleaning of the molding, as well as the hands, is recommended between the individual grinding stages to prevent carryover of coarse grits, which will cause deep scratches, to the next stage.

The grinding is performed with a steady hand at a constant speed and pressure in both directions. The following four steps are necessary:

Step 1: Abrasive paper number 240 is used and the molding is ground at an angle of 45 degrees to the specimen (board). Grinding continues until the turning grooves disappear and the cut nearly reaches the plane through the centerlines of the holes to be examined.

The location of the cut is checked with a magnifier; just a little of the outer side of the hole wall should be visible (A, Figure 25.14). A film of water on the surface eases the inspection.

Step 2: Abrasive paper number 320 is now used, and the grinding continues until the scratches from Step 1 disappear. Turning the molding 90 degrees makes it quite easy to see when the transverse scratches have disappeared.

The cut should now be in place because subsequent grinding stages do not remove much material. An indication of a correct location is that no external part of the hole wall is visible when viewed perpendicularly to the cut. See B, Figure 25.14.

Step 3: Abrasive paper number 400 is used and the grinding continues following the same guidelines (turning the molding 90 degrees and removing the previous scratches).

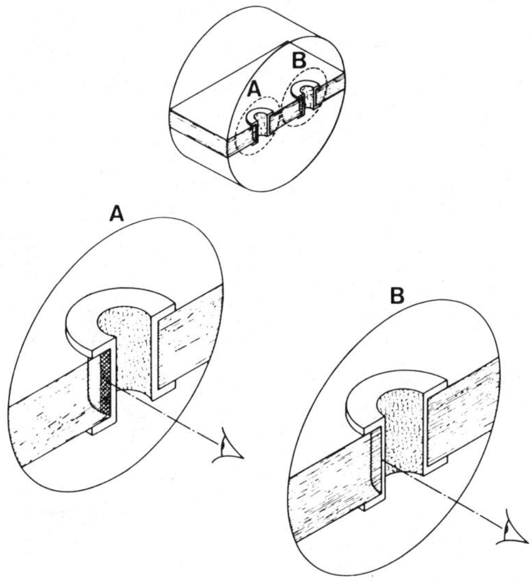

Fig. 25.14. Criteria of grinding. A: The cut should not be exactly in the right place after grinding step 1, since some material must be left for step 2. The outer side of the copper plated hole wall should be visible. B: The cut should now be in the right place after grinding step 2, since the subsequent grinding steps do not remove much material. No external side of the plated hole wall can be seen.

Step 4: Abrasive paper number 600 is used and the grinding is performed as described above, but very lightly.

The grinding is now finished. If the operator has taken care to avoid transferring grits, the surface of the cut should now be very even and exhibit only faint traces of the last grinding.

25.2.6 Polishing the Cut

The surface of the cut must now be polished in order to remove the last traces left from the final grinding stage. The objective is to get a clear surface structure permitting all details to be studied under a microscope, even at high magnification.

The principle is quite simple. The polishing compound is applied to a polishing disc having a suitable surface. All individual grains of the polishing compound must be available for polishing the surface while the molding is moved across the polishing disc.

25.2.6a Polishing Compounds

Polishing compounds of aluminum oxide or diamond powder are available in various grain sizes. Since a slight difference between the grain sizes of the existing brands may be expected, the following discussion indicates typical orders of magnitude. Minor deviations are of no importance to the result.

Aluminum oxide is available either as a powder, which is prepared as a suspension in distilled water by the operator, or as a commercial ready-to-use product. Commonly used grain sizes for polishing purposes are 5, 1, 0.25, and 0.1 μm, possibly also 0.05 μm.

Diamond powder is available either as a paste, applied with small syringes, or as a suspension in spray cans. Common grain sizes used for polishing are 7 (or 6), 1, and 0.25 μm.

Although diamond polishing is more expensive than aluminum oxide polishing, it is recommended. The polishing effect is greater and the rounding of the specimen edges is less. It is, however, important to use a good brand with a reasonable distribution of the grain sizes.

While some variation is unavoidable, the distribution should be as narrow as possible because the larger grains will scratch the surface and the smaller ones will be ineffective. See Figure 25.15. Also, the shape of the grains is important, with sharp-edged grains being the best.

25.2.6b Applying the Polishing Compound

Application of the diamond powder to the polishing disc is quite simple. The operator shakes the spray can for a few seconds, then holds it bottom up with the

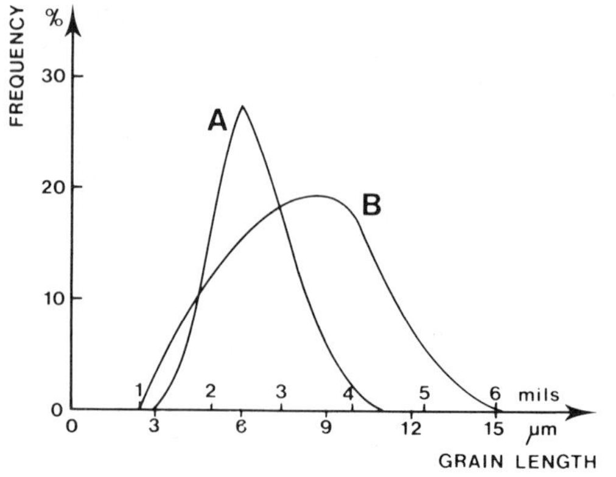

Fig. 25.15. Distribution curve of the grain sizes of a polishing compound.
A: Satisfactory brand with even distribution of the grain sizes.
B: Inferior brand with very uneven distribution of the grain sizes.

spray nozzle approximately 4 in. (10 cm) from the polishing disc while spraying.

He sprays the disc in a circular motion for a few seconds to distribute the diamond powder evenly over the surface. Later, when the polishing effect decreases from the loss of diamond powder, the polishing disc can be refurbished with a new application.

It is important to keep the polishing disc wet so that the polishing compound does not clog the disc. When polishing with diamond powder, the operator normally uses an alcohol-based wetting agent; distilled water is adequate for aluminum-oxide polishing.

25.2.6c The Polishing Disc

The surface character of the polishing disc is also important. The grains of the polishing compound must adhere without becoming embedded in the polishing disc and reducing the polishing action.

For coarse polishing, the polishing disc should be napless so that it will not round the specimen edges. A polishing disc with a silk- or nylon-woven surface is commonly used.

For both the fine polishing and the buff polishing, the polishing disc should be soft; the surface weave should have a relatively short nap, frequently made of a synthetic material.

Since polishing discs normally are mounted on the rotating wheel of a polishing machine, they come in various sizes, with and without a self-adhesive backing.

25.2.6d Manual Polishing

When only a few microsections are required, they can be polished manually. However, the result is not quite as fine as from polishing on a rotating wheel, and manual polishing is also more troublesome. A good technique is to stick a self-adhesive polishing disc (cloth) to an aluminum sheet so that it becomes completely flat and firm.

Correct polishing pressure is also important to the quality of the finished surface. Excessive pressure usually causes rounded edges. The polishing pressure should be restricted to about ½ lb. (250 g) for coarse polishing, and somewhat lower for fine polishing. During buff polishing a very low pressure should be used.

Manual polishing is performed by moving the molding over the polishing disc concentrically to the rim (Figure 25.16).

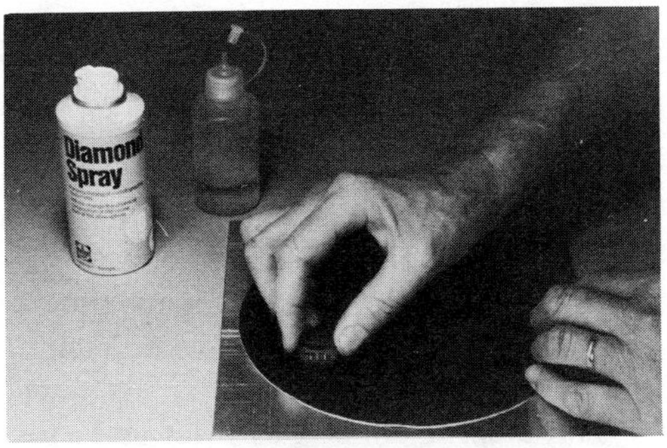

Fig. 25.16. Simple setup for manual polishing of the molding. A self-adhesive polishing disc is attached to an aluminum sheet, and the polishing is performed by a steady circular movement.

It is not necessary to turn the molding since the circular motion continuously changes the direction of polishing.

The operator should take care that the pressure is evenly distributed perpendicular to the molding. Otherwise, some areas may be inadequately polished. Since it is difficult to maintain a completely even pressure with manual motion, the operator should turn the molding a little between the fingers so that any tendency to uneven polishing is distributed over the whole surface.

25.2.6e Polishing Procedure

Polishing is frequently performed in two, and sometimes three, steps: First, a coarse polishing followed by a fine polishing, possibly followed by a buff polishing. The process is elaborated below.

The polishing is performed as follows:

Step 1: Coarse-polishing is performed with 6 μm diamond powder. The molding is polished until any scratches from the previous grinding step disappear. This takes a few minutes.

The molding is cleaned of all diamond powder, after which it is checked with a microscope to verify that the scratches from the previous grinding stage have been removed.

Step 2: Fine polishing is performed with 1 μm diamond powder. The procedure is as described in Step 1. The polishing time should be as short as possible to avoid rounding the specimen edges.

Step 3: Buff polishing is performed with aluminum oxide, grain size 0.1 or 0.05 μm, and very light pressure. This polishing step is necessary only for taking high-quality photographs of the microsections. In most cases, buff polishing is unnecessary.

It is extremely important for the operator to take great care to clean the molding thoroughly between steps so that coarser grains are not transferred to the next polishing stage. For the same reason, all polishing discs should be carefully stored—for example, in plastic bags—to exclude dust and other contaminants.

25.2.7 Etching the Cut

The softness of tin/lead allows a very thin layer of tin/lead to smear over the copper plating during the polishing. The result is a somewhat blurred boundary line between the tin/lead and copper platings, causing inaccurate measurement of the layer thickness.

In order to get a clear view, the operator must remove the smeared layers, which he can do in at least two ways. In many instances, a fine polishing on a vibratory polisher with 0.05 μm aluminum oxide (alumina) provides a clear enough view for most typical observations. However, the most popular method seems to be a very brief etching. The etchant must attack tin/lead as well as copper in order to clearly reveal the boundary line and the layer structure.

25.2.7a Swab Etching

A frequently recommended etchant for swab etching can be prepared as follows:

One part ammonium hydroxide, NH_4OH

One part distilled water, H_2O

Two parts 3 percent hydrogen peroxide, H_2O_2 water solution

The ammonium hydroxide is diluted with distilled water, after which hydrogen peroxide is added. The polished surface is wetted with a swab stick dipped in the etchant, which must be fresh. The molding is then rinsed in water, sprayed with methanol (methyl alcohol) or ethanol (ethyl alcohol) and blow-dried with warm air. The microsection is then ready for examination under a microscope.

25.2.7b Immersion Etching

A weaker etchant allows etching by immersion, which provides a far more uniform etching of the entire surface:

Six parts ammonium hydroxide, NH_4OH

Six parts distilled water, H_2O

One part 3 percent hydrogen peroxide, H_2O_2 water solution

The sample is etched for approximately 30 seconds, followed by the same rinse and dry procedure described above.

Another etchant, which perhaps better reveals the layer structure, is made up as follows:

One part glacial acetic acid, CH_3COOH
One part nitric acid, NHO_3

The operator should avoid breathing the fumes while mixing the two ingredients. Nitric acid fumes are especially dangerous and good ventilation is required if no fume cupboard is available. Because etchant drops on the skin can cause unpleasant wounds similar to burns, the operator must be very careful when etching the microsection.

The etching time is as brief as ½ second. The etching is performed most safely with a disposable hypodermic syringe filled with the etchant. The operator holds the

Fig. 25.17. Etching of the polished surface. The molding is held with a pair of tongs, the ends of which are bent a little to ensure a better grip around the molding.

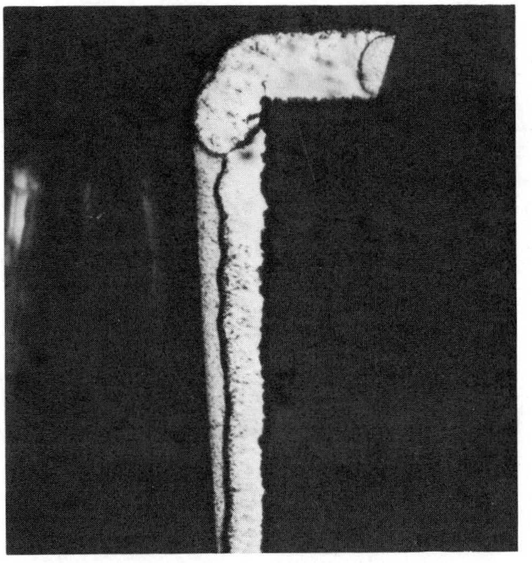

Fig. 25.18A. Example of unetched microsection. The photograph shows the transition between the hole wall and the solder pad, but the structure is not particularly clear.

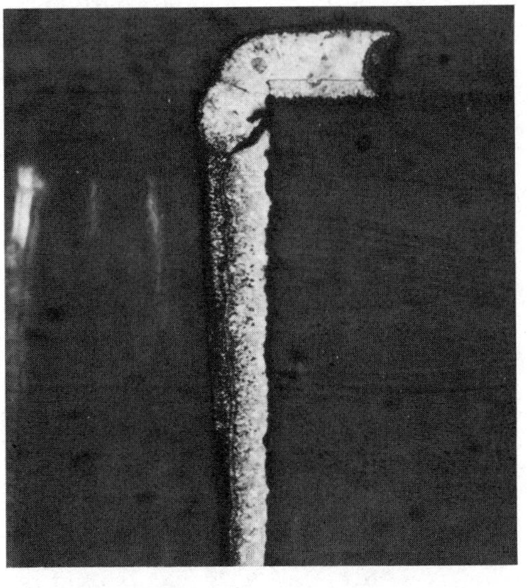

Fig. 25.18B. Same microsection as shown in Fig. 25.18A, but in an etched condition. The structure is now quite clear: The distinction between the copper foil of the solder pad and the copper plating is obvious. The tin/lead plating is clearly distinguishable from the copper plating. Note that the structure of the transition zone between the solder pad and the hole wall is not satisfactory.

molding with a pair of tongs (preferably made of nickel) over a sink, and squirts a few drops of the etchant over the polished surface. After etching for ½ second, the molding is flushed under tap water (Figure 25.17).

Finally, the molding is sprayed with methanol or ethanol and blow-dried with warm air. Figure 25.18 shows an example of a microsection prepared as previously described, then etched with the mixture of glacial acetic acid and nitric acid.

A less active etchant, safer to use because of its dilution, is prepared as follows:

Three parts glacial acetic acid, CH_3COOH

Four parts nitric acid, NHO_3

Sixteen parts distilled water, H_2O

WARNING:

When diluting an acid with water, the acid must always be added to the water. If water is added to the acid, excessive heat will develop within small volumes and cause such violent boiling that the solution splatters in all directions, exposing the operator to serious injury.

The sample is immersed for ½ to 2 seconds and immediately rinsed with running water followed by the same rinse and dry process as stated above. Etching by immersion instead of swabbing provides a better control due to the lower concentrations of active constituents.

25.2.7c Molding Materials Considerations

Certain types of molding materials do not resist such etchants as glacial acetic acid and nitric acid. The surface is attacked, resulting in cracking. A trial etching is recommended before using new molding materials.

25.2.7d Buff Polishing after Etching

A very light and short buff polishing of the cut after etching is, in the author's experience, not recommended, because the highest points of the cut become bright while the lower points remain untouched, presenting a false picture of the structure.

25.2.7e Other Metals

When microsections of a sample plated with other metals are to be made, a suitable etchant must be chosen. The publication ANSI/ASTM E407-70, "Standard Methods for Microetching Metals and Alloys," gives recipes for a number of etchants suited for various metals and alloys.

25.3 The Professional Method

The procedure previously described can be fairly time-consuming and a little difficult to perform. We shall now discuss a more practical and professional procedure. This procedure is based on the same fundamental principles of molding, grinding, and polishing, but the improvements are achieved by use of more suitable equipment.

Since the individual improvements are independent of each other, the necessary investments can be made one at a time. An expedient order of investment is, in the author's experience, the procurement of:

1. A combination grinding and polishing machine
2. A low-speed precision saw

25.3.1 Mechanical Grinding and Polishing

The operator will not only find it most convenient to machine-grind and polish the molding, but he will also achieve a far better result. There are many machines available on the market for grinding and polishing, as well as combination machines suited for both purposes, since a variety of wheels for abrasive paper and polishing discs can be mounted. The choice of a suitable machine is left to the reader.

Since microsectioning is not normally assigned to the incoming inspection department of an electronics company, a combination grinding and polishing machine is recommended. The conditions are quite different in the case of a PCB manufacturer or a metallurgical laboratory, where it is of utmost importance to prepare microsections in the most efficient way. In those cases, machines dedicated to individual tasks are preferred.

25.3.1a Grinding

The combination grinding and polishing machine has interchangeable wheels. One of the wheels is usually intended for abrasive paper, having a recess in which the abrasive paper is placed. Abrasive paper can be purchased in disc sizes corresponding to the recess and in the grit sizes listed in Paragraph 25.2.5.

The wheel of the machine is wetted with water to allow adhesion of the abrasive paper disc to the wheel, after which a metal ring is installed to prevent lifting of the rim of the abrasive paper. Interchanging an abrasive paper disc—for example, when going from one grinding step to the next—simply requires removing the metal ring and peeling the disc from the recess.

Grinding is performed as shown in Figure 25.19. The rotation speed can, for example, be 125 or 250 rpm. A jet of water against the disc keeps it wet and washes off ground material as well as detached abrasive grains.

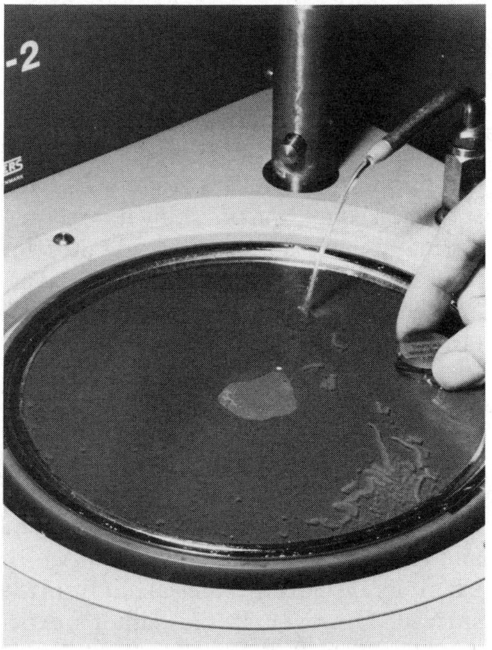

Fig. 25.19. Mechanical grinding of the molding.

The operator applies moderate pressure to hold the molding against the rotating abrasive paper disc. He can follow the procedure described in Paragraph 25.2.5 and keep the molding in such a position that the grinding traces form an angle of 45 degrees to the specimen. The molding is turned 90 degrees at the next grinding step.

It is extremely important that the wheel as well as the molding be carefully flushed with water before the next grinding step, so that detached abrasive grains are not transferred.

It is recommended that the operator carefully choose a suitable combination of grinding time and pressure, particularly when using the coarser abrasive papers. Since the process proceeds rapidly, he can very easily remove too much material. A common fault is that the surface of the cut is not perpendicular relative to the specimen, usually because of uneven pressure.

A low rotation speed is recommended until the operator has gained some experience. He should inspect the grinding progress from time to time.

It should be remembered that it is still an advantage to turn the molding in a lathe as described in Paragraph 25.2.4, instead of grinding the raw molding at the start.

25.3.1b Polishing

Since wheels for polishing are frequently not provided with a recess, it is practical to use polishing discs with a self-adhesive backing. The polishing follows quite closely the description given in Paragraph 25.2.6, and the same types of polishing discs and polishing compounds are used.

When polishing the molding as shown in Figure 25.20, the operator slowly moves

Fig. 25.20. Mechanical polishing of the molding.

the molding against the direction of rotation in order to avoid unilateral polishing effects in the form of flowing material, also called "comet-tailing." The operator can also hold the molding still, and just turn it a little between the fingers from time to time.

When the operator has to polish many microsections, he can save time by fitting the machine with a sample holder for three or six samples. See Figure 25.21. In this way it is possible to keep a constant pressure on the moldings since they are loaded with weights.

Frequently, the sample holder can be moved automatically across the polishing disc, and as the moldings can turn freely in the holder, comet-tailing effects are avoided. Finally, it is possible to supply the wetting agent automatically—for example, one drop per 5 seconds.

The final result depends to a certain degree on the rotation speed, which should neither be too high nor too low. The polishing pressure follows the guidelines given in Paragraph 25.2.6. In order to avoid comet-tailing as well as rounding of the specimen edges, the polishing time should normally not exceed 2 to 3 minutes for both the coarse and fine polishing steps. The buff polishing, if carried out, should last an even shorter time.

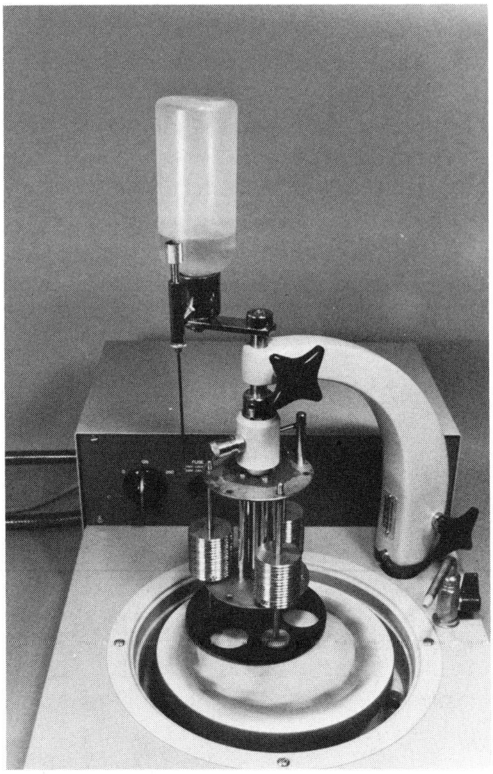

Fig. 25.21. Automatic polishing of several moldings at the same time by means of a sample holder.

25.3.2 Cutting the Specimen

It cannot be denied that turning the molding in a lathe, as described in Paragraph 25.2.4, may seem a little heavy-handed and risk damaging the finer details—for example, by tearing material out from the surface. However, in the author's experience, this risk is very small when preparing microsections from printed circuit boards.

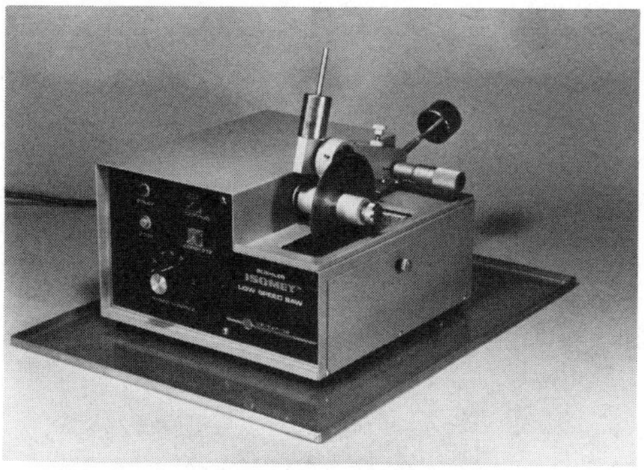

Fig. 25.22. Precision low-speed saw with a diamond wafering blade.

25.3.2a Cutting by Means of a Saw

It can be impractical to send the moldings to a machine shop to have them turned on a lathe. There may be long waiting times, but more important, communication problems may arise concerning how far the turning should go.

As a result, many companies choose to invest in precision cutting equipment, which is placed in the laboratory. An example of a precision low-speed saw is shown in Figure 25.22. The saw blade is a disc—for example, with a diameter of 4 in. (100 mm) and a thickness of 0.012 in. (0.3 mm). Since the rim is covered with diamond grains, the saw blade is also called a diamond wafering blade.

25.3.2b Cutting Procedure

The molding is suspended in a clamp on a movable arm. See Figure 25.23. The weight of the arm is balanced with a counterweight, and the pressure against the saw blade can be adjusted with suitable weights which load the arm—for example, with 2 to 4 oz. (50 to 100 g).

The lower part of the saw blade runs through a vessel containing a lubricant, which is drawn up by the saw blade. This lubricating film keeps the cutting area lubricated and cool.

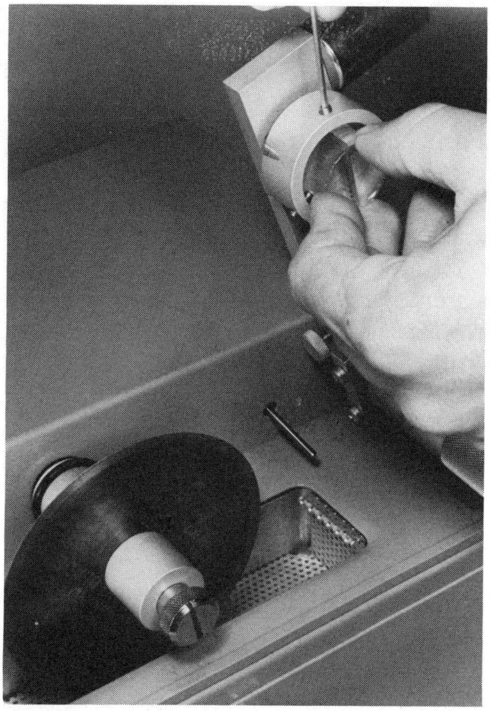

Fig. 25.23. Clamping the molding in the precision low-speed saw shown in Fig. 25.22.

The lubricant should have a suitable viscosity so that the molding is lifted a little with respect to the saw blade. This creates low friction, which reduces heat development and improves the cutting action. Too high a viscosity will produce a condition corresponding to aquaplaning and reduce the cutting action. Too low a viscosity may cause the diamond grains to break or cause an uneven surface. For this reason, diamond wafering blades should never be run while dry, and they should be lubricated with the prescribed lubricant.

The arm carrying the clamped molding can be displaced relative to the saw blade, usually by means of a micrometer screw so that the cut can be placed very accurately. The speed of the saw blade can be varied. For preparing microsections of printed circuit boards, a relatively low speed—for example, 150 to 200 rpm—should be used.

The precision cutting equipment offers two ways of preparing microsections. The first corresponds closely to the methods described in Paragraphs 25.2.1 to 25.2.4. Here, the largest problem is to locate the cut correctly, since the operator can use only visual alignment. An incorrectly started cut cannot be rectified.

In any case, the molding must be transparent, which precludes the use of a hard filler. It should also be noted that the saw blade has to cut through a lot of material, which means that the cutting process may last for as long as 10 minutes. This is too long, particularly when there are many moldings to cut.

25.3.2c Alternative Procedure for Preparing Microsections

A modified procedure is based on the fact that the cut obtained with a diamond wafering blade is very clean with sharp edges. Furthermore, the method is not destructive to the details which are to be examined.

The procedure is to clamp the bulk-cut specimen (Paragraphs 25.2.1 and 25.2.2) in the cutting equipment as shown in Figure 25.24. The flat shape makes it reasonably easy to locate the cut correctly, and the cutting takes only a few minutes.

The cut is placed just outside the centers of the holes—for example, 0.004 in. (0.1 mm)—leaving a little material for the following grinding and polishing. Possible burrs are removed with a few strokes across a piece of No. 600 abrasive paper.

The specimen is now ready for embedding (molding) as described in Paragraph 25.2.3. The advantage of this method is that the specimen is placed in the mold with the cut-through holes turning downwards, reducing the subsequent grinding considerably.

It is not absolutely necessary to adjust the length of the specimen to the inner diameter of the mold. As shown in Figure

Fig. 25.24. Cutting the specimen before encapsulation.

25.25, it is possible to use a small clip to hold the specimen upright. Unless a hard filler is used, the operator should, in the author's experience, avoid using supporting clips made of steel because they make grinding and polishing more difficult.

It is also possible to use an adhesive tape or a single-sided adhesive cardboard

Fig. 25.25. Using a small clip for supporting the specimen.

backing material, on which the specimen is stuck in an upright position. It is advantageous to first grind the edge to be taped with No. 240 or 320 abrasive paper before applying tape or backing.

The remaining treatment of the molding (grinding and polishing) follows the guidelines in Paragraphs 25.3.1a, 25.3.1b, and 25.2.7.

25.3.3 Edge Retention

When preparing microsections, it is very important to retain the specimen edge definition, or the microscopic examination will yield an incorrect result.

Since the epoxy molding material is softer than the PCB specimen, the epoxy inevitably wears away more rapidly than the specimen itself. This causes the specimen edges to become somewhat rounded, depending on the duration and intensity of the polishing. See Figure 25.26.

During the microscopic examination the rounded edges will deflect the light rays somewhat, distorting the image and causing an incorrect measurement of the layer thickness.

In order to retain the edges, the polishing time should be kept as short as possible. The microsection shown in Figure 25.18 shows that one can prepare usable microsections with no precautions other than selecting a suitable compromise between a short polishing time and an imperfect polishing.

When a more perfectly polished surface without any edge rounding is needed, the operator must employ other methods such as those described below.

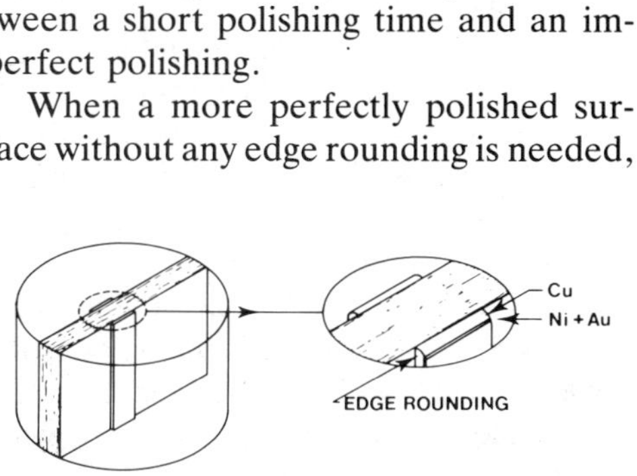

Fig. 25.26. Rounding of the specimen edges. Cu = copper foil + copper plating. Ni + Au = nickel and gold platings.

25.3.3a Hard Filler

An obvious solution is to increase the hardness of the epoxy until it corresponds to that of the PCB specimen. This is achieved by mixing a hard filler with the epoxy molding material. The filler usually consists of aluminum oxide in the form of very small, possibly porous, spherical grains. The advantage of this is that the epoxy can penetrate the grains and retain them better after curing.

The procedure is to mix hard filler with enough of the epoxy molding material to fill the lower part of the mold—for example, to a depth of 0.20 in. (5 mm). This requires filler equivalent to approximately 20 percent of the volume of the molding. The rest of the epoxy is then poured in on top.

The hard filler can also be mixed into the total volume of the epoxy, but it will inevitably sink to the bottom of the mold and not increase the hardness in the upper part of the molding. Since it is uncertain that the hard filler will completely fill the holes, the first procedure is recommended. It is also understood that it is necessary to place the specimen in the mold with the holes to be examined facing downwards.

Figure 25.27 shows a microsection with hard filler. When using a hard filler, the operator should be aware of the following:

1. The molding cannot be turned in a lathe because it will quickly wear out the tool.
2. The grinding must start with a very coarse abrasive paper—for example, No. 80, followed by No. 150, after which the grinding continues as described in Paragraph 25.2.5.

   It will be necessary to renew the very coarse abrasive papers quite

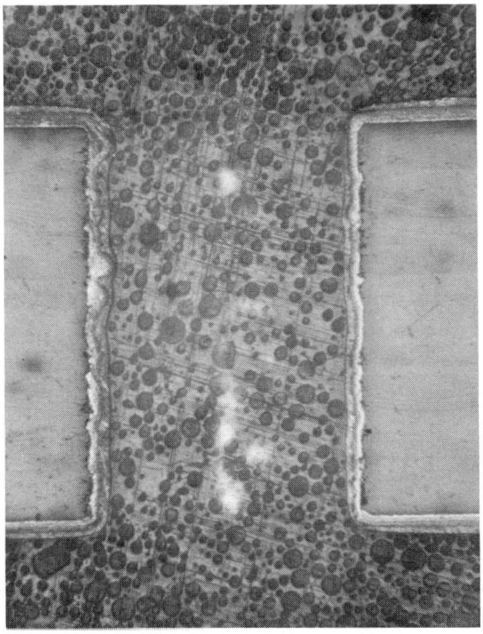

Fig. 25.27. Microsection with a hard filler.

frequently since the wear is greater than when grinding moldings without a hard filler.

25.3.3b Nickel Plating

In the case of gold-plated specimens it is also possible to reinforce the gold layer by depositing electroless nickel over the gold layer. The very hard nickel provides good protection against rounding the edges of the very thin gold layer.

Nickel can be deposited only on metals with an electromotive force less than that of nickel (+0.23 V). In practice, deposits are made on copper (−0.34 V) and gold (−1.50 V).

An electroless nickel bath is available as a kit consisting of a bottle of nickel salt solution, a bottle of reducing agent, a bottle of activator, and directions for use. The basic steps are:

1. Cleaning and degreasing the specimen
2. Activating the specimen
3. Electroless deposition of approximately 0.0002 in. (5 μm) of nickel. The desposition rate can be as much as 0.004 in. (10 μm) of nickel per hour. A thermostatically controlled bath is necessary for a correct deposition.

The subsequent preparation of the microsection (molding, grinding, and polishing) is performed as previously described.

25.4 Inspection of Microsections

A microscope is an absolute necessity for thoroughly examining a microsection. The inspector can use two different types of microscopes:

1. A stereo microscope
2. A metallurgical microscope

The choice depends on how detailed an examination of the microsection the inspector wants to perform. The following paragraphs discuss both types of microscopes.

25.4.1 Stereo Microscope

A stereo microscope as shown in Figure 25.28 has certain advantages over the metallurgical microscope, including a large depth of field and a relatively low price. On the other hand, there are disadvantages regarding the maximum magnification and image distortion.

The stereo microscope is universally applicable to so many other types of inspection that it is actually indispensable. Paragraph 7.3.3c describes the use of a stereo microscope for inspecting hole walls for voids.

A stereo microscope is also very useful for preparing microsections, since the operator can easily check the surface quality

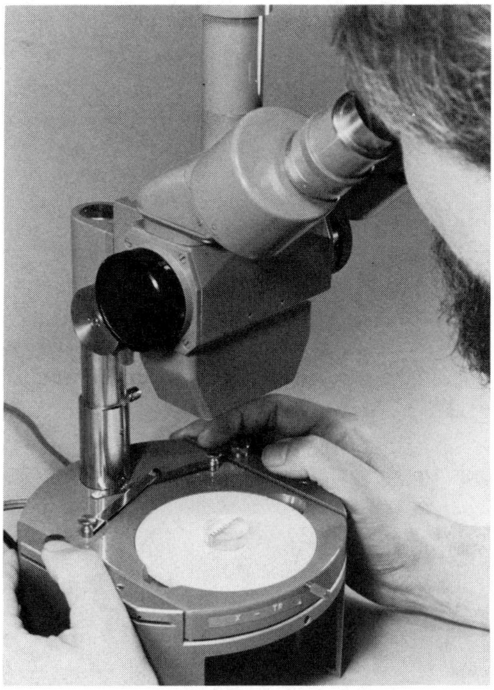

Fig. 25.28. Examination of a microsection by means of a stereo microscope.

during the grinding and polishing steps. He can, for example, verify that transverse scratches from the preceding stage have been removed. Here it is most expedient to view a rather large part of the surface at once, so the magnification should not be too high.

The stereo microscope has insufficient magnification for a thorough examination of the finished microsection. The maximum magnification is usually about 100X. However, the inspector can effectively study many interesting details, such as unevenness of the hole drilling, plating unevenness, formation of nodules and large cracks. He can also ascertain the copper plating thickness with acceptable accuracy.

25.4.1a Eyepiece Micrometer

An effective measurement requires that one of the eyepieces be provided with a micrometer (measuring graticule). The inspector can calibrate the eyepiece at a given magnification by placing a precision-ruled stage micrometer on the microscope stage.

When observed under the microscope, the layer thickness rarely matches the divisions on the micrometer eyepiece. The inspector therefore has to interpolate and take into consideration all the implied inaccuracies. See the detailed discussion of stage micrometers and micrometer eyepieces in Paragraph 25.4.2.

25.4.1b Micrometer Stage

If the stereo microscope is or can be fitted with a micrometer stage, it is easier to measure the plating thickness. A micrometer stage is a stage which can be moved mechanically in the X and Y directions by calibrated micrometer screws.

The stage carrying the microsection is now displaced in the X direction so that the two borderlines of the plated layer are successively aligned with the vertical line of the crosshair. The thickness is the difference between the two micrometer readings.

The inspector can avoid measuring errors by adjusting to the same side of the crosshair line. The expected resolution is about 0.00004 in. (1 μm) when using the vernier of the micrometer screw.

25.4.1c Microsection Illumination

The microsection must be illuminated separately because stereo microscopes normally have no built-in light sources. There are several possible choices:

1. Spot illumination, based on either a low-voltage incandescent lamp (tungsten) or a halogen lamp
2. Fiber optics for high-intensity light, derived from a halogen lamp, over small areas

3. Ring light, which is a low-wattage fluorescent lamp that fits around the body of the microscope

The rather limited magnification range prevents study of the important finer details, such as epoxy smear, corner cracking (cracking at the transition from the hole barrel to the solder pad), barrel cracking (cracking of the plated hole wall), nodules, and plating pockets. Such examinations require much higher magnification, which means using a metallurgical microscope.

25.4.1d Photographing Microsections through a Stereo Microscope

In addition to the normal eyepiece, many stereo microscopes have a vertical phototube to allow attaching a camera for photomicrography. An instant camera is a very practical accessory for this purpose, since it provides a finished photograph in a few minutes.

The photograph can be used for measuring the thickness of the copper-plated layer. There are several possibilities for measuring the thickness; for example, the inspector can photograph an object micrometer and use the photograph when measuring dimensions.

The micrometer can also be laid over the microsection and photographed with it. The object micrometer consists of a glass plate with an etched scale—for example, graduated in intervals of 0.001 in. (25 μm). The scale should face the microsection. Finally, the dimensions can be checked on the basis of a known dimension, such as the hole diameter.

The inspector should realize that a stereo microscope is not an accurate measuring instrument. Some distortion occurs in the optical system and causes small measuring errors.

25.4.2 Metallurgical Microscope

Sooner or later, the inspector will need a better microscope—for example, when measuring the thickness of the gold layer on an edge connector tab. It is therefore recommended to invest in a metallurgical microscope.

The microscope should have flatfield parfocal achromatic objectives, which provide a sharp overall viewing field without color dispersion. The magnification should be adjustable in steps up to at least 1000X.

An inverted type of microscope, which has the optical system below the stage, is easier to operate. The operator places the microsection over a small aperture in the stage, polished side down (Figure 25.29). This method has the advantage that the image remains in focus when the stage is moved, particularly when the magnification is high.

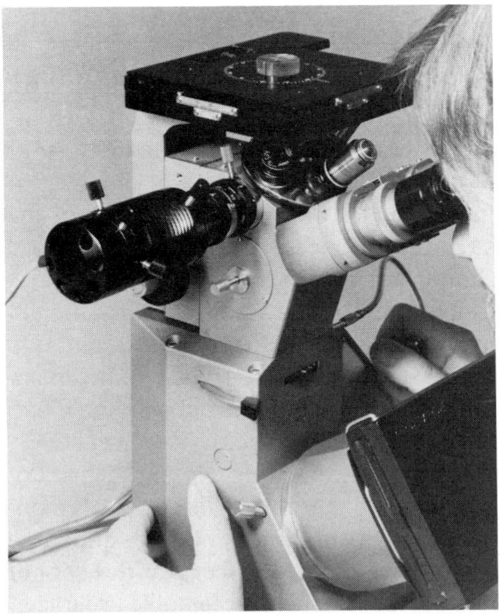

Fig. 25.29. Examination of a microsection by means of a metallurgical microscope. The microsection is placed on the stage at the top, polished side downwards.

In the conventional construction of microscopes, the stage is below the optical system, and the microsection is placed on the stage with the polished surface up. If the polished surface slants slightly with respect to the bottom of the molding, movement of the stage brings the image out of focus, even when the magnification is low. This is of course caused by the limited depth of field.

25.4.2a Special Microsection Holder

It is possible to make a special holder in which the sample can be fixed. In principle, the holder consists of a circular plate with three supporting legs. The plate is provided with an aperture a little smaller than the molding, the aperture being recessed on the under side, corresponding to the diameter of the molding.

When the molding is suspended against the recess by a flat spring, the holder can be moved across the stage, or the stage can be moved, without bringing the image out of focus.

25.4.2b Mechanical Stage

Metallurgical microscopes are usually provided with a mechanical stage so that it is easy to move the microsection. However, the stage is rarely moved by calibrated precision micrometer screws, which are normally provided with measuring microscopes. Instead, the stage has the usual stage screw knobs; however, their reading accuracy of 0.004 in. (0.1 mm) with a vernier, or possibly 0.002 in. (0.05 mm) with extra operator attentiveness, is quite insufficient.

25.4.2c Filar Micrometer Eyepiece

The measuring problem is best solved by a filar micrometer eyepiece which is inserted into the eyepiece tube instead of the normal eyepiece. It consists of a small housing containing a fine crosshair, which can be moved across the viewing field by a micrometer screw.

The distance between two positions of the crosshair—for example, corresponding to the boundaries of a plated layer—is the difference between the two readings of the micrometer screw. The crosshair should always be adjusted to the same side of the hairline to avoid measuring errors caused by the thickness of the hairline.

The filar micrometer eyepiece must be calibrated according to the actual magnification. This is done with a stage micrometer (object micrometer), which is placed on the stage. The stage micrometer bears a precision rule—for example, 0.04 in. (1 mm) long with divisions of 0.0004 in. (0.01 mm). It is quite simple to ascertain how many divisions on the micrometer drum correspond to a given length on the stage micrometer.

With the filar micrometer, the inspector can measure plating thicknesses as small as about 0.00004 in. (1 μm), the limit being determined by the wavelength of the light.

25.4.2d Eyepiece Micrometer

A considerably cheaper, but less accurate, solution is to use an eyepiece micrometer which is inserted into one of the eyepieces. The eyepiece micrometer is a circular glass disc ruled with subdivisions of 0.004 in. (0.1 mm), or finer. Calibration to the magnification actually used is performed by means of the aforementioned stage micrometer.

25.4.2e Illumination

Most metallurgical microscopes have a built-in illumination system which allows insertion of monochromatic filters as well as color temperature compensation filters. Both a polarizer for polarized light and a polarization filter can be built in.

25.4.2f Photographing Microsections through a Metallurgical Microscope

If the inspector wants to record his findings, he must be able to photograph the microsections. Metallurgical microscopes often have a built-in photographic eyepiece and a shock-proof shutter with various speeds. Camera backs for either 35 mm films or instant films are available as accessories.

With a viewfinder inserted in place of one of the normal eyepieces, the inspector can adjust the stage so that the microsection will be positioned correctly on the photograph. To facilitate microsection placement, the viewfinder eyepiece shows a frame corresponding to the film size.

Another solution is to attach a single-lens reflex camera back. The advantage is that the operator can see the image and adjust the stage. Another advantage is that the exposure is adjusted automatically by the camera's built-in exposure meter. This ensures correctly exposed photographs even when the operator is not an experienced photographer.

Dimensions on the finished photographs are measured as described in Paragraph 25.4.1d.

25.5 Assessment of Microsections

Microsection assessment is based on the inspector's background knowledge of failure mechanisms, acquired from reading the preceding chapters. Since a visual assessment is always subjective, he must sometimes supplement the assessment with measurements. These are best based on a photograph of the microsection.

The following discussion clearly shows that a thorough quality assessment of printed circuit boards requires examining the microsections at a fairly high magnification. Such examinations make it possible to assess various external and internal characteristics of the board which otherwise would be impossible.

25.5.1 Microsections of Conductors

Microsectioning conductors makes it possible to study a number of fairly elementary conditions as set out below:

- Copper foil thickness
- Copper plating thickness
- Tin/lead layer thickness
- Tin/nickel plating thickness
- Nickel and gold plating thickness
- Undercut
- Overhang
- Conductor edge coverage

25.5.1a Copper Foil Thickness

See the cross-sectional drawing in Figure 25.30. It is implied that the microsection is etched so that the boundary line between the copper foil and the superjacent layers appears clearly.

In principle, the copper foil must be of the specified thickness. In many cases, it is of little importance whether the thickness is ½ oz. = 0.0007 in. (17.5 μm) or 1 oz. = 0.0014 in. (35 μm). A waiver for overly thin copper foil requires the inspector to check whether special electrical requirements are imposed on the resistance of any conductors. If not, he can issue a waiver and rate the fault as a B-defect.

A thin copper foil offers certain advantages when etching. These advantages can improve the PCB manufacturer's production yield.

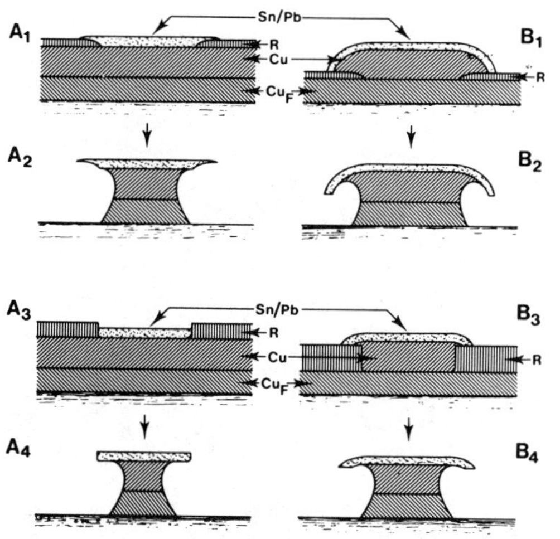

Fig. 25.30. Cross-sectional drawings of tin/lead plated conductors. The drawings show how the conductor cross section primarily depends upon the manufacturing method employed: panel plating versus pattern plating, and screen printing versus photo printing (dry film). Secondary details depend upon the thickness of the plating resist and the etchant used. The conductors are shown in a nonreflowed condition.
A_1: Panel plating and screen printing.
A_2: Etched conductor. A_3: Panel plating and dry film. A_4: Etched conductor.
B_1: Pattern plating and screen printing. B_2: Etched conductor. B_3: Pattern plating and dry film. B_4: Etched conductor.
Cu_F = copper foil. Cu = copper plating including electroless copper. Sn/Pb = tin/lead plating.
R = resist.

Some PCB manufacturers prefer to use a thinner copper foil than specified—for example, ½ oz. = 0.0007 in. (17.5 μm) copper instead of 1 oz. = 0.0014 in. (35 μm) copper. They then compensate for the reduced copper by a heavier copper plating. This especially applies to boards used in power supplies where the copper foil thickness can be specified as 2 oz. = 0.0028 in. (70 μm) copper.

The inspector should adopt a functional attitude and consider the total copper thickness (copper foil plus copper plating). When there are no reasons against an exemption, the fault is rated as a B-defect. Otherwise the fault is rated as an A-defect and the batch is rejected.

25.5.1b Copper Plating Thickness

See the cross-sectional drawing in Figure 25.30. The thickness, including the electroless copper, must fulfill the plating requirements for copper in the plated-through holes, usually 0.001 in. (25 μm), on the average.

If the requirements are not met at the conductors, the inspector can expect serious problems regarding the thickness of the copper plating in the plated-through holes. The final assessment of the plating should normally be based on the copper plating in the holes. Paragraph 25.5.3a elaborates on this.

25.5.1c Tin/Lead Layer Thickness

See the cross-sectional drawings in Figures 25.30 and 25.35. Usually, no thickness is specified for the tin/lead layer thickness after reflowing or for the thickness of hot-air leveled solder coatings.

The surface is slightly convex over the cross section. The inspector should verify that the tin/lead reaches right out to the conductor edges. The thickness at the edges can be as little as 0.00002 to 0.00004 in. (0.5 to 1 μm).

In the case of roller-tinned boards, special requirements regarding the thickness can be specified (Paragraph 9.6.2). Although failing to meet such requirements is, in principle, rated as an A-defect, the inspector's practical experience with solderability should be decisive when he considers an exemption.

25.5.1d Tin/Nickel Plating Thickness

See the cross-sectional drawing in Figure 25.31. The inspector should verify that the layer thickness meets the specification.

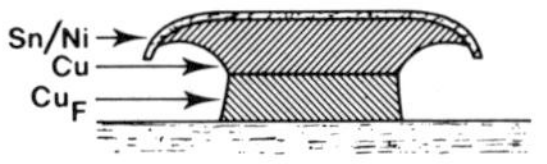

Fig. 25.31. Cross-sectional drawing of conductor with tin/nickel plating. Cu_F = copper foil. Cu = copper plating including electroless copper. Sn/Ni = tin/nickel plating. Note that in certain cases a tin/lead plating can occur on top of the tin/nickel plating, but usually only in the plated-through holes and on the solder pads.

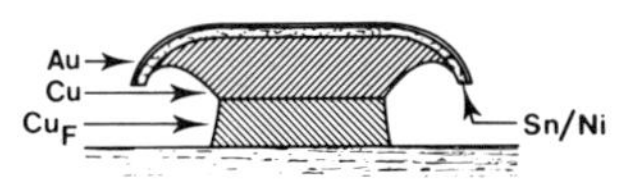

Fig. 25.32. Cross-sectional drawing of conductor with gold plating on top of a tin/nickel plating. Cu_F = copper foil. Cu = copper plating including electroless copper. Sn/Ni = tin/nickel plating. Au = gold plating.

Failure to do so is rated as an A-defect, and the batch is rejected.

25.5.1e Nickel and Gold Plating Thickness

See the cross-sectional drawing in Figure 25.32. The inspector should verify that the layer thicknesses are according to the specification. Failure to meet the requirement is rated as an A-defect and the batch is rejected.

25.5.1f Undercut

See the cross-sectional drawing in Figure 25.33. Since the etching factor must normally be greater than 1, the sidewise etching must be less than the total copper thickness. In this connection, the inspector should base his assessment on the existing edge and not on the nominal edge as determined on the basis of the master film.

Undercutting can seriously reduce the adhesion of narrow conductors. A 0.012 in. (0.3 mm) wide conductor consisting of 1 oz. = 0.0014 in. (35 μm) copper foil with a copper plating of 0.001 in. (25 μm)

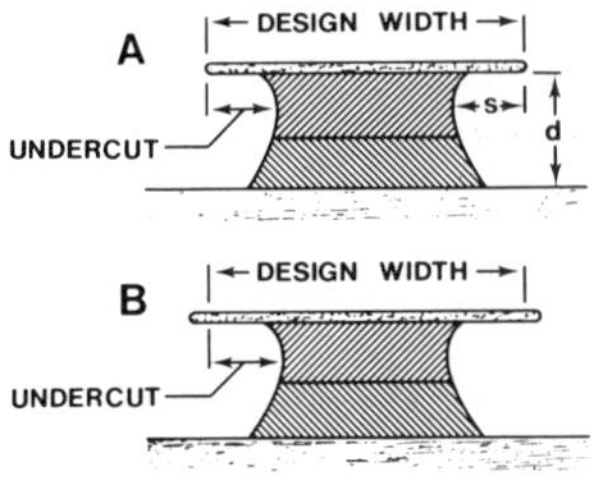

Fig. 25.33. Cross-sectional drawing of conductor with undercut. The etch factor d/s should be greater than 1; i.e., the sidewise etching should be less than the etching in depth. d = etching in depth. s = sidewise etching.

+ 50 percent has a total copper thickness of approximately 0.003 in. (75 μm). Undercutting implies, because of the sidewise etching, that the conductor adheres over a width of only 0.0006 in. (0.15 mm), which halves the nominal adhesion.

An undercut larger than the total copper thickness at any edge should be rated as an A-defect, and the batch is rejected. The reason is that today's printed circuit boards almost always have some very narrow conductors.

This example clearly illustrates why fine-line boards are necessarily based on thin copper foils—for example, ¼ oz. = 0.00035 in. (9 μm) copper—or on additive techniques.

When the pattern is coarser—that is, all conductors and annular rings are wider than 0.02 in. (0.5 mm), the inspector can safely issue a waiver for larger undercuts. In such cases, the limit can be set to 1.5 times the total copper thickness, implying that the adhesion never falls to less than half the nominal value.

25.5.1g Overhang

See the cross-sectional drawing in Figure 25.34. Reflowed boards should have no overhang. Overhang can break off or, even worse, partially break off. Overhang is cause for rejection and consequently rated as an A-defect (Paragraph 21.6).

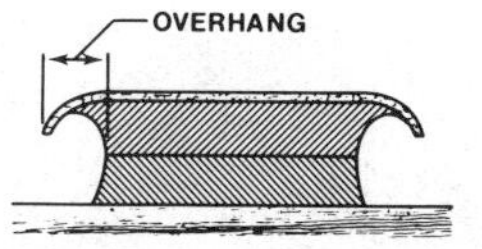

Fig. 25.34. Cross-sectional drawing of conductor with overhang. The overhang should be less than 0.002 in. (0.05 mm).

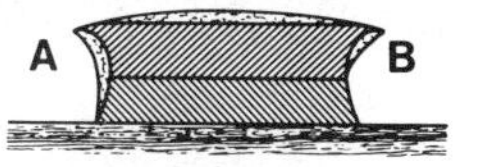

Fig. 25.35. Cross-sectional drawing of conductor showing various degrees of edge coverage. A: Full edge coverage. B: Partial edge coverage.

Nonfusible surface platings such as nickel/tin and gold over nickel should not exhibit overhang larger than 0.002 in. (0.05 mm) per edge. A larger overhang is cause for rejection and rated as an A-defect.

25.5.1h Conductor Edge Coverage

See the cross-sectional drawing in Figure 25.35. Some PCB specifications demand full coverage of all pattern edges after the tin/lead plating has been reflowed, whereas others do not. The best way of verifying with certainty that the edges are covered is to study some microsections of the conductors.

Failure to meet a requirement for full coverage shows itself very clearly and should, in principle, be rated as an A-defect. Consequently, the batch can be rejected. See Paragraphs 8.5.1d and 18.9.4, for discussions regarding possible exemptions.

25.5.1i Microsection Examples

The cases just described are illustrated with some practical examples of microsections. See Figures 25.36-1 through 25.36-7. The captions provide detailed explanations.

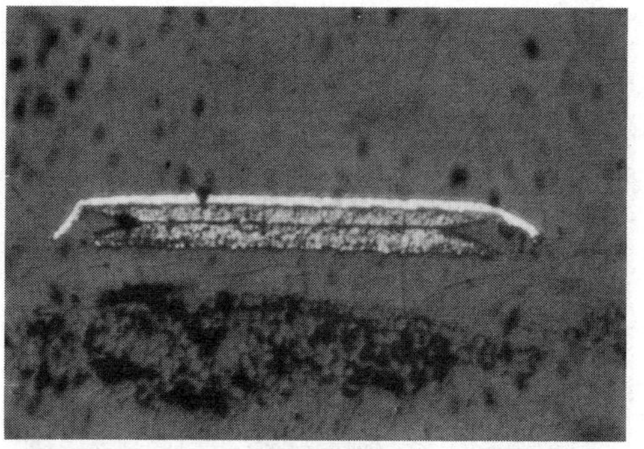

Fig. 25.36–1. Microsection of tin/lead plated conductor in a nonreflowed condition. The tin/lead overhang is clearly seen.

Fig. 25.36–2. Microsection of tin/lead plated conductor in a reflowed condition. Full edge coverage is achieved. However, the tin/lead layer is fairly thin at the edges.

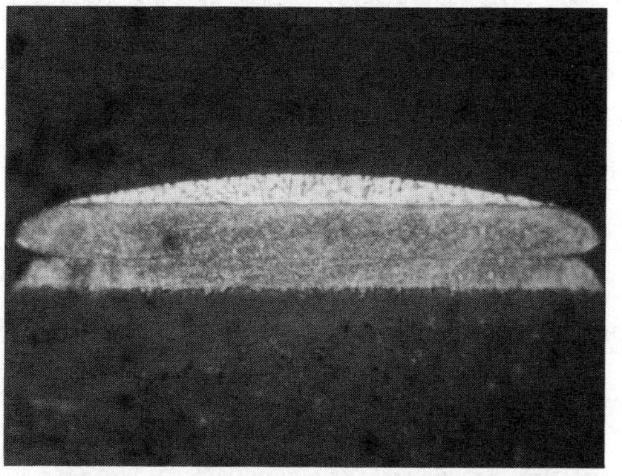

Fig. 25.36–3. Microsection of tin/lead plated conductor in a reflowed condition. Full edge coverage is not achieved.

Fig. 25.36–4. Microsection of tin/nickel plated conductor. Note the overhang.

Fig. 25.36–5. Microsection of conductor plated with nickel and gold.

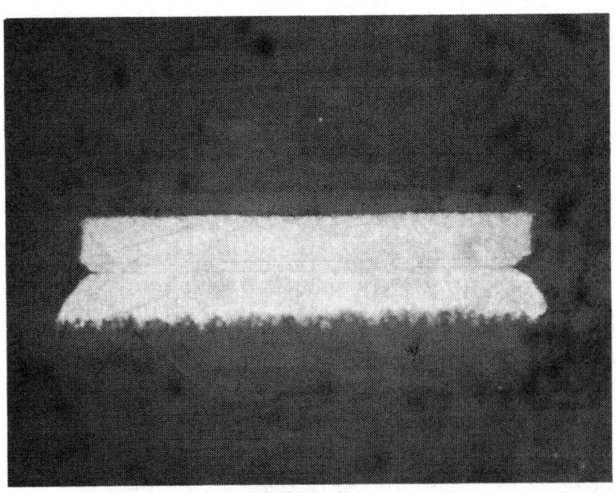

Fig. 25.36–6. Microsection of copper plated conductor. The protective coating is achieved by applying solder and insulation masks. In the actual case the mask is a liquid-film mask.

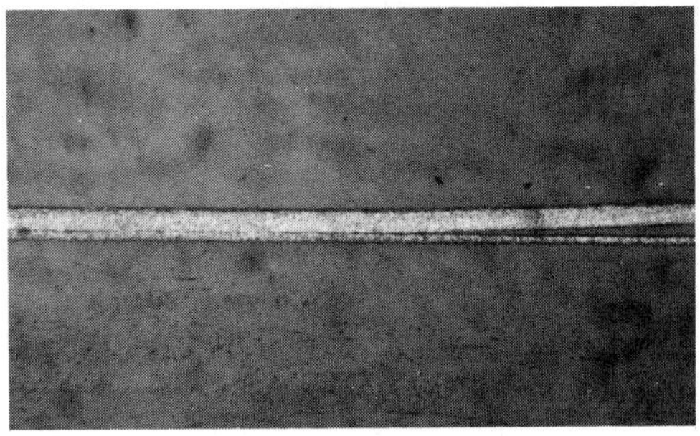

Fig. 25.36–7. Microsection of delaminated copper plating.

25.5.2 Microsections of Solder and Insulation Masks

In addition to the inspection of the primary details of the pattern described in Paragraph 25.5.1, a microsection can give deeper information regarding the properties of the solder mask, as described in the following paragraphs. The considerations apply equally to insulation masks, and it will often be the same microsections as inspected in Paragraph 25.5.1.

Now we shall discuss microsections of:

- Screen-printed solder masks
- Dry-film solder masks

25.5.2a Screen-Printed Solder Masks

See A of the cross-sectional drawing in Figure 25.37. A relatively moderate layer thickness on the order of 0.0006 to 0.0008 in. (15 to 20 μm) is characteristic of most screen-printed solder masks. It is very unusual to specify the thickness, and when the thickness is specified, it is very unlikely that the most critical area, at the conductor edge, will be mentioned.

As a result of surface tensions in the uncured ink, the solder mask will exhibit a convex profile across the conductor, leaving a very thin layer at the edge, perhaps only 0.00008 to 0.00016 in. (2 to 4 μm). The PCB specification may possibly

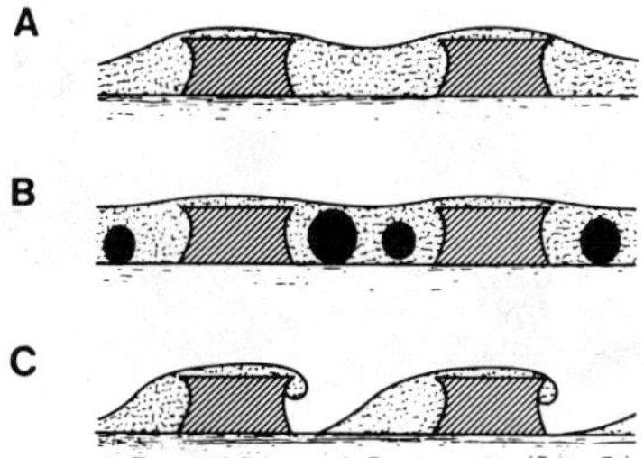

Fig. 25.37. Cross-sectional drawing of conductor covered with a screen-printed solder mask. A: Characteristical appearance. B: Occurrence of small air bubbles. C: Incomplete edge coverage.

require full coverage of the conductor edges by the solder mask.

When no specification exists, the inspector can use the following criteria, which are so modest that they can be met by any respectable PCB manufacturer:

1. Thickness at the middle of the conductor: 0.0004 in. (10 μm)
2. Thickness at the conductor edges: 0.00012 in. (3 μm)

When these requirements are not met, or one edge is not covered, the screen printing is so inferior that the boards should be returned for reprinting.

See the discussion in Paragraph 18.14.3 regarding coverage of the conductor sides where the conductor runs between two IC solder pads.

Concurrently with this inspection, the inspector should check whether the solder mask contains air bubbles (B, Figure 25.37). Since it is not always possible to avoid air bubbles when screen-printing, the inspector must accept them to a certain extent. Up to half the insulation distance may show air bubbles which must be closed cavities.

Whether failure to meet these requirements should be regarded as cause for rejection (an A-defect) depends on the use of the board, including the operating voltages and the environmental conditions.

The inspector must also check how the conductor sides are covered by the solder mask (C, Figure 25.37). Although this can be a problem, particularly in conjunction with adjacent conductors, the inspector should not repeat the assessment of unintentional openings and holes in the solder mask as described in Paragraph 18.14.4e.

25.5.2b Dry-Film Solder Masks

See the cross-sectional drawing in Figure 25.38. The layer thickness measured across a conductor will be considerably larger than in the case of screen-printed solder masks. As mentioned in Paragraph 12.7, the layer thickness depends on the pattern height and the thickness of the dry film used. See the curves shown in Figure 12.9.

In the case of dry-film solder masks, it is also relatively rare that a definite thickness is specified. When it is not specified, the inspector can use the following criteria when assessing the mask:

> The thickness across the whole conductor width shall be at least 0.001 in. (25 μm) for a pattern height of 0.0028 in. (70 μm) or less.

This requirement is met not only by the dry film thicknesses shown in Figure 12.9, but also by a frequently used dry film 0.002 in. (53 μm) thick.

Another criterion is that the spacing between adjacent parallel conductors must be filled with dry film.

If these criteria are not met, which is quite unlikely, the failure can be caused

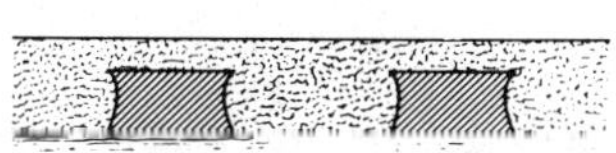

Fig. 25.38. Cross-sectional drawing of conductor covered with a dry-film solder mask.

only by the dry film. Because of process errors and/or lack of final inspection, the film manufacturer has delivered a dry film which is too thin.

Since the dry-film solder mask presumably was chosen for these quality criteria, such a fault is considered cause for rejection—that is, the fault is rated as an A-defect.

25.5.2c Microsection Examples

The aforementioned cases are illustrated with some practical examples of microsections in Figures 25.39-1 through 25.39-7. The captions give detailed explanations.

25.5.3 Microsections of Plated-Through Holes

Presumably, the best source of information about the board quality is microsections of its plated-through holes. They clearly reveal how well the PCB manufacturer has mastered processes decisively important to the board reliability, primarily hole drilling and plating. The following paragraphs clarify in more detail various conditions which the inspector should be aware of when assessing microsections:

- Copper plating thickness
- Copper plating evenness
- Tin/lead layer thickness
- Voids in hole wall
- Cracks in hole wall
- Etchback
- Epoxy smear
- Burrs and nail heading
- Plating contact
- Plating pockets
- Nodule formation
- Step-plating
- Adhesion to drilled hole wall
- Delamination within base material
- Blowholes

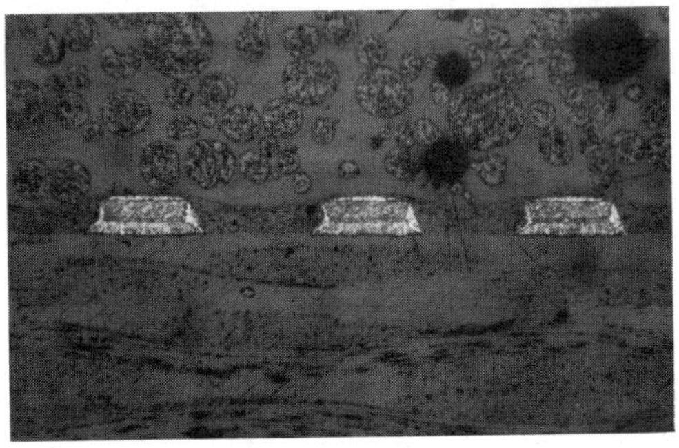

Fig. 25.39–1. Microsection of conductor with a screen-printed solder mask. Note the tin/lead coverage of the conductor edges.

Fig. 25.39–2. Microsection of conductor with a screen-printed solder mask, highly magnified. The thickness of the solder mask is 0.0003 to 0.0004 in. (8 to 10 μm) on top of the conductors. This is easiest assessed on the basis of a copper foil thickness of ½ oz. = 0.0007 in. (17.5 μm). The solder mask covers the conductor edges satisfactorily.

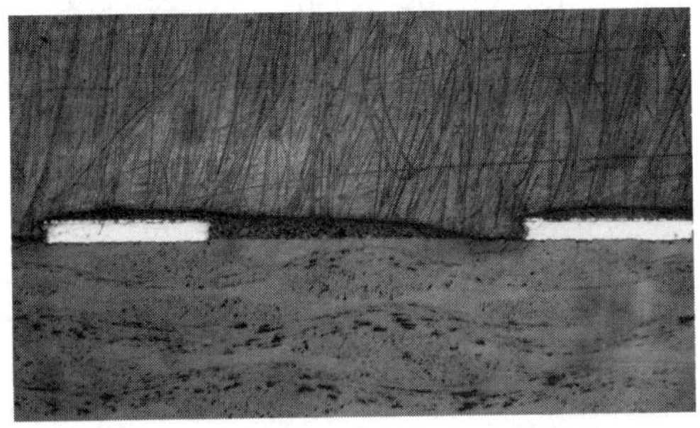

Fig. 25.39–3. Microsection of conductors with a screen-printed solder mask. An incomplete coverage of one edge of the conductors is seen, probably because the screen printing is performed crosswise with respect to the conductors.

Fig. 25.39–4. Microsection of conductors with a screen-printed solder mask. The black areas within the solder mask are small air bubbles which reduce the effective insulation between the conductors.

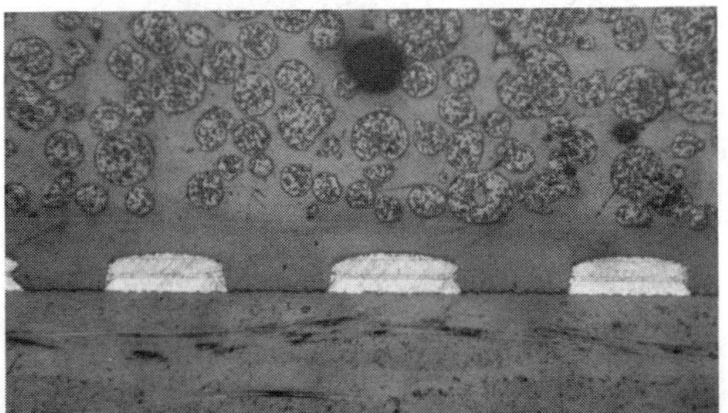

Fig. 25.39–5. Microsection of conductors with a dry-film solder mask. Note the large layer thickness of approximately 0.002 in. (50 μm) on top of the conductors. Note furthermore that the conductors are not covered completely with tin/lead on both edges.

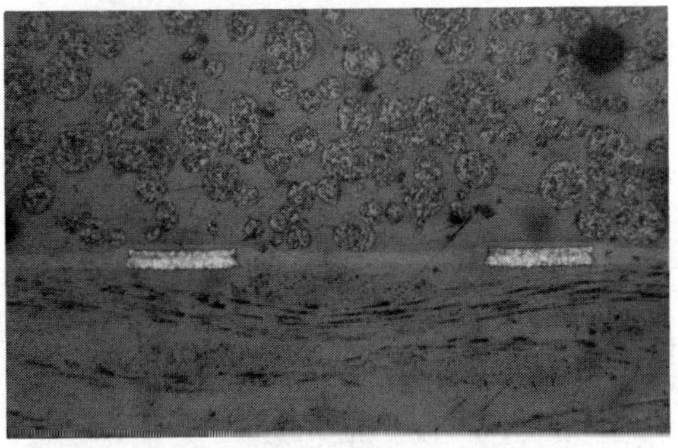

Fig. 25.39–6. Microsection of conductors with a liquid-film solder mask.

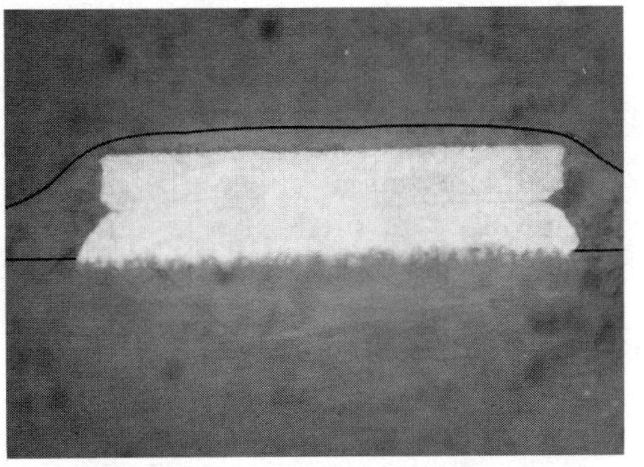

Fig. 25.39–7. Microsection of conductor with a liquid-film solder mask, highly magnified. The thickness of the solder mask is 0.0003 to 0.0004 in. (8 to 10 μm) on top of the conductors. (Due to a relatively poor contrast, the contour of the solder mask is inked with drawing ink.)

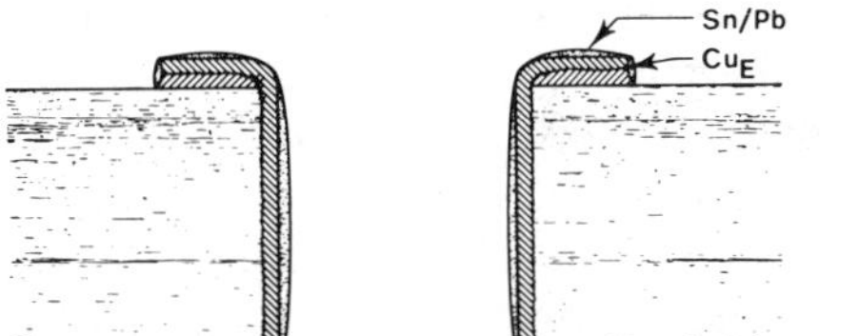

Fig. 25.40. Cross-sectional drawing showing the layer buildup of a plated-through hole. The drawing does not follow the drafting rules completely since two horizontal lines corresponding to the surface of the solder pads are missing. However, the appearance is in agreement with what can be seen in a metallurgical microscope which has a limited definition in depth. Note that in the subsequent cross-sectional drawings, all material indications are omitted as they show the same layer buildup as in this drawing. For the sake of clarity, all drawings show a nonreflowed condition. Cu_F = copper foil. Cu_E = electroless copper. Cu_P = copper plating. Sn/Pb = tin/lead plating.

25.5.3a Copper Plating Thickness

See the cross-sectional drawing in Figure 25.40. The most common requirement is an average copper plating thickness of 0.001 in. (25 μm), possibly supplemented by a requirement for a minimum thickness of 0.0006 in. (15 μm).

The inspector has to measure the copper plating thickness on the microsection. As mentioned in Paragraph 7.3.1b, he measures three to five places at each side of the microsection, afterwards calculating the average of the six to ten measurements. He should avoid local thin or thick areas. In the case of multilayer boards, he should measure just outside any internal pads. He then determines the minimum thickness.

It is a well-known fact that the plating thickness depends to some degree on the location of the holes on the board. It is therefore recommended that the microsection be taken from the middle of the board, primarily from dense areas.

In the case of multilayer boards, test coupons frequently accompany the boards so that microsections can be prepared without having to sacrifice an expensive board. The inspector should note that test coupons are usually located on the production panel just outside the board contour.

Even when a current robber or thief—which reduces the current density in adjacent areas during plating—is near the test coupon, there is a marked tendency towards a heavier copper deposit in the test coupon holes than in the holes located in the middle of the board.

Unfortunately, the inspector does not know the plating profile—that is, the thickness of the copper deposit in the holes as a function of the hole location. The plating profile depends primarily on the actual board—among other things, the pattern density, including the balancing of the pattern, and the ratio between the hole diameter and the board thickness. The plating profile depends to a very small extent on the location of the production panel in the plating bath, especially with respect to the anodes.

Some PCB manufacturers provide for 0.0014 in. (35 μm) copper in the holes of the test coupon which should ensure 0.001 in. (25 μm) copper in the holes in the center area of the board. If the inspector has asked for a scrap board, it can be used for microsectioning holes located in the center area of the board. Naturally, the board should not be rejected because of plating faults.

Failure to meet the requirements regarding plating thickness is, in principle, cause for rejection and consequently rated as an A-defect. The inspector can, however, issue a waiver for a lower plating thickness, provided the hole drilling is even. See Paragraph 7.3.1c for further details.

25.5.3b Copper Plating Evenness

See the cross-sectional drawing of Figure 25.41. The evenness of the copper plating depends considerably on the evenness of

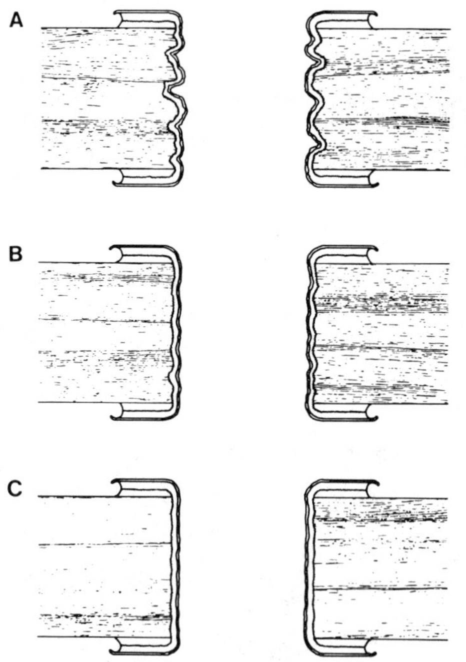

Fig. 25.41. Cross-sectional drawing showing various degrees of evenness of the copper plating in a plated-through hole. A: A very coarse and unacceptable hole wall. B: A fairly uneven but acceptable hole wall. C: A smooth and even hole wall.

the hole drilling, which is assessed at the same time. Uneven drilling can result in various defects in the copper plating, such as plating pockets, voids, and a tendency to crack.

It is, in practice, impossible to quantify, in a well-arranged way, how uneven a hole drilling may be before it is unacceptable. The inspector is instead referred to the illustrations in Figures 25.55-1 through 25.55-38 of Paragraph 25.5.3p. Various faults in the copper plating are described in some of the following paragraphs.

25.5.3c Tin/Lead Layer Thickness

See the cross-sectional drawing in Figure 25.40. The thickness of the tin/lead layer in the hole itself is not so decisive, provided that the tin/lead has not run down into the hole so that the diameter tolerance cannot be met.

A more interesting place to study the thickness of the reflowed tin/lead is at the transition from the hole barrel to the solder pad. Normally the layer thickness at this particular spot is quite modest—for example, 0.00002 to 0.00004 in. (0.5 to 1 μm).

Only a few PCB specifications explicitly state the thickness at the transition zone. See the remarks in Paragraph 8.5.2a concerning this problem.

25.5.3d Voids in Hole Wall

See the cross-sectional drawing in Figure 25.42. The causes of voids are elaborated in Paragraph 7.3.2. In order to reliably ascertain whether the microsection contains voids, the inspector can select solder holes in conjunction with the inspection described in Paragraph 7.3.3b.

It is an advantage, not least in conjunction with the following Requirement 1, that the microsection be transparent so that the entire hole wall can be inspected.

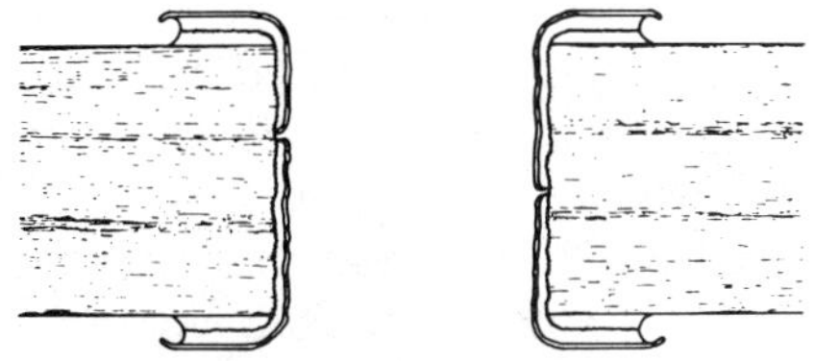

Fig. 25.42. Cross-sectional drawing showing voids in the hole wall of a plated-through hole.

Since this implies a reasonable depth of definition, the inspector must use a stereo microscope.

It is possible to find quite a number of specifications defining how many voids are acceptable in a solder hole, as well as the maximum dimension of such voids. All specifications, however, suffer from the weakness that they are not usable in conjunction with microsections which show only half of the solder hole.

The following requirements are adapted to the special conditions valid for microsections. The inspector should, however, check his own specification.

1. The microsection examined shall contain no more than two voids and the total area shall be less than 5 percent of the visible area of the hole wall.
2. The transition area between the hole and the solder pad shall contain no voids.
3. In multilayer boards, no voids shall occur on a level with internal pads and planes.
4. No voids shall occur at the same level on both sides of the cross section. (Such voids could, in theory, constitute the ends of a circumferential hole in the part of the barrel which has been removed.)

When one of these requirements is not met, there is a risk that the holes in the batch are unsatisfactory. The examination should therefore be extended to more microsections. If the requirements are still not met, the batch should be rejected and returned. The fault is rated as an A-defect.

The assessment of the areas mentioned in Requirement 1 is difficult and can only be performed as an estimate. In some specifications the requirement is expressed in lengths instead of areas, which makes estimating a little easier.

In other specifications, the largest dimension of a void is limited to 25 percent of the board thickness or of the hole circumference. In the author's experience, there is no reason to be too severe, because plated-through holes generally follow the rule that either they are all good or they are all bad.

25.5.3e Cracks in Hole Plating

Any tendency of plated-through holes to crack is aggravated by stress during soldering. This tendency shows up very clearly when the inspector makes microsections of soldered holes. The effect of soldering should not exceed the specified soldering conditions.

All cracking must be considered potentially dangerous with respect to reliability. Although it is possible to completely fill a large percentage of the solder holes with solder, the solder can be stopped at a circumferential crack; in such cases there is a definite risk of intermittent open circuits.

Cracking must be regarded as a serious fault requiring examination of several microsections. When all microsections indicate cracking, the fault must be regarded as cause for rejecting the entire batch, which is returned. The fault is, of course, rated as an A-defect.

It can be somewhat problematic to show the material structure clearly enough when etching with the etchant mentioned in Paragraph 25.2.7 (glacial acetic acid plus nitric acid). The solder, which is attacked quickly, forms a sludge layer, even when the etching time is as short as half a second.

A nonprofessional, but quite efficient, method of removing the sludge layer is to rub it with a finger under running water.

The inspector can then examine the microsection for cracks and, at the same time, turn his attention to the following types of faults:

25.5.3e1 Corner Cracking:

See A in the schematic drawing in Figure 25.43. Because of stress, the place most subject to cracking is the transition zone between the barrel and solder pad. The two surfaces of the fracture will usually fit together, but the crack can be filled with solder.

25.5.3e2 Barrel Cracking:

See B in the schematic drawing in Figure 25.43. It is not unusual to find cracks in the hole plating itself (barrel cracking). The cause of such cracks is often a delamination of the base material in the region around the solder hole. Delamination will, as a rule, cause circumferential cracks.

Cracks on both sides of the microsection can be interpreted as a circumferential crack. Since the crack can fill with solder, the inspector may have difficulty seeing it in the hole wall itself.

25.5.3e3 Cracks at Nodules:

See C in the schematic drawing in Figure 25.43. Around irregular nodules, large stresses can occur in the copper deposit, resulting in cracking. The cracks will frequently start at the foot of the nodules and continue through the copper plating. Under worst-case conditions, particularly during soldering, such cracks can develop into circumferential cracks.

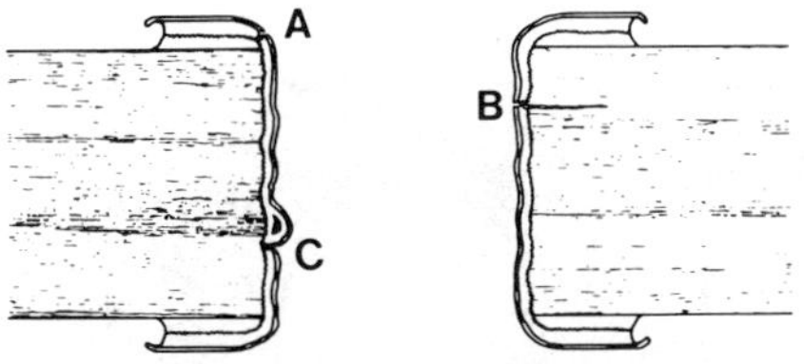

Fig. 25.43. Cross-sectional drawing showing different types of cracks in the hole wall of a plated-through hole. A: Corner cracking. B: Barrel cracking. C: Crack at a nodule.

25.5.3f Etchback

See the cross-sectional drawing in Figure 25.44. Etchback is used in the case of multilayer boards to create a better contact between the hole plating and the pads, preferably the internal pads. See A in Figure 25.44.

As mentioned in Paragraph 7.2.6c there are quite different opinions concerning the real value of an etchback, and some PCB manufacturers prefer not to perform etchback unless the customer requires it.

The inspector first checks whether etchback is specified, and if so, the limits of the etchback. Common limits are maximum values of 0.003 in. (75 μm) or 0.002 in. (50 μm), and a minimum value of 0.0003 in. (7.5 μm).

When a heavy etchback has been performed, the holes in multilayer boards having many layers will exhibit a wavy hole wall, even after reflowing. The inspector should preferably select such holes for his investigation.

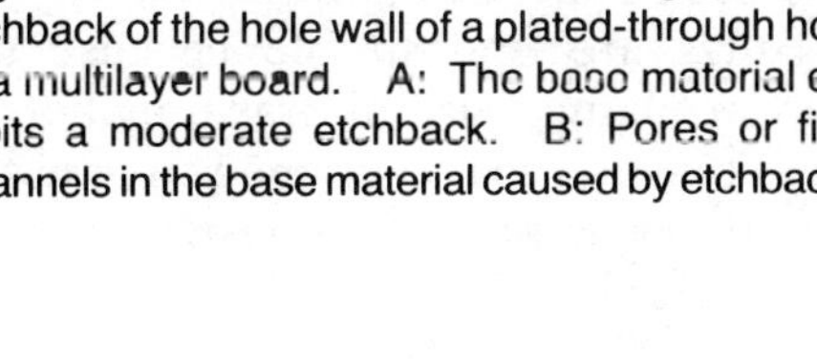

Fig. 25.44. Cross-sectional drawing showing etchback of the hole wall of a plated-through hole in a multilayer board. A: The base material exhibits a moderate etchback. B: Pores or fine channels in the base material caused by etchback.

Failure to meet the requirements is rated as an A-defect. Before rejecting the batch, the inspector should investigate additional microsections. In the case of excessive etchback, he should not issue a waiver since there is a possibility of serious stress in the copper deposit, resulting in a risk of cracking after heating, such as occurs during soldering.

On the other hand, the inspector should be able to issue a waiver in case of too light an etchback, or possibly no etchback at all. The waiver depends, of course, on the end user's specifications, but this is frequently not the case in civilian electronics production.

An unfortunate consequence of a heavy etchback can be the formation of pores or fine channels in the base material. Such defects can in some cases reach to the copper of an inner layer—for example, a ground plane. See Figure 25.44.

When copper plating the board, such pores or channels can become filled with copper, resulting in hard-to-find short circuits. If the inspector finds such defects during his inspection, he has every reason to inspect the batch very carefully.

The inspection can be performed easily as an electrical check for short circuits, followed by a thorough examination of microsections of defective holes. The fault is naturally cause for rejection and consequently rated as an A-defect.

25.5.3g Epoxy Smear

See the cross-sectional drawing in Figure 25.45. As mentioned earlier, epoxy smear is a thin layer of epoxy deposited internally in the drilled hole. This implies that epoxy can be present on the edge of internal as well as external pads. Epoxy smear is caused by poor hole drilling.

Most PCB manufacturers do not consider epoxy smear in ordinary double-sided plated-through boards a serious

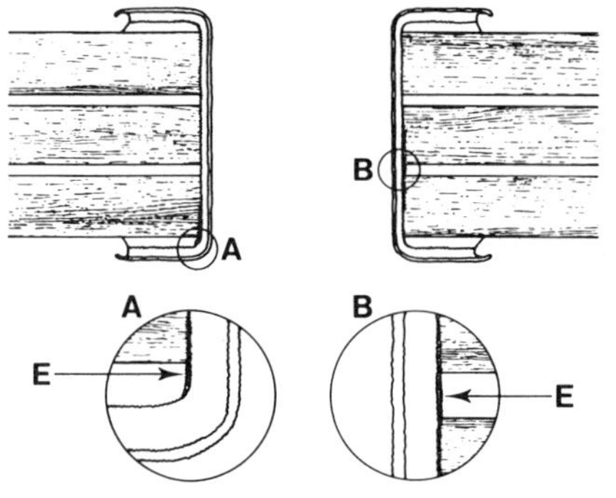

Fig. 25.45. Cross-sectional drawing of a multilayer board showing a plated-through hole with epoxy smear. A: Epoxy smear (E) at copper foil. B: Epoxy smear (E) at inner layer of board.

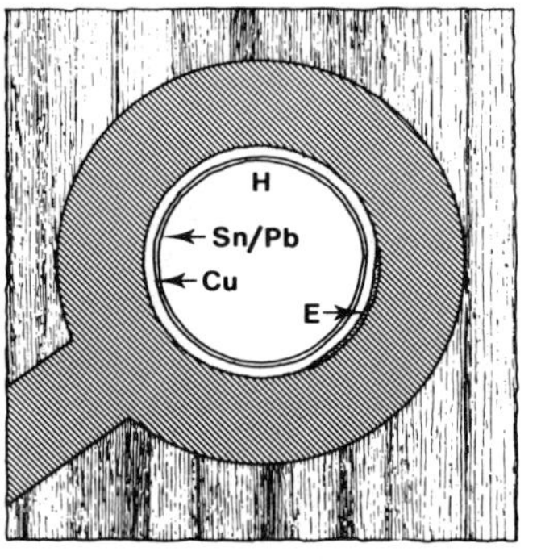

Fig. 25.46. Cross-sectional drawing of a multilayer board showing a plated-through hole with epoxy smear. Contrary to Fig. 25.45, the cut is placed in parallel with the board through one of the inner layers. E: Epoxy smear between the inner pad and the copper plating. Cu = copper plating. Sn/Pb = tin/lead plating. H = hole.

defect, since the hole plating under all circumstances has good contact with the plating on the top side of the pad. Therefore, they do not remove any epoxy smear from such boards, presumably because epoxy smear on the relatively thin edges is considered unimportant.

On the other hand, epoxy smear in multilayer boards is a very serious fault, and all PCB manufacturers take care to prevent epoxy smear and to remove all possible traces of it.

In the microsection, epoxy smear shows up as a separation between the copper deposit on the hole wall and the copper foil of the internal and external pads. See Figure 25.45. Since epoxy smear does not necessarily occur over the entire hole wall, it can be safer to prepare microsections perpendicular to the hole.

Such microsections show the annular ring of the internal pad. See Figure 25.46. Possible epoxy smear is now seen as a separation between the annular ring and the hole plating occurring over a shorter or longer part of the circumference.

In multilayer boards, epoxy smear is an indisputable cause for rejection. When the microsection shows epoxy smear, there is a considerable risk that the fault is present in the whole batch. Therefore, an examination of more microsections is recommended. If these also exhibit epoxy smear, the batch should be rejected. The fault is rated as an A-defect.

25.5.3h Drilling Deformation

See the cross-sectional drawing in Figure 25.47. The inspector should examine the hole for unevenness where the copper layers have been drilled through. A poor hole drilling can easily result in unevenness at such places.

25.5.3h1 Burrs

Burrs are irregular raised rings at the inner rims of the drilled holes. Paragraph 7.2.4 addresses the causes.

Particularly in the case of plated-through boards (ordinary double-sided boards as well as multilayer boards), the burr can be pressed back into the hole. This causes a thickening of the plating at the corner of the hole, also called a shoulder. See A and C, Figure 25.47.

Burrs are undesirable because the heavier plating current drawn by burred areas causes a coarse-grained structure in the copper deposit. The resulting stress

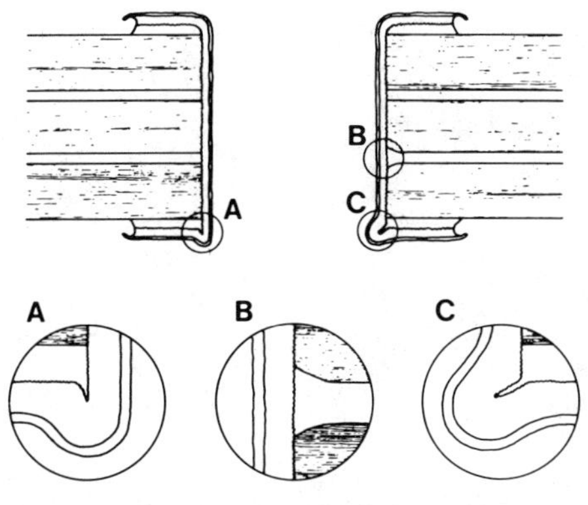

Fig. 25.47. Cross-sectional drawing of a multilayer board showing a plated-through hole with burrs and nail heading. A: Plating thickening at the hole rim, caused by a drilling burr. The burr is erect as produced during drilling. B: Nail heading showing up as a thickening of the inner layer (pad) due to incorrect drilling parameters. C: Plating thickening at hole rim, caused by a drilling burr. The burr is pressed back during the brushing prior to imaging.

effects can be relieved during soldering and cause cracking. In case of a bare-copper board, a large burr can be the origin of cracks in the solder fillet itself.

PCB specifications rarely mention burrs at all, much less a maximum value. The inspector can use the following guidelines when assessing burrs:

Nonplated-Through Boards:
A maximum burr height of 0.001 in. (25 μm) is accepted, whereas a height between 0.001 and 0.002 in. (25 and 50 μm) is regarded as a B-defect. A height exceeding 0.002 in. (50 μm) is cause for rejection and rated as an A-defect.

Plated-Through Boards:
A burr height up to 50 percent of the foil thickness, but no more than 0.001 in. (25 μm) is acceptable. Heights exceeding this are cause for rejection and rated as an A-defect.

The height of a burr is the dimension from the foot of the burr to its top. These limits apply even if the burr is pressed back into the hole.

25.5.3h2 Nail Heading

Nail heading occurs in multilayer boards as a mechanical deformation of the hole rim of the internal pads. See B, Figure 25.47, and the discussion in Paragraph 7.2.5b. The deformation causes a harder copper, resulting in a risk of cracking. In less fortunate circumstances, the copper at the hole rim can be twice its normal thickness. When inspecting for nail heading, the inspector can use the following criterion:

> Nail heading shall not increase the copper thickness at the hole rim by more than 50 percent. This is accepted as a matter of course.
>
> A thickness increase exceeding 50 percent is regarded as cause for rejection and rated as an A-defect. However, it is considered reasonable to examine further microsections, and if they also exhibit nail heading, the batch should be rejected.

25.5.3i Plating Contact

See the cross-sectional drawing in Figure 25.48. When examining the microsection, the inspector should assess the appearance of the metallic contact—also called the plating contact—between the hole plating and the inner and outer pads.

25.5.3i1 Plating Contact to Inner Layers

In multilayer boards, it is of the greatest importance to achieve full plating contact between the hole plating and the hole rim of the inner pads. Here it is required that contact to the pads across the full hole rim be attained, as shown at A, Figure 25.48.

The examples at B and C, Figure 25.48, show unacceptable contact conditions,

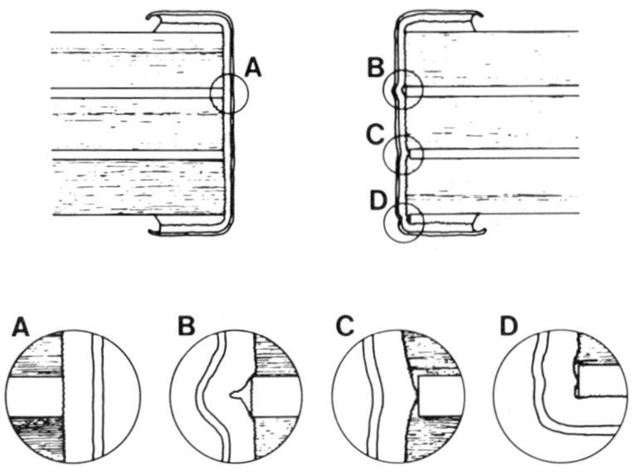

Fig. 25.48. Cross-sectional drawing of a multilayer board showing various degrees of metallic contact (plating contact) between the hole plating and the inner/outer pads. A: Metallic contact across the entire width of an inner pad. B and C: Partial contact to an inner pad is not acceptable. D: Partial contact to an outer pad is acceptable.

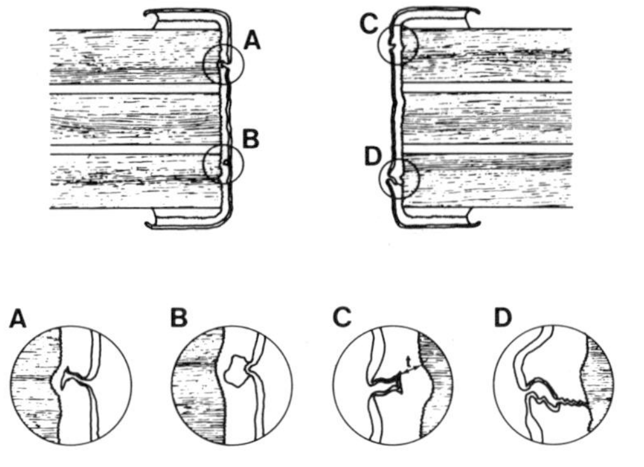

Fig. 25.49. Cross-sectional drawing of hole wall with plating pockets. A: Unacceptable plating pocket (open type). B: Unacceptable plating pocket (closed type). C: Acceptable plating pocket (t $\geqslant$ 0.0008 in. = 20 μm). D: Unacceptable plating pocket (with a crack).

which must be regarded as cause for rejection and consequently rated as A-defects.

25.5.3i2 Plating Contact to Outer Layers

It is not imperative that full plating contact be achieved between the hole plating and the hole rim of the external pads (solder pads). Under all circumstances, the hole plating will continue around the corner and make contact with the surface of the solder pad. Therefore, a partial plating contact as shown at D, Figure 25.48, is acceptable.

25.5.3j Plating Pockets

See the cross-sectional drawing in Figure 25.49. As mentioned in Paragraph 7.3.6, the occurrence and nature of plating pockets can be verified only by an examination of microsections. Plating pockets can be caused by an uneven and rough drilling. The structure of the hole plating, however, must be continuous and smooth. The following criteria can serve as guidelines for the inspector:

1. The area around plating pockets shall not contain cracks.
2. In multilayer boards, plating pockets shall not occur on a level with the individual layers.
3. The specified copper thickness (plating thickness) shall be present from the bottom of the plating pockets to the base material.

When requirements 1 and 2 are not met, the hole must be considered unacceptable. When Requirement 3 is not met, the plating pocket can be regarded as a void, subject to the assessment criteria given in Paragraph 25.5.3d.

Before rejecting the whole batch, the inspector should examine additional microsections to obtain an overall idea of the distribution and nature of the plating pockets. Usually, the worst problems occur in multilayer boards; the test coupons accompanying the boards prove their usefulness here.

25.5.3k Nodule Formation

See the cross-sectional drawing in Figure 25.50. Formation of nodules in plated-through holes can, to a certain degree, be

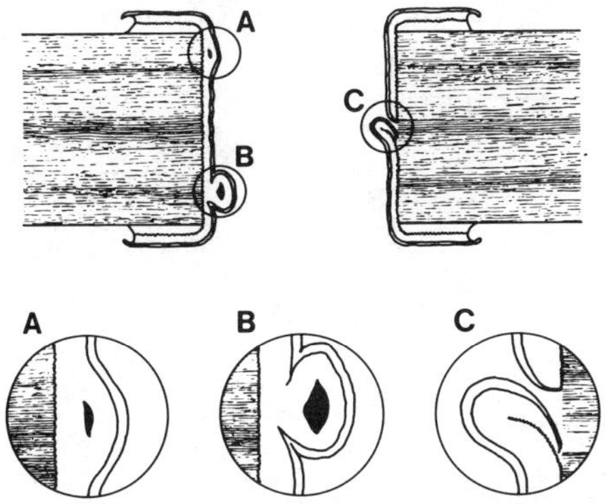

Fig. 25.50. Cross-sectional drawing of hole wall with nodules. A: Nodule on the surface of the hole wall, caused by plating on top of a foreign particle adhering to the hole wall. The nodule is acceptable because of the even structure. B: Large nodule which is unacceptable because of the irregular structure around the nodule. C: Formation of a nodule around a small flake of electroless copper which has worked loose. The nodule is unacceptable.

recognized by visual inspection. It is therefore possible to select holes with interesting nodules for microsectioning. The nodules can be divided into two groups:

Harmless Nodules:
These nodules, shown at A, Figure 25.50, are characterized by:

1. The projection outwards from the plated hole wall is so small that the diameter tolerance of the hole is met—that is, the projection is about 0.004 to 0.008 in. (0.1 to 0.2 mm).
2. The structure of the plating is even and smooth in the area around the nodule without any signs of cracking.
3. The nodule is typically located on the surface of the plating.

Nodules of this kind can be accepted routinely.

Detrimental Nodules:
Nodules as shown in B, Figure 25.50, are characterized by:

1. The projection outwards from the plated hole wall is so large that the diameter tolerance of the hole cannot be met—that is, the projection is about 0.008 to 0.012 in. (0.2 to 0.3 mm), or more.
2. The plating structure is very uneven in the area around the nodule. A plating gap or a plating void can possibly occur, or signs of cracking can be found.
3. The nodule can start down in the plated layer, but it can also be located on the surface of the plating.

A closer examination of the center area of the nodules can give some information regarding the cause of the nodules. If, for example, the electroless copper flakes in the area concerned, possibly because of an unstable support, a small nodule can form around the tiny flake concurrently with a plating gap. See C, Figure 25.50. At B, the nodule forms around an organic particle—for example, a small flake originating from the plating resist (Paragraph 7.3.7).

These examples clearly show that a microsection can give a good picture of some of the problems the PCB manufacturer encountered while manufacturing the boards in question. In the long run, such examinations permit a more accurate assessment of the PCB manufacturer's capabilities.

25.5.3l Step-Plating

See the cross-sectional drawing in Figure 25.51. A very exceptional fault is step-plating, also called skip-plating. The cause is usually loss of control of the copper deposition. Paragraph 7.3.9 gives detailed assessment criteria.

25.5.3m Adhesion to Drilled Hole Wall

See the cross-sectional drawing in Figure 25.52. Certain types of electroless copper

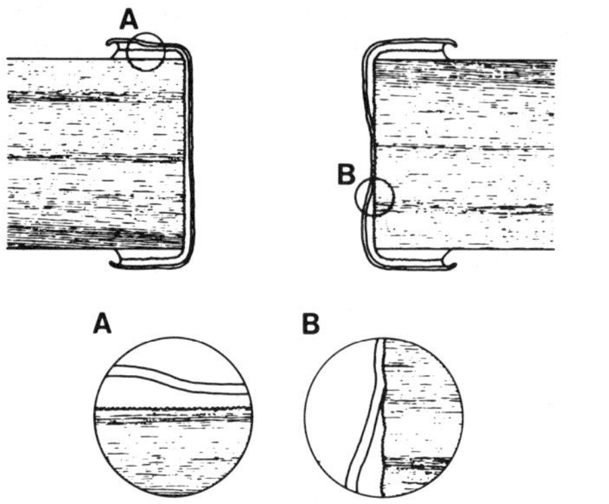

Fig. 25.51. Cross-sectional drawing of plated-through hole with step-plating. A: A very distinct plating step is seen in the plated copper layer. Since the very thin layer proceeds down through the hole, the presence of the plating-step is unacceptable. B: No copper plating is seen in a zone down in the hole. The tin/lead layer is bridging the zone and masks the fault which can be revealed by microsectioning only. The fault is unacceptable.

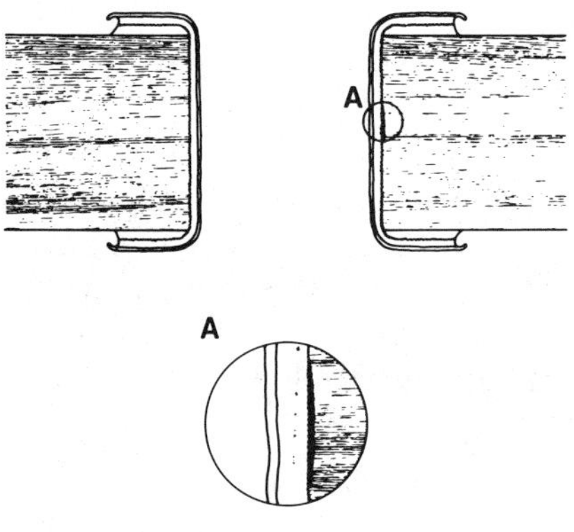

Fig. 25.52. Cross-sectional drawing of a plated-through hole exhibiting inferior adherence of the plating to the drilled hole wall. A possible tendency to inferior adhesion will be accentuated by a thermal shock caused when soldering the hole. A: Reduced adherence is characterized by a cavity behind the copper plating.

used in the basic metallization process (copper-reduction process) adhere poorly to the base material.

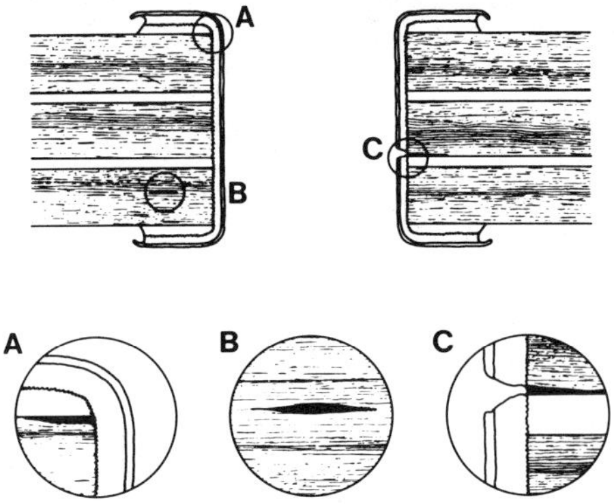

Fig. 25.53. Cross-sectional drawing of board exhibiting delamination of the base material. A: Delamination between the base material and an outer pad (solder pad). B: Delamination in the base material. C: Delamination between the base material and an inner pad.

This reveals itself in the microsection as a small space between the copper layer (electroless copper plus electroplated copper) and the base material. The fault is obviously cause for rejection and rated as an A-defect.

25.5.3n Delamination in Base Material

See the cross-sectional drawing in Figure 25.53. Drilling delamination, which is addressed in Paragraph 7.2.8, is an indisputable cause for rejection. The cavity can absorb moisture or electrolyte during the plating process. This results in blowholes in the solder fillets after mass soldering and/or reduced insulation resistance.

Delamination can occur between the base material and a solder pad, and in the case of multilayer boards, also between the base material and an inner pad. In addition, delamination can occur in the base material itself, possibly as a prolongation of a crack in the hole plating.

In multilayer boards, delamination can also be caused by a poor pressing technique. This can result in the same type of

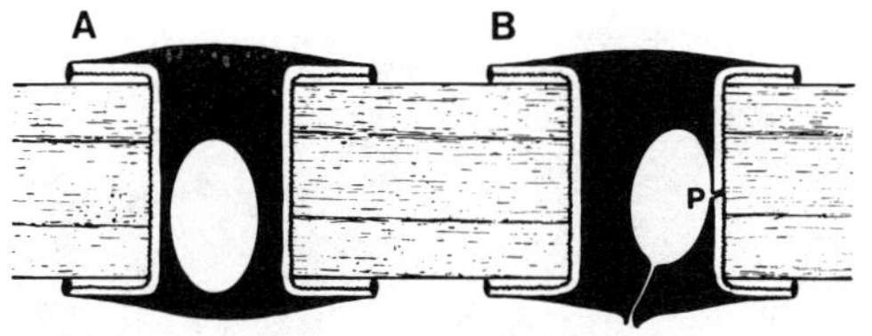

Fig. 25.54. Cross-sectional drawing of trial-soldered board exhibiting blowholes in the solder fillets. A: Closed blowhole. B: Open blowhole with a pore (P) in the copper plating.

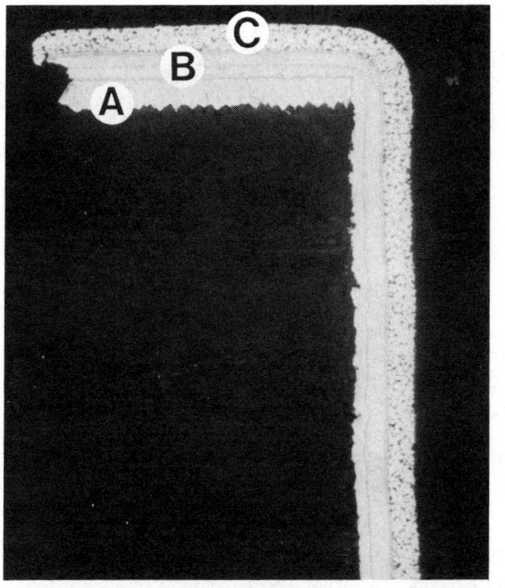

Fig. 25.55–1. Microsection of plated-through hole, very suitable for measuring plating thicknesses when the magnification is known. The photograph shows a section of the solder pad and the hole wall. The copper foil of the base material is seen as the first or lowest layer (A), top left. On top of the copper foil is seen the deposited copper (B), the dividing line indicating the limit between a relatively thick layer of electroless copper and a relatively thin layer of electrolytically deposited copper. On top of this comes a fairly thick layer of tin/lead (C) in a nonreflowed condition (courtesy Alpha Metals, Inc.).

defects as the short circuits caused by excessive etchback (Paragraph 25.5.3f). Sometimes the defect does not show up until after the board has been assembled and placed in use under severe environmental conditions.

Delamination is cause for rejection because of the aforementioned consequences, and the fault is rated as an A-defect.

25.5.3o Blowholes

See the cross-sectional drawing in Figure 25.54. Paragraph 10.7.4a describes the causes of blowholes and their assessment criteria. Since the inspector may well find both closed and open blowholes, microsections of soldered holes with blowholes in the solder fillets can be both interesting and instructive. See A and B, Figure 25.54.

If the inspector is very lucky in locating the cut, he can see a tiny pore in the copper plating through which the vapor or gas causing the blowhole has forced its way from the base material into the solder fillet. See C, Figure 25.54.

25.5.3p Microsection Examples

The cases described above are illustrated with some practical examples of microsections in Figures 25.55-1 through 25.55-38. The captions describe the microsections in detail.

Fig. 25.55–2. Microsection of a very even and smooth hole drilling (courtesy Alpha Metals, Inc.).

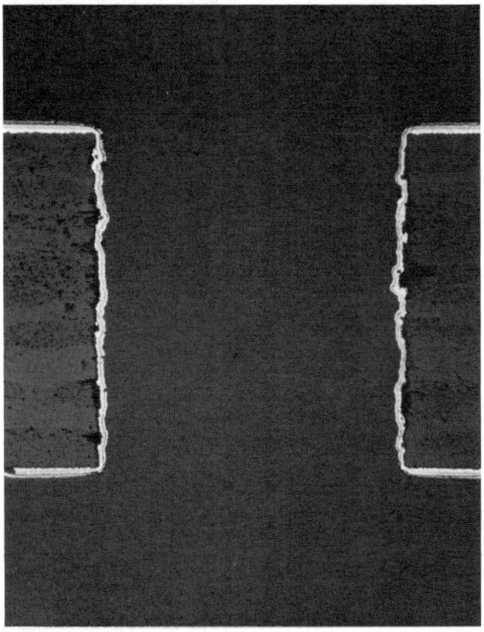

Fig. 25.55–3. Microsection of a somewhat uneven but acceptable hole drilling (Courtesy Alpha Metals, Inc.).

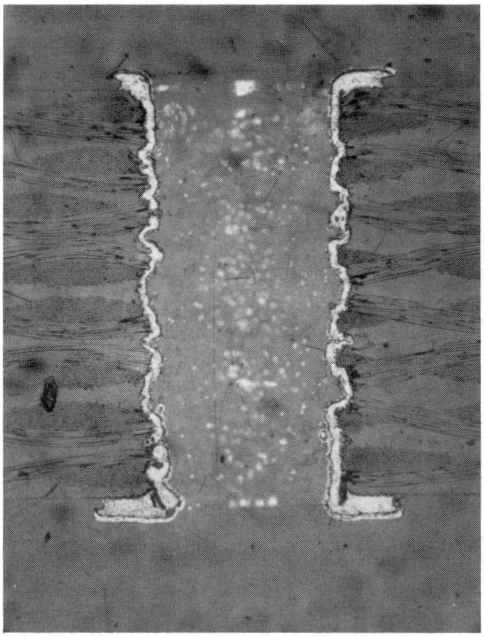

Fig. 25.55–4. Microsection of a very uneven and entirely unacceptable hole drilling.

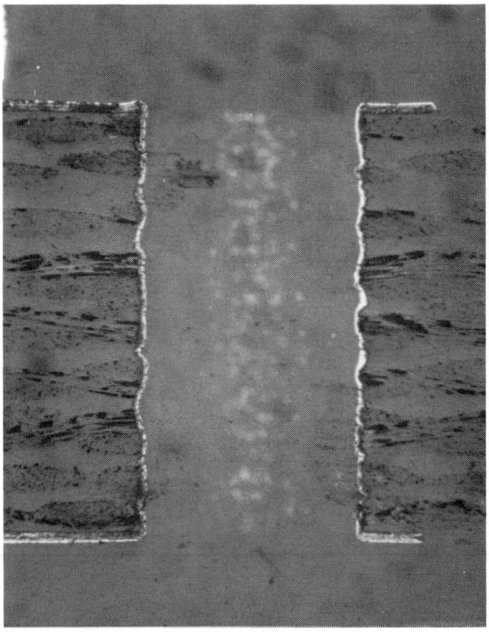

Fig. 25.55–5. Microsection of a hole with an unacceptably thin copper plating.

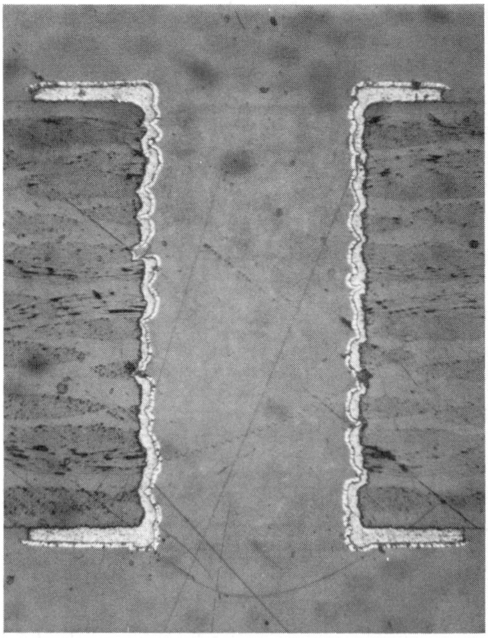

Fig. 25.55–6. Microsection of a hole with an unacceptably uneven copper plating.

Fig. 25.55–7. Microsection of an unevenness in the copper plating of a hole corresponding to Fig. 25.55–6, highly magnified.

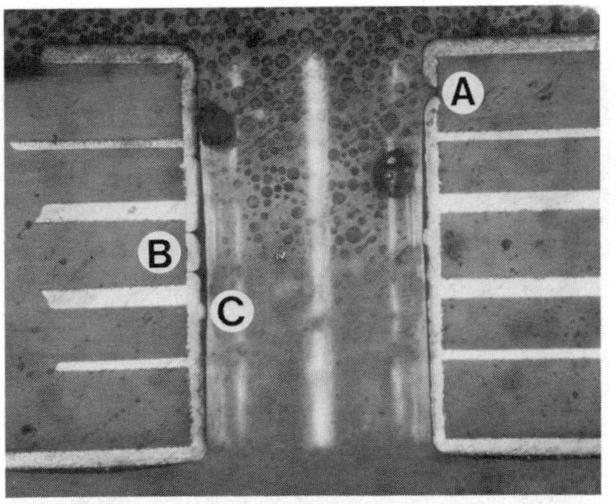

Fig. 25.55–9. Microsection of hole in a multilayer board. The hole plating contains a number of defects in the form of: A: Void. B: Pinholes. C: Nodules.

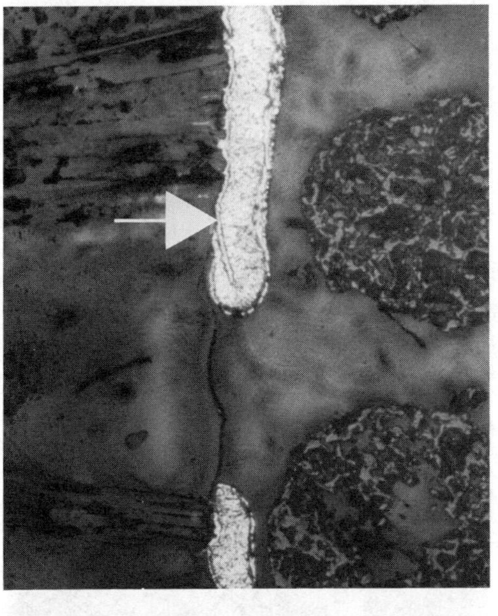

Fig. 25.55–8. Microsection of hole wall exhibiting a plating gap owing to the lack of electroless copper. At the arrow, the electroless copper has flaked off and later has been plated on the rear side also. The copper plating, however, has not been able to bridge the gap.

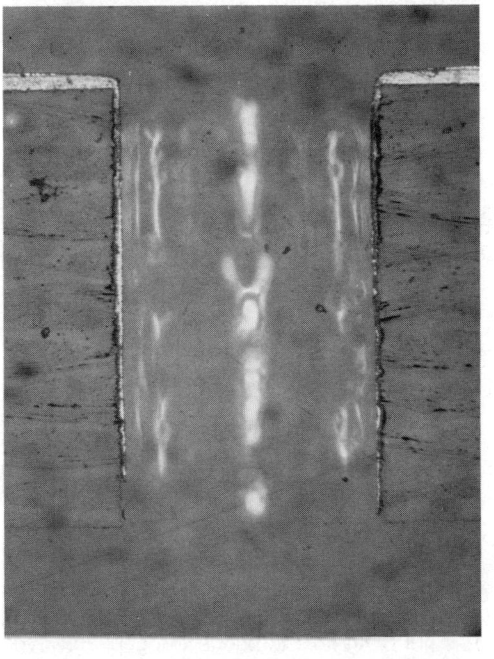

Fig. 25.55–10. Microsection of landless hole with incomplete hole plating at the landless side.

Fig. 25.55–11. Microsection of hole with corner cracking, i.e., a crack in the transition zone between the solder pad and the hole wall. Note that the surfaces of the fracture fit together.

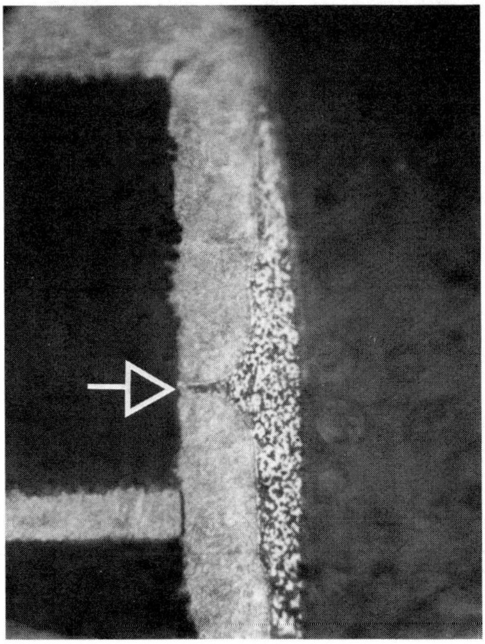

Fig. 25.55–13. Microsection, highly magnified, of a void in the copper plating of a hole in a multilayer board. The void is closed with tin/lead.

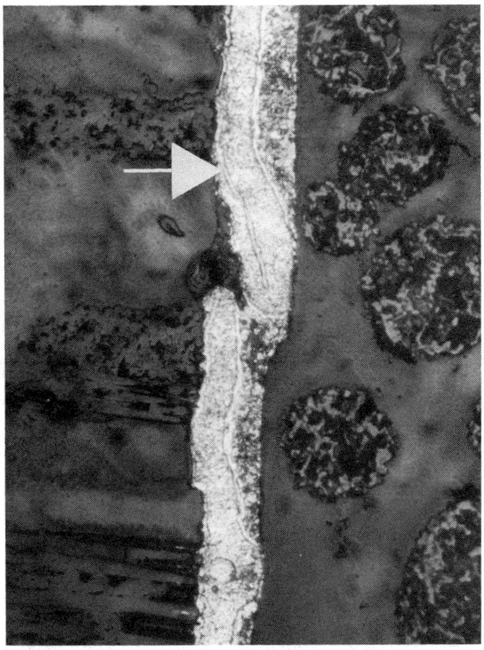

Fig. 25.55–12. Microsection of a hole wall with a crack in the copper plating. At the arrow the electroless copper has flaked off and later has been plated on the rear side. The copper plating has not been able to bridge the gap. However, the tin/lead plating has closed the gap which therefore cannot be revealed during a visual inspection of the hole.

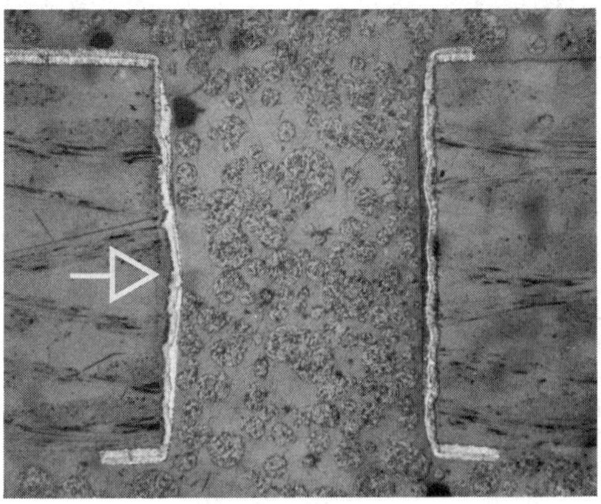

Fig. 25.55–14. Microsection of a hole where the plating has separated from the drilled hole wall.

Fig. 25.55–15. Microsection of hole with incomplete plating. The drilling is defective at one side of the board, and the plating contains many nodules.

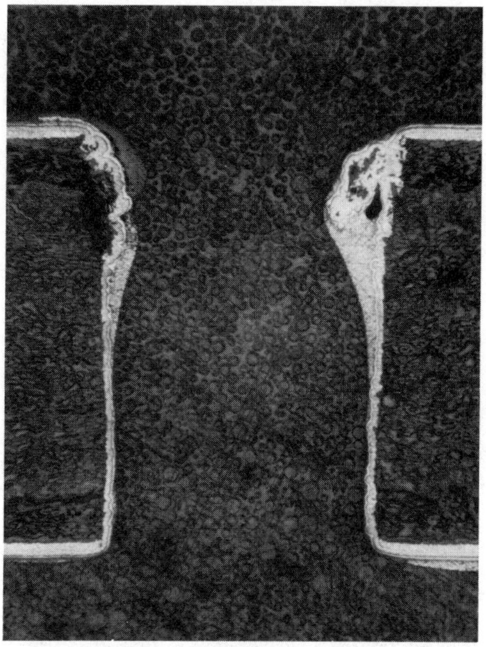

Fig. 25.55–17. Microsection of a hole with large nodules at one side of the board. The tolerance on the hole diameter is not met, and insertion of a component lead will probably damage the hole plating severely.

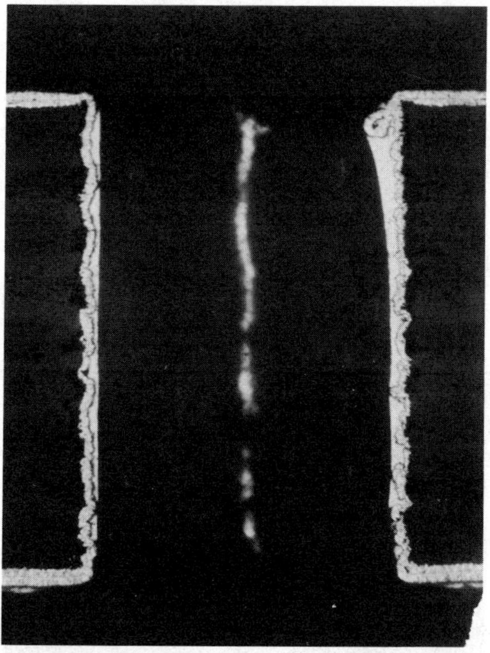

Fig. 25.55–16. Microsection of a hole exhibiting a large plating nodule at the transition zone between the solder pad and the hole wall. Note that the very uneven copper plating is concealed under an even layer of tin/lead as a result of reflowing.

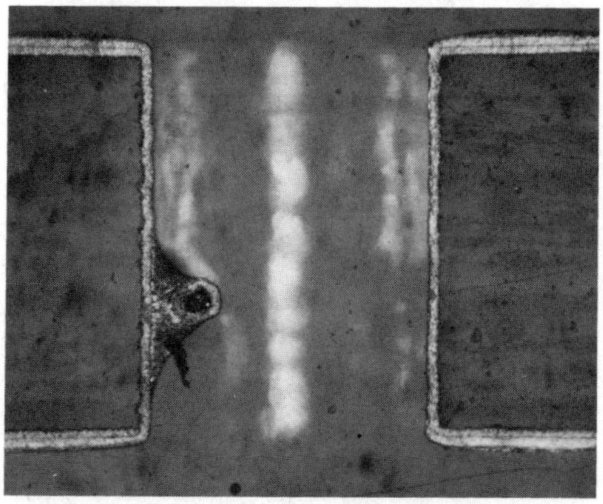

Fig. 25.55–18. Microsection of hole with a plating nodule located on the hole wall. The projection is about 0.008 in. (0.2 mm) and therefore unacceptable.

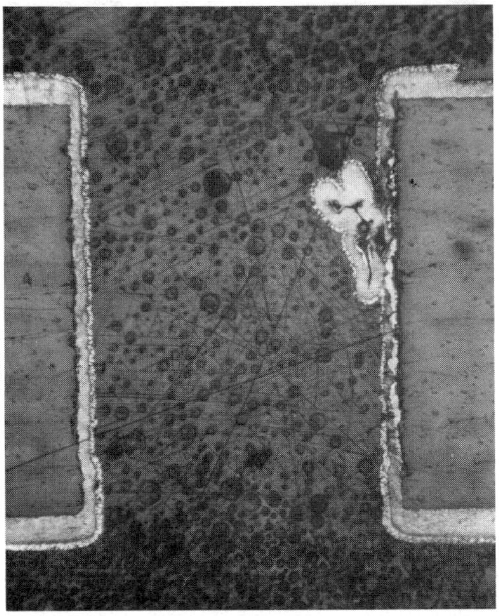

Fig. 25.55–19. Microsection of hole wall with a plating nodule located on the hole wall. The reason is a foreign particle fastened to the hole wall prior to the electroless copper deposition. Tearing-off the nodule will probably damage the hole plating seriously.

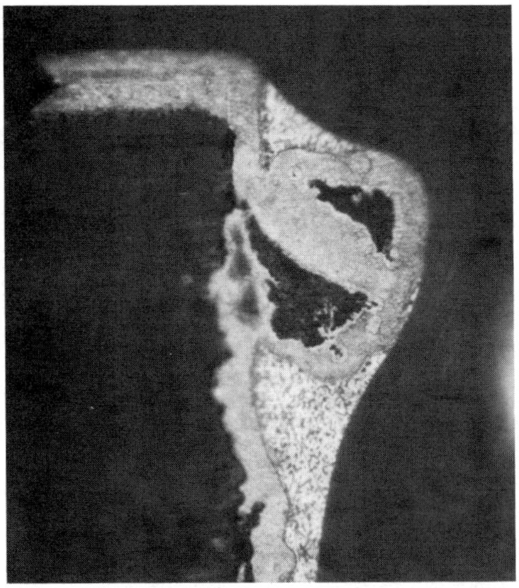

Fig. 25.55–20. Microsection, highly magnified, of a nodule located on the hole wall at the transition zone between the solder pad and the hole wall.

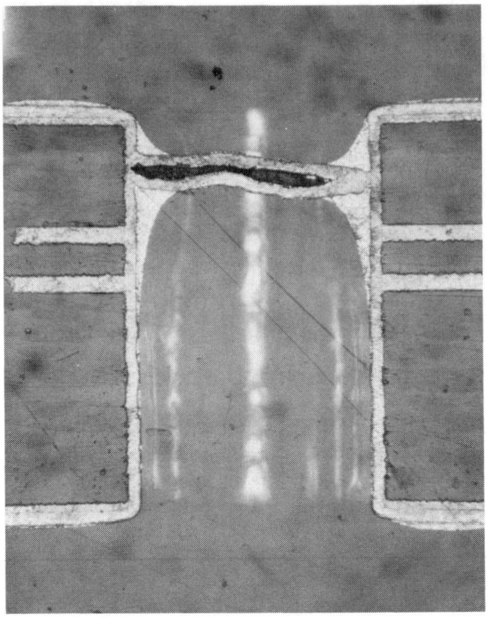

Fig. 25.55–21. Microsection of hole in a multilayer board. The thread in the hole in all probability originates from the brushing/scrubbing process prior to the electroless copper deposition.

Fig. 25.55–22. Microsection of hole with epoxy smear at the transition zone between the solder pad and the hole wall.

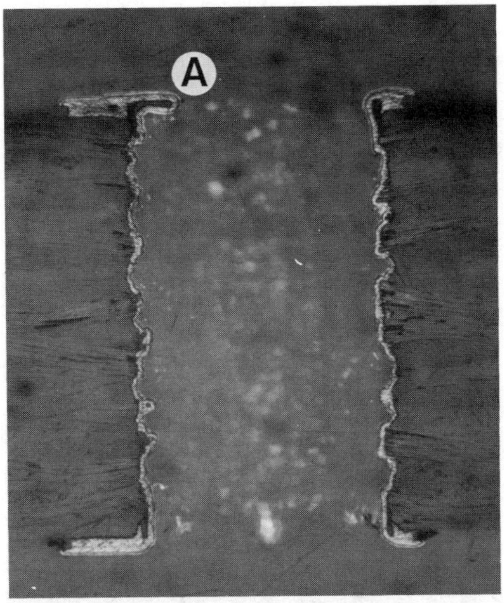

Fig. 25.55–23. Microsection of hole wall with epoxy smear at all transition zones between the solder pads and the hole wall. At (A) the epoxy smear is drawn out over the hole rim as a result of the brushing following the drilling. The subsequent plating creates a collar at the hole rim.

Fig. 25.55–25. Microsection of hole in a four-layer board exhibiting such heavy etchback that the inner and outer layers show up as distinct projections in the hole wall (Courtesy Alpha Metals, Inc.).

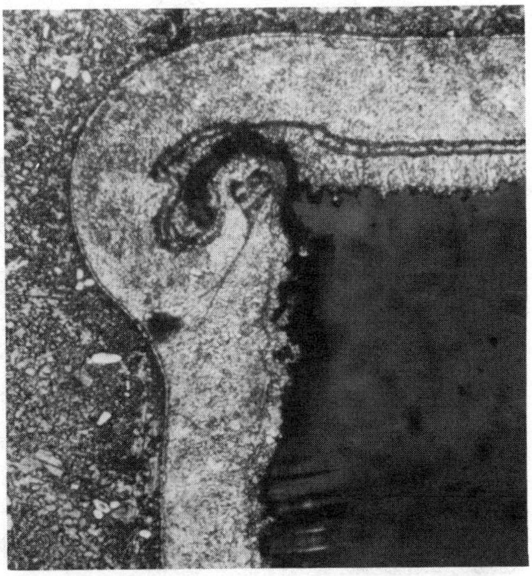

Fig. 25.55–24. Microsection, highly magnified, of hole wall with drilling burr and epoxy smear. Tho drilling burr with epoxy smear has been pressed back during the subsequent brushing, for which reason a shoulder is created during plating.

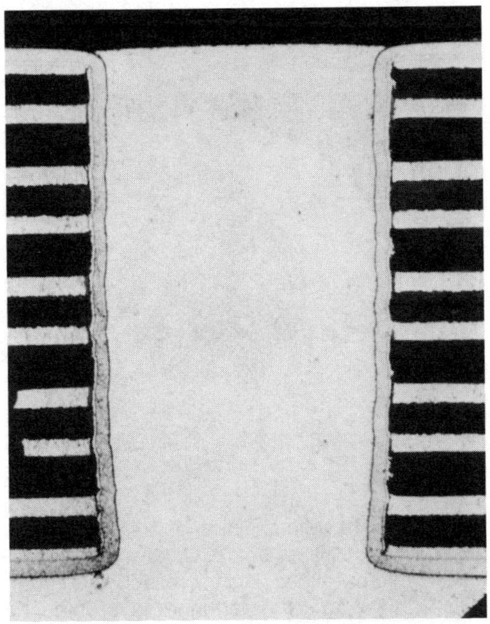

Fig. 25.55–26. Microsection of hole in a ten-layer board. The hole plating is quite even but pronounced epoxy smear occurs (Courtesy Alpha Metals, Inc.).

Fig. 25.55–27. Microsection, highly magnified, of epoxy smear between inner layer (right) and hole plating (left).

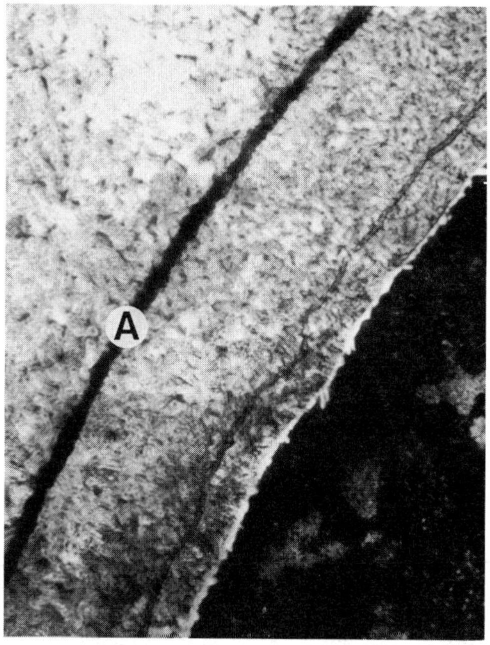

Fig. 25.55–29. Microsection, highly magnified, of hole with epoxy smear in a multilayer board. The photograph shows a section corresponding to Fig. 25.55–28, and the epoxy smear is clearly seen as the dark line (A).

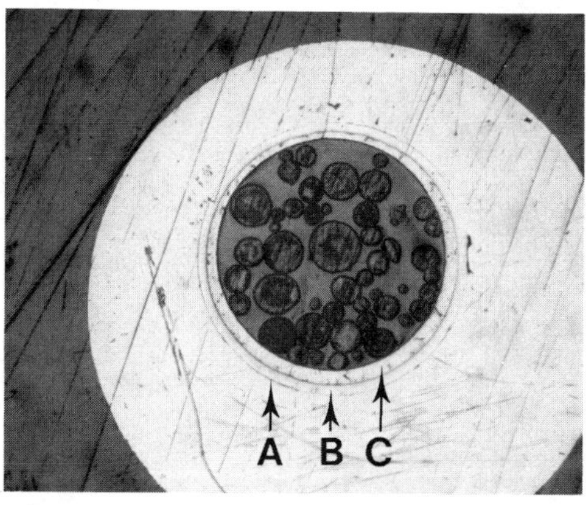

Fig. 25.55–28. Microsection of hole with epoxy smear in a multilayer board. Contrary to the preceding microsections, this microsection is in parallel with the board. As an inner layer is only 1 oz. = 0.0014 in. (35 μm) thick, the grinding must be performed with great care to avoid removing the pad. The photograph shows an inner pad with hole drilling. A: Epoxy smear. B: Copper plating. C: Tin/lead plating.

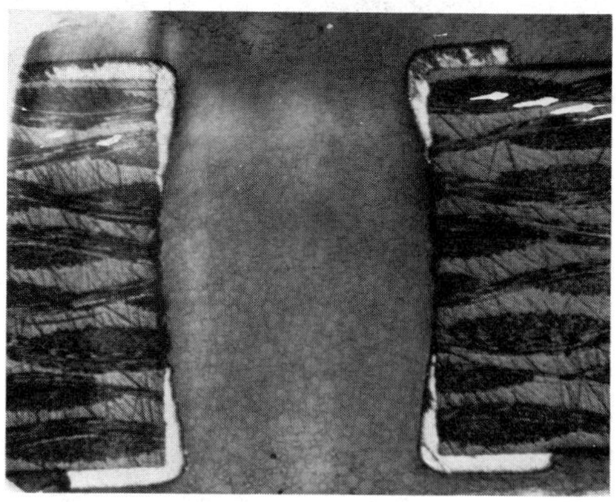

Fig. 25.55–30. Microsection of solder hole from a board which does not give complete filling of the solder holes when mass soldered. All plating is missing in the central region of the hole wall. The reason is excessively thin etch resist (tin/lead), resulting in etching of the copper deposition.

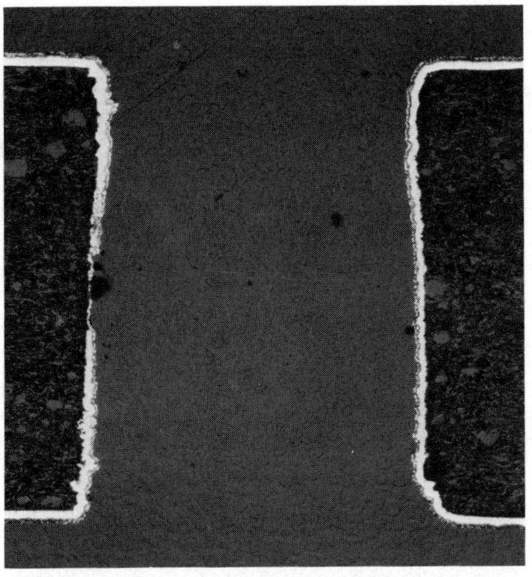

Fig. 25.55–31. Microsection of hole exhibiting step-plating. The copper plating is missing in the center region of the hole wall which is covered by tin/lead only (Courtesy Alpha Metals, Inc.).

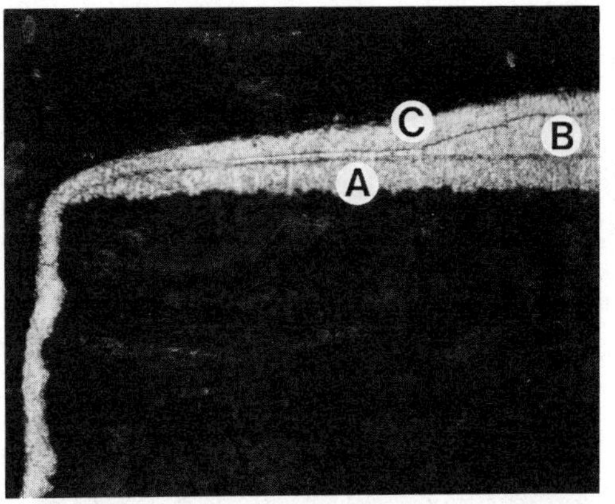

Fig. 25.55–33. Microsection of hole exhibiting step-plating, highly magnified. (A) indicates the copper foil of the base material, and (B) indicates the plated copper deposit, the thickness of which is suddenly decreased so that a lower plateau occurs on the solder pad. No copper deposit at all is found on the hole wall. (C) shows that the tin/lead layer continues through the hole. The electroless copper deposit must have been present on the hole wall.

Fig. 25.55–32. The photograph shows the defective area of Fig. 25.55–31, highly magnified. It is clearly seen that the fault is not the same as shown in Fig. 25.55–30 because the tin/lead plating continues with unchanged thickness across the defective area. Here the electroless copper deposit must have been present (courtesy Alpha Metals, Inc.).

Fig. 25.55–34. Microsection of soldered via hole exhibiting a closed blowhole. At (A) a tiny hole in the copper deposit can be seen dimly. During soldering, gases from the base material have forced their way through the tiny hole into the solder and thereby created the blowhole.

Fig. 25.55–35. Microsection of soldered via hole exhibiting an open blowhole. Note that there is not much solder in the hole. The side-to-side connection depends entirely upon the quality of the hole plating.

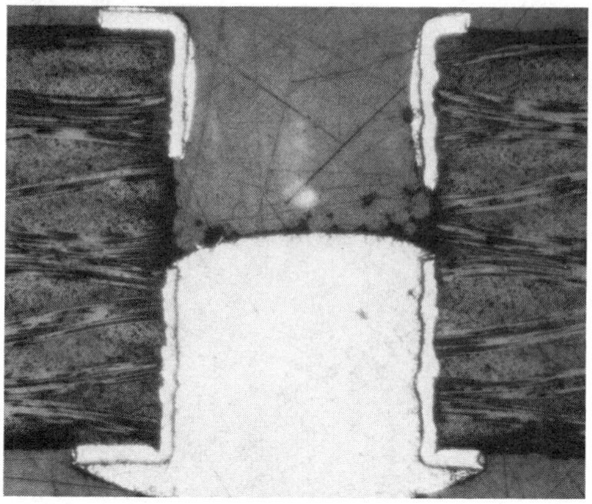

Fig. 25.55–37. Microsection of soldered via hole with interrupted side-to-side connection. The solder has not been able to fill the hole because of a plating or etching fault as shown in Fig. 25.55–30.

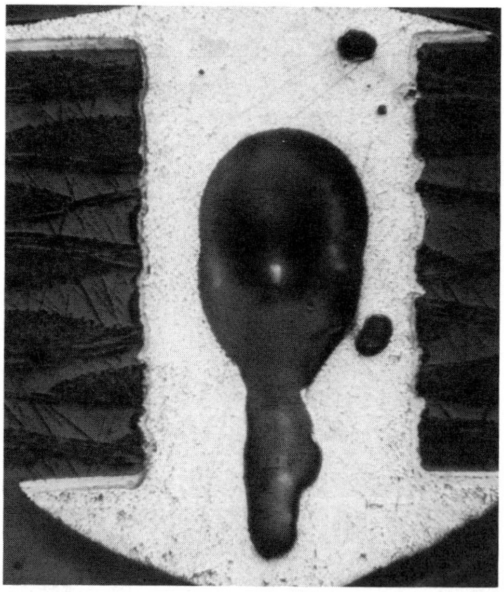

Fig. 25.55–36. Microsection of soldered via hole exhibiting a blowhole. A long channel leads from the blowhole to the open air, but it cannot be seen because of the location of the cut. Contrary to Fig. 25.55–35, the side-to-side connection is ensured owing to the reasonable filling of the hole with solder.

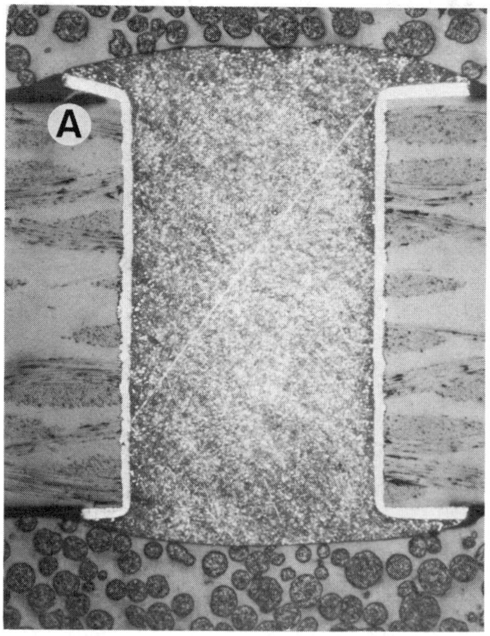

Fig. 25.55–38. Microsection of soldered via hole with lifting of the pad at (A). The filling, however, is characterized as very satisfactory.

25.5.4 Microsections of Multilayer Board Buildup

The buildup of a multilayer board can be studied by means of microsections. Next we shall address the following problems:

- Layer registration
- Minimum width of inner annular rings
- Layer spacing

25.5.4a Layer Registration

It can be important to examine the registration between the inner and outer pads of a multilayer board. The layers are frequently very thin, and under improper process conditions they can shift slightly with respect to each other during the lamination of the board.

Registration faults can, to a certain degree, be revealed by X-raying the board, but only a few electronics companies have the necessary equipment. A solution is to prepare microsections and measure the misregistration on photographs of the microsections. This method implies some problems, which are discussed below.

Since it is impossible to know in which direction the layers have shifted, the inspector must examine microsections of two cuts perpendicular to each other. The cuts must be located in the same area—for example, a test coupon—to ensure that both the examined areas are subject to the same misregistration.

The theoretical basis of the determination appears in Figure 25.56A. View A shows an outer solder pad with a concentric hole drilling and, with dotted lines, an inner pad. The misregistration along the X- and Y-axes, which correspond to the length and width of the board, are Δx and Δy, respectively. The total misregistration M is calculated as

$$M = \sqrt{\Delta x^2 + \Delta y^2}$$

View B shows a microsection placed in the X-axis direction. For reasons of clarity, the drawing shows only one hole. The centerlines through the two pads are drawn on the photograph (drawing), after which the lengthwise displacement Δx is measured. The centerlines are determined on the basis of the pad rims. The two pads should be circular, but not necessarily of the same size.

The crosswise displacement Δy is measured in a similar manner, with which the total misregistration M_1 occurring at the spot selected is calculated according to the formula given above. The calculated misregistration is not necessarily representative of the board.

For example, it is possible for the misregistration to increase linearly—that is, it is negligible at one end of the board and large at the other end. It is therefore recommended to measure the misregistration M_2 elsewhere, preferably at the opposite corner.

Incidentally, it is more practical if multilayer boards are provided with a test coupon at each of the two opposite corners, so that the investigation can be made on test coupons.

When assessing the misregistration, the inspector uses the largest of the values M_1 and M_2 found. The acceptable misregistration depends on the design conditions of the board, which can vary from company to company. A commonly used maximum value is 0.007 in. (0.18 mm), understood as the maximum shifting of the layers. The inspector should, however, check with the PCB designer before using this value.

The above method of measuring the misregistration seems somewhat safer than the method prescribed by the IPC in “Test Methods Manual IPC-TM-650,” Paragraph 22.10. This depends on differences regarding the hole locations and the number of microsections. In the author’s opinion the IPC method generally gives misregistration values which are too small.

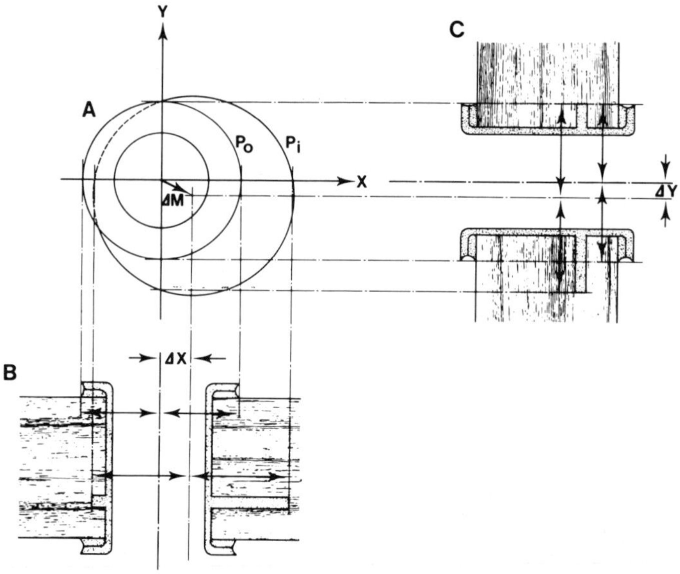

Fig. 25.56A. Determination of misregistration between the inner and outer layers of a multilayer board. A: Board seen from above. B: Microsection placed in the X-axis direction. C: Microsection placed in the Y-axis direction. P_o = outer pad (solder pad) with concentric hole drilling. P_i = inner pad. ΔM = total misregistration between inner and outer pads. Δx = misregistrationlong the X-axis direction. Δy = misregistration along the Y-axis direction.

Another approach would be to place, at each corner of each layer, copper areas as shown in Figure 25.56B, in exact register. After a light grinding of the corners of the finished board, it is possible to study the registration at all four corners without having to perform a destructive test. The only problem is the space occupied by the test areas.

A variation of the above principle can be used for checking the correct buildup of the board. By placing a small copper area along the edge, regularly displaced for each layer, a correct buildup is characterized by a regular staircase as shown in Figure 25.56C, whereas an incorrect buildup shows up as an irregular staircase.

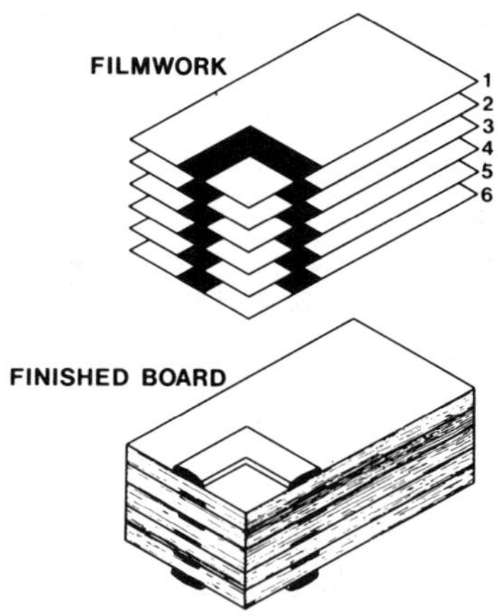

Fig. 25.56B. Determination of misregistration based on copper areas placed at each board corner.

25.5.4b Minimum Width of Inner Annular Rings

Most official standard specifications for multilayer boards give the minimum width of the inner annular rings as 0.001 in. (25 μm). The inspector should routinely check that the annular rings have a width of not less than 0.001 in. (25 μm). It is, however, not particularly probable that the cuts are

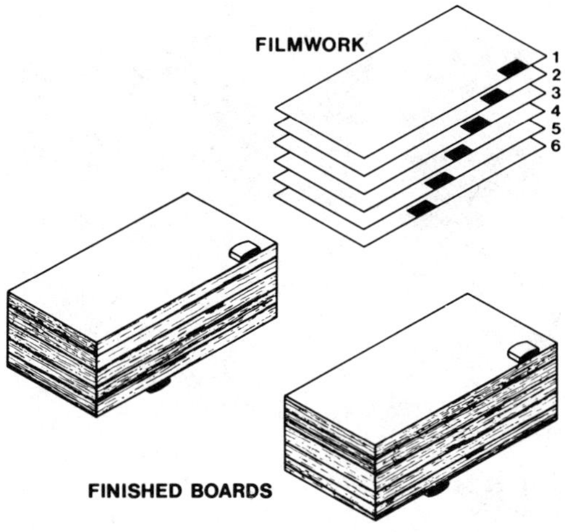

Fig. 25.56C. Checking the correct buildup by means of copper areas along the board edge, regularly displaced on the artwork/filmwork. Left: a regular staircase indicates correct buildup. Right: an irregular staircase indicates incorrect buildup.

placed in such a way that the microsections show the minimum width.

25.5.4c Layer Spacing

See the cross-sectional drawing in Figure 25.57. The inspector routinely verifies that the distance between the copper layers—for example, between the conductors of two adjacent layers—complies with the specification.

A commonly used minimum value is 0.004 in. (0.102 mm), but more exacting classes require minimum values of 0.003 in. (0.076 mm) and 0.002 in. (0.051 mm).

If the multilayer board comprises one or more transmission lines with a defined characteristic impedance—for example, 90 ohms—the impedance depends on the conductor width, the dielectric separation to the adjacent layers, and the dielectric constant.

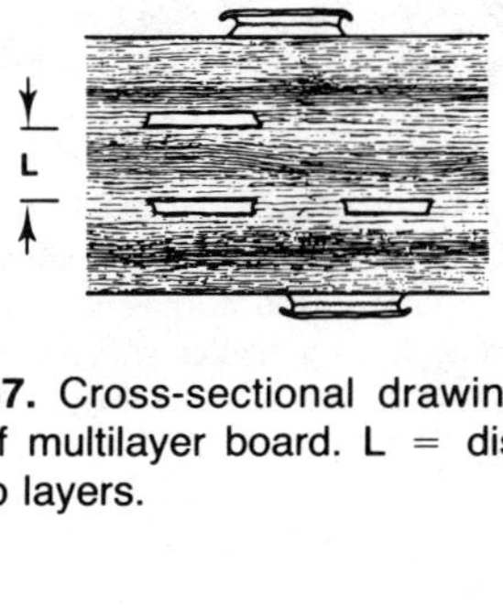

Fig. 25.57. Cross-sectional drawing of layer buildup of multilayer board. L = distance between two layers.

It will be most expedient if the test coupon contains a similar transmission line so that the inspector can use a microsection to check the mechanical dimensions decisive to the impedance.

For the inspector's information, commonly accepted tolerances are given below for the distance L between layers and the conductor width W, valid for the lowest tolerance class. For closer tolerances, the inspector is referred to IPC-ML-910, "Design and End Product Specification for Rigid Multilayer Printed Wiring Boards."

Tolerance of Distance L between Layers:

0.002 < L ≤ 0.0043 in.:	± 0.0010 in.
0.05 < L ≤ 0.11 mm:	± 0.025 mm
0.0043 < L ≤ 0.006 in.:	± 0.0015 in.
0.11 < L ≤ 0.15 mm:	± 0.038 mm
0.006 < L ≤ 0.012 in.:	± 0.0020 in.
0.15 < L ≤ 0.30 mm:	± 0.051 mm
0.012 < L ≤ 0.020 in.:	± 0.0025 in.
0.30 < L ≤ 0.51 mm:	± 0.064 mm
0.020 < L ≤ 0.030 in.:	± 0.0030 in.
0.51 < L ≤ 0.76 mm:	± 0.076 mm

Tolerance on Conductor Width W:

Nonplated conductor:	+0.004, −0.005 in. (+0.102, −0.127 mm)
Plated conductor:	+0.008, −0.005 in. (+0.204, −0.127 mm)

Failure to meet these tolerances should be regarded as cause for rejection, and the fault is rated as an A-defect, provided that the PCB manufacturer has been informed about the tolerances and has accepted them.

25.5.4d Microsection Examples

The cases described above are illustrated with some practical examples of microsections. See Figures 25.58-1 through 25.58-4. The captions provide details of the microsections.

25.5.5 Microsections of Edge Connectors

To a certain degree, visual inspection can reveal details such as the nature of the transition zone, while other details, such as plating thicknesses, cannot be assessed at all. By making microsections lengthwise and crosswise with respect to the contact tabs, the inspector can expose all conditions.

Since the gold layer is very thin and soft, there is a risk of edge rounding while the microsection is being polished, so a layer of nickel should be applied for protection (Paragraph 25.3.3b). We shall next address the following problems:

- Transition zone
- Plating thickness
- Coverage of edges

25.5.5a Transition Zone

As mentioned in Paragraph 14.6, the transition from the connector tabs to the conductors can be produced in several different ways as illustrated in Figures 14.9, 14.11, and 14.12. A lengthwise cut is made through the contact tab in question. Care should be taken that the cut runs approximately through the middle of the associated conductors.

When inspecting the microsection under high magnification, the inspector can assess the conditions in the transition zone. It is also possible to determine the sequence in which the various platings have been applied.

The inspection criteria are exactly the same as stated in Paragraph 18.16.9. The inspector should be particularly aware of

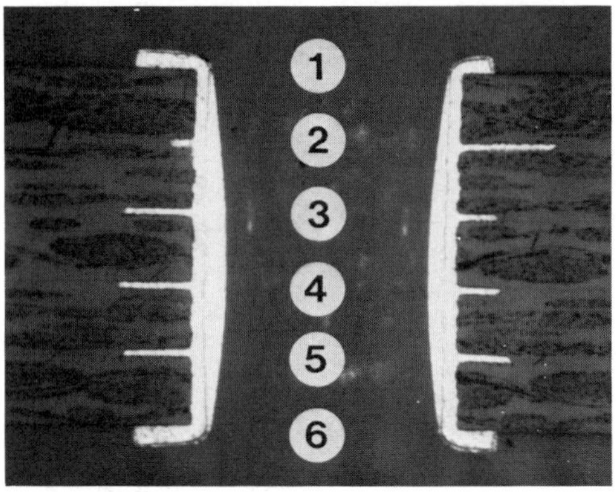

Fig. 25.58–1. Microsection of plated-through hole in a six-layer board, suited for measuring the misregistration between the individual layers. The largest misregistration is approximately 0.0075 in. (190 μm). The thickness of the copper foil on the outer and inner layers is 1 oz. = 0.0014 in. (35 μm) and 1/2 oz. = 0.0007 in. (17.5 μm), respectively.

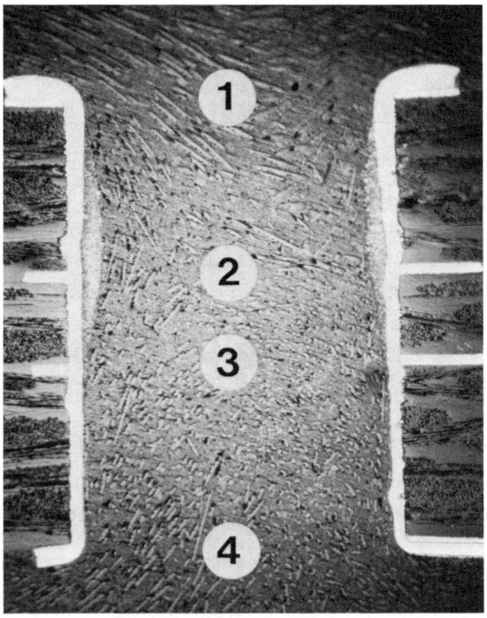

Fig. 25.58–2. Microsection of plated-through hole in a six-layer board. The misregistration between layer 1 and layer 3 is approximately 0.002 in. (50 μm). The hole drilling is concentric with layer 1 and consequently eccentric with respect to the other layers. The smallest width of the annular rings occurs at layer 3 and is about 0.004 in. (100 μm).

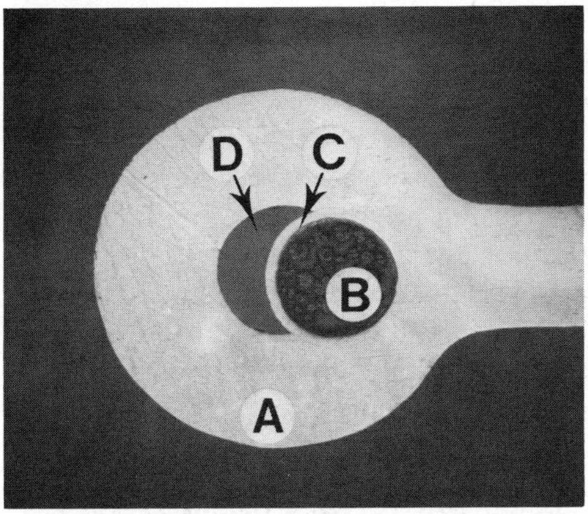

Fig. 25.58–3. Microsection of plated-through hole in a multilayer board. The microsection is located in parallel with the board and shows an inner pad (A) with hole drilling (B) and hole plating (C). For some unknown reason a large aperture (D) occurs in the inner pad. This type of microsection gives more information about the total amount of misregistration than the microsection in Fig. 25.58–1. In return it is only possible to see the misregistration of a single layer with respect to the hole drilling. See the extraordinarily large misregistration of approximately 0.016 in. (450 μm).

Fig. 25.58–4. Microsection of an eight-layer board, well suited for determining the tolerance of the conductor width and mutual distance. The buildup comprises three double-sided inner layers separated by prepregs. Inner layers are easily recognized as the conductors are widest at the surface of the base material. The thickness of the copper foil on the outer and inner layers is 2 oz. = 0.0028 in. (70 μm) and 1 oz. = 0.0014 in. (35 μm), respectively.

bare copper, which is an indisputable cause for rejection. The fault is rated as an A-defect.

25.5.5b Plating Thickness

When measuring plating thicknesses the inspector should be aware of the problems caused by the fundamental difference between a measurement according to the beta-backscatter method and a direct gauging based on a microsection (Paragraph 14.3.5). Independent of the density of the gold plating structure, the inspector can measure the lineal plating thickness by means of a microsection.

Figure 14.4 shows how the plating thickness can vary across the contact tabs of the edge connector, which means that the inspector should examine contact tabs taken from the center of the edge connector.

The cut is placed in parallel with the connector edge so that the microsection includes cross sections of a number of contact tabs—for example, eight tabs. To check the distribution of the layer thickness, the inspector can prepare microsections of the contact tabs at the connector ends.

The buildup of the layers is shown schematically in Figure 14.7, and the inspector checks that the nickel and gold layers meet the minimum requirements in the PCB specification. Overly thin layers are regarded as cause for rejection, and the fault is rated as an A-defect.

In particular, the thickness of the gold layer may cause difficulties when making a complaint, and the inspector is advised to consider the remarks given in Paragraph 14.3.5, especially Cases 1 and 2.

25.5.5c Edge Coverage

Many PCB specifications require that the edges of the contact tabs be covered with nickel and gold. Fortunately, the microsections used for determining the plating thickness (Paragraph 25.5.5b) are suitable for this investigation. The contact tabs should preferably be taken from the central area of the edge connector where the plating thickness is smallest (Paragraphs 14.3.2 and 14.3.3 and Figure 14.4).

The inspector checks whether the edges are covered. Failure to meet this requirement is regarded as cause for rejection, and the fault is rated as an A-defect.

The inspector cannot routinely reject boards without edge coverage of the contact tabs unless the PCB specification specifies such coverage. The PCB manufacturer will, in all probability, be disinclined to accept a return of the boards. See the discussion in Paragraph 14.5.1 concerning alternative methods resulting in noncovered edges. Application of such methods requires prior approval by the customer if the PCB specification requires coverage of the edges.

The author is acquainted with some very depressing cases of a new PCB supplier using an alternative method which did not ensure the required edge coverage. It is therefore recommended that the plating conditions be discussed in detail so that the first delivery does not entail unpleasant surprises.

25.5.5d Microsection Examples

The above cases are illustrated by some practical examples of microsections. See Figures 25.59-1 through 25.59-10. The captions give a detailed explanation of the microsections.

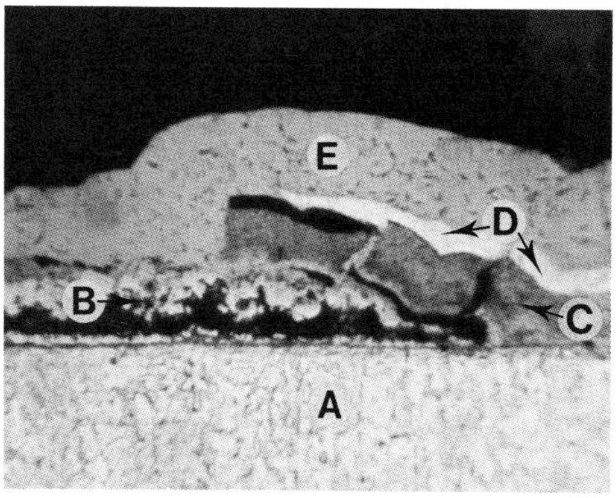

Fig. 25.59–1. Microsection of transition zone of an edge connector tab with gold overlapping, cf. Para 14.6.1. The photograph shows a lengthwise cut through the contact tab. (A) is the underlying plated copper layer, and (B) is the tin/lead layer. (C) and (D) indicate the nickel and gold layers, respectively. Note that the latter two layers exhibit a crack where they are lying on top of the tin/lead layer. This can result in delamination of the gold and nickel layers in the transition zone. (E) is a protective layer of nickel, cf. Para. 25.3.3b.

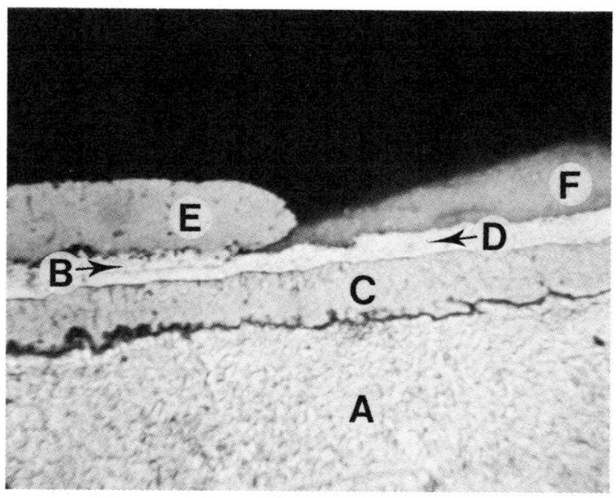

Fig. 25.59–2. Microsection of transition zone of an edge connector tab with epoxy sealing, cf. Para 14.6.2. The photograph shows a lengthwise cut through the contact tab. (A) is the underlying plated copper layer. (C) and (D) indicate the nickel and gold layers, respectively. The gap between the conductor and the contact tab is not seen, but (B) indicates tin/lead which has overflowed the contact tab a little during reflowing. At this place the tin/lead has formed an alloy with the gold. (F) indicates the epoxy sealing which by accident has been displaced in a direction away from the transition area and therefore is of no effect. (E) is a protective layer of nickel.

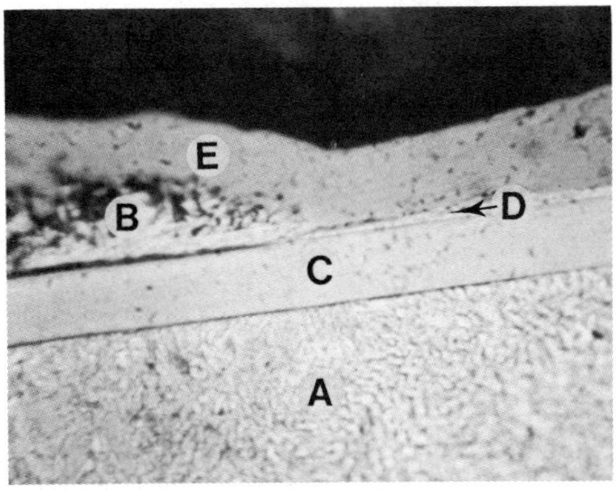

Fig. 25.59–3. Microsection of transition zone of an edge connector tab, made by an alternative technique. The photograph shows a lengthwise cut through the contact tab. (A) is the plated copper. (C) and (D) indicate the nickel and gold layers, respectively. (B) indicates the tin/lead layer. It is seen clearly that the edge connector has been plated with nickel and gold before the board itself has been plated with tin/lead. (E) is a protective layer of nickel.

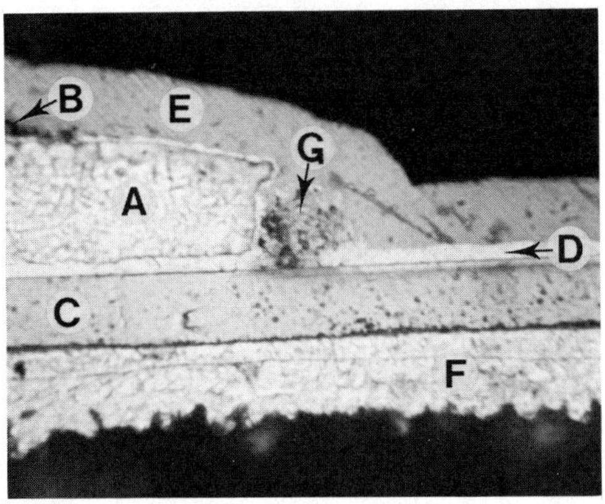

Fig. 25.59–4. Microsection of transition zone of an edge connector tab, made by an alternative technique. The photograph shows a lengthwise cut through the contact tab. (A) is the plated copper. (C) and (D) indicate the nickel and gold layers, respectively, which are deposited directly on top of the copper foil. (D) indicates the tin/lead, and at (G), a little tin/lead has flowed down and formed an alloy with the gold. (E) is a protective layer of nickel.

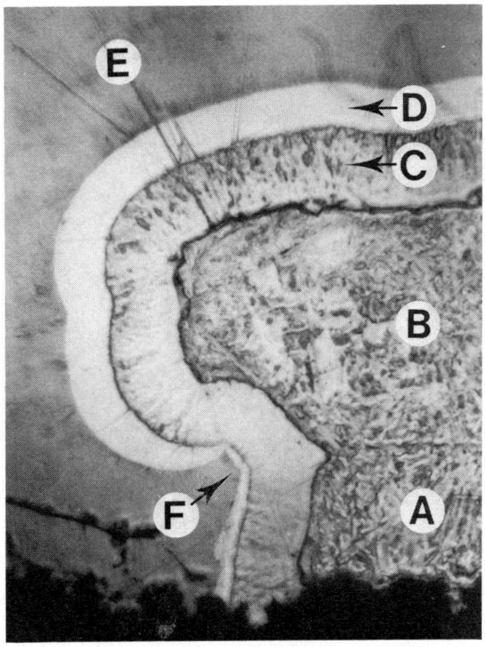

Fig. 25.59–5. Microsection of one of the edges of an edge connector contact tab, plated in accordance with the gold overlapping method. (A) indicates the copper foil of the base material with a thickness of ½ oz. = 0.0007 in. (17.5 μm), and (B) indicates the plated copper with a thickness of approximately 0.0012 in. (30 μm). (C) and (D) indicate the nickel layer of 0.0003 to 0.0004 in. (7.5 to 10 μm), and the gold layer of 0.00016 to 0.00024 in. (4 to 6 μm), respectively. Note that the nickel plating has a leveling effect which does not apply to the gold plating, cf. (D). Here the thickness of the gold layer is approximately 0.00004 in. (1 μm). (E) indicates a protective layer of nickel.

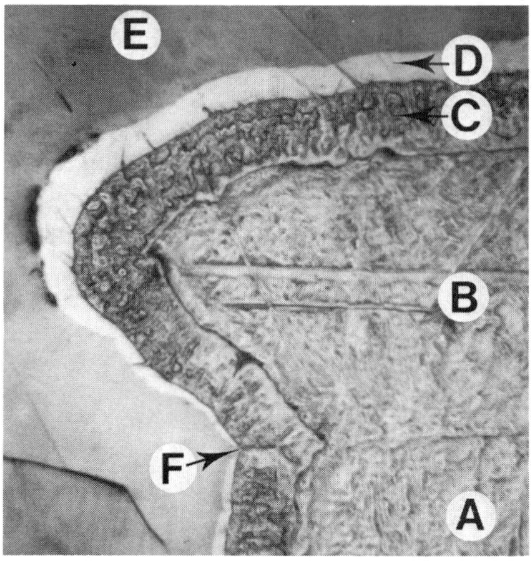

Fig. 25.59–6. Microsection of one of the edges of an edge connector contact tab, plated in accordance with the gold overlapping method. The letter indications are as stated in Fig. 25.59–5. Note the somewhat smaller thickness of the nickel and gold layers, and note in particular the very moderate thickness of the gold layer at (F), less than 0.00004 in. (1 μm). A plating of this kind can very well result in the cases of edge corrosion shown in Fig. 27.7.

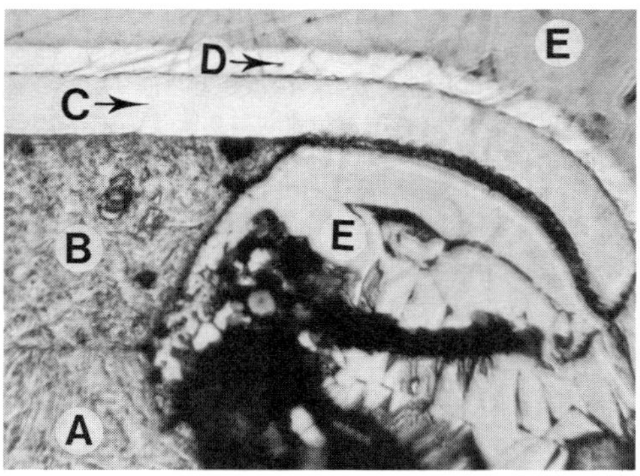

Fig. 25.59–7. Microsection of one of the edges of an edge connector contact tab plated in the same way as shown in Fig. 25.59–3. (A) indicates the copper foil at the base material, and (B) indicates the plated copper layer. (C) and (D) indicate a nickel layer of 0.0003 in. (7.5 μm) and a gold layer of 0.00012 to 0.00016 in. (3 to 4 μm), respectively. Note the large overhang and also that the edge of the contact tab has received no surface protection. (E) indicates a protective layer of nickel.

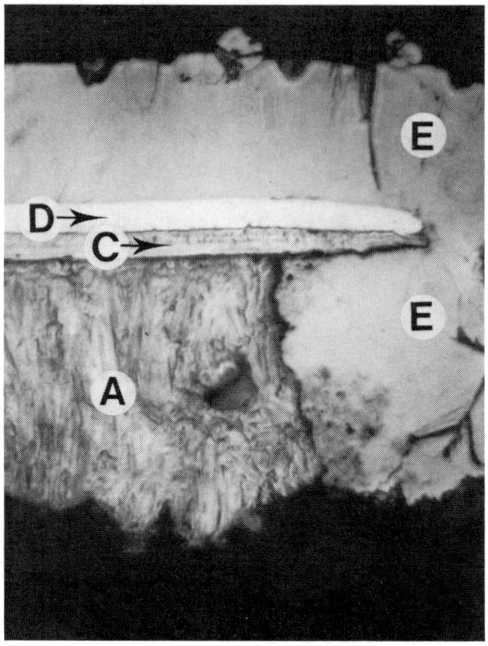

Fig. 25.59–8. Microsection of one of the edges of an edge connector contact tab plated in the same way as shown in Fig. 25.59–4; i.e., the nickel layer (C) of approximately 0.00016 in. (4 μm) and the gold layer (D) of approximately 0.00016 in. (4 μm) are deposited directly on the copper foil of the base material. Note the large overhang and also that the edge of the contact tab has received no surface protection. (E) indicates a protective layer of nickel.

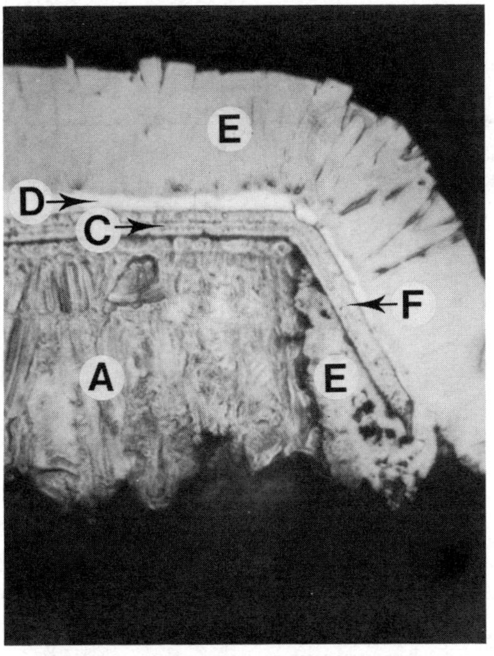

Fig. 25.59–9. Microsection of the same type of edge connector contact tab as shown in Fig. 25.59–8. Note that the overhang is deformed mechanically and also that the gold layer exhibits a crack where the overhang has been bent.

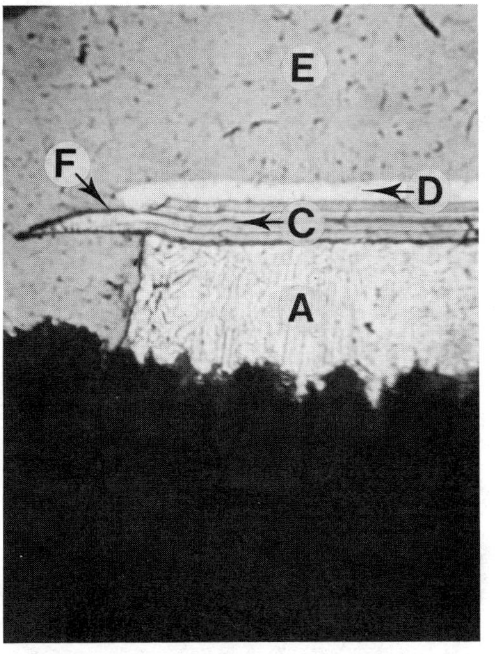

Fig. 25.59–10. Microsection of the same type of edge connector contact tab is shown in Fig. 25.59–8. At (F) the gold layer has flaked off. Note the very clear stratification of the nickel layer, which is not unusual for nickel platings.

Chapter 26

Solder Mask Thickness Measurement

26.1 Introduction

An often disputed point is the solder mask thickness, which is quite important to the mask stability during mass soldering. The thickness can, in principle, be measured in several different ways:

1. A measurement based on microsections, as detailed in Paragraph 25.5.2. However, this is a destructive method unsuitable for routine tests.
2. A measurement based on eddy currents. This measurement uses the change in the high-frequency energy in a small measuring probe as it approaches a conducting nonferrous surface.

   The eddy current loss is a measure of the distance between the conducting surface and the probe. When the probe is placed on the surface of the solder mask, the distance is determined by the thickness of the mask. The thickness is therefore indicated by the change in the eddy current loss.

   In practice, the measuring equipment is not very easy to use on printed circuit boards because the probe requires a fairly large measuring area—for example, a circular area with a diameter of 0.5 in. (13 mm). Obviously, it is not possible to measure the thickness where the mask lies over a conductor. In the author's experience, even measurements made on large areas can be somewhat unreproducible.
3. A measurement with a betascope. A prerequisite of this measuring method is that standards for the adjustment of the betascope are available. These standards must be based on the same combination of mask ink and base material as in the actual board, and the thickness of the mask must be known.

   According to the measuring principle, the betascope records the mass of the mask ink under a certain area. Therefore, the measured layer thickness corresponds to the "compressed" layer thickness and not to the actual layer thickness.

   The type of mask ink on the board is frequently unknown to the inspector since the PCB manufacturer always has a free choice among a number of approved brands. The pigmentation of the mask ink (the addition of dyes and fillers) inevitably affects the measuring result. Also, the incidence of small air bubbles in the finished solder mask affects the measuring result. Under worst-case conditions, a measuring error of ±25 percent, possibly up to ±50 percent, can occur.

4. A measurement with a light-sectioning microscope. The next paragraph details this method.

26.2 Measuring with a Light-Sectioning Microscope

A light-sectioning microscope as shown in Figure 26.1 is particularly suitable for nondestructive measurements of greater thicknesses. The principle appears in Figure 26.2, which shows a highly magnified conductor cross section with a solder mask applied.

A very narrow light beam, created by a slot and directed towards the solder mask surface at a 45 degree angle, is reflected at the same angle. It is seen in the eyepiece of the microscope as a narrow band of light across the surface at right angles to the viewer (through point A, perpendicular to the paper).

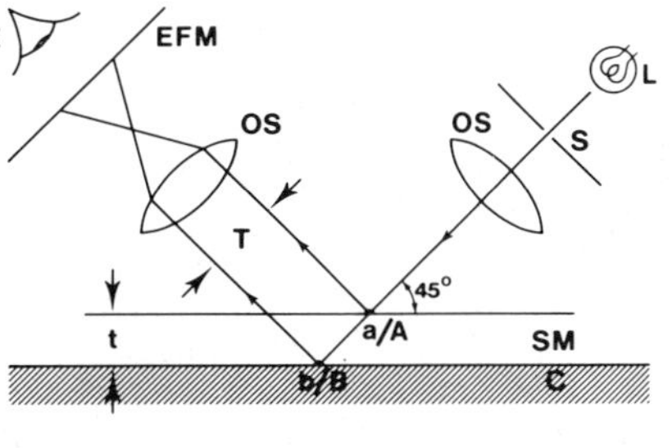

Fig. 26.2. Path of rays in a light-sectioning microscope.
SM = solder mask.
L = light source.
E = eye.
C = conductor.
S = slot.
EFM = eyepiece filar micrometer.
OS = optical system.
a = point of reflection on upper side of solder mask.
b = point of reflection on upper side of conductor.
A = band of light (perpendicular to the paper, through a).
B = band of light (perpendicular to the paper, through b).
T = distance between bands of light, A and B.
t = thickness of solder mask.

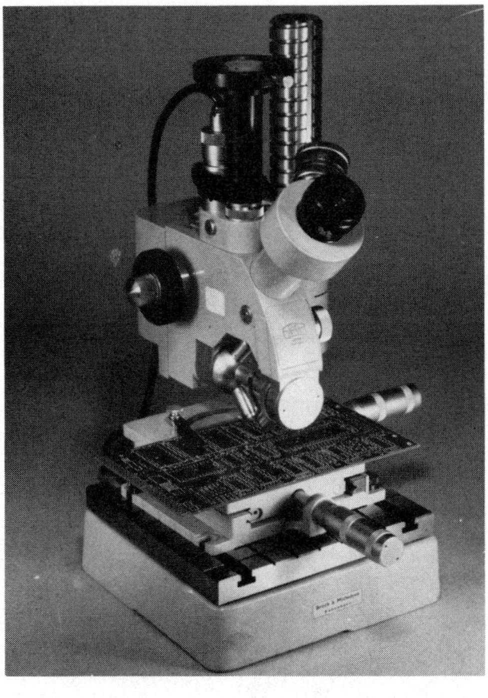

Fig. 26.1. Example of a light-sectioning microscope being used for measuring the thickness of a solder mask. See also Fig. 26.4.

In order to measure the solder mask thickness, the inspector must achieve a reflection of the light beam from the underside of the solder mask, which is the same as the upper side of the conductor. The simplest method is to perform the measurement at an edge of the solder mask—for example, at the rim of the mask aperture around a solder pad. See Figure 26.3. Here we have a step, and the light beam is reflected as a band of light A from the upper side of the solder mask and as a displaced band of light B from the upper side of the conductor attached to the pad.

In the cross-sectional drawing shown in Figure 26.2, the two bands of light A and B pass through points a and b. In the eyepiece, band B is seen displaced with

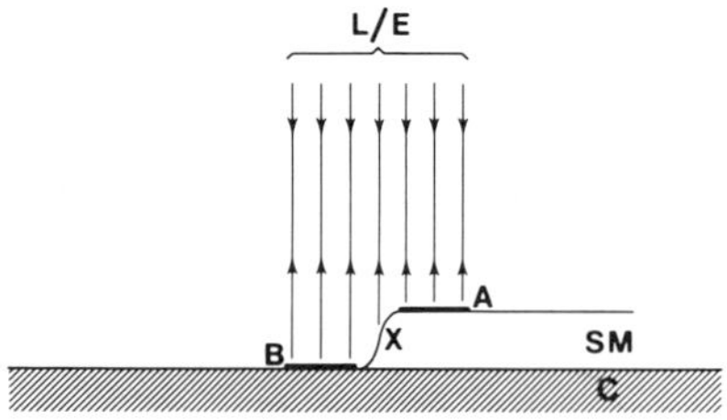

Fig. 26.3. Cross-sectional drawing perpendicular to the cross-sectional drawing in Fig. 26.2.
SM = solder mask.
L = light source.
X = step.
A = band of light through a, Fig. 26.2.
B = band of light through b, Fig. 26.2.
C = conductor.
E = eye.

respect to band A. The distance T between bands A and B is measured with an eyepiece filar micrometer (EFM), the crosshair being moved from one band of light to the next by the micrometer. The thickness t of the solder mask is calculated by the formula

$$t = (A - B) \sin 45° = 0.707 \text{ T}$$

Under suitable magnification the inspector can achieve a total measuring range of 0.00004 to 0.004 in. (1 to 100 μm), depending on the sharpness of the reflected bands of light, which in turn depends on the surface character. Figure 26.4 shows how the thickness of a solder mask is ascertained with a light-sectioning microscope.

If the solder mask is transparent, the two bands of light will be seen even without using a step. It is, however, necessary to consider the refractive index of the solder mask. This implies, as shown in Figure 26.5, that the distance T measured between the two bands of light seen in the eyepiece will be too small. The result is that the calculated thickness of the solder mask becomes too thin.

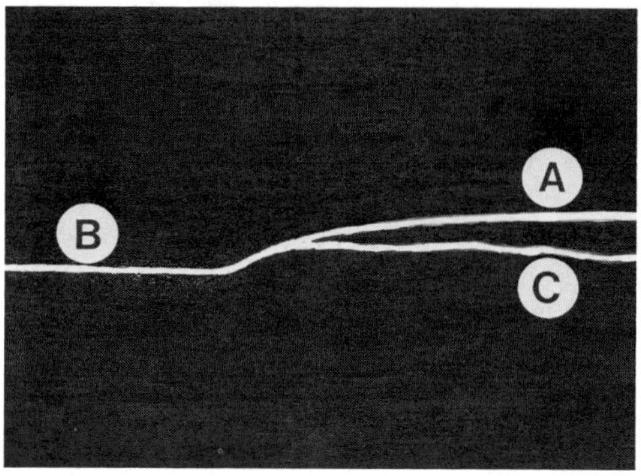

Fig. 26.4. Photograph of the bands of reflected light obtained by light-sectioning a solder mask. The band of light (A) is reflected from the upper side of the solder mask, and the band of light (B) is reflected from the tin/lead surface of the conductor. The band of light (C) is an unwanted reflection from the tin/lead surface immediately below the solder mask. By means of the eyepiece filar micrometer the thickness (A) - (B) of the solder mask is determined as approximately 0.0008 in. (20 μm).

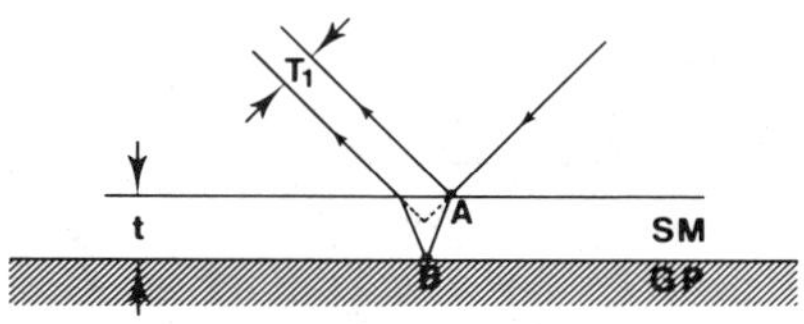

Fig. 26.5. Changed path of rays when light-sectioning a transparent mask without using a step. The change is caused by the refraction factor of the mask.
T_1 = reduced distance between the bands of light (A) and (B).
GP = ground plane.
Remaining designations: see Fig. 26.2.

The inspector must apply a correction factor, which implies that he knows the refractive index of the solder mask. Because he probably does not know the refractive index, which can be as much as 2, it is safer to measure across a step. Since the field of vision of the light-sectioning microscope is about 0.04 in. (1 mm), the step should not be wider than 0.02 in. (0.5 mm).

Chapter 27

Edge Connector Porosity Test

27.1 Introduction

Paragraph 25.5.5b described the use of microsectioning to measure the thickness of the nickel/gold platings of edge connector contacts. Meeting the PCB specification does not ensure that the contact tabs are sufficiently resistant to a corrosive atmosphere.

Paragraph 18.16 detailed the thorough visual inspection of nickel/gold-plated edge connectors, and Paragraph 18.16.5 discussed pinholes. Unfortunately, a visual inspection cannot reveal pores or microcracks in the plated layers. The density of these layers depends partly on the nature of the substrate and partly on the plating processes. Therefore, the inspector must occasionally check the plating tightness.

Nickel/gold plated contact tabs corrode through pores, pinholes, and microcracks in the plated layers. This process, often called point corrosion, involves two different mechanisms. The first one is the corrosive elements in the atmosphere attacking the underlying copper and nickel directly through the holes, and the second one is an electrochemical process which takes place at the bottom of the holes.

The copper and nickel turn into dissociated salts which, in the form of sulfides, sulfates, carbonates, and oxides, grow up through pores and microcracks and then spread over the contact surface.

The corrosion products damage the surface of the contact tabs and increase the contact resistance, possibly enough to cause an electrical discontinuity. The inspector should therefore perform a corrosion test to ascertain whether the plating is dense enough—that is, with no pores, pinholes, or microcracks. See the considerations in Paragraph 3.7.2, Reference A12, Reading List, regarding corrosion of gold plated contact tabs and recommended layer thicknesses.

It is impossible in practice to test the sample board(s) for corrosion under the future environmental conditions to which the boards will be subjected, because the exposure time would be too long. It is therefore necessary to use accelerated test methods, which expose the boards to highly corrosive environmental conditions. By exposing the sample board(s) to high concentrations of corrosive gasses at increased temperature and humidity, the inspector can reduce the exposure time to weeks, days, or even hours.

27.2 Test Methods

The inspector can test porosity in several ways, all of them destructive, at least when the porosity test reveals pores in nickel and gold platings. Pores do not become smaller after the test, and gas or acid tests damage the remaining part of the pattern.

It is more practical and economical to perform the test on a board which the PCB manufacturer has rejected and included in the delivery for use in this test. The board must not have been rejected for plating faults. Below we shall elaborate some of the most common test methods:

- Electrographic test
- Gas exposure test with sulfur dioxide and hydrogen sulfide
- Gas exposure test with nitric acid
- Acid test with nitric acid
- Gas exposure test with hydrogen sulfide

There is good reason to warn against misinterpretation of the results of accelerated test methods. The sole purpose of a porosity test is to ascertain the presence of pores, pinholes, and microcracks in the overplated layers, frequently expressed as the number of these defects per contact tab or per unit area.

The assessment should remain unaffected by the degree of corrosion—that is, the extent to which corrosion products in the form of point or surface contamination of the contact tab occur. The intensity and duration of the exposure are decisive factors, and it is not possible to routinely convert the test results to the actual environmental conditions of the end product. Figures 27.1 to 27.5 show corrosion of gold plated contact tabs.

A consequence of the severe porosity requirements in many official PCB specifications is the use of fairly thick gold platings, which seems to be the only true safe method of reducing the number of pores and pinholes.

However, with the current high price of gold, it seems preferable to match the thickness of the gold plating with the actual conditions to which the boards will be exposed. This requires a very careful investigation of the corrosion, which is

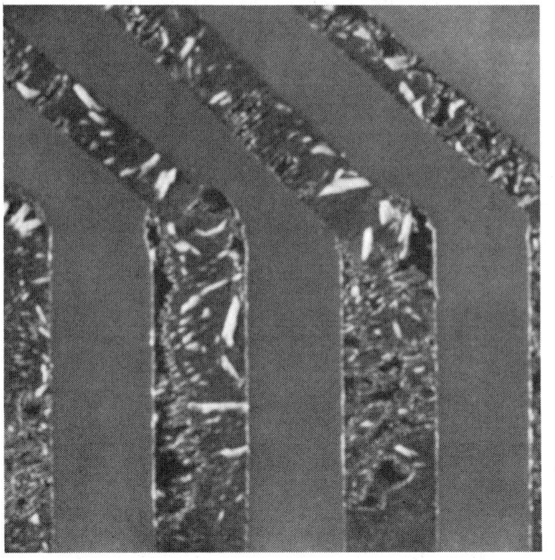

Fig. 27.1. The corrosion products are seen as green crystalline flakes. The corrosion is spreading over the majority of the surface, but it is a far-advanced point corrosion which started in pinholes and microcracks. The test conditions are a sulfur dioxide atmosphere with a concentration of 100 ppm SO_2, a relative humidity of 87% and a temperature of 40°C, for 96 hours (Courtesy of the Danish Research Center for Applied Electronics).

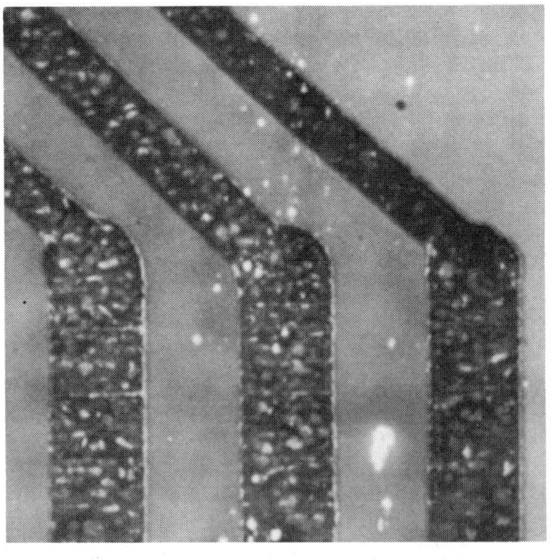

Fig. 27.2. This is a dense but weakly developed point corrosion. There is no marked formation of crystals, only a slight discoloration around each cell of corrosion. Test conditions are as given in Fig. 27.1 (Courtesy of the Danish Research Center for Applied Electronics).

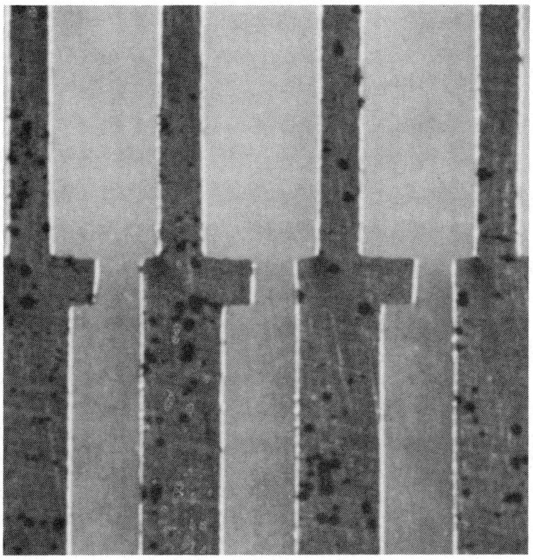

Fig. 27.3 The picture shows a weak but unmistakable point corrosion resulting from pinholes. The test conditions are a hydrogen sulfide atmosphere holding a concentration of 10 ppm H_2S, a relative humidity of 93% and a temperature of 40°C, for 48 hours (Courtesy of the Danish Research Center for Applied Electronics).

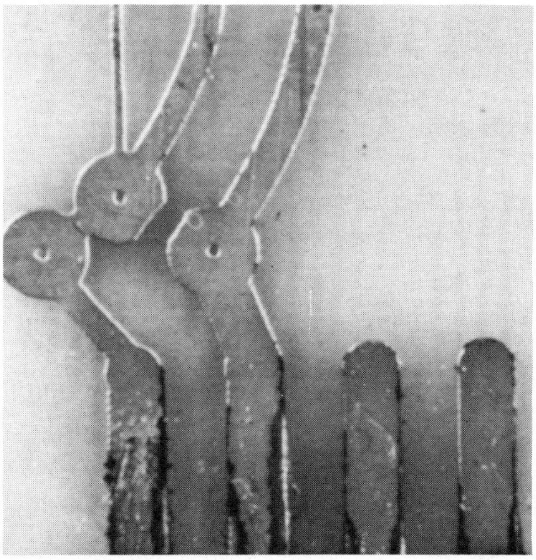

Fig. 27.4. An unmistakable corrosion of the edge of the contact tabs can be seen. The corrosion is due to the gold plating on the edges of the contact tabs being too thin. Test conditions are as given in Fig. 27.3 (Courtesy of the Danish Research Center for Applied Electronics).

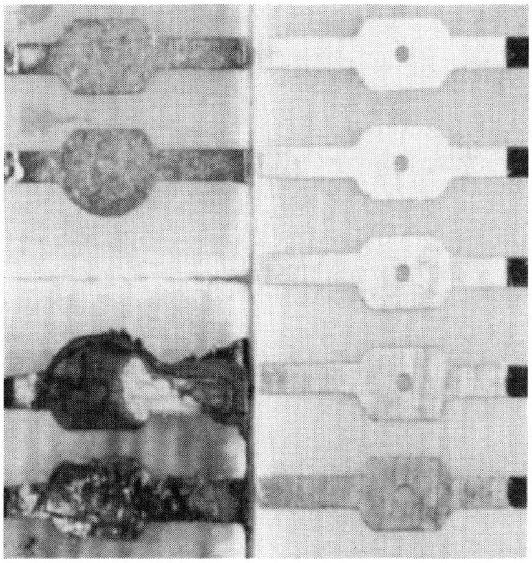

Fig. 27.5. Top left: Bare-copper board before the test.
Bottom left: Same board after exposure to a hydrogen sulfide atmosphere. A very heavy and detrimental surface corrosion is evident. The surface is completely covered with a crystalline structure.
Right: This picture shows good gold plating with no traces of corrosion after the test (Courtesy of the Danish Research Center for Applied Electronics).

not usually the duty of the incoming inspection department, but rather that of a reliability department.

27.2.1 Electrographic Test

Since this test method is described in IEC 326-2, BS 9760, and IPC-TM-650, only a short overview is presented here.

As the name suggests, an electrogram is produced by electrolysis on specially impregnated indicator paper, which is placed on the contact tabs. Any pores show up as colored stains on the indicator paper, which is very useful as test documentation.

The process is somewhat complicated since the inspector must place a piece of suitably wetted photographic blotting paper over the indicator paper, and cover it with a high-purity aluminum plate. This

Fig. 27.6. Execution of an electrographic test. The edge connector area of the board is placed in the clamping device which is compressed by means of the handle.

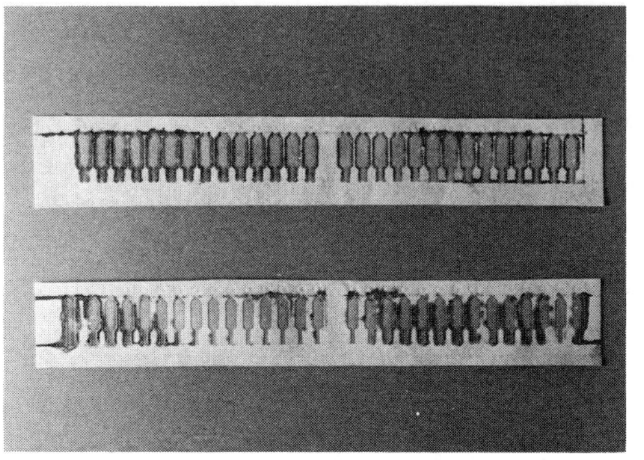

Fig. 27.7. Indicator paper exhibiting stains as indications of pores and pinholes. A solitary pore is seen on one of the contact tabs. A great deal of corrosion along the edges of the contact tabs is evident, but this is quite common to most edge connectors.

assembly is compressed with a clamping device so that the pressure between the indicator paper and the contact tabs is between 200 and 250 lb./sq. in. (14 to 17 kg/cm^2). See Figure 27.6.

A dc current is now passed between the contact tabs (the anode) and the aluminum plate (the cathode) for 30 seconds, at an intensity of 50 mA/sq. in. (0.08 mA/mm2) of anode area. After the indicator paper is removed and dried, it is examined for stains. See Figure 27.7. The size of the pores cannot be measured from the size of the stains since the stains tend to spread over the indicator paper.

The aforementioned publications detail the procedure for impregnating the indicator paper. For electrograms of gold plated on copper, the impregnation is performed with cadmium chloride, and with sodium sulfide after drying. Pores show up as brown stains.

For electrograms of gold on nickel, the paper is impregnated with nioxime and ammonia vapor. Pores in the gold layer show up as purple stains, while pores extending through the nickel layer to the underlying copper show up as brown/green stains.

BS 9762 states a requirement of no more than one pore per contact tab. A modification, however, allows the presence of two pores on 10 percent of the contact tabs examined.

Most PCB manufacturers use electrographic testing. To the inexperienced user, this test method may appear somewhat troublesome, and perhaps also a little questionable. The shelf life of the impregnated indicator paper is only 4 to 6 weeks. However, the inspector does not need to prepare the indicator paper, since it is commercially available.

27.2.2 Gas Exposure with Sulfur Dioxide and Hydrogen Sulfide

Another type of porosity test can be performed by exposing the edge connector to a sulfur dioxide atmosphere (SO_2) for 24 hours followed by a hydrogen sulfide atmosphere (H_2S) for 24 hours. This is an accelerated test because the gas concentrations far exceed the conditions to which the board is likely to be exposed in practice.

The objective is to reveal possible pores and pinholes only. It is, on the other hand, conceivable that pores and pinholes would not appear during use under normal atmospheric conditions.

Since IEC 326-2 and BS 9760 describe this test, it is only briefly described here. The method is quite simple: The degreased sample is placed on a perforated and glazed ceramic plate in a conventional 2 gallon (10 liter, European size) desiccator vessel with ground-in lid. The operator dispenses 0.2 fl. oz. (approximately 5 ml) distilled water over the base of the chamber beneath the ceramic plate.

A 4 fl. oz. (100 ml) glass jar is filled with sulfur dioxide gas from a liquid gas cylinder, the air being displaced upwards. The filled glass jar is then placed on the ceramic plate, after which the desiccator is closed. See Figure 27.8. After 24 hours the desiccator is opened and the glass jar removed.

The desiccator with the sample is now left open for 1 hour under normal room conditions. Then the glass jar is filled with hydrogen sulfide and, as before, placed in the desiccator, which is again closed and left for 24 hours. The desiccator is then opened and the sample is inspected under a 10X magnification for stains and corrosion products on the contact tabs. The same approval criteria as stated in Paragraph 27.2.1 also apply to this test.

The test requires a process time of 48 hours. Since it may be difficult to obtain the gasses, this method is not commonly used.

Fig. 27.8. Desiccator vessel with inserted sample.

WARNING:

1. Breathing concentrated hydrogen sulfide can be fatal. The first and only sign of danger is loss of the sense of smell, so that the operator no longer notices the very unpleasant smell of hydrogen sulfide. At this point there is little time left; medical assistance should be obtained without delay.
2. Sulfur dioxide is also dangerous in high concentration, but since it causes persistent twinges in the throat, the operator can quickly recognize the danger.
3. A fume cupboard must be used.

27.2.3 Gas Exposure with Nitric Acid

The simplest way to perform a porosity test is to expose the edge connector to fumes of concentrated nitric acid (HNO_3). A nitric acid concentration of 60 to 65 percent is poured into the bottom chamber of a desiccator, after which the sample is placed on the perforated ceramic plate. The lid is placed on the desiccator, which is left for 1 hour. The sample is then removed and dried for 1 hour at 230°F (110°C) in an oven.

The operator must not touch the microfine drops on the plated surface while transfering the sample from the desiccator to the oven. Otherwise, it will be difficult to localize the pores and pinholes. After drying, the sample is inspected under a 10X magnification for stains and corrosion products. Figure 27.9 shows an example. The approval criteria stated in Paragraph 27.2.1 also apply here.

WARNING:

1. Nitric acid fumes must never be inhaled. They are very dangerous and will cause lung injury if inhaled.
2. Nitric acid is a powerful etchant which is dangerous if it comes in contact with the skin. The operator should exercise the greatest care.
3. A fume cupboard must be used.

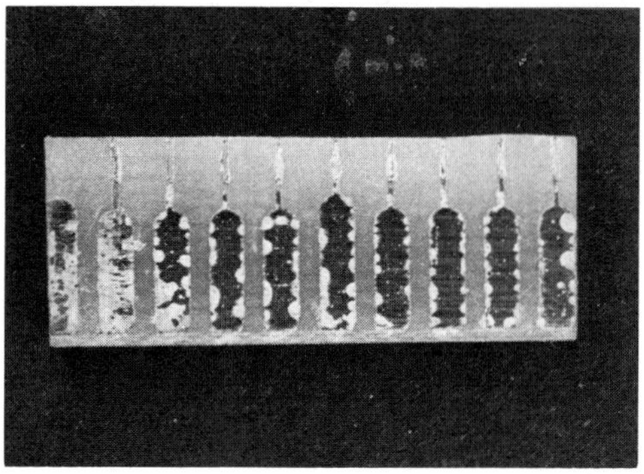

Fig. 27.9. Edge connector exhibiting point corrosion after a gas test with nitric acid.

Because nitric acid is so dangerous, the author discourages use of this test by personnel without chemical training. The company must also have the necessary laboratory facilities.

27.2.4 Acid Test with Nitric Acid

A variant of the preceding gas exposure test with nitric acid is an acid test in which the operator places the edge connector in a beaker with nitric acid (60 to 65 percent) for one minute. The beaker should be placed in a fume cupboard with exhaust (Figure 27.10). Pores and pinholes show up as colored stains. At places with uncovered copper—for example, at the milled end of the contact tabs—no coloring usually occurs since the exposed copper is simply etched away. The approval criteria stated in Paragraph 27.2.1 also apply here.

Fig. 27.10. Execution of an acid test in a fume cupboard.

WARNING:

The same warnings as given in Paragraph 27.2.3 apply to this test.

27.2.5 Gas Exposure Test with Hydrogen Sulfide

In regard to the degree of corrosion, this test is estimated to correspond to common and moderate environmental conditions over a period of approximately 10 years, and is recommended when there are no special requirements regarding test methods.

This test is easy to perform, and the etchant is harmless to the skin. The resulting hydrogen sulfide gas has such an unpleasant smell, even at very low concentrations, that the operator is very unlikely to expose himself to dangerous concentrations. This contrasts sharply with nitric acid fumes, which give little warning.

WARNING:

Breathing hydrogen sulfide in high concentrations can be fatal. The operator should exercise great caution if an accident happens during the test or preparation.

A desiccator is used as in the other gas exposure tests previously described. The operator pours 4 fl. oz. (100 ml) of water into the bottom chamber of the desiccator, and then adds 4 to 5 drops of ammonium sulfide $(HN_4)_2S$ to the water.

Unless a fume cupboard is used, extremely good ventilation must be provided.

The sample is placed in the desiccator—for example, by placing it at an angle against the glass wall—after which the lid is placed on the desiccator for 24 hours. The sample is then removed for a short drying period before it is examined for stains with a 10X magnifier (Figure 27.11). The approval criteria stated in Paragraph 27.2.1 also apply here.

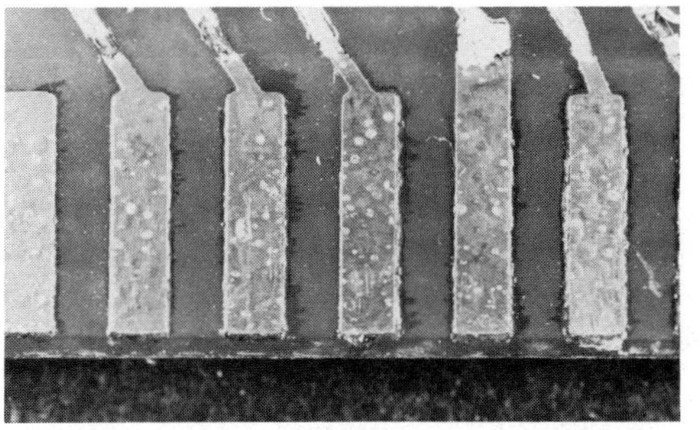

Fig. 27.11. Edge connector exhibiting point corrosion after a gas test with hydrogen sulfide.

Chapter 28

Insulation Resistance Measurement

28.1 Introduction

Paragraph 22.4 mentioned insulation resistance measurements. This chapter discusses the best way of performing these measurements. Since the insulation resistance is frequently in the order of 10^8 or 10^9 megohms, insulation measurements place exacting demands on both the equipment and the measuring setup.

Thorough investigation requires that the board first be subjected to environmental conditioning—for example, humid heat, meaning a relative humidity of 92 percent at 104°F (40°C). The test duration can be 96 hours, although 10 and 21 day durations may also be required, depending on the severity of the standards followed.

Normally, an incoming inspection department lacks time for such prolonged tests, and the boards are tested as received, possibly after a short drying period. In spite of the limitations, this approach gives a good indication of the magnitude of the insulation resistance.

A precise measurement is seldom needed, since the main objective is to identify catastrophic failures, where the insulation resistance drops dramatically, usually as the result of adverse production processes.

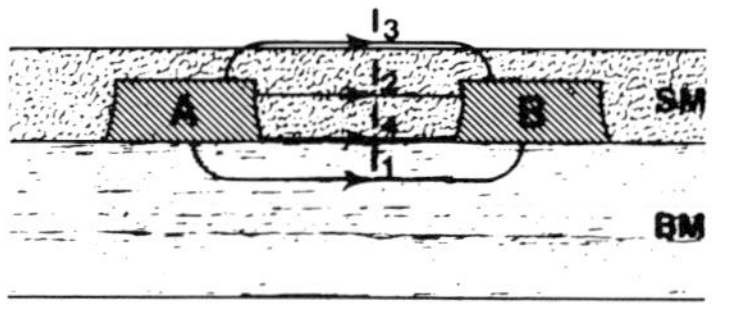

Fig. 28.1. Cross-sectional drawing of printed circuit board showing leakage current paths.

A and B = conductors.
BM = base material.
SM = solder mask.
I_1 = leakage current in base material.
I_2 = leakage current in solder mask.
I_3 = leakage current on surface of solder mask.
I_4 = leakage current in border area between base material and solder mask.

28.2 Components of Insulation Resistance

Before discussing measuring methods, it is expedient to see how the insulation resistance is composed. To aid the understanding, we will consider leakage currents instead of insulation resistances, and will discuss the four different types of leakage currents I_1 to I_4. See the cross-sectional drawing in Figure 28.1, where A and B indicate two conductors on a board with a solder mask.

28.1.1 Leakage Current I_1

The value of the leakage current I_1, which passes through the base material itself, depends on several factors. The primary factor is the fundamental insulation properties of the base material, including its structure and the materials from which it was made.

Another factor is the effect of relative humidity, since the board always absorbs some moisture. This absorption, which can be measured as the increased board weight, takes place slowly over a long time—for example, two months—after which a stationary condition is reached. The result is that the leakage current I_1 increases during this period.

It should be noted that many official standards prescribe, for example, exposure periods of 21 days, reflecting the absorption process. However, an exposure of 4 days (96 hours) is used in many measurements since most absorption takes place during this interval. The absorption is about 0.25 percent for epoxy glass boards and 1 percent for phenolic paper boards.

The leakage current can also rise with the temperature due to the higher ion mobility and degree of dissociation. Within certain temperature limits, we can often anticipate a doubling of the leakage current per 14 to 18°F (8 to10°C) increase in temperature.

Each of these parameters affects the other. When, for example, a humid board is exposed to a temperature rise, the insulation resistance first falls due to the increased ion mobility and degree of dissociation (I_1 rises). Later, however, the insulation resistance rises again due to the drying of the board (I_1 falls). In this way the insulation resistance reaches a minimum value within a certain time.

28.2.2 Leakage Current I_2

The leakage current I_2 passes through the solder mask and, in principle, behaves in just the same manner as the leakage current I_1 through the base material.

28.2.3 Leakage Current I_3

The leakage current I_3 is actually a surface current, but it has to pass through the solder mask twice. However, the surface can be contaminated by dust and fingerprints, possibly in conjunction with a film of moisture. Therefore, I_3 is determined primarily by the volume resistance of the solder mask where it passes through the solder mask. Although a screen-printed solder mask can be very thin, particularly at the conductor edges, it is possible, with good-quality printing ink, to achieve satisfactory insulation resistance (the leakage current becomes quite low).

28.2.4 Leakage Current I_4

The leakage current I_4, which flows through the border area between the base material and the solder mask, depends on the cleaning of the base material surface before solder mask application. This contamination can, in worst-case conditions, be ion contamination (Paragraph 4.6.6c) caused by an inadequate cleaning and neutralization after the board is etched.

Since the solder mask behaves like a membrane through which humidity can penetrate, the insulation resistance in the border area decreases, and I_4 consequently increases.

28.2.5 Mealing

The inspector should check whether small white spots, called mealing, are present under the solder mask. They occur mostly where there are voids or cavities in the solder mask, as a consequence of pumping action.

Moisture absorbed by the solder mask migrates as vapor towards cavities where dissociated salts are precipitated, and in time forms the white spots. The fault, which is considered cause for rejection and consequently rated as an A-defect, should not be mistaken for measling. The latter is a defect in the structure of the base material (Paragraph 4.5.1).

Since mealing usually takes some time to form, the inspector rarely sees signs of mealing. An insulation measurement can reveal the concealed defect since the insulation resistance may decrease by several orders of magnitude due to salt residues. In order to promote the dissociation of possible salt residues, it is expedient to wet the surface of the solder mask with distilled water.

When measuring insulation resistance, it is not possible to distinguish between the various components. The above discussion illustrates how different effects take place as a function of external influences.

28.3 Megohmmeter

The inspector measures insulation resistance with a megohmmeter (Figure 28.2). The measuring range should include 10^9 megohms. Because of the large differences in insulation resistance exhibited by printed circuit boards, it is advantageous to use instruments having a logarithmic scale covering a range of nine decades in order to not have to switch between measuring ranges. The lower reading accuracy associated with a nine-decade scale is immaterial to this type of measurement. The test voltage should be 100 V dc, although 500 V dc is sometimes used.

Most megohmmeters have a guard circuit which neutralizes all external and unwanted leakage paths which could affect the measurement, so that only the insulation resistance is measured. The resistances in question are shown in the diagram of Figure 28.3. R_x indicates the insulation resistance to be measured between the conductors A and B. R_A and R_B represent the leakage resistance from the conductors A and B to the surroundings, ultimately to ground.

The leakage currents pass along the surface of the board and through the board to the supporting points S and over the surface and through these points to the ground. By means of the guard system it is possible to isolate the leakage resistances R_A and R_B as shown in Figure 28.4. R_A loads the test voltage supply V_T, but if R_A is greater than 1 megohm, which is normally the case, it can be ignored.

Fig. 28.2. Megohmmeter with a logarithmic measuring range from 1 MΩ to 10^9 MΩ, suitable for measuring insulation of printed circuit boards (RE Instruments A/S).

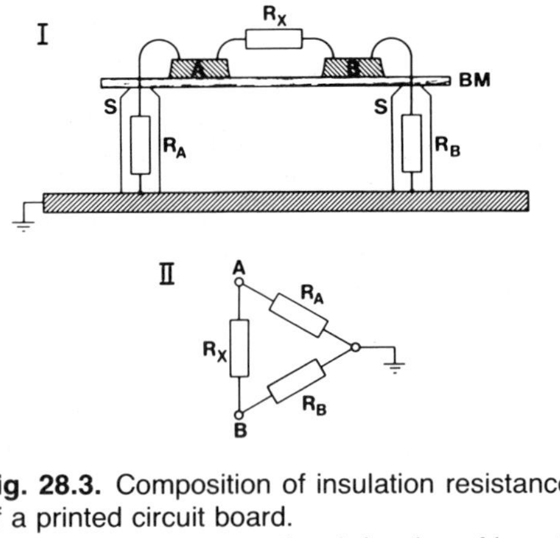

Fig. 28.3. Composition of insulation resistance of a printed circuit board.

I	=	cross-sectional drawing of board.
II	=	equivalent diagram.
A and B	=	conductors.
R_x	=	insulation resistance to be determined.
R_A and R_B	=	leakage resistance to ground.
BM	=	base material.
S	=	supporting stud.

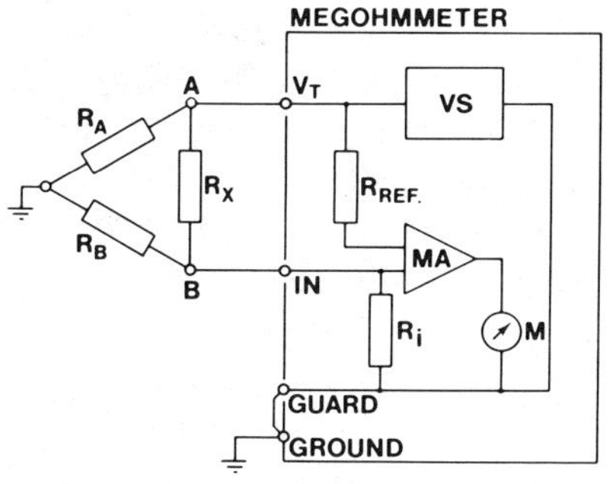

Fig. 28.4. Elimination of leakage resistances by means of a guard circuit.
V_T = test voltage.
V_S = voltage supply.
MA = measuring amplifier.
M = meter.
IN = input terminal of measuring amplifier.
R_1 = input resistance of measuring amplifier.
R_{REF} = internal reference resistance.
Remaining designations: See Fig. 28.3.

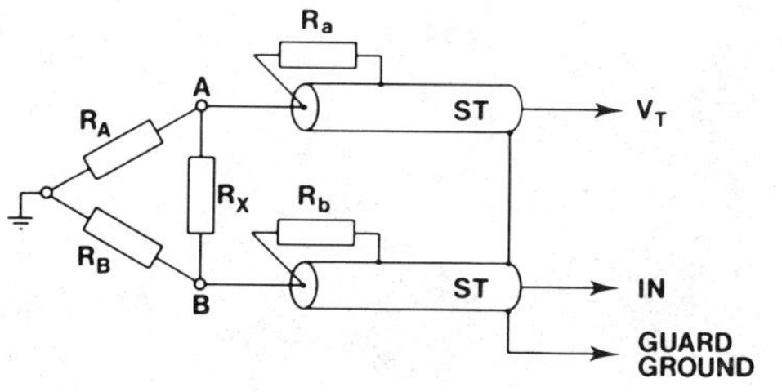

Fig. 28.5. Use of shielded test leads.
R_a and R_b = leakage resistance of test leads.
ST = shielded test lead.
Remaining designations: See Fig. 28.3.

R_B loads the input of the measuring amplifier. If the input resistance R_I, for example, is 1 megohm, a leakage resistance of 100 megohms will cause only 1 percent error, which is negligible.

The inspector is recommended to very carefully follow the instrument manufacturer's instructions on using the guard and ground terminals of the megohmmeter. An erroneous connection can, for example, put the instrument chassis at the test potential, which may be hazardous.

Since the insulation resistances R_A and R_B are usually very high, the measuring setup is sensitive to external fields. Long measuring leads can disturb the measurements. Shielded test leads can be easily employed when the guard circuit is correctly used. See Figure 28.5. R_a and R_b represent the leakage resistance from the inner conductor to the shield of the test cables, and by connecting the shields to the guard circuit, R_a and R_b are placed in parallel with R_A and R_B, respectively, eliminating their influences from the resistance reading.

28.4 Measuring Setup

A simple measuring setup is shown schematically in Figure 28.6 and illustrated in Figure 28.7. A large sheet of aluminum placed below the setup shields against external fields. A shielding housing placed over the board during the measurement may also be necessary. The shielding housing consists of three sides and a top, and can be easily made from an aluminum sheet.

Both shields are connected to the guard terminal of the megohmmeter. The guard terminal is connected to the ground terminal, which of course is connected to the ground. The board is placed on a test fixture, which can simply be a sheet of aluminum having four supporting studs made

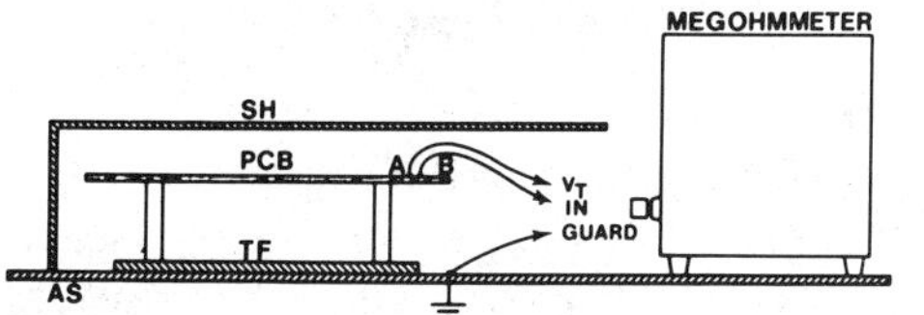

Fig. 28.6. Schematic drawing of measuring setup.
A and B = measuring points on conductors.
PCB = printed circuit board.
SH = shielded housing.
AS = aluminum sheet.
TF = test fixture.
V_T – test lead to test voltage terminal.
IN = test lead to input terminal of measuring amplifier.
GUARD = test lead to guard terminal.

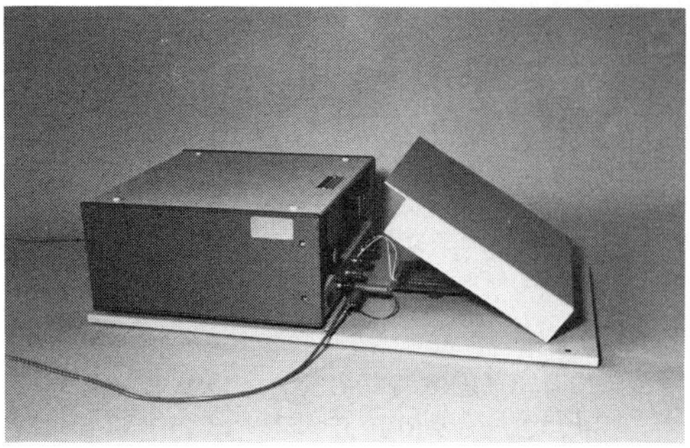

Fig. 28.7. Practical setup of measuring equipment.

from an insulating material, such as perspex.

The selected measuring points are connected to the test voltage and measuring input terminals of the megohmmeter with insulated test leads which do not touch each other or the shielding arrangement. The leads can be connected to the measuring points on the board by soldering, but the operator must take care not to contaminate the board with flux during soldering, since this may cause an inaccurate measurement.

It is also possible to solder a small tapered pin to each lead so that the operator can connect them by carefully pressing the pins into solder holes on the board. This method eliminates flux contamination.

28.5 Resistance Measurement

The board insulation resistance is measured with the board in an "as received" condition—that is, the board is not cleaned before the measurement. The inspector must therefore be careful to not touch the board surface with bare fingers, since fingerprints can reduce the insulation resistance. Incidentally, it is always good practice to hold a printed circuit board by the edges only.

The inspector should now select two closely spaced conductors—for example, 0.012 in. (0.3 mm)—running in parallel. The length should be at least 1 in. (25 mm). The two conductors are then connected to the megohmmeter with test leads as described in Paragraph 28.4. The board is placed on the text fixture and the shielding housing is put in place. The test voltage is adjusted as follows:

For conductor spacings
up to 0.04 in. (1 mm): 100 ±15 V
For conductor spacings
over 0.04 in. (1 mm): 500 ±50 V

The test voltage is turned on, and the measuring result is read after a 1 minute stabilization period.

The inspector is now facing the problem of choosing a resistance value of R_I to be used as an acceptance criterion. According to IEC 326-3 and DIN 48303, the value is determined by the following requirement:

$$R_I = 160 \times R_M \times \frac{S}{L}$$

where R_M indicates a material constant based on certain test conditions, S is the spacing between the conductors, and L is the length of the conductors.

When the spacing varies over parts of the lengths, the effective ratio S/L valid for the total length can be calculated with the expression

$$\frac{1}{\frac{S}{L}} = \frac{1}{\frac{S_1}{L_1}} + \frac{1}{\frac{S_2}{L_2}} + \cdots + \frac{1}{\frac{S_n}{L_n}}$$

The material constant R_M is specified as the surface resistance under different test conditions, for example:

Humid heat:
Duration:
96 hours
Relative humidity:
92 percent

Temperature:
104°F (40°C)

Post-treatment:
Duration:
1 to 2 hours
Relative humidity:
73 to 77 percent
Temperature:
59 to 95°F (15 to 35°C)

Typical values of R_M after post-treatment:
Epoxy glass boards:
$R_M > 10^6$ megohms
Phenolic paper boards:
$R_M > 5 \times 10^5$ megohms

Dry heat:
Duration:
1 hour
Relative humidity:
20 percent
Temperature:
Epoxy glass boards:
257°F (125°C)
Phenolic paper boards:
212°F (100°C)

Typical values of R_M after dry heat:
Epoxy glass boards:
$R_M > 10^5$ megohms
Phenolic paper boards:
$R_M > 10^4$ megohms

Some examples of ascertaining the insulation resistance R_I under different test conditions are given below:

Example 1:
Conditioning with humid heat implies a delay of 4 days before the board insulation resistance can be measured. In the case of an epoxy glass board having conductors 2 in. (50 mm) long with a spacing of 0.012 in. (0.3 mm) we find:

$$R_I = 10^6 \times 160 \times \frac{0.012}{2} \simeq 10^6 \text{ megohms}$$

Example 2:
Conditioning with dry heat implies that the measurement can be made fairly quickly. It is, however, a prerequisite that the measurement be performed at a high temperature—that is, with the board placed in a hot oven. It is therefore necessary to place the measuring setup with the shielding sheet and housing within the oven and the megohmmeter outside. The test leads brought out from the oven should be Teflon-insulated coaxial cables. See Figure 28.5

For the same board as discussed in Example 1 we find:

$$R_I = 10^5 \times 160 \times \frac{0.012}{2} \cong 10^5 \text{ megohms}$$

The difference between the two examples is primarily due to the high ion mobility caused by the high temperature.

Example 3:
The problem of measuring within the oven can be alleviated by removing the board after baking and letting it cool down to ambient temperature (77°F, ~25°C). This means a temperature drop of some 212°F (~100°C). The material constant R_M must, of course, be converted to the lower temperature.

Under the assumption that the insulation resistance is doubled per 8°C (~14.4°F) decrease in temperature, the insulation resistance increases $2^{12.5}$ times—that is, approximately 5000 times. The material constant R_M now becomes 5×10^8 megohms, and for the same board as before, we find:

$$R_I = 5 \times 10^8 \times 160 \times \frac{0.012}{2}$$

$$\cong 5 \times 10^8 \text{ megohms}$$

This corresponds, in reality, to a well-dried board, and it seems quite reasonable to find that the insulation resistance is now 5000 times higher.

As mentioned in Paragraph 4.6.5e, it is possible to introduce a safety factor of 100, so that the rejection value R_R of the insulation resistance is 100 times lower than the value of R_I found from the formulas.

The most important matter when checking the insulation resistance is to determine whether a serious contamination has occurred—for example, as a result of salt residues originating from the etching. With some experience, the inspector can quickly identify disastrous cases, which means boards with a greatly reduced insulation resistance. Such boards should be rejected, and consequently the fault is rated as an A-defect.

Examples 1 to 3 also illustrate what the inspector can expect to find, under local climatic conditions, as the lowest insulation resistance when testing boards as received—that is, without any preconditioning. An acclimatization for 1 hour on the inspection premises is, however, always assumed. The inspector can expect results between the values found in Examples 1 and 3—that is, $R_I \cong 10^7$ megohms

To give a practical illustration of the influence of various test conditions on the test result, it is most expedient to perform the measurements on a test board as shown in Figure 28.8. Here the insulation resistance is measured on a comb-shaped pattern section with a conductor spacing of 0.012 in. (0.3 mm) and a total conductor length of 20 in. (500 mm) The values found in the preceding examples would therefore be 10 times lower, with all other properties unchanged. The results are as follows:

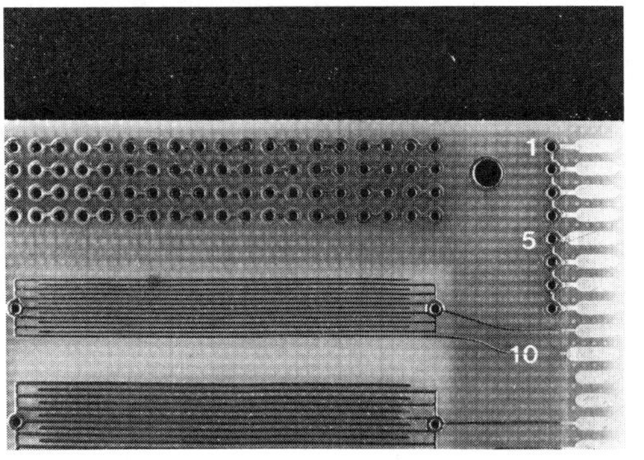

Fig. 28.8. Example of epoxy glass board being used for the practical tests stated in Examples No. 1 to 3. The insulation resistance of the comb is measured between contact Nos. 9 and 10. The test voltage is 100 V dc.

Test condition 1:
(after humid heat, cooled down)
Calculated value:
$R_{I,C} = 10^5$ megohms
Rejection value:
$R_R = 10^3$ megohms
Measured value:
$R_{I,M} = 5 \times 10^5$ megohms

Test condition 2:
(heat, 257 °F ≅ 125°C)
Calculated value:
$R_{I,C} = 10^4$ megohms
Rejection value:
$R_R = 10^2$ megohms
Measured value:
$R_{I,M} = 2 \times 10^3$ megohms

Test condition 3:
(after dry heat, cooled down)
Calculated value:
$R_{I,C} = 5 \times 10^7$ megohms
Rejection value:
$R_R = 5 \times 10^5$ megohms
Measured value:
$R_{I,M} = 10^9$ megohms

Obviously, the base material used has a material constant R_M which is far better than the typical value stated under the test conditions for humid heat and dry heat.

Chapter 29

Hardness Test of Copper Plating in Holes

29.1 Introduction

Cracking in any of the plated-through holes casts serious doubt on the board reliability. In order for the customer to assess the PCB manufacturer's capability of producing reliable plated-through holes, the inspector must ascertain the tendency of the copper plating to crack.

Paragraph 7.3.4e mentioned that some boards more than others tend to crack in the plated-through holes, and outlined a certain correlation between the hardness of the plated copper and the tendency to crack. An indirect test of the plating hardness is the change of resistance when the hole is subjected to thermal cycling, as outlined below.

The coefficient of thermal expansion in the Z-axis direction (perpendicular to the board) is three to four times higher for epoxy glass base materials than for copper. The influence of heat, such as occurs during reflowing, puts a significant tension on the hole walls of plated-through holes, stretching the hole wall. The harder the copper, the greater the risk of cracking, which usually shows up as a circumferential crack.

The tendency to crack is increased by a rough hole drilling, partly because this can create stress regions in the plating and partly because the copper wall can be very thin in places.

When the board cools after the heat treatment, the base material as well as the hole wall contracts. However, if the copper in the hole wall is hard and has a low coefficient of elasticity, the hole wall will not return to the original dimensions and the electrical resistance of the hole increases slightly. This permanent resistance change is used to indicate the copper hardness.

29.2 Principle of the Test

Although the test is described in IEC 326-2A (1980)—the first supplement to IEC 326-2—it is outlined below for the reader's convenience. Paragraph 29.3 describes the test used in actual practice.

The copper plating hardness test is based on the resistance variation which occurs in plated-through holes subjected to thermal cycling. A board with a number of holes connected in series is used as a sample. IEC 326-5 and 326-6 show a special test pattern, specimen D being intended for such measurements since it has 30 holes connected in series.

The specimen is heated for 20 seconds in an oil bath at a temperature of 500°F (260°C). The specimen is then transferred to another oil bath with a temperature of 77°F (25°C). The test may include, for example, five temperature cyclings, with

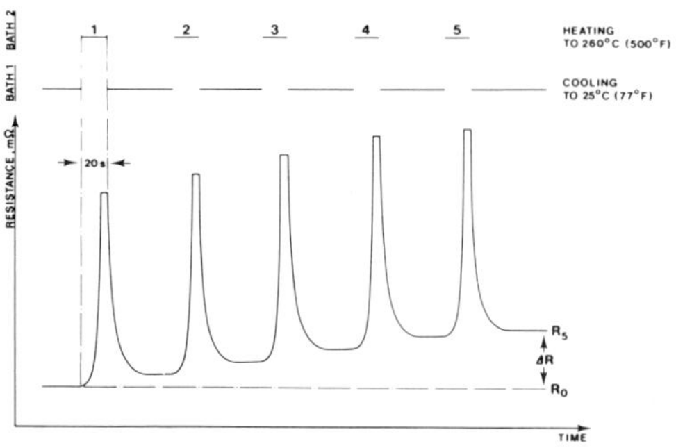

Fig. 29.1. Measuring sequence corresponding to 5 heating and cooling cycles.
R_o = resistance at the start of the test.
R_5 = resistance at the end of the test.
ΔR = permanent change in resistance.

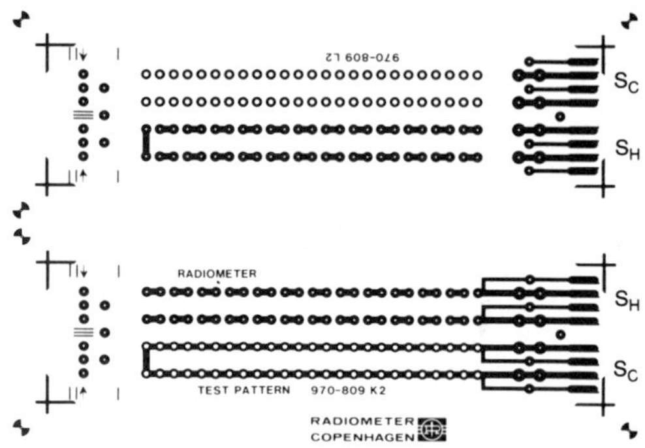

Fig. 29.2. Example of test coupon with extended test pattern. The filmwork shows the front and rear sides of the test coupon. Note that in one of the sections the plated-through holes are not interconnected on one side of the test coupon.

the temperature being continuously monitored throughout the test.

The low resistance values make it necessary to use a four-terminal measuring method. A power supply feeds a constant 100 mA current through the holes, and a chart recorder records the voltage drop across the holes, producing the curve shown in Figure 29.1.

The curve shows a certain resistance R_o at the beginning of the test and a slightly higher resistance R_5 at the end. The difference ΔR represents the permanent resistance change of the holes, with the assumption that the resistance of the interconnecting conductor segments does not change. It is normal practice to calculate the relative resistance change with respect to the initial value by

$$\frac{\Delta R}{R_o} \times 100\ \%$$

In the author's opinion, the conditions are somewhat uncertain because the resistance of the interconnecting conductor segments has the same magnitude as the resistance of the holes. Changing the copper thickness from 1 oz. = 0.0014 in. (35 μm) to ½ oz. = 0.0007 in. (17.5 μm) or even to ¼ oz. = 0.00035 in. (9 μm) affects the result, and since the inspector may not always know the foil thickness, he can be misled.

It is possible to modify the test a little by extending the test pattern with an extra section (Section S_C) which has plated-through holes like the first section (Section S_H), but these holes are interconnected on one side only. The holes act both as a heat sink and as an extra compensation, giving the section a resistance which corresponds as closely as possible to the total resistance of the interconnecting conductor segments of Section S_H. Figure 29.2 shows the test coupon filmwork.

By performing the test as a bridge measurement as shown in Figure 29.3, the inspector can cancel the resistance of the interconnecting conductor segments so well that the resistance reading represents the resistance of only the holes.

A four-terminal measurement offers the advantage that the voltage drop in the leads and current terminals does not affect the measurement because the voltage measurement excludes the voltage drop. See Figure 29.4.

This approach is valid in theory only. In practice, the two sections have slightly different time constants for heating and

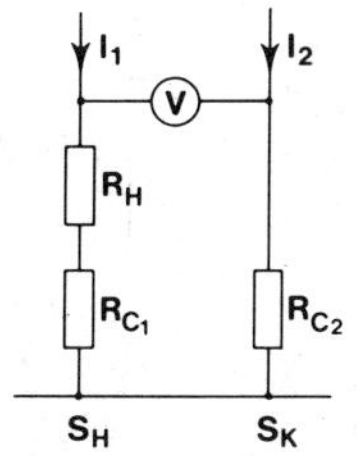

Fig. 29.3. Schematic diagram of measuring principle.
$I_1 = I_2$ = constant currents.
V = voltmeter.
R_H = resistance of holes connected in series.
R_{C1} = resistance of conductor section S_H.
R_{C2} = resistance of conductor section S_C.
S_H = section of test coupon with holes.
S_C = section of test coupon compensation.

cooling. The recorded ΔR curve can therefore exhibit peculiar transients during the temperature cycling.

Instead, the extra section is used to obtain an approximate value of the cold resistance of the interconnecting conductor sections ($R_{c1} \approx R_{C_2}$, see Figure 29.3). The corrected resistance R'_o is calculated as

$$R'_o = R_o - R_{C_2}$$

and the relative change in resistance is calculated as

$$\frac{\Delta R}{R'_o} \times 100\%$$

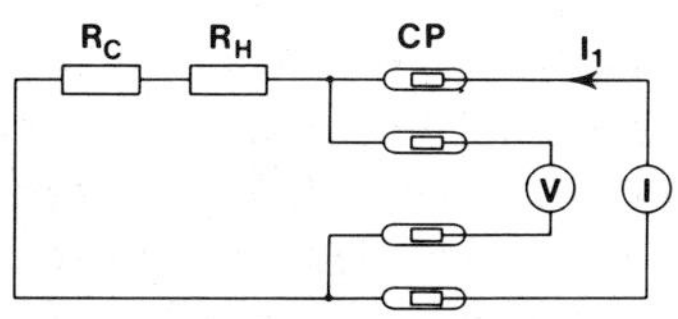

Fig. 29.4. The execution of a 4-terminal measurement.
I = constant current generator.
V = voltmeter.
I_1 = constant current.
R_H = resistance of holes connected in series.
R_C = resistance of conductor section.
CP = contact points.

On the whole, this means no interaction from the conductor segments which interconnect the holes in test coupon Section S_H.

Unfortunately, the author cannot quantify the test results in the form of a correlation between the relative change in resistance and the tendency to crack. However, his experience shows that a relatively large change in resistance indicates an increased risk of cracking.

If a hole wall cracks because of the thermal stress, this will show up very clearly as a sharp deflection in the recorded curve. A second sharp deflection will be recorded if the hole wall contracts during cooling so that the electrical continuity is restored.

The defective hole can be located by a simple voltage measurement. Since there is an open circuit, the full generator voltage appears across the hole with the cracked hole wall. After locating the defective hole, the inspector should perform a microsection.

29.3 The Practical Test

We shall now describe how the test is performed in practice. The test is actually carried out on a test coupon with a pattern shown in Figure 29.2. The customer should reach an agreement with the PCB manufacturer that every six months he delivers a number of test coupons produced simultaneously with an actual order.

29.3.1 Tin/Lead Removal

The first step is to remove the tin/lead layer, unless the PCB manufacturer has done this. The etchant must not affect the copper plating. A solution for treating the test coupon can be:

100 parts nitric acid, 60% (HNO_3)
1 part fluoroboric acid (HBF_4)
200 parts deionized water (H_2O)

WARNING:

The safety precautions given in Paragraph 27.2.3 regarding the use of nitric acid apply here.
When diluting an acid with water, the acid must always be added to the water. If water is added to the acid, excessive heat will develop within small volumes and cause such violent boiling that the solution splatters in all directions, exposing the operator to serious injury.

Another etchant, which is easier to obtain and widely used for stripping tin/lead from edge connectors, is described below. It should be noted that the hydrogen peroxide deteriorates quickly, so the solution should not be too old.

8 parts glacial acetic acid (CH_3COOH)
1 part hydrogen peroxide (H_2O_2)
24 parts water (H_2O)

Because of the four-terminal method of measurement, the edge connector contact tabs do not need to be gold plated; a tin/lead plating is adequate. When stripping the tin/lead, the inspector should mask the contact tabs with tape in order to avoid stripping the tin/lead from the contact tabs.

29.3.2 Measuring Setup

Figure 29.5 shows the details of the measuring setup. The constant-current generator must have two outputs, each capable of delivering 100 mA. It is easiest to use a voltage supply of 5 volts, with series resistors of 50 ohms, ½ to 1 watt. These resistors make the currents I_1 and I_2 effectively independent of the resistance of the holes and the interconnecting segments. Since the resistance of Section 1 is approximately 35 milliohms (50 holes @ 0.4 milliohm according to Figure 7.18 plus a total interconnection resistance of

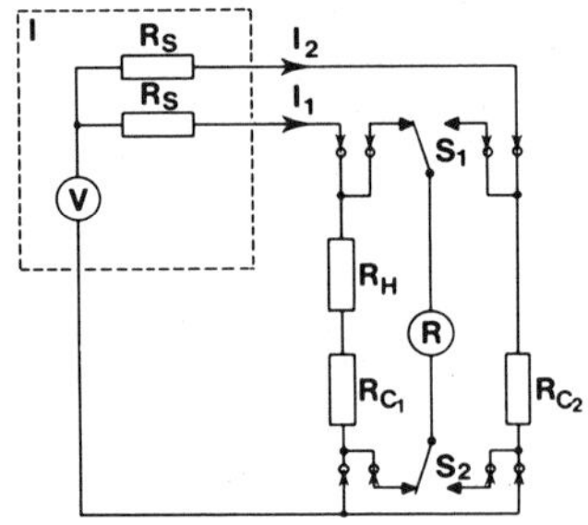

Fig. 29.5. Schematic diagram of measuring circuit.

I = constant current generator.
V = voltage supply (5 V).
R_S = series resistor (50 Ω).
I_1 and I_2 = constant currents, each 100 mA.
R = recorder.
R_H = resistance of holes connected in series.
R_{C1} and R_{C2} = resistance of conductor segments.
S_1 and S_2 = double switch.

15 milliohms) and the temperature coefficient of the copper resistance is 0.004 percent per °C, the 35 milliohm value increases to approximately 70 milliohms when the test coupon is heated to 260°C (500°F). The corresponding voltage drop is 7 mV, which the recorder must be sensitive enough to indicate.

The inspector measures the cold resistance of the two sections before starting the temperature cycling, and uses the switch S_1-S_2 to conveniently switch between the measuring points. During this and the following measurements, the zero point of the recorder should correspond to the reference line of the recorder paper so that the recorder shows no zero offset. This is easily verified by short-circuiting the recorder input terminals.

29.3.3 Specimen Heating and Cooling

A 1 or 2 pint (0.5 to 1 liter) beaker of Pyrex or similar glass is used for heating the specimen. The heating medium should be silicone oil with a flame point above

570°F (300°C). The inspector places the beaker on an electric hot plate and immerses a suitably scaled thermometer halfway down in the oil to verify that the oil temperature is within 500 to 509°F (260 to 265°C).

WARNING:

1. Ordinary mineral oil must not be used because the flame point is too low, creating a fire hazard.
2. Suitable measures should be taken to avoid accidentally overturning the beaker. Hot oil at 500°F (260°C) can inflict severe burns.

A smaller beaker filled with silicone oil is used for cooling the specimen. After the specimen cools to ambient temperature (68 to 77°F, ~20 to 25°C), the oil must be removed from the holes so that they are subjected to the full temperature shock when the specimen is dipped in the hot oil. Absorbent paper, such as paper towels, is recommended for removing the remaining oil.

The test coupon is connected to the measuring setup, either with a receptacle or by leads soldered to the connector tabs. The test coupon is dipped far enough into the oil to cover all holes, but the edge connector is kept above the oil. The test coupon is moved back and forth to ensure good heat transfer.

The recorder paper speed should be set to 1 in. (25 mm) per minute, and the sensitivity to 10 millivolts full deflection, depending on the resistance of Section 1. The test can start once this is done.

The test coupon is dipped in the cold oil and the cold resistances of Sections 1 and 2 are recorded successively. The test coupon must be kept in the cold oil until the recorder deflection stabilizes. The distance between the two recorded curves determines

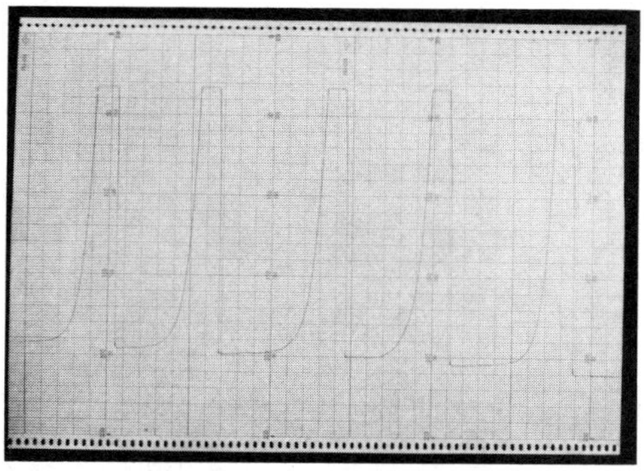

Fig. 29.6. Measuring result shown on a recorder paper.

$$R'_o = R_o - R_{C_2}$$

which is the approximate value of the cold resistance of the holes connected in series. This can be checked by the curve shown in Figure 7.18.

After removing the remaining cold oil from the holes—for example, with a paper towel—the inspector dips the test coupon in the hot oil for 20 ±1 seconds. Then the test coupon is moved quickly back to the cold oil until the recorder deflection restabilizes. The switch S_1-S_2 must be set to select Section 1 during the restabilization period and the subsequent cycling.

The test includes five heating and cooling cycles. This gives the inspector a curve as shown in Figure 29.6, from which he can ascertain the permanent change in resistance of the holes as

$$\Delta R = R_5 - R_o$$

The relative change in the resistance can now be calculated as

$$\frac{\Delta R}{R'_o} \times 100\%$$

Reading List

As previously mentioned, a very limited number of courses addressing PCB technology exists, and a person wanting training or follow-up education within this field must rely on informal study. Obviously, this must be based on the PCB literature available. The author has built up his own PCB library over several years, and selectively lists below the most useful of his books and magazines dealing with PCB technology.

A. Books on PBC Technology

A1 *Handbook of Electronic Packaging*
Charles A. Harper
1969, McGraw-Hill, Inc.

A2 *Plastic Coatings for Electronics*
James J. Licari
1970, McGraw-Hill, Inc.

A3 *Handbook of Materials and Processes for Electronics*
Charles A. Harper
1970 McGraw-Hill, Inc.

A4 *Printed Circuit Boards, A Guide to Specifying and Designing*
C. Thomas
1971 Publications Department Eindhoven
Philips Product Book

A5 *Handbook of Wiring, Cabling and Interconnecting for Electronics*
Charles A. Harper
1972, McGraw-Hill, Inc.

A6 *Electronic Drafting and Design*
Nicholas M. Raskhodoff
1972, Prentice-Hall, Inc.

A7 *Electronic Drafting Workbook*
Cyrus Kirshner and Kurt M. Stone
1973, McGraw-Hill, Inc.

A8 *Printed Circuit Troubleshooting*
H. R. Shemilt
1974, Electrochemical Publications Ltd.

A9 *Printed Circuit Design and Drafting*
Jessi S. Cook III
1976, Tad Institute, Cambridge
Massachusetts, U.S.A.

A10 *Printed Circuit Technology*
Norman S. Einarson
1977, Printed Circuit Technology
P.O. Box 334
Burlington, Massachusetts, U.S.A.

A11 *Leiterplatten* (In German)
Günther Herrmann
1978, Eugen G. Leuze Verlag

A12 *Generation of Precision Artwork for Printed Circuit Boards*
(2nd. Edition: See A20)
Preben Lund
1978, John Wiley & Sons Ltd.

A13 *Printteknik, 2nd. Edition* (In Danish)
Preben Lund
1979, Teknisk Forlag A-S

A14 *The Design & Drafting of Printed Circuits*
Darryl Lindsey
1979, Bishop Graphics, Inc.

A15 *Präzisionsvorlagen für Leiterplatten* (In German)
Preben Lund
1979, Carl Hanser Verlag

A16 *Printed Circuits Handbook* (2nd. Edition)
Clyde F. Coombs, Jr.
1979, McGraw-Hill, Inc.

A17 *Printed Circuit Boards for Microelectronics, 2nd. Edition*
J. A. Scarlett
1980, Electrochemical Publications Ltd.

A18 *Handbook of Printed Circuit Design, Manufacture, Components & Assembly*
Giovanni Leonida
1981, Electrochemical Publications Ltd.

A19 *How to Design Printed Circuit Boards for UL Recognition*
Preben Lund
1983, Bishop Graphics, Inc.

A20 *Generation of Precision Artwork for Printed Circuit Boards*
(2nd. Edition of A12)
Preben Lund
1985, Bishop Graphics, Inc.

B. Magazines

B1 ***Dedicated PCB Magazines***

Circuits Manufacturing
Morgan-Grampian Publishing Co.
1050 Commonwealth Avenue
Boston, Massachusetts 02215, U.S.A

Electronics
Lake Publishing Corporation
17730 West Peterson Road
Libertyville, Illinois 60048, U.S.A.

PC Design
1790 Hembree Road
Alpharetto, Georgia 30201, U.S.A.

Printed Circuit Fabrication
1790 Hembree Road
Alpharetto, Georgia 30201, U.S.A.

Circuit World
8 Barns Street
Ayr KA7 1XA
Scotland

B2 ***Magazines with a Regular PCB Section.***

Galvanotechnik
Eugen G. Leuze Verlag
D-7968 Saulgau/Württ.
West Germany

B3 ***Magazines Frequently Carrying PCB Articles.***

Electronic Packaging and Production
Cahners Publishing Co.
270 St. Paul Street
Denver, Colorado 80206, U.S.A.

Electronic Production
KP Publishing Ltd.
322 St. John Street
London EC1V 4QH
England

Index of Design Recommendations

Many descriptions of defect types and their causes also stated design recommendations for preventing the defects, thereby drawing the designer's attention to the effects of certain design flaws on the finished board—for example, its reliability or producibility. To help the designer retrieve the design recommendations, they have been indexed with key words and phrases.

Since it is impossible to list all preventive measures, the author strongly recommends that the designer obtain a reasonable understanding of the more common defect types and their failure mechanisms. Such understanding is an important prerequisite to designing producible boards, free of production traps.

Index

Note: The index references refer to numbered entries within chapters. Each entry number consists of a chapter number followed by the number of the entry within the chapter.